Earth Science and Geology

Earth Science and Geology

Edited by Joe Carry

SYRAWOOD
PUBLISHING HOUSE

New York

Published by Syrawood Publishing House,
750 Third Avenue, 9th Floor,
New York, NY 10017, USA
www.syrawoodpublishinghouse.com

Earth Science and Geology
Edited by Joe Carry

International Standard Book Number: 978-1-68286-601-6 (Hardback)

Cataloging-in-Publication Data

Earth science and geology / edited by Joe Carry.
    p. cm.
Includes bibliographical references and index.
ISBN 978-1-68286-601-6
1. Earth sciences. 2. Geology. I. Carry, Joe.
QE26.3 .E27 2018
550--dc23

TABLE OF CONTENTS

Preface...VII

Chapter 1 **Inferring the Effects of Compositional Boundary Layers on Crystal Nucleation, Growth Textures and Mineral Chemistry in Natural Volcanic Tephras through Submicron-Resolution Imaging**...1
Georg F. Zellmer, Naoya Sakamoto, Shyh-Lung Hwang, Nozomi Matsuda, Yoshiyuki Iizuka, Anja Moebis and Hisayoshi Yurimoto

Chapter 2 **The Feeding Habits of Mesosauridae**..8
Rivaldo R. Silva, Jorge Ferigolo, Piotr Bajdek and Graciela Piñeiro

Chapter 3 **An Empirical Approach for Estimating Stress-Coupling Lengths for Marine-Terminating Glaciers**..26
Ellyn M. Enderlin, Gordon S. Hamilton, Shad O'Neel, Timothy C. Bartholomaus, Mathieu Morlighem and John W. Holt

Chapter 4 **Scaling Precipitation Input to Spatially Distributed Hydrological Models by Measured Snow Distribution**..38
Christian Vögeli, Michael Lehning, Nander Wever and Mathias Bavay

Chapter 5 **Exploring Google Earth Engine Platform for Big Data Processing: Classification of Multi-Temporal Satellite Imagery for Crop Mapping**.....................53
Andrii Shelestov, Mykola Lavreniuk, Nataliia Kussul, Alexei Novikov and Sergii Skakun

Chapter 6 **The Influence of Reactive Oxygen Species on Local Redox Conditions in Oxygenated Natural Waters**...63
Andrew L. Rose

Chapter 7 **A Plastic Network Approach to Model Calving Glacier Advance and Retreat**.................................74
Lizz Ultee and Jeremy N. Bassis

Chapter 8 **New Advances in Dial-Lidar-Based Remote Sensing of the Volcanic CO_2 Flux**..86
Alessandro Aiuppa, Luca Fiorani, Simone Santoro, Stefano Parracino, Roberto D'Aleo, Marco Liuzzo, Giovanni Maio and Marcello Nuvoli

Chapter 9 **Microbial and Biogeochemical Dynamics in Glacier Forefields are Sensitive to Century-Scale Climate and Anthropogenic Change**....................99
James A. Bradley, Alexandre M. Anesio and Sandra Arndt

Chapter 10 **Stress Controls of Monogenetic Volcanism**...118
Joan Martí, Carmen López, Stefania Bartolini, Laura Becerril and Adelina Geyer

Chapter 11 **Initial Opening of the Eurasian Basin, Arctic Ocean**..................................135
Kai Berglar, Dieter Franke, Rüdiger Lutz, Bernd Schreckenberger and Volkmar Damm

Chapter 12   **Modeling the Controls on the Front Position of a Tidewater Glacier in Svalbard**..149
Jaime Otero, Francisco J. Navarro, Javier J. Lapazaran, Ethan Welty, Darek Puczko and Roman Finkelnburg

Chapter 13   **Transformations and Decomposition of MnCO$_3$ at Earth's Lower Mantle Conditions**..160
Eglantine Boulard, Yijin Liu, Ai L. Koh, Mary M. Reagan, Julien Stodolna, Guillaume Morard, Mohamed Mezouar and Wendy L. Mao

Chapter 14   **Orientation of the Eruption Fissures Controlled by a Shallow Magma Chamber in Miyakejima**...169
Nobuo Geshi and Teruki Oikawa

Chapter 15   **Molecular and Optical Properties of Tree-Derived Dissolved Organic Matter in Throughfall and Stemflow from Live Oaks and Eastern Red Cedar**......................178
Aron Stubbins, Leticia M. Silva, Thorsten Dittmar and John T. Van Stan

Chapter 16   **Determining the Stress Field in Active Volcanoes using Focal Mechanisms**...........................191
Bruno Massa, Luca D'Auria, Elena Cristiano and Ada De Matteo

Chapter 17   **The Stoichiometry of Nutrient Release by Terrestrial Herbivores and its Ecosystem Consequences**...204
Judith Sitters, Elisabeth S. Bakker, Michiel P. Veldhuis, G. F. Veen, Harry Olde Venterink and Michael J. Vanni

Permissions

List of Contributors

Index

PREFACE

The main aim of this book is to educate learners and enhance their research focus by presenting diverse topics covering this vast field. This is an advanced book which compiles significant studies by distinguished experts in the area of analysis. This book addresses successive solutions to the challenges arising in the area of application, along with it; the book provides scope for future developments.

Earth science is the study of the planet Earth. It includes different sub-fields namely soil science, physical geography, hydrology, atmospheric science, geophysics, biogeography, mineralogy, sedimentology, etc. It also includes the study about Earth's magnetic field and Earth's atmosphere. The aim of this text is to present researches that have transformed this discipline and aided its advancement. It will provide in-depth analysis about the various techniques and concepts used in this subject. It aims to serve as a resource guide for students and experts alike and contribute to the growth of this discipline.

It was a great honour to edit this book, though there were challenges, as it involved a lot of communication and networking between me and the editorial team. However, the end result was this all-inclusive book covering diverse themes in the field.

Finally, it is important to acknowledge the efforts of the contributors for their excellent chapters, through which a wide variety of issues have been addressed. I would also like to thank my colleagues for their valuable feedback during the making of this book.

Editor

Inferring the Effects of Compositional Boundary Layers on Crystal Nucleation, Growth Textures, and Mineral Chemistry in Natural Volcanic Tephras through Submicron-Resolution Imaging

Georg F. Zellmer[1]*, Naoya Sakamoto[2], Shyh-Lung Hwang[3], Nozomi Matsuda[2], Yoshiyuki Iizuka[4], Anja Moebis[1] and Hisayoshi Yurimoto[2]

[1] *Volcanic Risk Solutions, Institute of Agriculture and Environment, Massey University, Palmerston North, New Zealand,* [2] *Isotope Imaging Laboratory, Department of Natural History Sciences, Hokkaido University, Sapporo, Japan,* [3] *Department of Materials Science and Engineering, National Dong Hwa University, Hualien, Taiwan,* [4] *Institute of Earth Sciences, Academia Sinica, Taipei, Taiwan*

Edited by:
Oliver Jagoutz,
Massachusetts Institute of
Technology, USA

Reviewed by:
Christy B. Till,
Arizona State University, USA
Fabio Arzilli,
University of Manchester, UK

*Correspondence:
Georg F. Zellmer
g.f.zellmer@massey.ac.nz

Specialty section:
This article was submitted to
Petrology,
a section of the journal
Frontiers in Earth Science

Crystal nucleation and growth are first order processes captured in volcanic rocks and record important information about the rates of magmatic processes and chemical evolution of magmas during their ascent and eruption. We have studied glass-rich andesitic tephras from the Central Plateau of the Southern Taupo Volcanic Zone by electron- and ion-microbeam imaging techniques to investigate down to sub-micrometer scale the potential effects of compositional boundary layers (CBLs) of melt around crystals on the nucleation and growth of mineral phases and the chemistry of crystal growth zones. We find that CBLs may influence the types of mineral phases nucleating and growing, and growth textures such as the development of swallowtails. The chemistry of the CBLs also has the capacity to trigger intermittent overgrowths of nanometer-scale bands of different phases in rapidly growing crystals, resulting in what we refer to as cryptic phase zoning. The existence of cryptic phase zoning has implications for the interpretation of microprobe compositional data, and the resulting inferences made on the conditions of magmatic evolution. Identification of cryptic phase zoning may improve thermobarometric estimates and thus geospeedometric constraints. In future, a more quantitative characterization of CBL formation and its effects on crystal nucleation and growth may contribute to a better understanding of melt rheology and magma ascent processes at the onset of explosive volcanic eruptions, and will likely be of benefit to hazard mitigation efforts.

Keywords: crystal nucleation, crystal growth, swallow tail textures, cryptic phase zoning, Ngauruhoe volcano, Ruapehu volcano

INTRODUCTION

Crystal nucleation and growth are first-order processes in the evolution of silicate melts during cooling or degassing (e.g., Cashman and Marsh, 1988; Marsh, 1988; Fokin et al., 2006; Toramaru et al., 2008; Hammer, 2009; Hammer et al., 2010; Melnik et al., 2011) and determine mineral and melt chemistry and their evolution, which form the foundation of thermobarometric and hygrometric constraints (e.g., Putirka, 2008; Lange et al., 2009; Waters and Lange, 2015). Minerals may also provide critical insights into the rates of magmatic processes occurring during magma evolution and ascent at the onset of volcanic eruptions, with analytical methods ranging from crystal size distribution studies (e.g., Cashman and Marsh, 1988; Marsh, 1988, 1998; Hammer et al., 1999; Piochi et al., 2005; Noguchi et al., 2006, 2008; Clarke et al., 2007; Toramaru et al., 2008; Melnik et al., 2011) to diffusion geospeedometry (e.g., Zellmer et al., 1999, 2003, 2011, 2016; Morgan et al., 2004; Costa and Dungan, 2005; Martin et al., 2008; Druitt et al., 2012; Ruprecht and Plank, 2013). Thus, understanding the processes that govern crystal nucleation and growth, as well as crystal morphology, structure, and composition, are key to unlocking volcanic hazards and many other compelling questions. Small scale variations in crystal chemistry have become apparent in the last few decades through advances in microanalytical techniques (Anderson, 1983; Wallace and Bergantz, 2002; Davidson et al., 2007; Jerram and Martin, 2008, and references therein). At present, the most advanced instruments provide submicroscopic resolution (down to nanometer scale) of crystal chemistry and crystal structure (Zellmer et al., 2015, 2016). With increasing magnification, the small-scale complexity of compositional and structural variations has become apparent, elucidating magmatic processes that were previously not resolvable. In natural tephras, the latest small-scale variations are likely acquired at the onset or during eruption of the samples, and understanding such variations may thus be crucial in the characterization and mitigation of volcanic hazards (Zellmer et al., 2016).

In this perspective article, we focus on the narrow compositional boundary layers (CBLs) of melt around growing crystals (Zhang, 2009) and their potential effects on crystal nucleation, growth, and compositional as well as phase zoning, using as examples some tephras from New Zealand's recent Central Plateau eruptions in the Southern Taupo Volcanic Zone (Moebis, 2010; Moebis et al., 2011). CBLs are the interface-melt layers around growing or dissolving crystals with composition different to that of the bulk melt. CBLs develop when crystal growth or dissolution rates exceed the rate of diffusion of the crystal-forming ions within the melt. Their thicknesses therefore depend on growth/dissolution rate of the crystal, diffusion rate of the ions, and on the relative motion of crystal and melt (Levich, 1962). We demonstrate here through semiquantitative imaging techniques that there are complex interdependencies between CBLs and crystal nucleation, growth, and zoning. A detailed quantification of these interdependencies is beyond the scope of this contribution, but we argue that their study is important and may ultimately improve volcanic hazard characterization and mitigation strategies.

METHODS

Sampling Details and Sample Preparation

Tephra samples from the Central Plateau were collected by one of us (AM) for tephrochronological work. In the present study, we report data from Mangatawai tephra 407-28 and Papakai tephra 606-30 sourced from Ngauruhoe volcano, with ages of c. 2800 and c. 3980 cal. years B.P., respectively (Moebis, 2010), and from Tufa Trig tephra 108-137 (TF13 of Donoghue et al., 1997) sourced from Ruapehu volcano, with an age of c. 540 cal. years B.P. Volcanic glass shards from individual tephra units were handpicked at Massey University. The particles were embedded in an epoxy plug (EPOTEK 301), then cut and polished with diamond pastes of successively finer grades. The final polish was made at the Institute of Earth Sciences (IES), Academia Sinica, using a vibration polisher (Buhler: Vibromet-2) with 0.3 μm alumina compounds for several hours. For electron microbeam work, the polished specimens were then coated by a layer of carbon (Q150TE, Quorum Technologies Ltd., UK). For subsequent analysis by secondary ion mass spectrometry (SIMS), the carbon coat was removed with ethanol, and samples were coated with a thin film (c. 70 nm) of gold (SC-701MC, Sanyu Electron Co., Ltd.) at the Isotope Imaging Laboratory (IIL) of Hokkaido University.

Electron Probe Microanalysis

Crystals within tephra shards were analyzed at IES using a JEOL JXA-8900R electron microprobe equipped with four wavelength-dispersive spectrometers. A 2 μm defocused beam was operated for analysis at an acceleration voltage of 15 kV with a beam current of 12 nA. Measured X-ray intensities were ZAF-corrected using the standard calibration of synthetic (s) and natural (n) chemical-known standard minerals with various diffracting crystals, as follows: diopside (s) or wollastonite (s) for Si with TAP crystal, rutile (s) for Ti with PET crystal, corundum (s) for Al (TAP), chromium oxide (s) for Cr (PET), fayalite (s), or hematite (n) for Fe with LiF crystal, tephroite (s) for Mn (PET), periclase (s) for Mg (TAP), wollastonite (s) for Ca (PET), albite (n) for Na (TAP), and adularia (n) for K (PET). Peak and both upper and lower baseline X-rays were counted for 10 s for each element, respectively. Standards run as unknowns yielded major oxide relative standard deviations for Si, Na, and K of less than 1%, and less than 0.5% for other elements. Detection limits, based on 3σ of standard calibration, were less than 600 ppm for all elements.

Electron Microscopy and Isotopography

Backscattered electron (BSE) images were obtained using a field emission scanning electron microscope (FE-SEM; JEOL JSM-7000F) at the Isotope Imaging Laboratory (IIL), Hokkaido University, equipped with a high sensitivity energy dispersive X-ray spectrometer (EDS; Oxford X-Max 150). However, high-magnification work on crystal rims could not be conducted with this technique due to edge effects. Therefore, the IIL isotope microscope system, a Cameca ims-1270 SIMS instrument equipped with a stacked CMOS-type active pixel sensor (SCAPS) for ion imaging, was applied to visualize at high magnification the elemental distribution on the sample surface ("isotopography," for details see Yurimoto et al., 2003). An O- primary beam of 23 keV was irradiated on the sample surface of approximately 60 × 60 μm^2 with beam currents ranging from c. 1.5 to 20 nA. The exit slit was narrowed enough to eliminate the contribution of interference ions to the isotopic images. The positive secondary ion images of ^{23}Na, ^{24}Mg, ^{27}Al, ^{28}Si, ^{40}Ca, and ^{56}Fe on the sample surface were collected in a SCAPS detector, with exposure times ranging between 200 and 750 s. A spatial resolution of about 300 nm was achieved for our samples. Ablation rates of only about 1 $\mu m/h$ allowed the successful acquisition of images at 5 s per frame of essentially the identical sample surface across all elements. Secondary ion intensity varied with primary beam intensity. Image gray scales were adjusted to balance these variations.

Transmission Electron Microscopy (TEM)

For TEM analyses, several samples of c. 50 nm thickness were prepared from the corona pyroxene overgrowing an olivine crystal of a selected Ruapehu tephra by applying the focused ion beam technique (FIB; SMI-3050). Selective area electron diffraction patterns and images were obtained using a transmission electron microscope (JEOL JEM-3010) operated at 300 kV. The instrument was equipped with an energy dispersive X-ray (EDX) spectrometer (Oxford EDS-6636) featuring an ultrathin window and a Si(Li) detector, capable of detection of elements from boron to uranium. EDX spectra were collected for 200 s. Semi-quantitative analysis was based on the Cliff-Lorimer thin film approximation with experimental k-factors obtained from natural minerals (Loretto, 1994).

RESULTS

Electron probe mineral chemical data are given in **Supplementary Table 1**. In summary, besides minor oxide phases that have not been analyzed but are present in all studied samples, Mangatawai tephra 407-28 contains crystals of plagioclase (An_{57-85}) and orthopyroxene ($En_{63-72}Fs_{23-27}Wo_{3-5}$); Papakai tephra 606-30 contains crystals of plagioclase (An_{45-85}) and orthopyroxene ($En_{67-73}Fs_{24-29}Wo_{3-5}$); and Tufa Trig tephra 108-137 contains crystals of plagioclase (An_{51-66}), orthopyroxene ($En_{63-84}Fs_{12-33}Wo_{3-5}$), clinopyroxene ($En_{44-55}Fs_{10-17}Wo_{35-43}$), and olivine ($Fo_{71-75}$).

The present study focusses on results from semi-quantitative high-resolution imaging. **Figures 1A–C** show SCAPS elemental maps of a large zoned plagioclase crystal, set in a glassy groundmass with microlites of plagioclase and pyroxene. A complexly zoned large plagioclase crystal experiencing synneusis (Vance, 1969; Dowty, 1980) with a microphenocryst (bottom right) is seen, with zoning particularly evident in the Na-image. Microlites show swallow-tail textures indicating rapid crystal growth. There is striking evidence of distinct CBLs of melt present around all crystals, about 1 micron in width. CBLs are enriched in Mg and depleted in Al and Na around plagioclase crystals, and vice versa around pyroxene microlites (cf. small pyroxene microlite in center right of image). Complex zoning of Na in plagioclase is evident even in the microlites, with

a wavelength of similar width to the CBL. The black arrow in c indicates plagioclase nanolites forming just outside the CBL of the large plagioclase crystal, the white arrow points to a magnesian nanolite forming in the CBL. **Figure 1D** is a BSE image of groundmass microlites and nanolites of another sample. Pyroxene microlites have a Fe-rich rim and are surrounded by Fe-poor CBLs, similar in grayscale as plagioclase microlites. Magnetite nanolites are seen distributed throughout the groundmass, but the largest are surrounding the plagioclase microlites.

Figure 2 provides an analysis of an olivine crystal with pyroxene corona texture. Panel (a) shows the BSE image of part of this crystal set in a glassy groundmass. SCAPS elemental maps of the corona texture are provided for Mg (panel b) and Ca (panel c). Variations in Ca concentration within the overgrowth are evident, as observed previously in other corona orthopyroxenes (Zellmer et al., 2016), with a very strong enrichment in the rim of the overgrowth, and concomitant depletion in the CBL. Weaker Ca-enrichment zones are evident within the overgrowth. Such calcic lamellae are ubiquitous in overgrowth orthopyroxenes from the southern Taupo Volcanic Zone, and also occur in some orthopyroxene microphenocrysts (cf. Zellmer et al., 2016). Panels (d) and (e) are TEM bright field images of FIB sections, the location of which are indicated in panel (b). The images show clear phase boundaries of nm-thin layers of Ca-rich clinopyroxenes within Ca-poor orthopyroxene overgrowth. Two clinopyroxene domains can be identified by semiquantitative EDX spectrometry: domain A adjacent to the glass, with Al-enrichment (panel f), and domain B, inside the overgrowth, low in Al. Neither domain is in Fe/Mg exchange equilibrium with their orthopyroxene host, the EDX spectrum of which is provided in **Supplementary Image 1**. In equilibrium, K_D(Fe-Mg) should be 1.09 ± 0.14 (Putirka, 2008). Domain A yields a K_D(Fe-Mg) of >2.00, while domain B yields a K_D(Fe-Mg) of <0.80. Two-pyroxene thermobarometry therefore cannot be conducted.

DISCUSSION

Crystal nucleation may initially be randomly distributed within the glass, i.e., may be homogeneous (Fokin et al., 2006). However, crystal growth and associated CBL development appears to result in a more favorable nucleation environment of magnesian phases (such as pyroxene) in magnesian CBLs, which are less favorable for nucleation of Mg-poor phases (such as plagioclase) that will nucleate away from these CBLs (**Figure 1C**). Mg-poor phases will preferentially nucleate in Mg-poor CBLs. For example, preferential plagioclase nucleation would be expected in the CBLs developing around growing orthopyroxene crystals (further discussed below, cf. **Figure 1D**). Inhomogeneous crystal nucleation has previously been described (e.g., Hammer et al., 2010), and has recently been associated with CBLs that developed during dissolution of natural olivine in SiO_2-rich melts (Zellmer et al., 2016).

CBLs will be less depleted in crystal-forming elements at crystal corners, and this may be the reason for the development of swallow-tail textures (Vernon, 2004), which form by more

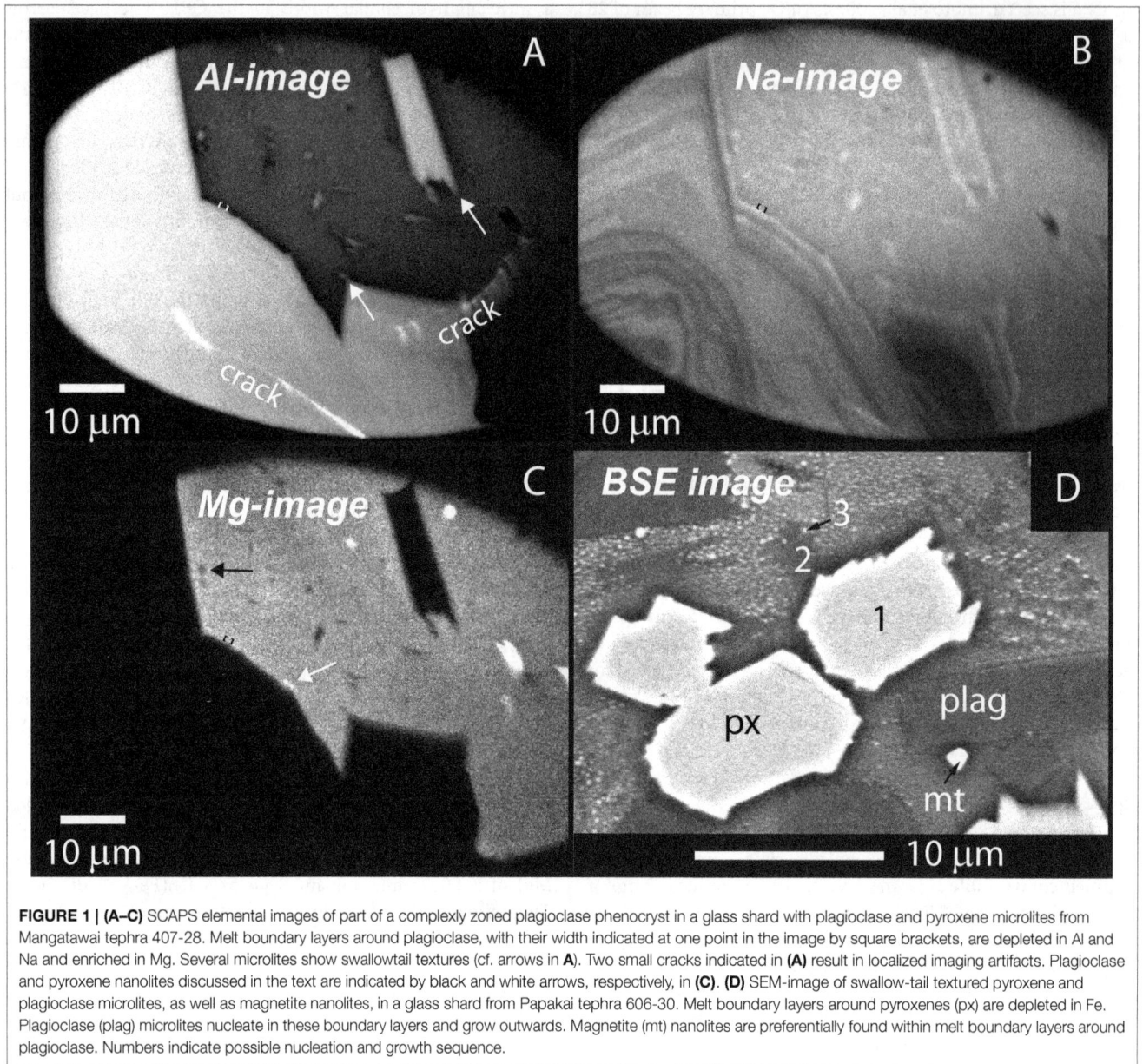

FIGURE 1 | (A–C) SCAPS elemental images of part of a complexly zoned plagioclase phenocryst in a glass shard with plagioclase and pyroxene microlites from Mangatawai tephra 407-28. Melt boundary layers around plagioclase, with their width indicated at one point in the image by square brackets, are depleted in Al and Na and enriched in Mg. Several microlites show swallowtail textures (cf. arrows in **A**). Two small cracks indicated in **(A)** result in localized imaging artifacts. Plagioclase and pyroxene nanolites discussed in the text are indicated by black and white arrows, respectively, in **(C)**. **(D)** SEM-image of swallow-tail textured pyroxene and plagioclase microlites, as well as magnetite nanolites, in a glass shard from Papakai tephra 606-30. Melt boundary layers around pyroxenes (px) are depleted in Fe. Plagioclase (plag) microlites nucleate in these boundary layers and grow outwards. Magnetite (mt) nanolites are preferentially found within melt boundary layers around plagioclase. Numbers indicate possible nucleation and growth sequence.

rapid crystal growth out from those corners in response to greater availability of crystal-forming elements within these local environments (**Figures 1A–C**). Crystal chemical zonation at circa 1 micron wavelength (**Figure 1B**) may be the response of CBL development at the same length scale, and subsequent CBL destruction, e.g., by movement of the crystal through the melt and concomitant erosion of the CBL, resulting in a regular local variations of ions available to the growing crystals. Such variations are even seen in microlites. Repetitive CBL development and destruction may thus represent an alternative way to form the fine-scale outer zoning in crystal composition observed in many volcanic phenocrysts, as opposed to being due to growth during magmatic rejuvenation and convection in subvolcanic reservoirs (cf. Shelley, 1993). Timescales predicted by

modeling diffusion at intracrystalline boundaries of incompatible element enriched zones may thus not always be associated with magmatic intrusion events, for example, but instead might be recording timescales of crystallization during cooling or decompression-induced degassing.

As a result of these complex interactions of crystal nucleation and CBL formation by crystal growth, small scale heterogeneities in the mineral distribution of the groundmass may result. **Figure 1D** provides an example where orthopyroxene microlite growth has apparently resulted in Ca and Al enrichment of developing CBLs, which served as nucleation environments for plagioclase microlites. These, in turn, generated Fe-rich CBLs that represented favorable environments for nucleation of growth of magnetite nanolites.

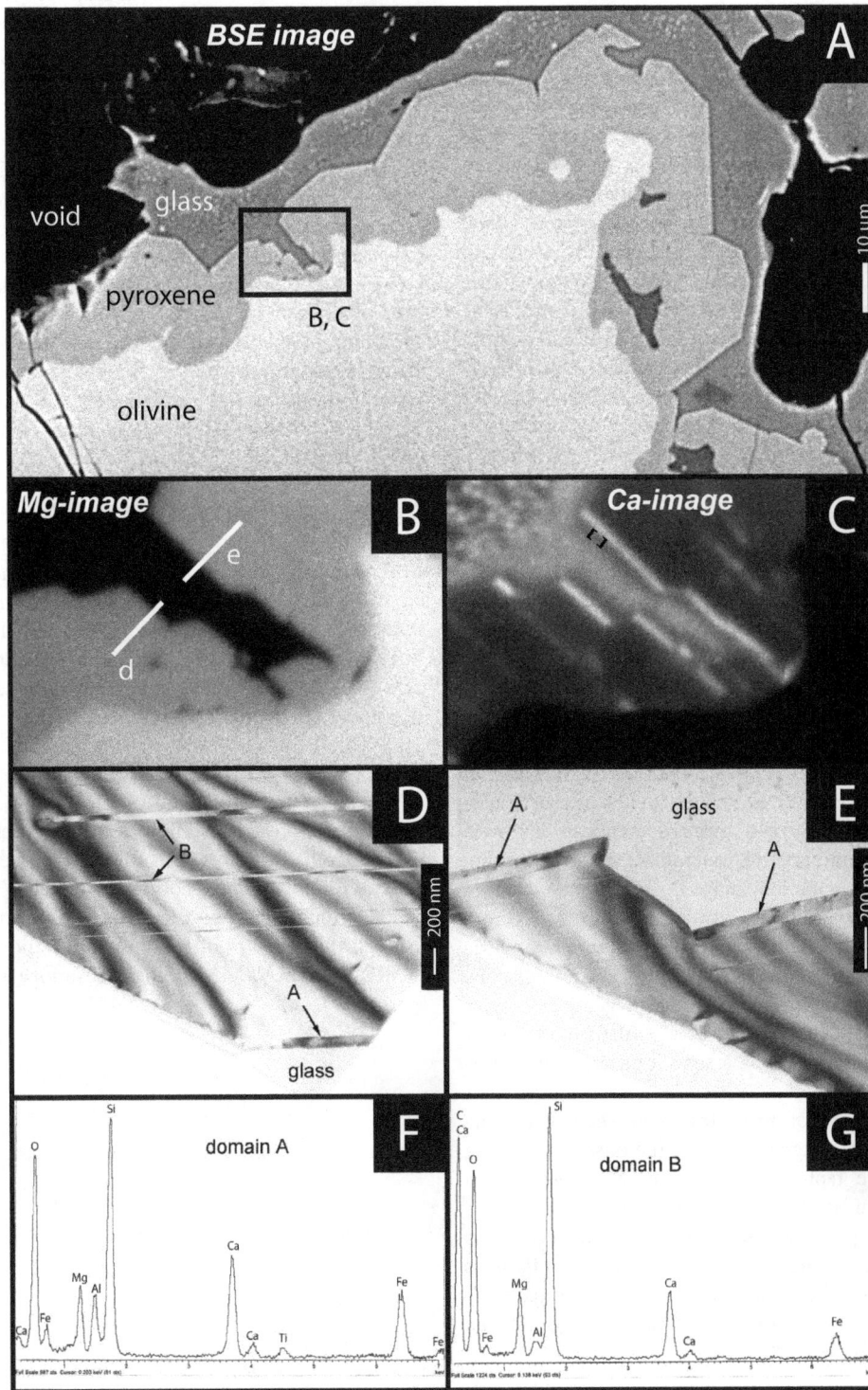

FIGURE 2 | Pyroxene corona textures on olivine displaying cryptic phase zoning in a glass shard from Tufa Trig tephra 108-137. (A) Backscattered electron image of part of the grain, indicating the area investigated by SCAPS imaging. **(B)** SCAPS Mg-image, showing the Mg-rich olivine (bright), the pyroxenes (gray), and the Mg-poor glass (dark). Position of focussed ion beam (FIB) sections for TEM work is indicated. **(C)** SCAPS Ca-image showing the Ca-poor olivine (dark), the dark-gray low-Ca pyroxene with some bright high-Ca bands, and the glass with variable Ca-content: low in the boundary layer (indicated by square brackets) right next to the outermost high-Ca growth zone, high in between the two pyroxene crystals, and highly variable away from the crystals. **(D,E)** TEM bright field images of FIB sections showing nm-scale bands of the high Ca-phase at the pyroxene rims (domain A), as well as inside the pyroxene crystal (domain B). Semi-quantitative analytical electron microscopy energy-dispersive X-ray spectra of high-Ca domains A and B are provided in **(F,G)**, respectively, yielding a higher-Ca, low Mg/Fe aluminous pyroxene in domain A and a lower-Ca, high Mg/Fe pyroxene in domain B. See **Supplementary Image 1** for the EDX spectrum of the orthopyroxene host crystal, provided for comparison.

The effect of CBLs may not only be restricted to variations in major (and presumably trace) element zonation in phases displaying solid solution, but may extend to triggering phase changes on microscopic to submicroscopic scales: **Figure 2** illustrates the submicroscopic phase changes within rapidly growing pyroxene crystals, which is what one may refer to as cryptic phase zoning. These lamellae are suggestive of exsolution textures, with clinopyroxene exsolving from orthopyroxene (e.g., Vernon, 2004). However, if this was the case, one would expect Fe/Mg exchange equilibrium between the host and exsolution zones. Although the EDX chemical analysis is only semiquantitative, given the narrow width of the clinopyroxene bands, the magnitude of the measured disequilibria of both clinopyroxene domains with their orthopyroxene host is likely too great to be attributable to analytical uncertainties. Thus, we prefer a scenario where rapid orthopyroxene growth results in a calcic CBL, which intermittently triggers brief precipitation intervals of clinopyroxene (domain B) instead of orthopyroxene. These intermittent intervals of clinopyroxene precipitation would then rapidly deplete the CBL in Ca, such that the system switches back to orthopyroxene as the preferred phase. Finally, domain A at the edge of the crystal likely represents growth related to quenching. In this scenario, cryptic phase zoning is related to crystal growth and associated chemical variations of the evolving CBL over time.

The presence of cryptic phase zoning in pyroxenes represents a challenge for two-pyroxene thermobarometry as well as pyroxene-melt thermobarometry (e.g., Putirka, 2008). Electron microprobe analysis (EMPA) is unable to resolve such sub-micron features and will yield average compositions with somewhat elevated amounts of calcium. Detailed imaging of submicron phase zonation may in future allow more successful targeting of crystal growth zones (cf. Zellmer et al., 2015) and thus may improve thermobarometric constraints. Temperature uncertainties still represent one of the principle limiting factors in diffusion geospeedometry (Costa and Morgan, 2010; Petrone et al., 2016), and tighter temperature constraints would be a significant advance in this respect.

Nucleation and growth of microlites are processes occurring during the last stage of magma ascent at the onset and during eruption of volcanic tephras (Hammer et al., 1999; Piochi et al., 2005; Noguchi et al., 2006, 2008; Clarke et al., 2007). Understanding these processes down to submicrometer scale is important to better characterize the crystallization of microlites in volcanic conduits, which have been used to estimate the timescales of magma ascent in order to characterize and mitigate the hazards of explosive volcanic eruptions (Zellmer et al., 2016). Our study shows that CBLs in natural magmas may significantly influence crystal nucleation and crystal growth, as well as the resulting crystal morphologies, crystal chemical zonation, and the distribution of small scale heterogeneities within the crystallizing groundmass. We have outlined some of the complex interplays of crystal nucleation, growth, CBL formation, and crystal chemical as well as phase zonation. Additional work will be required to properly quantify these processes, including the characterization of potential variations in the width of CBLs between different samples, and what this might reveal in terms of relative movement of crystals and melt, and thus about potential variations in melt rheology and melt ascent processes. Given the growing importance of the role of igneous petrology in informing volcanic hazards and their mitigation (Saunders et al., 2012), we anticipate significant advances in this field in the near future.

AUTHOR CONTRIBUTIONS

GFZ designed the project, assisted with analytical work, interpreted the data and wrote the paper. SH undertook TEM analyses. NS and NM undertook SEM and SIMS imaging work. YI undertook EPMA work. NS and HY checked SEM and SIMS data quality. AM collected and prepared the samples. All authors contributed to the discussion of results.

ACKNOWLEDGMENTS

We thank Fred Davis and Bob Stewart for useful discussions and Oliver Jagoutz for editorial handling of this contribution. We are grateful for constructive comments of Fabio Arzilli and Christy B. Till, which improved the script. GFZ acknowledges funding through the Ministry of Business, Innovation and Employment (MBIE, grant MAUX1507), and the National Geographic Society (grant 9577-14).

REFERENCES

Anderson, A. T. Jr. (1983). Oscillatory zoning of plagioclase: Nomarski interference contrast microscopy of etched polished sections. *Am. Mineral.* 68, 125–129.

Cashman, K. V., and Marsh, B. D. (1988). Crystal size distribution (CSD) in rocks and the kinetics and dynamics of crystallisation. II: Makaopuhi lava lake. *Contrib. Mineral. Petrol.* 99, 292–305. doi: 10.1007/BF00375363

Clarke, A. B., Stephens, S., Teasdale, R., Sparks, R. S. J., and Diller, K. (2007). Petrologic constraints on the decompression history of magma prior to Vulcanian explosions at the Soufrière Hills volcano, Montserrat. *J. Volcanol. Geother. Res.* 161, 261–274. doi: 10.1016/j.jvolgeores.2006.11.007

Costa, F., and Dungan, M. (2005). Short time scales of magmatic assimilation from diffusion modeling of multiple elements in olivine. *Geology* 33, 837–840. doi: 10.1130/G21675.1

Costa, F., and Morgan, D. J. (2010). "Time constraints from Chemical Equilibration in Magmatic Crystals," in *Timescales of Magmatic Processes: From Core to Atmosphere*, eds A. Dosseto, S. P. Turner, and J. A. Van Orman (Chichester: John Wiley & Sons, Ltd.), 125–159.

Davidson, J. P., Morgan, D. J., and Charlier, B. L. A. (2007). Frontiers in textural and microgeochemical analysis: isotopic microsampling of magmatic rocks. *Elements* 3, 253–259. doi: 10.2113/gselements.3.4.253

Donoghue, S. L., Neall, V. E., Palmer, A. S., and Stewart, R. B. (1997). The volcanic history of Ruapehu during the past 2 millennia based on the record of Tufa Trig tephras. *Bull. Volcanol.* 59, 136–146. doi: 10.1007/s004450050181

Dowty, E. (1980). Synneusis reconsidered. *Contrib. Mineral. Petrol.* 74, 75–84. doi: 10.1007/BF00375491

Druitt, T. H., Costa, F., Deloule, E., Dungan, M., and Scaillet, B. (2012). Decadal to monthly timescales of magma transfer and reservoir growth at a caldera volcano. *Nature* 482, 77–80. doi: 10.1038/nature10706

Fokin, V. M., Zanotto, E. D., Yuritsyn, N. S., and Schmelzer, J. W. P. (2006). Homogeneous crystal nucleation in silicate glasses: a 40 years perspective. *J. Non-Crystal. Solids* 352, 2681–2714. doi: 10.1016/j.jnoncrysol.2006.02.074

Hammer, J. E. (2009). Capturing crystal growth. *Geology* 37, 1055–1056. doi: 10.1130/focus112009.1

Hammer, J. E., Cashman, K. V., Hoblitt, R. P., and Newman, S. (1999). Degassing and microlite crystallization during pre-climactic events of the 1991 eruption of Mt. Pinatubo, Philippines. *Bull. Volcanol.* 60, 355–380. doi: 10.1007/s004450050238

Hammer, J. E., Sharp, T. G., and Wessel, P. (2010). Heterogeneous nucleation and epitaxial crystal growth of magmatic minerals. *Geology* 38, 367–370. doi: 10.1130/G30601.1

Jerram, D. A., and Martin, V. M. (2008). "Understanding crystal populations and their significance through the magma plumbing system," in *Dynamics of Crustal Magma Transfer, Storage and Differentiation*, eds C. Annen and G. F. Zellmer (London: Geological Society), 133–148.

Lange, R. A., Frey, H. M., and Hector, J. (2009). A thermodynamic model for the plagioclase-liquid hygrometer/thermometer. *Am. Mineral.* 94, 494–506. doi: 10.2138/am.2009.3011

Levich, V. G. (1962). *Physicochemical Hydrodynamics*. Englewood Cliffs, NJ: Prentic-Hall.

Loretto, H. M. (1994). *Electron Beam Analysis of Materials*. London: Chapman & Hall.

Marsh, B. D. (1988). Crystal size distribution (CSD) in rocks and the kinetics and dynamics of crystallisation. I: Theory. *Contrib. Mineral. Petrol.* 99, 277–291. doi: 10.1007/BF00375362

Marsh, B. D. (1998). On the interpretation of crystal size distributions in magmatic systems. *J. Petrol.* 39, 553–599. doi: 10.1093/petroj/39.4.553

Martin, V. M., Morgan, D. J., Jerram, D. A., Caddick, M. J., Prior, D. J., and Davidson, J. P. (2008). Bang! Month-scale eruption triggering at Santorini volcano. *Science* 321, 1178. doi: 10.1126/science.1159584

Melnik, O. E., Blundy, J. D., Rust, A. C., and Muir, D. D. (2011). Subvolcanic plumbing systems imaged through crystal size distributions. *Geology* 39, 403–406. doi: 10.1130/G31691.1

Moebis, A. (2010). *Understanding the Holocene Explosive Eruption Record of the Tongariro Volcanic Centre*. PhD thesis, Massey University, Palmerston North.

Moebis, A., Cronin, S. J., Neall, V. E., and Smith, I. E. (2011). Unravelling a complex volcanic history from fine-grained, intricate Holocene ash sequences at the Tongariro Volcanic Centre, New Zealand. *Q. Int.* 246, 352–363. doi: 10.1016/j.quaint.2011.05.035

Morgan, D. J., Blake, S., Rogers, N. W., De Vivo, B., Rolandi, G., Macdonald, R., et al. (2004). Time scales of crystal residence and magma chamber volume from modelling of diffusion profiles in phenocrysts: Vesuvius 1944. *Earth Planet. Sci. Lett.* 222, 933–946. doi: 10.1016/j.epsl.2004.03.030

Noguchi, S., Toramaru, A., and Nakada, S. (2008). Relation between microlite textures and discharge rate during the 1991-1995 eruptions at Unzen, Japan. *J. Volcanol. Geother. Res.* 175, 141–155. doi: 10.1016/j.jvolgeores.2008.03.025

Noguchi, S., Toramaru, A., and Shimano, T. (2006). Crystallization of microlites and degassing during magma ascent: constraints on the fluid mechanical behaviour of magma during the Tenjo Eruption on Kozu Island, Japan. *Bull. Volcanol.* 68, 432–449. doi: 10.1007/s00445-005-0019-4

Petrone, C. M., Braschi, B., and Tommasini (2016). Pre-eruptive magmatic processes re-timed using a non-isothermal approach to magma chamber dynamics. *Nat. Commun.* 7:12946. doi: 10.1038/ncomms12946

Piochi, M., Mastrolorenzo, G., and Pappalardo, L. (2005). Magma ascent and eruptive processes from textural and compositional features of Monte Nuovo pyroclastic products, Campi Flegrei, Italy. *Bull. Volcanol.* 67, 663–678. doi: 10.1007/s00445-005-0410-1

Putirka, K. D. (2008). "Thermometers and barometers for volcanic systems," in *Reviews in Mineralogy and Geochemistry*, Vol. 69, eds K. D. Putirka and F. Tepley (Chantilly, VA: Mineralogical Society of America), 61–120.

Ruprecht, P., and Plank, T. (2013). Feeding andesitic eruptions with a high-speed connection from the mantle. *Nature* 500, 68–72. doi: 10.1038/nature12342

Saunders, K., Blundy, J., Dohmen, R., and Cashman, K. (2012). Linking petrology and seismology at an active volcano. *Science* 336, 1023–1027. doi: 10.1126/science.1220066

Shelley, D. (1993). *Igneous and Metamorphic Rocks under the Microscope: Classification, Textures, Microstructures and Mineral Preferred Orientation*. London: Chapman & Hall.

Toramaru, A., Noguchi, S., Oyoshihara, S., and Tsune, A. (2008). MND (microlite number density) water exsolution rate meter. *J. Volcanol. Geother. Res.* 175, 156–167. doi: 10.1016/j.jvolgeores.2008.03.035

Vance, J. A. (1969). On synneusis. *Contrib. Mineral. Petrol.* 24, 7–29. doi: 10.1007/BF00398750

Vernon, R. H. (2004). *A Practical Guide to Rock Microstructure*. Cambridge, New York, NY; Madrid, Cape Town; Singapore, Sao Paulo, New Delhi: Cambridge University Press.

Wallace, G. S., and Bergantz, G. W. (2002). Wavelet-based correlation (WBC) of zoned crystal populations and magma mixing. *Earth Planet. Sci. Lett.* 202, 135–145. doi: 10.1016/S0012-821X(02)00762-8

Waters, L. E., and Lange, R. A. (2015). An updated calibration of the plagioclase-liquid hygrometer-thermometer applicable to basalts through rhyolites. *Am. Mineral.* 100, 2172–2184. doi: 10.2138/am-2015-5232

Yurimoto, H., Nagashima, K., and Kunihiro, T. (2003). High precision isotope micro-imaging of materials. *Appl. Surface Sci.* 203–204, 793. doi: 10.1016/S0169-4332(02)00825-5

Zellmer, G. F., Blake, S., Vance, D., Hawkesworth, C., and Turner, S. (1999). Plagioclase residence times at two island arc volcanoes (Kameni islands, Santorini, and Soufriere, St. Vincent) determined by Sr diffusion systematics. *Contrib. Mineral. Petrol.* 136, 345–357. doi: 10.1007/s004100050543

Zellmer, G. F., Hawkesworth, C. J., Sparks, R. S. J., Thomas, L. E., Harford, C., Brewer, T. S., et al. (2003). Geochemical evolution of the Soufrière Hills volcano, Montserrat, Lesser Antilles Volcanic Arc. *J. Petrol.* 44, 1349–1374. doi: 10.1093/petrology/44.8.1349

Zellmer, G. F., Hwang, S.-L., Sakamoto, N., Iizuka, Y., Harada, S., Kimura, J.-I., et al. (2015). "Interaction of arc magmas with subvolcanic hydrothermal systems: insights from compositions and metasomatic textures of olivine crystals in fresh basalts of Daisen and Mengameyama, Western Honshu, Japan," in *The Role of Volatiles in the Genesis, Evolution and Eruption of Arc Magmas*, eds G. F. Zellmer, M. Edmonds, and S. M. Straub (London: Geological Society; Special Publications), 219–236.

Zellmer, G. F., Rubin, K. H., Dulski, P., Iizuka, Y., Goldstein, S. L., and Perfit, M. R. (2011). Crystal growth during dike injection of MOR basaltic melts: evidence from preservation of local Sr disequilibria in plagioclase. *Contrib. Mineral. Petrol.* 161, 153–173. doi: 10.1007/s00410-010-0518-y

Zellmer, G. F., Sakamoto, N., Matsuda, N., Iizuka, Y., Moebis, A., and Yurimoto, H. (2016). On progress and rate of the peritectic reaction Fo + SiO_2 → En in natural andesitic arc magmas. *Geochim. Cosmochim. Acta* 185, 383–393. doi: 10.1016/j.gca.2016.01.005

Zhang, Y. (2009). *Geochemical Kinetics*. Princeton, NJ: Princeton University Press.

Conflict of Interest Statement: The authors declare that the research was conducted in the absence of any commercial or financial relationships that could be construed as a potential conflict of interest.

2

The Feeding Habits of Mesosauridae

Rivaldo R. Silva[1], Jorge Ferigolo[2], Piotr Bajdek[3] and Graciela Piñeiro[4]*

[1] Laboratório de Paleontologia da Universidade Luterana do Brasil, Torres, Brazil, [2] Museu de Ciências Naturais da Fundação Zoobotânica do Rio Grande do Sul, Porto Alegre, Brazil, [3] Independent Researcher, Aleja Najświętszej Maryi Panny 20/20A, Częstochowa, Poland, [4] Facultad de Ciencias, Instituto de Ciencias Geológicas, Montevideo, Uruguay

Mesosauridae comprises the oldest known aquatic amniotes which lived in Gondwana during the Early Permian. Previous work in the Uruguayan mesosaur-bearing Mangrullo Formation suggested that mesosaurids lived in an inland water body, inferred as moderately hypersaline, with exceptional preservational conditions that justified describing these strata as a Fossil-Lagerstätte. Exquisitely preserved articulated mesosaur skeletons, including gastric content and associated coprolites, from the Brazilian Iratí Formation in the State of Goiás (central-western Brazil) indicate excellent conditions of preservation, extending the Konservat-Lagerstätte designation to both units in the Paraná Basin. The near-absence of more resistant fossil remains, like actinopterygian and temnospondyl bones, demonstrates the faunistic poverty of the mesosaur-bearing "salty sea." Our studies of the alimentary habits of mesosaurids through the use of stereoscopic microscopy, light and electronic microscopy, and X-ray diffractometry suggest that the diet of mesosaurids was predominantly composed of pygocephalomorph crustaceans (possibly not exceeding 20 mm in length). However, the presence of bones and bone fragments of small mesosaurs in the gastric content, cololites, coprolites, and possible regurgitalites may also indicate cannibalistic and/or scavenging habits. Cannibalism is relatively common among vertebrates, particularly during conditions of environmental stress, like food shortage. Likewise, the apparent abundance of pygocephalomorph crustacean fossils in the Iratí and Mangrullo Formations, outside and within the studied gastric, cololite, and coprolite contents, might have to do with environmental stress possibly caused by volcanic activity, in particular ash spread into the basin during the Early Permian. In this context, casual necrophagy on the dead bodies of small mesosaurs and large pygocephalomorphs might have been an alternative alimentary behavior adopted for survival in mesosaurs.

Keywords: Mesosauridae, Early Permian, Iratí Formation, Mangrullo Formation, Fossil-Lagerstätte, diet, bromalites

Edited by:
Corwin Sullivan,
Institute of Vertebrate Paleontology
and Paleoanthropology (CAS), China

Reviewed by:
David Marjanović,
Museum für Naturkunde, Germany
Eric M. Roberts,
James Cook University Townsville,
Australia

*Correspondence:
Graciela Piñeiro
fossil@fcien.edu.uy

Specialty section:
This article was submitted to
Paleontology,
a section of the journal
Frontiers in Earth Science

INTRODUCTION

Fossilized feces, gastric contents, and regurgitations ("bromalites," sensu Hunt et al., 1994; see also Hunt and Lucas, 2012) can tell us a lot about the dietary behavior of extinct organisms (Chin and Kirkland, 1998; Chin et al., 2008; Bajdek et al., 2014; Qvarnström et al., 2016). They mostly fossilize *in situ* (Hu et al., 2010; Bajdek, 2013), as isolated coprolites (Eriksson et al., 2011; Dentzien-Dias et al., 2012; Niedźwiedzki et al., 2016) or associated with the producer's skeleton (Hone et al., 2015; Wang et al., 2016). However, the latter is a far less frequent situation and the coprolite producers are usually very hard or impossible to identify. The mesosaur coprolites described herein represent one of the rare cases where the producer can be identified with confidence, because no other tetrapod or vertebrate is found in the mesosaur-bearing strata (Piñeiro, 2006; Piñeiro et al., 2012b). Also, the present study represents an exceptional case where coprolites, gastric contents, cololites, and regurgitalites of a single animal species, i.e., all the "basic" bromalite types, are described. Lower Permian vertebrate coprolites have been briefly described in several papers (e.g., Hunt and Lucas, 2005a,b; Hunt et al., 2005a,b; Shelton, 2013), but more detailed studies were until now restricted to Middle/Upper Permian materials (Smith and Botha-Brink, 2011; Dentzien-Dias et al., 2012, 2013; Owocki et al., 2012; Bajdek et al., 2016; Niedźwiedzki et al., 2016).

Mesosauridae comprises the oldest known aquatic reptiles which lived in Gondwana during the Early Permian. Most authors agree that Mesosauridae are basal sauropsids (e.g., Laurin and Reisz, 1995; Piñeiro et al., 2016) or basal parareptiles (Modesto, 1999; Tsuji and Müller, 2009), including three monotypic species: *Mesosaurus tenuidens*, *Stereosternum tumidum*, and *Brazilosaurus sanpauloensis*. All three species are found in Brazil, whereas only *M. tenuidens* has been reported from Uruguay (Piñeiro, 2004; Morosi, 2011; Piñeiro et al., 2012a). Nevertheless, recent studies would suggest the presence of not more than two species in the Paraná Basin (Piñeiro, 2004, 2006; Piñeiro et al., 2015a, 2016), and the taxonomic diversity of the mesosaurids from the African Karoo Basin remains uncertain (Piñeiro et al., 2015a).

Previously, the diet of mesosaurs was only inferred from indirect evidence. MacGregor (1908) suggested that mesosaurids preyed on fish because of their long snout and very peculiar tooth morphology. Romer (1966) and Carroll (1982) followed Frech's (1897–1902) suggestion that the long and horizontalized teeth of mesosaurs could indicate a sludge filter-type habit. Later, Oelofsen and Aráujo (1983), based on the distribution of the fauna in the "Iratí-Sea," proposed that *Stereosternum* and *Brazilosaurus*, inhabitants of shallow waters, had a crustacean-based diet (e.g., *Liocaris*). Araújo (1976), Araújo-Barberena (1994), and Carroll (1988) followed the proposition that *Stereosternum* and *Brazilosaurus* were predators. Piñeiro (2002, 2004, 2006) questioned the hypothesis of MacGregor (1908) because of the absence of "fishes" in the mesosaur-bearing strata of the Lower Permian Mangrullo Formation of Uruguay and probably in the correlative Brazilian Iratí and South African Whitehill formations (Piñeiro, 2002, 2004). However, according to some Brazilian paleontologists, additional biostratigraphic studies will be needed to confirm this (Marina Bento-Soares, personal communication). Accordingly, Modesto (1996, 2006) considered that invertebrates of medium or large size would be preferred prey items of mesosaurs. In Uruguay, it is clear that isolated actinopterygian, acanthodian and coelacanthid scales, teeth and bones are present in levels that overlie and underlie those containing mesosaurs (see also Piñeiro et al., 2012b). Some bromalites, very probably produced by "fishes," are found in the layers dominated by actinopterygian, coelacanthid, and acanthodian remains (Piñeiro, 2002, 2008; Piñeiro et al., 2012b). They are three-dimensional, sub-spherical structures containing tiny fragments of thick ganoid scales; despite their poor preservation, they are clearly spiral. Thus, because of their very

different morphology with respect to the mesosaur coprolites, they are good stratigraphic markers in incompletely preserved sections of the Mangrullo Formation that do not contain mesosaur remains (but see critical notes on coprostratigraphy in Bajdek et al., 2014; Niedźwiedzki et al., 2016).

Recently, Pretto et al. (2014), based on histological studies of the tooth microstructure in *Stereosternum tumidum*, reported the presence of a tripartite alveolar periodontium that increased the resistance of the teeth against bite forces; they used this evidence to suggest that mesosaurids could have been predatory animals, feeding on crustaceans and juvenile mesosaurids.

The first direct evidence about mesosaurid dietary preferences was provided by Raimundo-Silva et al. (1997) and Raimundo-Silva (1999) through the study of an almost complete specimen assigned to *Brazilosaurus sanpauloensis* which preserves a cololite and some associated coprolites. Moreover, several isolated, well-preserved coprolites that could belong to the same species were also examined. Later, Piñeiro et al. (2012b) reported the discovery of similar coprolites from the Mangrullo Formation preserved in isolation or associated with adults, and also gastric contents in specimens assigned to *Mesosaurus tenuidens*. The material in both the gastric contents and the coprolites was identified as carapaces and abdomens of pygocephalomorph crustaceans, in some cases mixed with bones of small mesosaurids and putative plant remains. Recently, Ramos (2015) analyzed the coprolite and gastric contents from the Mangrullo Formation of Uruguay, concluding that mesosaurids could have been selective predators which fed on prey items smaller than 20 mm in length and that fragmentary, small mesosaur bones might have been accidentally ingested while catching pygocephalomorphs—taking into account the evidence presented therein (and in this contribution) that crustaceans could have been scavengers of mesosaur carcasses.

This paper aims to present an updated overview of the existing knowledge about the alimentary habits of Mesosauridae. Gastric contents, cololites, coprolites, and regurgitalites from the Lower Permian Iratí (Brazil) and Mangrullo (Uruguay) formations are described in detail. The trophic and taphonomic differences between these two paleoenvironments in the Paraná Basin and the mesosaurid species *Brazilosaurus sanpauloensis* and *Mesosaurus tenuidens* are discussed.

MATERIALS AND METHODS

Material

The materials used in this study come from Lower Permian deposits from Brazil (Iratí Formation) and Uruguay (Mangrullo Formation; **Figure 1**). The specimens from the Mangrullo Formation were found at the Paso del Cuello locality, Tacuarembó County, northern Uruguay, and are housed at the Vertebrate Collection of the Facultad de Ciencias (FC-DPV), Universidad de la República Oriental del Uruguay (**Figures 1A–C,E–G**), while the samples from the Iratí Formation were collected from outcrops in the municipality of Perolândia (Goiás State, Brazil) and are deposited in the Museu de Ciências Naturais of the Fundação Zoobotânica do Rio Grande do Sul, Paleovertebrate Collection (MCN-PV; **Figures 1D,H,I**). The

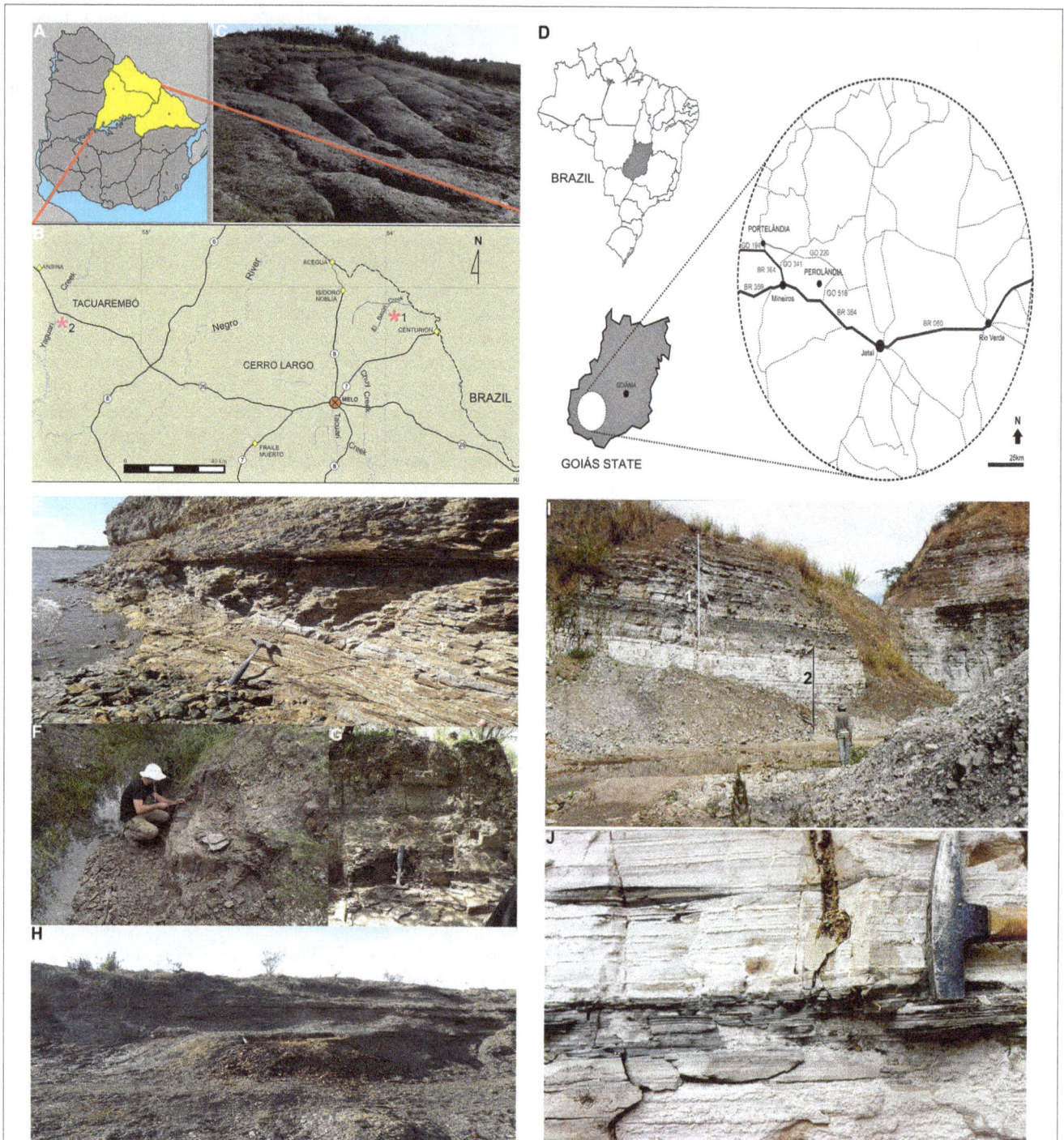

FIGURE 1 | Geographic location of the Uruguayan Mangrullo and the Brazilian Iratí Lower Permian Konservat-Lagerstätten, and principal outcrops.
(A) Map of Uruguay showing the area of outcrops of the Mangrullo Formation; **(B)** Main localities where the studied Uruguayan specimens were found; **(C)** Photograph of the Mangrullo Formation section at the El Barón locality; **(D)** Geographic location of outcrops of the Early Permian Iratí Formation that provided the studied specimens from Brazil; **(E)** Lithostratigraphic section at the El Barón locality (Cerro Largo County, Uruguay) showing the basalmost shales where mesosaurid, pygocephalomorph, and plant remains have appeared; **(F)** Limestones and siltstones at the Arroyo Yaguarí locality (Tacuarembó County, Uruguay) which yielded mesosaurid and pygocephalomorph specimens as the most frequent association; **(G)** Detail of the profile depicted in **(F)**; **(H)** Bituminous shales at the El Barón locality (Cerro Largo County) intercalated by several bentonitic layers and levels of gypsum crystals; **(I)** General view of the litostratigraphic section at the SUCAL quarry (Perolândia, Goiás State, Brazil), which consists of limestones intercalated with shales at the middle and the top (1) and yellowish limestones at the base (2); **(J)** Detail of the basal limestone showing the gray, fossiliferous levels in Goiás. All the localities shown have yielded the mesosaur coprolites described herein but only Arroyo Yaguarí has provided specimens preserving gastric contents.

specimens (**Figures 2–9**) consist of: (1) partially articulated mesosaurid individuals preserving gastric and gut contents, plus associated coprolites (**Figure 2**), (2) partially articulated individuals preserving only gastric contents (**Figure 6**), (3) isolated coprolites (**Figures 3,4,7,8**), and (4) regurgitalites (**Figure 9**). Some other mesosaurid materials from São Paulo State, belonging to the fossil collection of the Instituto de Geociências, Universidade de São Paulo, Brazil (GP-2E) were also studied by GP. The complete list of the examined 136 specimens is provided as Supplementary Information. Skeletons from Brazil have been mostly preserved as original bones, in place or slightly displaced. Partial skeletons from the Mangrullo Formation are mostly preserved as molds.

All identifiable skeletal remains from Brazil here studied were attributed to *Brazilosaurus sanpauloensis* by Raimundo-Silva et al. (1997) and Raimundo-Silva (1999) based on the criteria established by Oelofsen and Aráujo (1983); Oelofsen and Araújo (1987): adults developing less pachyostotic ribs, and hemal arches which are U-shaped due to their marked pachyostosis. However, these diagnostic characters turned out not to be useful, considering that the ribs are in fact wider than the radius, ulna, tibia, or fibula in specimens assigned to *Brazilosaurus*, and that U-shaped hemal arches can be present

along with V-shaped ones in the tail of a single individual (see Piñeiro et al., 2015a for a revision of the *Brazilosaurus* status and general mesosaurid diversity). Most of the materials were collected *ex-situ* in abandoned and active quarries, when rock debris and blocks formed as the result of mining activities. There are also specimens that were collected *in situ*. The materials collected from the debris of quarries often represent single blocks fragmented into several parts, which are here referred to as portions A, B, C of the same specimen. Moreover, most of the studied specimens have split into part and counterpart. They consist of two complementary plates that will be referred to as A/A′, B/B′ and so on. For instance, in MCN-PV 2254, a partially articulated adult individual, part, and counterpart (MCN-PV 2254 A and A′; **Figure 2**) are differentiated in that one (the part) shows bones and the other (the counterpart) just molds. Other specimens consist of molds in both slabs due to loss by weathering.

Methods

The isolated bones, and the dispersed material around the adult individuals preserving gut contents, were measured and compared to the bones of the specimens studied by Sedor (1994) and Piñeiro et al. (2016) in order to determine their relative age and the minimum number of juvenile individuals associated with the adults in each of the samples (from Brazil and from Uruguay).

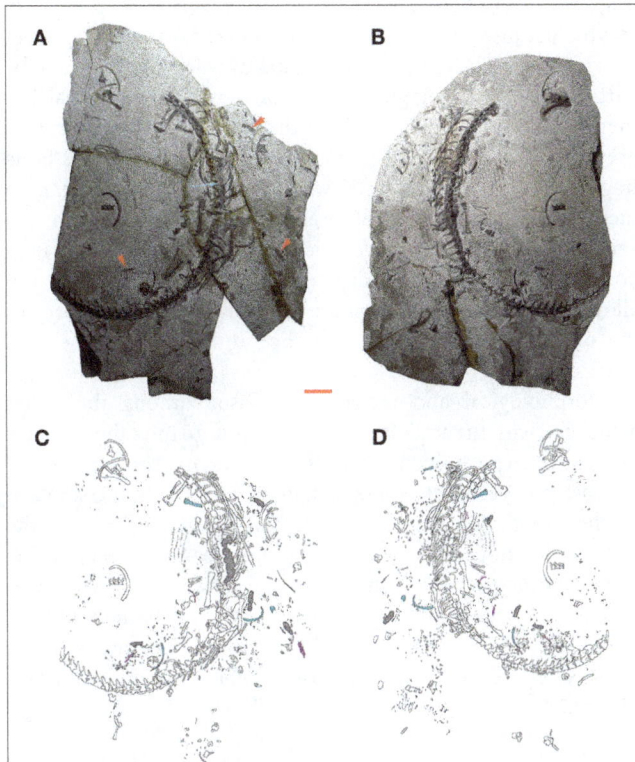

FIGURE 2 | MCN-PV 2254. Part (A) and counterpart (B) of a mesosaur specimen assigned to the species *Brazilosaurus sanpauloensis* from Goiás State. Red arrows point to the small coprolites placed around the articulated skeleton. Blue arrow shows the location of the cololite. (C,D) Interpretive drawings of (A,B), respectively. Isolated bones from immature individuals of different ontogenetic stages associated with the adult are in blue and pink; cololite and coprolites are in gray. Scale bar: 10 mm.

FIGURE 3 | Mesosaur coprolites from the Iratí Formation (Lower Permian of Brazil). (A–C) (MCN-PV 2220, 2210, 2231), isolated coprolites showing the different morphologies found. (D) MCN-PV 2222, coprolite assemblage containing specimens of different sizes. (E) MCN-PV 2104. Coprolite from the Iratí Formation (Lower Permian, Brazil) showing a general morphology that mirrors that of a pygocephalomorph crustacean. This specimen seems to have preserved part of the bases of the antennae. Scale bars: 10 mm.

FIGURE 4 | Mesosaur coprolites from the Iratí Formation (Lower Permian of Brazil). (A) Small coprolite showing bone inclusions which is part of an assemblage (MCN-PV 2218) of coprolites that otherwise lack such inclusions **(B)**. **(C)** Specimen MCN-PV 2102 clearly showing that its original three-dimensional shape has been weathered and its contents dispersed. Scale bars: 10 mm.

After this, the samples were placed in an oven at 60°C for 24 h, in order to remove any moisture. The samples were glued on stubs with silver paste, metalized with 24 carat gold, and examined and photographed under a Jeol Scanning Electron Microscope (SEM), JSM-5200, under 25 kV with variable magnification. Other samples, mostly those specimens (coprolites and gastric contents) from the Uruguayan Mangrullo Formation, were also analyzed under SEM, but they remained uncoated.

All measurements are given in millimeters. The measurements taken were anteroposterior length and dorsoventral diameter of the cololites, and minimum and maximum diameter of the coprolites. They were taken using a Mitutoyo caliper with a span of 15 cm opening and accuracy of 0.05 mm. Bones were measured following Sedor (1994) and fragmented bones were measured by combining the measurements taken from parts and counterparts (A, A′).

RESULTS

MCN-PV 2254 (Figure 2)

MCN-PV 2254: Represented Mesosaur Individuals

Inside and around the adult individual MCN-PV 2254, indeed in almost every area of both slabs (A, A′), there is a large number of small crustacean fragments, bones, and bone fragments, all of them disarticulated and mostly badly preserved, maybe because they had already suffered different degrees of digestion. Thus, the very fragmentary elements could be part of the gastric or gut content of the adult individual but they also could have come from degradation of the coprolites preserved around the skeleton (**Figure 2**). About 25 bones/bone fragments and various teeth were identified as belonging to mesosaurs; others could not be identified because they are too small, and still poorly ontogenetically differentiated. Concerning crustacean remains, the degree of fragmentation does not allow identification, but most are probably pygocephalomorphs, the only crustacean group described thus far from the Iratí Formation.

Morphological and metric comparison among the bones associated with the adult individual allows us to infer the presence of at least two juvenile stages. These bones do not show the degree of erosion expected for bones originally retained in the stomach or the intestinal tract. Other tiny bones scattered around the adult skeleton, possibly from hatchling mesosaurs, seem not be related to the previously mentioned juvenile individuals due to fractures, missing parts, or excessive weathering because of their immaturity. These remains might represent elements formerly retained in the gastrointestinal tract of the adult mesosaur (see **Figure 2**).

MCN-PV 2254: Macroscopic Description of the Cololite

A macroscopic analysis of the cololite in MCN-PV 2254, which is associated with the vertebral column of the adult individual, shows that when the vertebral column is hyperextended, the cololite lies next to the vertebrae from the anterior portion of the 10th dorsal vertebra (D10) until the third caudal vertebra (3 Cd; see **Figure 2**). Three distinct types of bromalite preservation are seen in the specimen. First, part of the last meal is still contained

Determination of the crustacean species in the gut contents and inside the coprolites was difficult because of their weathering and high degree of fragmentation, but they might be juvenile pygaspids (Dr. Irajá Damiani Pinto, personal communication).

The methods for a more profound study of the gut and coprolite contents were restricted to scanning electron microscopy and the X-ray diffraction. Further, both gut and coprolite contents were compared under normal light microscopy, stereoscopic microscopy, and scanning electron microscopy. Thin sections were made at the laboratory of the Instituto de Geociências, Universidade Federal do Rio Grande do Sul (UFRGS), and the Paleontological Laboratory at the Museu de Ciências e Tecnologia, Pontifícia Universidade do Rio Grande do Sul (PUCRS) and photographed under a Zeiss microscope, at magnifications between 80 and 500 times. The thin sections were made following standard histological techniques (e.g., Ferigolo, 1985).

For study under scanning electron microscopy, cololite/coprolite contents were removed from the slabs using electronic drills, dental probes, stylets, and tweezers. Some samples were etched in 5% hydrochloric acid for 10 min, whereas others were treated with phosphoric acid for a variable time.

FIGURE 5 | Mesosaur coprolites and cololites from the Lower Permian Iratí Formation: morphology and structure. (A) MCN-PV 2218 showing the different coprolite types (see text for a more detailed description). **(B)** MCN-PV 2220, type-4 coprolite bearing a very small tooth (arrow). **(C)** Detail showing the good preservation of the tooth included in FC-DPV 2220, even after the digestive process. Scale bars: 10 mm in **(B)** and 5 mm in **(C)**. **(D)** Photograph of the cololite in MCN-PV 2254 to show the presence of bone inclusions (red arrow) with different degrees of preservation. **(E)** SEM image of the same cololite illustrated in **(D)**, showing crustacean remains as part of its content. r, ribs; dv, dorsal vertebrae of the adult individual bearing the studied cololite.

in a mass in the intestinal tube (i.e., a cololite). Some remnants of the last meal are oval, reddish-brown, and of a slightly lighter shade than the bones of the adult mesosaur. The cololite lies between the medial surfaces of the right and left series of ribs and ventral to the vertebral column, indicating an intra-abdominal position. Second, numerous fragmented and weathered small bones and carapaces, representing part of the gastric contents, are dispersed around the skeleton of the adult. They lie close to the vertebral column, spanning the last dorsal and the sacral segments. Third, at least five coprolites are preserved close to the adult individual, here interpreted as the most probable producer (see **Figures 2C,D**).

MCN-PV 2254: Cololite Studied under Stereoscopic Microscopy

The cololite in specimen MCN-PV 2254 was studied under a stereoscopic microscope. A lot of crustacean fragments disrupted by the digestive processes can be seen inside the content, along with several small fragments of bones (**Figure 5B**). We interpret these fragmentary carapaces and bones as belonging to small pygocephalomorph crustaceans and very young mesosaurids respectively, but because they appear to be partially digested, no more precise taxonomic identification can be made. Nevertheless, one fragment of bone was removed for study under scanning electron microscopy.

MCN-PV 2254: Cololite Studied under Scanning Electron Microscopy (SEM)

Examination of the cololite under SEM has shown the crustacean fragments to be elongated bodies, twisted or very deformed, very homogeneous microscopically and without structure, except that multiple layers of exoskeleton are visible in some cases (**Figure 5E**). They also contain bones that appear to be not as weathered as the crustaceans. One bone fragment was separated from the cololite of MCN-PV 2254 for SEM study. It was identified as a bone fragment on the basis of its typical bone tissue microstructure, meaning concentric layers around vascular canals, and probable osteocyte vacuities.

X-ray diffraction

Of the analyzed cololite samples from MCN-PV 2254, only one showed peaks indicating a pure apatite composition; the others showed apatite and dolomite or apatite, dolomite and quartz. Apatite indicates the presence of bone debris in the cololite, since this mineral is not present in the dolomitic limestone that contains the specimen. It was also present in vertebral fragments of the skeleton of MCN-PV 2254, which were analyzed for comparison. The dolomite and quartz may be present as a result of fossilization processes and contamination of samples, since it is very difficult to remove cololite fragments without taking along small portions of the surrounding rock. However, the fossilized fecal matter often

FIGURE 6 | Mesosaurid gastric contents from the Mangrullo Formation (Lower Permian, Uruguay). (A) FC-DPV 2616 and **(B)** FC-DPV 2608, in partially articulated individuals; **(C)** FC-DPV 2636 and **(D)** FC-DPV 2651, identified based on the presence of well-preserved skeletal elements of an adult individual which were possibly separated from the rest of the skeleton by the excavation process using heavy machinery. **(E)** Close-up of the gastric content showing the presence of possible plant remains or an insect wing. Scale bars: 10 mm, with the exception of scale bar in **(E)** which is 5 mm. **(F)** FC-DPV 2650 showing a partial large individual preserving gastric content, associated with a small one (hatchling individual), whose body remains were possibly scavenged after death (red arrows), producing total disarticulation, and loss of most of bones. Scale bar: 10 mm.

contains a few quartz grains and other mineral clasts (i.e., gastroliths) which may have been swallowed deliberately or incidentally with food and water, as exemplified by analyses of crocodile stomach contents (Pauwels et al., 2007; Wings, 2007).

The crystallographic parameters "a" and "c" (**Tables 1, 2**) suggest that the apatite present in the cololite of MCN-PV 2254 is virtually identical to that found in bone. The minimal and insignificant differences in chemical composition that exist are explained by recrystallization of apatite around the end of void-fillings (zeolitic channels) and alteration by factors such as pH, temperature, and exposure to chemicals, all resulting from digestive processes. The analysis did not find a difference between the carbonate of the dolomitic limestone and that found in the crustacean shells. This suggests that a diagenetic precipitation of calcium carbonate took place, and that the crustacean carapaces were not originally mineralized (Piñeiro et al., 2012c).

MCN-PV 2254: Coprolites

Several ellipsoid coprolites can be seen around the adult specimen. Their shape and appearance are similar to those of other mesosaur coprolites (**Figure 2**).

X-ray diffraction

Coprolites were analyzed using X-ray diffraction. The results obtained were very similar to those for the cololite. The sample

of MCN-PV 2254 only presented apatite in contrast with other specimens (e.g., MCN-PV 2222 and 2229) which showed apatite, dolomite, and quartz.

The crystallographic parameters "a" and "c" (**Table 2**) indicate that the apatite type in the cololite is similar to that found in the sampled vertebrae of MCN-PV 2254 and that the small differences are due to the digestive process. As its contents show, the cololite is the product of partial digestion, whereas coprolites represent the end of the digestive process. For the carbonate we propose the same explanation given in the case of the cololite.

Study of Isolated Coprolites from the Iratí Formation (Figures 3–5)

Most of the specimens have an elongate ellipsoid, oval, or slightly curved shape (see **Figures 3A,B,D,E, 5A**). The specimen MCN-PV 2226 has a more rounded shape (**Figure 3C**). As a rule, coprolites have a reddish-brown color, and most of the exceptions have a lighter shade (**Figure 4**). Some, possibly due to diagenesis, have a slightly bluish (e.g., MCN-PV 2212 A/A′; MCN-PV 2220 A/A′) or brown-black hue (e.g., MCN-PV 2208, MCN-PV 2218 A/A′/B) and some show small bone inclusions (**Figure 4A**).

The coprolites are usually dispersed, or sometimes agglomerated, in slightly different planes. In the specimens in which there are a larger number of coprolites most are more or less parallel to each other, while some are oriented obliquely

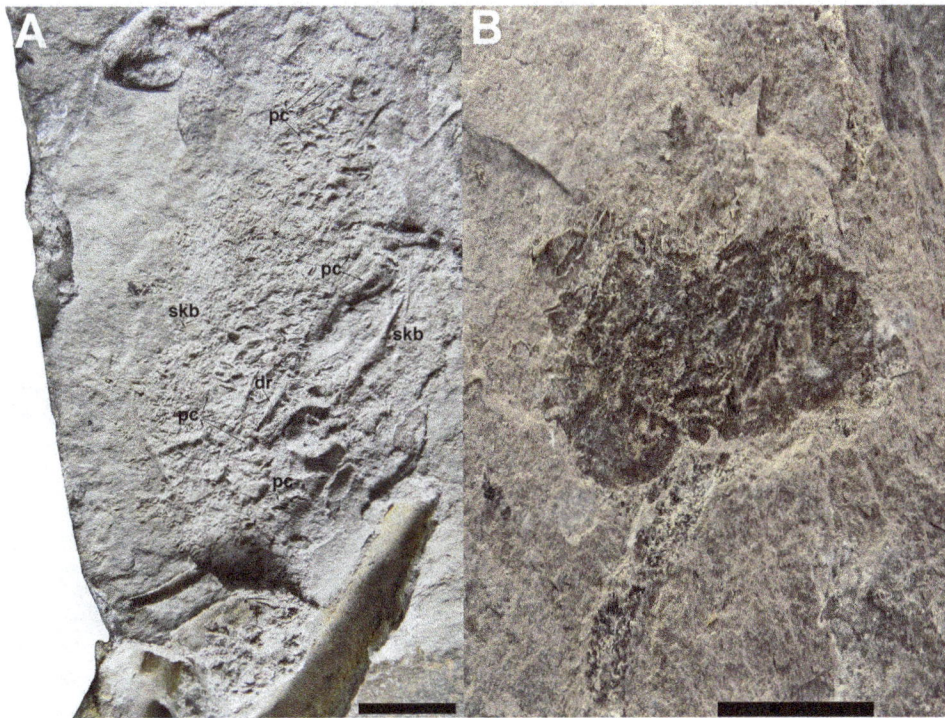

FIGURE 7 | Mesosaur gastric contents and coprolites contrasted. (A) FC-DPV 2156, gastric content found in the Lower Permian Mangrullo Formation. **(B)** FC-DPV 2651, irregular coprolite from the Lower Permian Mangrullo Formation. dr, dorsal ribs; pc, pygocephalomorph crustaceans; skb, skull bones. Scale bars; 10 mm.

to the others (**Figures 3D, 5A**). Some specimens show coprolites overlapping each other (**Figure 4B**).

Some coprolites are complete (e.g., MCN-PV 2158 A/A′; MCN-PV 2254 A/A′), but some are broken and incomplete (MCN-PV 2158; MCN-PV 2210 A/A′; MCN-PV 2254). Most of them were split lengthwise by the fracture between the part and counterpart slabs (**Figure 3**). Some complete coprolites are well-preserved and visible on both slabs (e.g., MCN-PV 2102; MCN-PV 2231; MCN-PV 2254 A/A′). Others have only the thin outer layer preserved, and still others were completely lost so that just a slight halo remains. A few of them were in an initial stage of decomposition when fossilized, because their content is dispersed (**Figures 3C, 4C**).

Of the 220 coprolites studied, it was possible to establish both diameters of 107; only the largest diameter of a further six; and only the smallest diameter of the remaining 107. In the 113 coprolites for which the largest diameter could be measured, this parameter ranged from 9.8 to 32.5 mm (average largest diameter: 18.12 mm). In the smallest 214 coprolites for which the minimum diameter could be measured, this parameter ranged from 2.3 to 9.9 mm (average smallest diameter: 5.15 mm). Most coprolites have a largest diameter of 15.0–17.0 mm and a smallest diameter of 4.0–6.0 mm.

There was variation in content of the casting, not only between the different coprolites but between these preserved in the same slab. Four basic types were established: Type 1, only amorphous content; Type 2, with some crustacean fragments;

Type 3, predominance of minuscule crustacean fragments; Type 4, crustacean and bone fragments and/or teeth (**Figures 3–5**).

MCN-PV 2220 is a coprolite containing an incomplete mesosaur tooth with a greatly altered enamel layer, which is opaque due to decalcification by the digestive process (**Figures 5B,C**).

Under microscopic stereoscopy, sometimes, the bone fragments appear very similar to fragmentary crustaceans, due to digestion. However, generally they are distinguished based on morphology, as the crustaceans are preserved as elongated, compact, fractured carapaces in continuity with each other. Other bone fragments are small polygonal pieces with rounded angles, due to digestion. Some of the best preserved of these were selected for further study under scanning electron microscopy, such as the specimen MCN-PV 2227 (**Figure 5D**).

Coprolites from the Iratí Formation under Scanning Electron Microscopy (SEM)

The mineralogical composition of coprolites from the Iratí Formation was analyzed under SEM equipped with EDS and the main component found was apatite (**Figure 10**). Some coprolites showed very small bone fragments with recognizable concentric bone layers. The fragments are much modified, bony microscopic structures, maybe suggesting they underwent a greater degree of digestion than the teeth (**Figures 5B,C**).

FIGURE 8 | The different types of coprolites found in the Mangrullo Formation (Lower Permian, Uruguay). (A) FC-DPV 2619, elliptical coprolite without bone inclusions; (B) FC-DPV 2647, elliptical coprolite with bone inclusions; (C) FC-DPV 2210, ellipsoid amorphous coprolites without bone inclusions; (D) FC-DPV 2807 irregularly shaped coprolites bearing bone inclusions and teeth (red arrow). Scale bars: 10 mm.

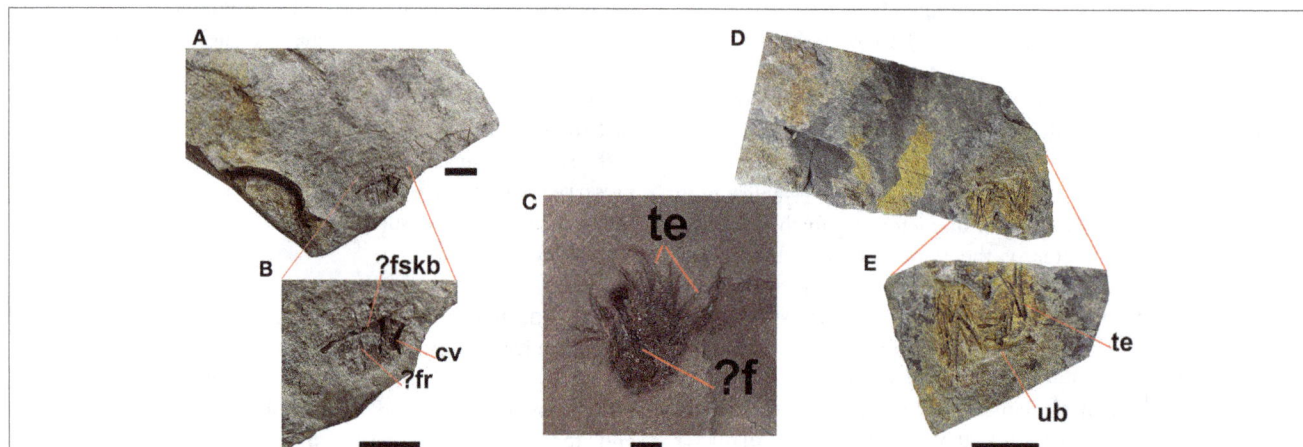

FIGURE 9 | Regurgitalites from the Mangrullo Formation (Lower Permian, Uruguay). (A,B) FC-DPV 2648; (C) GP-2E 19A; (D,E) FC-DPV 2649. Scale bars: 10 mm. cv, cervical vertebra; f, femur; fr, fragmentary rib; fskb, fragmentary skull bone; te, teeth; ub, unidentified bone.

TABLE 1 | Values of crystallographic parameters "a" and "c" for the apatite in the cololite.

	Parameter "a" (Å)	Parameter "c" (Å)
Vertebra	9,3381	6,8840
Sample 1	9,3350	6,8908
Sample 2	9,3575	6,9014
Sample 3	9,3458	6,8908

TABLE 2 | Values of crystallographic parameters "a" and "c" for the apatite in the vertebra of MCN-PV 2254 and the coprolites found in the samples.

	Parameter "a" (Å)	Parameter "c" (Å)
Vertebra	9,3381	6,8840
Coprolite 2254	9,3423	6,8840
Coprolite 2229	9,3461	6,8896
Coprolite 2222	9,3600	6,8954

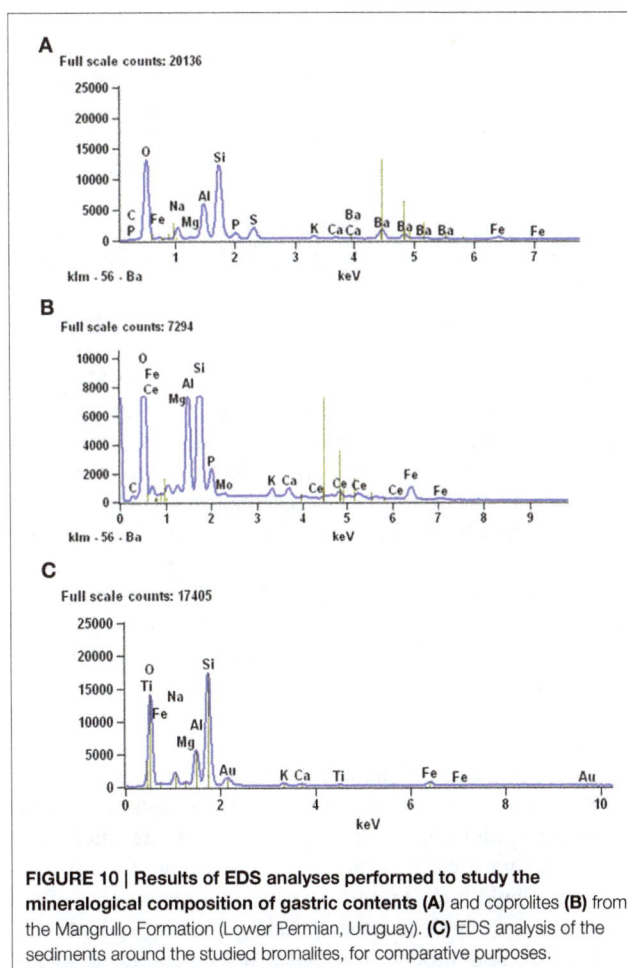

FIGURE 10 | Results of EDS analyses performed to study the mineralogical composition of gastric contents (A) and coprolites (B) from the Mangrullo Formation (Lower Permian, Uruguay). (C) EDS analysis of the sediments around the studied bromalites, for comparative purposes.

Study of the Bromalites from the Mangrullo Formation (Lower Permian) of Uruguay

Gastric Contents

Partially articulated skeletons of large adults from the Mangrullo Formation Konservat-Lagerstätte preserve gastric contents (Piñeiro et al., 2012b; Ramos, 2015; **Figure 6**). The main difference with respect to the cololite described from the Iratí Formation is that the elements show a lower digestive degradation of carapaces and bones by the acids, meaning that they were eaten a shorter time before the death of the animal. Although they are difficult to identify more precisely, we can identify isolated mesosaur bones (fragmentary humerus, vertebrae, ribs, mandibles, and skull; **Figure 6A**), different parts of pygocephalomorph crustacean bodies (e.g., thoracic and abdominal segments, and very probably also small carapaces; **Figure 6C**), and putative insect wings or plants (**Figures 6B,D,E**), as being part of the mesosaur gastric contents. These specimens demonstrate the presence of different-sized prey in the contents, as well as different degrees of degradation due to the digestion of the carapaces and bones. Thus, it may be possible to determine how much time passed between the last meal and death. Notably, crustacean remains are more degraded than the bones in all the examined specimens; this can be explained by the absence of mineralization of pygocephalomorph carapaces (see Piñeiro et al., 2012b,c; Ramos, 2015). Although chitin is well-resistant to digestion, many sauropsids secrete chitinases to hydrolyze it and also for defense against many pathogens (Marsh et al., 2001; Siroski et al., 2014). While we cannot know whether such enzymes were present in mesosaurs based on the fossilized specimens, we can say that the crustacean carapaces were almost completely digested in most of the coprolites found, with recognizable fragments present in just a few cases. The bones within the contents are mostly very small and moderately better-preserved vertebrae, ribs and other skeletal elements perhaps belonging to hatchling or juvenile mesosaurids. However, bones

of large, adult mesosaurs can also be observed, even in the coprolites (see **Figures 8B,D**).

Coprolites

Most of the preserved mesosaur coprolites in the Mangrullo Formation (**Figures 7, 8**) are more or less cylindrical with rounded ends, but some are anisopolar or irregular in shape (**Figure 8D**). Their morphology cannot be described in detail because of the nearly two-dimensional form of preservation caused by diagenetic compression (see also Hone et al., 2015; Wang et al., 2016). The putative coprolites differ from gastric contents in shape and in being masses constrained by better-defined regular or irregular margins. The state of preservation of remains in the amorphous groundmass differs too (see **Figure 8**). Coprolites from the Mangrullo Konservat-Lagerstätte can be classified as (1) long elliptical, groundmasses containing probably completely macerated crustacean remains and no bone inclusions; (2) short elliptical, groundmasses with putative bone inclusions; (3) more or less spherical forms with amorphous texture and some inclusions of uncertain origin; and (4) irregular bodies that can contain bones or teeth (**Figures 8B,D**). As expected for tetrapod coprolites, none show evidence of being spiral (Hunt et al., 1994; Niedźwiedzki et al., 2016). The major axis of the elliptical ones ranges from 20 to 40 mm and the minor one from 5 to 10 mm.

The flattened condition of the coprolites is possibly due to compression of the shale and fine siltstone layers that contain them. Most were preserved as part and counterpart, making it possible to see their interiors and the texture of their contents. Otherwise, three-dimensional, not so well-preserved coprolites seem to have been weathered faster than the surrounding matrix and their exposed surfaces may have been destroyed fairly rapidly (see **Figure 4C**).

Coprolites can be associated with partially articulated skeletons or bones of adult individuals and with scattered, altered bones, and fragments of different-sized hatchling and juvenile mesosaurs (as described for MCN-PV 2154). Also, they can be associated with complete and fragmentary remains of pygocephalomorph crustaceans. As in the Iratí Formation, there are no bones that can be attributed to vertebrates other than mesosaurids in the levels that yielded the coprolites described herein. The gastric contents may be preserved mainly as phosphatized structures (**Figure 10**); their preservation was favored by the extreme conditions in the Mangrullo and Iratí formations, including low oxygen rates and increased salinity (see **Figure 10C**).

Regurgitalites

Small accumulations of moderately-preserved bones and well-preserved teeth are often found in the mesosaur-bearing levels of the Mangrullo Formation (**Figure 9**). They are interpreted as most likely mesosaur regurgitalites, consisting of indigestible parts that would cause intestinal occlusion if not expelled. They comprise relatively small bones revealing a lower degree of acid etching than bones found in cololites and coprolites, as well as large teeth which lack evident etching marks (Ramos, 2015). In fact, regurgitation may result in only slight chemical and structural alterations on bones, as exemplified by owl pellets (Dauphin et al., 2003). As in the mesosaur coprolites, there is some kind of groundmass surrounding the elements, but the groundmass gives the impression of having been less dense or more liquid in the regurgitalites. This is suggested by the absence of a well-defined outer border in the regurgitalites (**Figure 9**), in contrast to the coprolites.

DISCUSSION

The mesosaur bromalites described in this study represent a rare case where the source animal can be identified with confidence, because no other tetrapod or vertebrate is found in the mesosaur-bearing strata (Piñeiro, 2006; Piñeiro et al., 2012a). Because all "fish" groups of the Permian period had a spiral valve and none of the coprolite specimens studied herein shows spiral morphology, they are interpreted as produced by tetrapods (see Gilmore, 1992; Hunt et al., 1994; Argyriou et al., 2016; Niedźwiedzki et al., 2016). The only tetrapods known from these particular strata are mesosaurids, represented by a large amount of bone remains. Moreover, the content of the coprolites does not differ substantially from the gut contents

observed in mesosaurid skeletons. Therefore, the coprolites are attributed to mesosaurids, although some irregular coprolites could be also assigned to the large arthropods which are present in the Mangrullo Lagerstätte (Piñeiro et al., 2012b). Indeed, some of the pygocephalomorph crustaceans found in the Mangrullo Formation are unusually large (more than 80 mm) and they may also have been scavengers of dead mesosaurs. They are found in intimate association with the mesosaur skeletons, as can be seen even in the holotypic specimen of *Mesosaurus tenuidens* housed in the Muséum national d'Histoire naturelle in Paris (**Figure 11**). Therefore, pygocephalomorphs could potentially have produced some of the coprolites. However, the coprolites intimately associated with mesosaur specimens (e.g., MCN-PV 2254) and those found isolated are quite similar in morphology, texture, and color. They are also fairly comparable in size, ranging from 6 to 35 mm in length (see **Table 3**). This range allows us to infer that coprolites were produced by animals of mesosaur size, i.e., 40–100 cm in length.

In MCN-PV 2254 the adult skeleton, along with the cololite, coprolites, and bones of juveniles, possibly was preserved in the margins of the water body inhabited by the group. A part of the coprolite and cololite contents was lost and scattered around the adult skeleton. We can thus suggest that the last meal of the adult individual contained tiny bone fragments as well as small crustacean remains which can be distinguished by their morphology. Rapid burial would ensure the preservation of such delicate material. Thus, it seems even possible that the event that disarticulated the individuals also buried them. The presence of microbial mats would also ensure the preservation of bromalites and other fossils, since bacteria may play an important role in the fossilization of fecal matter (Hu et al., 2010; Bajdek et al., 2016).

The remains of a smaller, possibly young mesosaur and other bones that may belong to a hatchling are preserved adjacent to the adult individual and led us to analyze three hypotheses: (1) the smaller bones correspond to animals that died well-before, and were disarticulated by long exposure to air; (2) the small bones were part of the gastric content of the adult individual preserved in the same slab; (3) the bones belong to juvenile mesosaurs that died at the same time and perhaps for the same reason as the adult.

We consider hypothesis 2 least likely on the basis of the following evidence:

(1) The bones do not show any evidence of acid etching produced by gastric fluids.
(2) The skeletons of the juvenile individuals are too large to have passed through the oral cavity or the digestive tract of the articulated adult specimen.
(3) These bones belong to at least two juveniles of different ages associated with the adult. All bones are in the same layer and show similar preservation, and thus, it is unlikely that they were disarticulated and deposited close to an articulated, well-preserved individual by a separate event.

Even though the first hypothesis cannot be disproved, associations of bones from more than one individual or

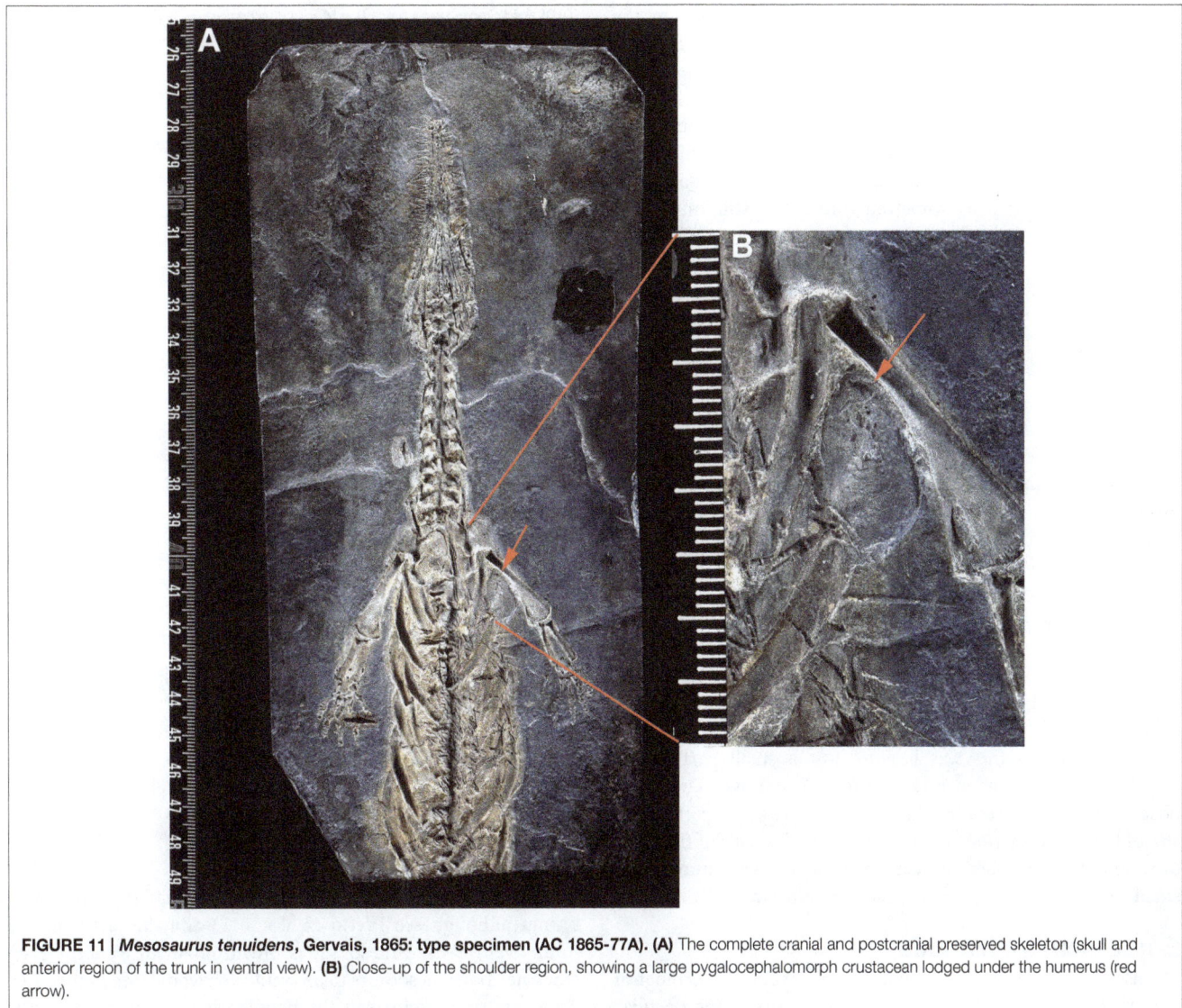

FIGURE 11 | *Mesosaurus tenuidens*, **Gervais, 1865: type specimen (AC 1865-77A). (A)** The complete cranial and postcranial preserved skeleton (skull and anterior region of the trunk in ventral view). **(B)** Close-up of the shoulder region, showing a large pygalocephalomorph crustacean lodged under the humerus (red arrow).

TABLE 3 | Measurements of coprolites from the Iratí (MCN-PV) and the Mangrullo (FC-DPV) formations.

Specimen	MCN-PV 2231	MCN-PV 2254	MCN-PV 2218	MCN-PV 2212	MCN-PV 2222	FC-DPV 1633	FC-DPV 2609	FC-DPV 2607	FC-DPV 2619
Diameter 1	18	23	10–23	28–19	6–25	19	10–15	9–35	25
Diameter 2	10	5	4–7	4–7	2–6	6,5	4–6	9–10	10

Diameter 1 corresponds to length and diameter 2 corresponds to width of the elliptical coprolites. Measurements are in mm.

even of partially articulated skeletons are very common in the Mangrullo Formation shales and could suggest parental care and/or aggregation in mesosaurs (see Piñeiro et al., 2012d, 2015b).

It is very probable that the scenario represented by this specimen is more complex than initially thought. We can interpret the slab as bearing one adult and two young individuals (maybe belonging to the same species) that apparently died and were buried at the same time. It is evident that most of the elements are *in situ*, but there is evidence that the assemblage was not immediately buried after the death of the animals, as some alteration can be seen in the adult skeleton: the skull and front limbs are missing or incomplete, and the hind limbs and tail are very incomplete and have been moved from their original anatomical position. It is probable that the animals died very close to the shore, where the decay of the soft tissues was quicker than in the deep, anoxic bottom waters of the basin. The specimens could have been scavenged by pygocephalomorphs and other mesosaurs, as is shown also by some specimens found in the Mangrullo Formation (**Figure 6E**).

The gut content spread away from the skeleton during the decay processes and, like the coprolites, was buried faster than the bones. The latter remained exposed for a prolonged time, accounting for the partial disarticulation of the adult skeleton. As explained above, there is enough evidence to support the presence of two juvenile individuals next to the adult. One juvenile might have been born a very short time before its death, and the other is somewhat older but still immature. Their bones were lighter and more easily swept away. However, another possible explanation for the presence of the isolated bones is that the original slab could have been broken by the bulldozers working in the quarry. This is a frequent occurrence in the Iratí and in the Mangrullo Formation as well. Indeed, the caterpillars are producing a taphonomic bias toward the young-adult associations but they do not explain why young adults represent the most frequently preserved ontogenetic stage in the two studied populations. Perhaps they were, along with the newborns, the most affected by the environmental stress produced by the evaporitic and progressive drought conditions in the basin engendered by continued volcanism that affected the region during the beginning of the Permian (López-Gamundi, 2006; Santos et al., 2006; Piñeiro et al., 2012d).

Interestingly, coprolite specimens from the Mangrullo and Iratí Formations are quite comparable in size in spite of the noticeably smaller body size of the Brazilian species *Brazilosaurus sanpauloensis*. This anomaly might suggest a rarity of coprolites of very mature mesosaurids in the Mangrullo Formation of Uruguay. Perhaps this was because just a small part of the population had the good luck to attain an old age. There is a positive but loose correlation between the feces size and the body size of the producer when comparing the individuals of the same or different species; big animals can produce large amounts of small feces (Chin, 1997; Chame, 2003; Flessa et al., 2012).

Ecological Inferences

Mesosaur bromalites tell us quite much about the ecological (trophic) interactions between the components of the peculiar "Mesosaur community" (Piñeiro et al., 2012b). As expected, mesosaurs fed on pygocephalomorph crustaceans, possibly as their preferred prey. They appear to have been size-selective, eating crustacean prey not exceeding 20 mm in length (Ramos, 2015). This can be asserted from almost all the coprolites and from all the studied gastric contents. We also show that small scattered mesosaur bones are present in the cololites, and exceptionally in coprolites (e.g., teeth), suggesting that these skeletal pieces must have been ingested. However, there is no evidence that large mesosaurs preyed on small ones, even though mesosaur bones are common in stomach contents and coprolites (see **Table 4**). Schwimmer et al. (2015) described partially articulated vertebrae of a baby turtle in a spiral bromalite (coprolite or cololite) of a shark. In contrast, bone elements present within cololites and coprolites in the material from the Mangrullo and Iratí Formations show strong acid etching and many fractures, and they are never articulated. In a hypothetical scenario of predation, we would expect to find more or less articulated skeletons in the gastric contents (where the digestive processes would still allow for a general identification of the

TABLE 4 | Specimens from the Mangrullo Lagerstätte (Uruguay) with preserved cololites having identifiable content (modified from Ramos, 2015).

Specimen	Coprolites		Gastric contents	
	Crustaceans	Mesosaurs	Crustaceans	Mesosaurs
FC-DPV 2611	x			
FC-DPV 2629			x	
FC-DPV 2210			x	
FC-DPV 2608				
FC-DPV 2609	x		x	
FC-DPV 2610			x	x
FC-DPV 2612				
FC-DPV 2156	x	x		
FC-DPV 2126				
FC-DPV 2613	x	x		
FC-DPV 2635			x	x (teeth)
FC-DPV 2607			x	x
FC-DPV 2616	x	x		
FC-DPV 2617	x			
FC-DPV 2619			x	
FC-DPV 2620			x	x
FC-DPV 2626	x			
FC-DPV 2627	x	x	x	
FC-DPV 2624			x	x
FC-DPV 2621			x	x
FC-DPV 2377			x	
FC-DPV 1633			x	
FC-DPV 2625			x	x

bones). Mesosaur teeth seem not to be adapted to powerful biting (although see Pretto et al., 2014) and the jaw aperture in an average-sized mesosaur is much too small to allow even small newborn mesosaurs to be swallowed whole (see **Figure 12**). Thus, we support instead the hypothesis that mesosaurs were scavengers and fed on soft tissues of smaller decaying individuals. Some bones could have been accidentally ingested along with soft tissue. This could be why some of them are so eroded that they are difficult to identify. As the soft tissues are partially decomposing, the teeth need not be so strong and fixed to the jaws, explaining the high number of them preserved separately from the rest of the skeleton and the frequent occurrence of toothless skulls. Another hypothesis to explain the presence of mesosaur bones in the gastric contents and coprolites is that they were eaten accidentally, during the capture of pygocephalomorph crustaceans that were scavenging on the carrion accumulated on the bottom.

Teeth are among the skeletal elements that are most resistant to dissolution, although extant crocodiles normally decalcify ingested teeth (Fisher, 1981a,b). This may explain why the teeth found in the regurgitalites and gastric contents of mesosaurs tend to be better-preserved than bones. Interestingly, if tooth fragments found in the samples derive from prey, mesosaurs must have swallowed at least some cranial elements during acts of predation. Such a behavior is known in modern crocodiles,

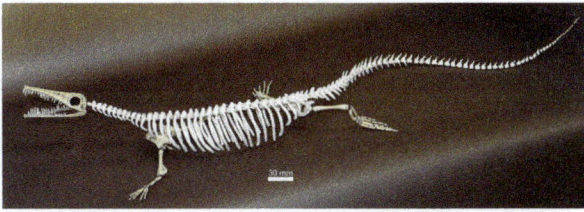

FIGURE 12 | Skeletal reconstruction of a young adult *Mesosaurus tenuidens* Gervais, 1865 based on comparative measurements and anatomical observation of more than 50 individuals (see Villamil et al., 2015). This reconstruction allows us to obtain a more accurate idea of the size of the prey that a mesosaur of standard size could have eaten.

FIGURE 13 | PIMUZ A III 192. Mesosaur specimen assigned to the species *Brazilosaurus sanpauloensis* showing a huge mass of fecal matter completely filling the distal part of intestine and some smaller feces already outside of the body (see text for a more detailed interpretation of the preservation).

although very large prey needs to be dismembered first (Grigg and Gans, 1993; Webb and Manolis, 2009). But, in the case of mesosaurs, the evidence suggests that partial decomposition of the consumed carrion facilitated the presence of teeth in gastric contents and coprolites. Alternately, the mesosaurs might have sometimes swallowed their own teeth when they broke during feeding. Such tooth damage is known in some archosaurs, which are often characterized by fast tooth replacement (Grigg and Gans, 1993) as mesosaurs were Pretto et al. (2014). Stone et al. (2000) described from the Upper Jurassic of the Morrison Formation a large bone-rich phosphatic mass interpreted as a coprolite, which contained a broken distal end of an *Allosaurus* tooth. The authors suggested that the tooth broke and was accidentally ingested when the theropod used its jaws to break bones into smaller portions (Stone et al., 2000). Some mesosaur coprolites were found to contain teeth that could have either belonged to very young individuals or been recently erupted replacement teeth of adults, which are also small and short.

Based on the analyzed material, it is hard to determine whether the mesosaurids consumed carrion of the same or another species. Distinguishing between mesosaur species is a very complex issue, even in the case of well-preserved individuals (see Piñeiro et al., 2015a). Therefore, we prefer to identify mesosaurs as scavengers rather than cannibals. The largest crustacean remains (e.g., abdomens and some thoracic segments) often appear to be more weathered than expected, and thus it seems possible that they were also scavenged from the bottom. Moreover, some possible coprolites/regurgitalites resemble the crustaceans in shape and some aspects of internal structure (**Figure 3D**). It would be conceivable that crustaceans larger than 20 mm in length were swallowed whole, but were soon regurgitated in an incompletely digested state because of their large size. The fossil record contains some previously documented examples of that kind of inappropriate predator-prey interaction, in which a predator and an excessively large prey animal both died after the former swallowed the latter (see Bieńkowska-Wasiluk, 2010). Alternatively, the crustaceans may have been expelled as feces (see **Figure 13**).

The presence of apparent bone inclusions in the coprolites (**Figure 4A**) is intriguing. Reptiles are characterized by a long period of digestion and, for example, crocodiles digest the bones they ingest practically completely (Milàn, 2012; see Owocki et al.,

2012). However, mesosaurs were fairly small; while large snakes may defecate even less than once a year, gracile arboreal snake species, which have low body mass, sometimes defecate within 24–48 h of ingesting a meal (Lillywhite et al., 2002). Relatively large bones are seen also in the putative regurgitalites, which also contain teeth of varied size and fragmentary bones that are difficult to identify, although they could be ribs or other long bones (**Figure 9**). Mesosaurs probably regurgitated bone fragments, perhaps consumed accidentally, which were too large to pass through the gastrointestinal tract. For example, raptor birds (e.g., Bond, 1936; Dauphin et al., 2003; Terry, 2004) and crocodiles (e.g., Grigg and Gans, 1993; Webb and Manolis, 2009; Semeniuk et al., 2011), as well as probably ichthyosaurs (Doyle and MacDonald, 1993; Nature News, 2002; Thies and Hauff, 2013), regurgitate most of the indigestible or hard-to-digest remains. Regurgitation might also have been caused by some kind of environmental stress, as suggested by the conditions under which the Mangrullo Formation is thought to have been deposited. Because digestion efficiency depends on body temperature in extant reptiles, undigested remnants of prey may be regurgitated during periods of unfavorable environmental temperature, as exemplified by crocodiles (Verdade et al., 1994) and snakes (Hailey and Davies, 1987), and regurgitation may also be caused by disease (Lock et al., 2003). Also, the presence of poorly digested remains in feces, caused by a short food-retention time in the gastrointestinal tract, may have to do with fluctuating food availability (see Chin et al., 2003).

Taphonomic Effects of the Environmental Conditions under Which the Iratí and Mangrullo Lagerstätten Were Deposited

Considering the sedimentary facies present in the Iratí and Mangrullo formations (**Figure 1**) and the data from the fossils described above, we can infer similar trophic conditions across the entire water body where mesosaurs lived (see Piñeiro et al.,

2012b). Yet, it seems that specimens with gastric contents are comparatively more frequently found in the Uruguayan Mangrullo Formation than coprolites, the latter being more abundant and better preserved in the Brazilian deposits. In the Mangrullo Formation the evidence is suggestive of near-shore conditions, where not only its autochthonous communities are represented but also the parautochthonous one (plants and insects). Such combination of environmental and taphonomic conditions might in part have contributed to the decomposition of feces (see Chin and Gill, 1996; Wahl et al., 1998; Chin et al., 2009). Material preserved within the gastrointestinal tract may have a better preservation potential than fecal matter expelled in the form of excrement (see Section Discussion in the Supplementary Material in Niedźwiedzki et al., 2016), resulting in better preservation of such materials in the Mangrullo Formation. The presence of increased salinity and stagnant conditions at the bottom, associated with the presence of algal mats (Piñeiro et al., 2012b), delayed the decay processes and produced a very favorable environment for preservation of very delicate soft materials. In contrast, in the facies studied at the State of Goiás, Brazil, fossilization occurred in deeper water conditions. Feces may disintegrate in water and hence some believe that subaerial conditions are more suitable for their preservation (Poinar and Boucot, 2006), as seen in many coprolites from young archeological contexts preserved via desiccation (Wood and Wilmshurst, 2014). However, the process of mineralization and/or lithification may require some initial humidity of the fecal mass (see Bajdek et al., 2016). The most crucial factors allowing the preservation of feces are rapid burial followed by rapid lithification (Eriksson et al., 2011). For example, Bajdek (2013) described a putative coprolite preserved beneath sediments of a turbidity current. However, in the Mangrullo and Iratí Formations the sedimentation rate was very low. Preservation of the fossils described herein was also favored by the periodic volcanic events that affected the basin (López-Gamundi, 2006; Santos et al., 2006; Piñeiro et al., 2012b), which facilitated rapid burial of the specimens. However, it seems likely that the excrement masses were covered by microbial mats which are common in the sediments (Piñeiro et al., 2012b), essentially as suggested by Hu et al. (2010). The role of bacteria in the fossilization of feces has already been discussed by some authors (Chin and Kirkland, 1998; Hollocher et al., 2001; Hollocher and Hollocher, 2012; Bajdek et al., 2016; Qvarnström et al., 2016).

Interestingly, the fecal matter in the putative coprolite specimen described by Bajdek (2013), buried beneath sediments of a turbidity current, was replaced by fine clastic material, only preserving abundant shell fragments as content and the general excrement-shaped morphology. In contrast, the coprolite (and cololite) specimens from the Mangrullo and Iratí Formations, likely preserved as a result of microbial activity, are composed of fossilized original phosphatic fecal matter, like most carnivore coprolites known from the fossil record (Chin, 2002). As expected, EDS analysis of mesosaur stomach contents and coprolite matrix from the Mangrullo Formation reveals that the latter is characterized by a higher content of phosphate (**Figure 10**). This is explained by a higher degree of decomposition of the ingested bone material in the final stage of

digestion. Finally, one of the interesting aspects of the material studied herein is the presence of some kind of groundmass surrounding the regurgitated elements, which nevertheless gives the impression of originally having been less dense or more liquid than in coprolites.

CONCLUSIONS

This study showed for the first time preserved residues from four distinct phases of the digestive process, which can all be assumed to belong to the same basal amniote: gastric contents, cololites, coprolites, and possibly regurgitalites. Coprolites are associated with mesosaur skeletons and bones and even those that were found isolated can be assigned to Mesosauridae, because of the absence of any other vertebrate in the studied outcrops. Therefore, this is one of the rare cases when coprolites can be directly ascribed to their producers.

Mesosaur diet included pygocephalomorph crustaceans as the main food item, corroborating some of the hypotheses previously suggested. However, contrary to other hypotheses based on indirect inferences, there is no evidence in support of the use of fish as a food item by any of the recognized mesosaurid species.

In addition, bone fragments occurring in cololites and gastric contents, and possibly also in coprolites and regurgitalites, are identified as belonging to juvenile Mesosauridae, which might suggest that mesosaurs developed a cannibalistic predatory behavior. However, the lack of any further evidence in support of this hypothesis, together with the morphology of the jaws and teeth of mesosaurids, leads us to suggest that mesosaurids were facultative scavengers, eating carrion of the same or different mesosaurid species. Both cannibalistic behavior and scavenging are quite typical under environmental stress, overcrowding and insufficient food resources. Also, the presence of undigested bone elements in the regurgitalites and feces of ectothermic reptiles sometimes results from environmental stress.

Perhaps the stress conditions were in part caused by the extended volcanism that began during the Early Permian. The volcanic events, as well as the presence of microbial mats on the bottoms of the water bodies, facilitated preservation of the fecal matter.

AUTHOR CONTRIBUTIONS

RS: Studied the Brazilian materials, wrote part of the manuscript, made **Tables 1, 2**, performed some of the analyses. JF: Wrote part of the manuscript and provided relevant hypotheses to the better understanding of some aspects of the manuscript. PB: Wrote part of the manuscript and helped with the English language and bibliography, made important contributions to the article related to his specialty on the coprolite and cololite study. GP: Studied the Uruguayan materials, wrote the manuscript, and prepared all the figures, made **Tables 3, 4** and performed part of the analyses detailed in the "Materials and Methods" Section.

FUNDING

This study was partially financed by ANII FCE_2011_6450 and by NGS grant 049714 to GP.

ACKNOWLEDGMENTS

We are very grateful to the personnel and students of the Departamento de Paleontologia of the São Paulo University and the Paläontologisches Institut und Museum, University of Zürich for allowing the study of some specimens included in this contribution. We also thank Silvia Villar and Alejandro Márquez (Servicio de Microscopía Electrónica, Facultad de Ciencias, Montevideo, Uruguay) for assistance in the analyses performed under SEM. Joselyn Falconnet (Muséum National d'Histoire Naturelle, Paris) provided the photographs for **Figure 11**. We are grateful to the academic editor Corwin Sullivan and to David Marjanović and Eric M. Roberts for their constructive and very much thorough revision of this manuscript. We are indebted to William Idiart Rodriguez for assistance in the construction of **Figure 1D**.

REFERENCES

Araújo, D. C. (1976). Taxonomia e relações dos Proganosauria da Bacia do Paraná. *Anais Acad. Brasil. Ciências* 48, 91–116.

Araújo-Barberena, D. C. (1994). "Mesossauros e Pareiassauros: dúvidas e controvérsias," in *Workshop de Integração da Geologia e Paleontologia de Vertebrados no RS*, Vol. 1 (Porto Alegre: Resumo das Comunicaçõ*es)*, 21–22.

Argyriou, T., Clauss, M., Maxwell, E. E., Furrer, H., and Sánchez-Villagra, M. R. (2016). Exceptional preservation reveals gastrointestinal anatomy and evolution in early actinopterygian fishes. *Sci. Rep.* 6:18758. doi: 10.1038/srep18758

Bajdek, P. (2013). Coprolite of a durophagous carnivore from the Upper Cretaceous Godula Beds, Outer Western Carpathians, Poland. *Geol. Q.* 57, 361–364. doi: 10.7306/gq.1094

Bajdek, P., Owocki, K., and Niedźwiedzki, G. (2014). Putative dicynodont coprolites from the Upper Triassic of Poland. *Palaeogeogr. Palaeoclimatol. Palaeoecol.* 411, 1–17. doi: 10.1016/j.palaeo.2014.06.013

Bajdek, P., Qvarnström, M., Owocki, K., Sulej, T., Sennikov, A. G., Golubev, V. K., et al. (2016). Microbiota and food residues including possible evidence of pre-mammalian hair in Upper Permian coprolites from Russia. *Lethaia* 49, 455–477. doi: 10.1111/let.12156

Bieńkowska-Wasiluk, M. (2010). Taphonomy of oligocene teleost fishes from the outer carpathians of Poland. *Acta Geol. Polonica* 60, 479–533.

Bond, R. M. (1936). Eating habits of falcons with special reference to pellet analysis. *The Condor* 38, 72–76. doi: 10.2307/1363552

Carroll, R. L. (1982). The early evolution of reptiles. *Ann. Rev. Ecol. Sys.* 13, 87–109. doi: 10.1146/annurev.es.13.110182.000511

Carroll, R. L. (1988). *Vertebrate Paleontology and Evolution.* New York, NY: W.H. Freeman and Company, 698.

Chame, M. (2003). Terrestrial mammal feces: a morphometric summary and description. *Mem. Inst. Oswaldo Cruz* 98, 71–94. doi: 10.1590/s0074-02762003000900014

Chin, K. (1997). "Coprolites," in *Encyclopedia of Dinosaurs*, eds P. J. Currie and K. Padian (New York, NY: Academic Press), 147–150.

Chin, K. (2002). "Analyses of coprolites produced by carnivorous vertebrates," in *The Fossil Record of Predation. Paleontological Society Special Papers* 8, eds M. Kowalewski and P. H. Kelley, 43–49.

Chin, K., Bloch, J., Sweet, A., Tweet, J., Eberle, J., Cumbaa, S., et al. (2008). Life in a temperate polar sea: a unique taphonomic window on the structure of a Late Cretaceous Arctic marine ecosystem. *Proc. R. Soc. B Biol. Sci.* 275, 2675–2685. doi: 10.1098/rspb.2008.0801

Chin, K., Eberth, D. A., Schweitzer, M. H., Rando, T. A., Sloboda, W. J., and Horner, J. (2003). Remarkable preservation of undigested muscle tissue within a late cretaceous tyrannosaurid coprolite from Alberta, Canada. *Palaios* 18, 286–294. doi: 10.1669/0883-1351(2003)018<0286:RPOUMT>2.0.CO;2

Chin, K., and Gill, B. D. (1996). Dinosaurs, dung beetles, and conifers: participants in a Cretaceous food web. *Palaios* 11, 280–285. doi: 10.2307/3515235

Chin, K., Hartman, J. H., and Roth, B. (2009). Opportunistic exploitation of dinosaur dung: fossil snails in coprolites from the Upper Cretaceous two medicine formation of Montana. *Lethaia* 42, 185–198. doi: 10.1111/j.1502-3931.2008.00131.x

Chin, K., and Kirkland, J. I. (1998). Probable herbivore coprolites from the Upper Jurassic Mygatt-Moore Quarry, Western Colorado. *Mod. Geol.* 23, 249–275.

Dauphin, Y., Andrews, P., Denys, C., Fernández-Jalvo, Y., and Williams, T. (2003). Structural and chemical bone modifications in a modern owl pellet assemblage from Olduvai Gorge (Tanzania). *J. Taphonomy* 1, 209–232.

Dentzien-Dias, P. C., de Figueiredo, A. E. Q., Horn, B., Cisneros, J. C., and Schultz, C. L. (2012). Paleobiology of a unique vertebrate coprolites concentration from Rio do Rasto Formation (Middle/Upper Permian), Paraná Basin, Brazil. *J. S. Am. Earth Sci.* 40, 53–62. doi: 10.1016/j.jsames.2012.09.008

Dentzien-Dias, P. C., Poinar, G. Jr., de Figueiredo, A. E. Q., Pacheco, A. C. L., Horn, B. L. D., and Schultz, C. L. (2013). Tapeworm eggs in a 270 million-year-old shark coprolite. *PLoS ONE* 8:e55007. doi: 10.1371/journal.pone.0055007

Doyle, P., and MacDonald, D. I. M. (1993). Belemnite battlefields. *Lethaia* 26, 65–80. doi: 10.1111/j.1502-3931.1993.tb01513.x

Eriksson, M. E., Lindgren, J., Chin, K., and Månsby, U. (2011). Coprolite morphotypes from the Upper Cretaceous of Sweden: novel views on an ancient ecosystem and implications for coprolite taphonomy. *Lethaia* 44, 455–468. doi: 10.1111/j.1502-3931.2010.00257.x

Ferigolo, J. (1985). "O uso da Microscopia Eletrônica de Varredura na sistemática dos mamíferos," in *Congresso Brasileiro de Paleontologia 8* (Rio de Janeiro). Coletβnea de trabalhos paleontológicos (Brasilia: DNPM), 43–49 (Série Geologia, 27; Secção Paleontologia e Estratigrafia).

Fisher, D. C. (1981a). Crocodilian scatology, microvertebrate concentrations, and enamel-less teeth. *Paleobiology* 7, 262–275. doi: 10.1017/S0094837300004048

Fisher, D. C. (1981b). *Taphonomic Interpretation of Enamel-Less Teeth in the Shotgun Local Fauna (Paleocene, Wyoming)*. Contributions from the Museum of Paleontology, The University of Michigan, Vol. 25, 259–275.

Flessa, K., Struthers, A., Fall, L. M., and Dexter, T. A. (2012). "Really crappy talk: fecal volume and body size in birds, mammals and dinosaurs," in *GSA Annual Meeting & Exposition 4–7 November 2012 (Paper No. 30774–206053)* (Charlotte, NC).

Frech, F. (1897–1902). *Lethaea Palaeozoica.* Bd II. Stuttgart.

Gilmore, B. (1992). Scroll coprolites from the Silurian of Ireland and the feeding of early vertebrates. *Palaeontology* 35, 319–333.

Grigg, G., and Gans, C. (1993). "Morphology and physiology of the Crocodylia," in *Fauna of Australia Vol 2A Amphibia and Reptilia* (Canberra, ACT: Australian Government Publishing Service), 326–336.

Hailey, A., and Davies, P. M. C. (1987). Digestion, specific dynamic action, and ecological energetics of Natrix maura. *Herpetol. J.* 1, 159–166.

Hollocher, K., and Hollocher, T. C. (2012). Early processes in the fossilization of terrestrial feces to coprolites, and microstructure preservation. *NMMNH Bull.* 57, 79–91.

Hollocher, T. C., Chin, K., Hollocher, K. T., and Kruge, M. A. (2001). Bacterial residues in coprolite of herbivorous Dinosaurs:

role of bacteria in mineralization of feces. *Palaios* 16, 547–565. doi: 10.1669/0883-1351(2001)016<0547:BRICOH>2.0.CO;2

Hone, D., Henderson, D. H., Therrien, F., and Habib, M. B. (2015). A specimen of Rhamphorhynchus with soft tissue preservation, stomach contents and a putative coprolite. *PeerJ* 3:e1191. doi: 10.7717/peerj.1191

Hu, S. X., Zhang, Q. Y., and Zhou, C. Y. (2010). Fossil coprolites from the Middle Triassic Luoping biota and ecological implication. *J. Earth Sci.* 21, 191–193. doi: 10.1007/s12583-010-0209-7

Hunt, A. P., Chin, K., and Lockley, M. G. (1994). "The paleobiology of vertebrate coprolites," in *The Paleobiology of Trace Fossils*, ed S. K. Donovan (London: John Wiley), 221–240.

Hunt, A. P., and Lucas, S. G. (2005a). A new coprolite ichnotaxon fromthe Early Permian of Texas. *NMMNH Bull.* 30, 121–122.

Hunt, A. P., and Lucas, S. G. (2005b). The origin of large vertebrate coprolites from the Early Permian of Texas. *NMMNH Bull.* 30, 125–126.

Hunt, A. P., and Lucas, S. G. (2012). Classification of vertebrate coprolites and related trace fossils. *NMMNH Bull.* 57, 137–146.

Hunt, A. P., Lucas, S. G., and Spielmann, J. A. (2005a). Early Permian vertebrate coprolites from north-central New Mexico with description of a new ichnogenus. *NMMNH Bull.* 31, 39–42.

Hunt, A. P., Lucas, S. G., and Spielmann, J. A. (2005b). Biochronology of Early Permian vertebrate coprolites of the American Southwest. *NMMNH Bull.* 31, 43–45.

Laurin, M., and Reisz, R. R. (1995). A reevaluation of early amniote phylogeny. *Zool. J. Linn. Soc. Lond.* 113, 165–223. doi: 10.1111/j.1096-3642.1995.tb00932.x

Lillywhite, H. B., de Delva, P., and Noonan, B. P. (2002). "Patterns of gut passage time and the chronic retention of fecal mass in viperid snakes," in *Biology of the Vipers*, eds G. W. Schuett, M. Höggren, and H. W. Greene (Traverse City, MI: Biological Sciences Press), 497–506.

Lock, B., Heard, D., Detrisac, C., and Elliott Jacobson, E. (2003). An epizootic of chronic regurgitation associated with chlamydophilosis in recently imported emerald tree boas (*Corallus caninus*). *J. Zoo Wildl. Med.* 34, 385–393. doi: 10.1638/02-065

López-Gamundi, O. (2006). Permian plate margin volcanism and tuffs in adjacent basins of west Gondwana: age constraints and common characteristics. *J. S. Am. Earth Sci.* 22, 227–238. doi: 10.1016/j.jsames.2006.09.012

MacGregor, J. H. (1908). "On Mesosaurus brasiliensis nov. sp. from the Permian of Brazil," in *Commissão dos Estudos das Minas de Carvão de Pedra do Brasil, Parte II*, ed I. C. White (Rio de Janeiro: National Press), 301–336.

Marsh, R. S., Moe, C., Lomnethc, R. B., Fawcett, J. D., and Place, A. (2001). Characterization of gastrointestinal chitinase in the lizard Sceloporus undulatus garmani (Reptilia: Phrynosomatidae). *Com. Biochem. Physiol. Part B Biochem. Mol. Biol.* 128, 675–682. doi: 10.1016/S1096-4959(00)00364-X

Milàn, J. (2012). Crocodylian scatology – a look into morphology, internal architecture, inter- and intraspecific variation and prey remains in extant crocodylian feces. *NMMNH Bull.* 57, 65–71.

Modesto, S. P. (1996). *The Anatomy, Relationships, and Palaeoecology of Mesosaurus tenuidens and Stereosternum tumidum (Amniota: Mesosauridae) from the Lower Permian of Gondwana*. dissertation Ph.D. thesis, University of Toronto, Toronto, ON.

Modesto, S. P. (1999). Observations on the structure of the early permian reptile *Stereosternum tumidum* Cope. *Palaeontol. Afr.* 35, 7–19.

Modesto, S. P. (2006). The cranial skeleton of the Early Permian aquatic reptile *Mesosaurus tenuidens*: implications for relationships and palaeobiology. *Zool. J. Linn. Soc. Lond.* 146, 345–368. doi: 10.1111/j.1096-3642.2006.00205.x

Morosi, E. (2011). *Estudio comparativo del cráneo en Mesosauridae de la Formación Mangrullo (Pérmico Temprano) de Uruguay*. Montevideo: Facultad de Ciencias; Universidad de la República.

Nature News (2002, February 12). Jurassic vomit comes up at meeting.

Niedźwiedzki, G., Bajdek, P., Qvarnström, M., Sulej, T., Sennikov, A. G., and Golubev, V. K. (2016). Reduction of vertebrate coprolite diversity associated with the end-Permian extinction event in Vyazniki region, European Russia. *Palaeogeogr. Palaeoclimatol. Palaeoecol.* 450, 77–90. doi: 10.1016/j.palaeo.2016.02.057

Oelofsen, B., and Araújo, D. C. (1987). *Mesosaurus tenuidens and Stereosternum tumidum* from the Permian Gondwana of both Southern Africa and South America. *S. Afr. J. Sci.* 83, 370–372.

Oelofsen, B., and Araújo, D. C. (1983). Palaecological implications of the distribution of mesosaurid reptiles in the Permian Irati sea (Paraná Basin), South America. *Rev. Brasil. Geociê. São Paulo* 13, 1–6.

Owocki, K., Niedźwiedzki, G., Sennikov, A. G., Golubev, V. K., Janiszewska, K., and Sulej, T. (2012). Upper Permian vertebrate coprolites from Vyazniki and Gorokhovets, Vyatkian regional stage, Russian Platform. *Palaios* 27, 867–877. doi: 10.2110/palo.2012.p12-017r

Pauwels, O. S. G., Barr, B., Sanchez, M. L., and Burger, M. (2007). Diet records for the dwarf crocodile, *Osteolaemus tetraspis* tetraspis in Rabi Oil Fields and Loango National Park, southwestern Gabon. *Hamadryad* 31, 258–264.

Piñeiro, G. (2002). *Paleofaunas del Pérmico-Eotriásico de Uruguay*. dissertation/master's thesis, PEDECIBA, Universidad de la República, Montevideo.

Piñeiro, G. (2004). *Paleofaunas del Pérmico y Permo-Triásico de Uruguay. Bioestratigrafía, Paleobiogeografía y Sistemática*. dissertation/PhD thesis, Universidad de la República, Montevideo.

Piñeiro, G. (2006). "Nuevos aportes a la paleontología del Pérmico de Uruguay," in *Cuencas Sedimentarias de Uruguay: Geología, Paleontología y Recursos Naturales – Paleozoico*, eds G. Veroslavsky, M. Ubilla, and S. Martínez (Montevideo: DI.R.A.C./FCien), 257–279.

Piñeiro, G. (2008). "Los mesosaurios y otros fósiles de fines del Paleozoico," in *Fósiles de Uruguay*, ed D. Perea (Montevideo: DIRAC, Facultad de Ciencias), 179–205.

Piñeiro, G., Ferigolo, J., Meneghel, M., and Laurin, M. (2012d). The oldest known amniotic embryos suggest viviparity in mesosaurs. *Hist. Biol.* 24, 620–630. doi: 10.1080/08912963.2012.662230

Piñeiro, G., Ferigolo, J., Ramos, A., and Laurin, M. (2012a). Cranial morphology of the Early Permian mesosaurid *Mesosaurus tenuidens* and the evolution of the lower temporal fenestration reassessed. *Comptes Rendus Palevol.* 11, 379–391. doi: 10.1016/j.crpv.2012.02.001

Piñeiro, G., Ferigolo, J., Núñez, P., Meneghel, M., and Laurin, M. (2015a). "Mesosaurid taxonomy and systematics: the evolutive significance of the mesosaurid lower temporal fenestra: an overview of more than a century of research," in *Biology and Evolutionary History of Mesosaurs, the Oldest Known Gondwanan Aquatic Reptiles, Topic Research, Frontiers in Earth Science*, eds G. Piñeiro and M. Laurin.

Piñeiro, G., Ferigolo, J., Núñez, P., Meneghel, M., Villar, S., and Laurin, M. (2015b). "Mesosaurid reproductive biology," in *Biology and Evolutionary History of Mesosaurs, the Oldest Known Gondwanan Aquatic Reptiles, Topic Research, Frontiers in Earth Science.* eds G. Piñeiro, and M. Laurin.

Piñeiro, G., Meneghel, M., and Núñez-Demarco, P. (2016). The ontogenetic transformation of the mesosaurid tarsus: a contribution to the origin of the amniotic astragalus. *PeerJ* 4:e2036. doi: 10.7717/peerj.2036

Piñeiro, G., Morosi, E., Ramos, A., and Scarabino, F. (2012c). Pygocephalomorph crustaceans from the Early Permian of Uruguay: constraints on taxonomy. *Rev. Brasil. Paleontol.* 15, 33–48. doi: 10.4072/rbp.2012.1.03

Piñeiro, G., Ramos, A., Goso, C., Scarabino, F., and Laurin, M. (2012b). Unusual environmental conditions preserve a mesosaur-bearing Konservat-Lagerstätte from Uruguay. *Acta Palaeontol. Pol.* 57, 299–318. doi: 10.4202/app.2010.0113

Poinar, G., and Boucot, A. J. (2006). Evidence of intestinal parasites of dinosaurs. *Parasitology* 133, 245–249. doi: 10.1017/S0031182006000138

Pretto, F. A., Cabreira, S. F., and Schultz, C. L. (2014). Tooth microstructure of the Early Permian aquatic predator *Stereosternum tumidum*. *Acta Palaeontol. Pol.* 59, 125–133. doi: 10.4202/app.2011.0121

Qvarnström, M., Niedźwiedzki, G., and Žigaitė, Ž. (2016). Vertebrate coprolites (fossil faeces): an underexplored *Konservat-Lagerstätte*. *Earth Sci. Rev.* 162, 44–57. doi: 10.1016/j.earscirev.2016.08.014

Raimundo-Silva, R. (1999). *Hábito Alimentar de Brazilosaurus Sanpauloensis (Reptilia, Mesosauridae) Formação Iratí, Estado de Goiás, com Base em Conteúdo Digestivo e Coprólitos*. dissertation/master's thesis, Instituto de Geociências, Universidade Federal do Rio Grande do Sul, Porto Alegre.

Raimundo-Silva, R., Ferigolo, J., and Sedor, F. A. (1997). "Primeiras evidências de conteúdo digestivo em *Brazilosaurus sanpauloensis* (Reptilia, Mesosauridae) da Formação Iratí, Bacia do Paraná," in *XV Congresso Brasileiro de Paleontologia* (São Pedro: Resumos), 85.

Ramos, A. (2015). *La Dieta de Los Reptiles Mesosauridae (Reptilia: Proganosauria) del Pérmico Temprano de Uruguay.* dissertation/graduation's thesis, Universidad de la República, Montevideo.

Romer, A. S. (1966). *Vertebrate Paleontology.* Chicago, IL: The University of Chicago Press, 468.

Santos, R. V., Souza, P. A., Alvarenga, C. J. S., Dantas, E. L., Pimentel, E. L., Oliveira, C. G., et al. (2006). Shrimp U—Pb zircon dating and palynology of bentonitic layers from the permian irati formation Parana basin, Brazil. *Gondwana Res.* 9, 456–463. doi: 10.1016/j.gr.2005.12.001

Schwimmer, D. R., Weems, R. E., and Sanders, A. E. (2015). A late cretaceous shark coprolite with baby freshwater turtle vertebrae inclusions. *Palaios* 30, 707–713. doi: 10.2110/palo.2015.019

Sedor, F. A. (1994). *Estudo Pós-Craniano de Brazilosaurus Sanpauloensis Shikama & Ozaki, 1966 (Anapsida, Proganosauria, Mesosauridae) da Formação Irati, Permiano da Bacia do Paraná, Brasil.* dissertation/master's thesis, Programa de Pós-graduação em Geociências, Universidade Federal do Rio Grande do Sul, Porto Alegre.

Semeniuk, V., Manolis, C., Webb, G. J. W., and Mawson, P. R. (2011). The saltwater crocodile, *Crocodylus porosus* Schneider, 1801, in the Kimberley coastal region. *J. R. Soc. West. Aust.* 94, 407–416.

Shelton, C. D. (2013). A new method to determine volume of bromalites: morphometrics of Lower Permian (Archer City Formation) heteropolar bromalites. *Swiss J. Palaeontol.* 132, 221–238. doi: 10.1007/s13358-013-0057-z

Siroski, P. A., Poletta, G. L., Parachu Marco, M. V., Ortega., H. H., and Merchant., M. E. (2014). Presence of chitinase enzymes in crocodilians. *Acta Herpetol.* 9, 139–146. doi: 10.13128/Acta_Herpetol-13237

Smith, R. M. H., and Botha-Brink, J. (2011). Morphology and composition of bone-bearing coprolites from the Late Permian Beaufort Group, Karoo Basin, South Africa. *Palaeogeogr. Palaeoclimatol. Palaeoecol.* 312, 40–53. doi: 10.1016/j.palaeo.2011.09.006

Stone, D. D., Crisp, E. L., and Bishop, J. T. (2000). A large meat-eating dinosaur coprolite from the jurassic morrison formation of Utah. *Geol. Soc. Am. Abstr. Programs* 32, 220.

Terry, R. C. (2004). Owl pellet taphonomy: a preliminary study of the post-regurgitation taphonomic history of pellets in a temperate forest. *Palaios* 19, 497–506. doi: 10.1669/0883-1351(2004)019<0497:OPTAPS>2.0.CO;2

Thies, D., and Hauff, R. B. (2013). A Speiballen from the lower jurassic posidonia shale of South Germany. *N. Jb. Geol. Paläont. Abh.* 267, 117–124. doi: 10.1127/0077-7749/2012/0301

Tsuji, L. A., and Müller, J. (2009). Assembling the history of the Parareptilia: phylogeny, diversification, and a new definition of the clade. *Fossil Rec.* 12, 71–81. doi: 10.1002/mmng.200800011

Verdade, L. M., Packer, I. U., Michelloti, F., and Rangel, M. C. (1994). "Thermoregulatory behavior of broad snouted caiman (*Caiman latirostris*) under different thermal regimes," in *Workshop Sobre Conservación y Manejo del Yacare Overo Caiman latirostris*, Vol. 4 (Santa Fé), 84–94.

Villamil, J., Meneghel, M., Blanco, R. E., Jones, W., Núñez Demarco, P., Rinderknecht, A., et al. (2015). Optimal swimming speed estimates in the Early Permian mesosaurid *Mesosaurus tenuidens* (Gervais, 1865) from Uruguay. *Hist. Biol.* 28, 963–971. doi: 10.1080/08912963.2015.1075018

Wahl, A. M., Martin, A. J., and Hasiotis, S. (1998). "Vertebrate coprolites and coprophagy traces, Chinle Formation (Late Triassic), Petrified Forest National Park, Arizona, in *National Park Service Paleontological Research*," in *National Park Service Geological Resources Division Technical Report NPS/NRGRD/GRDTR-98/01*, eds V. L. Santucci and L. McClelland (Lakewood, CO: National Park Service; Geological Resources Division), 144–148.

Wang, M., Zhou, Z., and Sullivan, C. (2016). A fish-eating enantiornithine bird from the early cretaceous of china provides evidence of modern avian digestive features. *Curr. Biol.* 26, 1–7. doi: 10.1016/j.cub.2016.02.055

Webb, G., and Manolis, C. (2009). *Green Guide to Crocodiles of Australia.* Sydney, NSW: New Holland Publishers.

Wings, O. (2007). A review of gastrolith function with implications for fossil vertebrates and a revised classification. *Acta Palaeontol. Pol.* 52, 1–16.

Wood, J. R., and Wilmshurst, J. M. (2014). Late Quaternary terrestrial vertebrate coprolites from New Zealand. *Quaternary Sci. Rev.* 98, 33–44. doi: 10.1016/j.quascirev.2014.05.020

Conflict of Interest Statement: The authors declare that the research was conducted in the absence of any commercial or financial relationships that could be construed as a potential conflict of interest.

3

An Empirical Approach for Estimating Stress-Coupling Lengths for Marine-Terminating Glaciers

Ellyn M. Enderlin[1,2]*, Gordon S. Hamilton[1,2], Shad O'Neel[3], Timothy C. Bartholomaus[4], Mathieu Morlighem[5] and John W. Holt[6]

[1] Climate Change Institute, University of Maine, Orono, ME, USA, [2] School of Earth and Climate Sciences, University of Maine, Orono, ME, USA, [3] Alaska Science Center, US Geological Survey, Anchorage, AK, USA, [4] Department of Geological Sciences, University of Idaho, Moscow, ID, USA, [5] Department of Earth System Science, University of California Irvine, Irvine, CA, USA, [6] Institute for Geophysics, University of Texas at Austin, Austin, TX, USA

Edited by:
Alun Hubbard,
Aberystwyth University, Norway

Reviewed by:
Jaime Otero,
Technical University of Madrid, Spain
Samuel Huckerby Doyle,
Aberystwyth University, UK

*Correspondence:
Ellyn M. Enderlin
ellyn.enderlin@gmail.com

Specialty section:
This article was submitted to
Cryospheric Sciences,
a section of the journal
Frontiers in Earth Science

Despite an increase in the abundance and resolution of observations, variability in the dynamic behavior of marine-terminating glaciers remains poorly understood. When paired with ice thicknesses, surface velocities can be used to quantify the dynamic redistribution of stresses in response to environmental perturbations through computation of the glacier force balance. However, because the force balance is not purely local, force balance calculations must be performed at the spatial scale over which stresses are transferred within glacier ice, or the stress-coupling length (SCL). Here we present a new empirical method to estimate the SCL for marine-terminating glaciers using high-resolution observations. We use the empirically-determined periodicity in resistive stress oscillations as a proxy for the SCL. Application of our empirical method to two well-studied tidewater glaciers (Helheim Glacier, SE Greenland, and Columbia Glacier, Alaska, USA) demonstrates that SCL estimates obtained using this approach are consistent with theory (i.e., can be parameterized as a function of the ice thickness) and with prior, independent SCL estimates. In order to accurately resolve stress variations, we suggest that similar empirical stress-coupling parameterizations be employed in future analyses of glacier dynamics.

Keywords: marine-terminating glaciers, force balance, stress-coupling, Columbia Glacier, Helheim Glacier, glacier dynamics

INTRODUCTION

Dynamic mass loss from Greenland's marine-terminating glaciers accounted for nearly half of the ice sheet's mass loss over the last two decades (Enderlin et al., 2014) yet the processes controlling the timing and magnitude of changes in dynamics are poorly understood (Vieli and Nick, 2011). Force balance estimates (van der Veen and Whillans, 1989), constrained with surface observations, potentially offer insight into the spatial and temporal variations in driving and resistive stresses that govern glacier flow. However, poor spatial coverage, low resolution, and large uncertainties in velocity and thickness have historically limited its application to a handful of glaciers (e.g., Whillans et al., 1989; O'Neel et al., 2005; Kavanaugh and Cuffey, 2009; Sergienko and Hindmarsh, 2013; van der Veen et al., 2014). Moreover, differences in the way that observational datasets have been incorporated into analyses have thus far prevented rigorous comparisons of how different glaciers respond to common environmentally-forced stress perturbations.

Recent advances in observational technology have increased the accuracy and spatial coverage of radar-derived bed elevation estimates and decreased the temporal repeat interval of surface velocity and elevation observations. These improvements have vastly expanded the potential to investigate differences in glacier dynamics using analytical and numerical modeling techniques. To take advantage of the increased availability of high-resolution surface observations, the processing of these input data must be carefully considered. A key consideration is the distance over which stresses are horizontally transferred within glacier ice (Kamb and Echelmeyer, 1986; Bahr et al., 1994; O'Neel et al., 2005; Maxwell et al., 2008), i.e., the stress-coupling length (hereafter referred to as the SCL). Because of stress coupling, glaciers act as low-pass filters, limiting the spatial scales over which stress perturbations at the bed will be expressed at the glacier surface (e.g., Armstrong et al., 2016). Under-estimation of the SCL, or failure to account for stress-coupling altogether (i.e., calculation of local strain rates), will lead to noisy results. Although the noise introduced by random observational errors decreases with increasing length of the smoothing window, smoothing over lengths greater than the SCL (or over-estimation of the SCL) can obscure the controlling processes of observed variability in ice flow (Kamb and Echelmeyer, 1986). To demonstrate the need for accurate SCL estimates when computing strain rates, we constructed a synthetic surface speed profile representing a fast-flowing tidewater glacier and calculated the strain rate (i.e., longitudinal speed gradient) over the SCL (**Figure 1**, black squares and linear trendline). We then added random errors (± 300 m year^{-1}) to the synthetic speed data (**Figure 1**, red diamonds) and calculated strain rates for the true, over-estimated, and under-estimated SCLs (**Figure 1**, solid, dashed, and dot-dashed lines, respectively). Under- or over-estimation of the SCL by $\sim$50% in **Figure 1** is associated with errors in strain rates of 40 and 52%, respectively, compared to the <3% error for the strain rate estimate obtained over the SCL. Thus, in order to maximize our ability to resolve glacier dynamics, and to determine the scales over which basal boundary conditions can be inferred from surface observations, a robust method is needed to estimate the SCL.

Theory suggests that the SCL varies as a function of glacier geometry (i.e., thickness and width), the difference between the bulk (i.e., vertically-averaged) effective viscosity and the effective viscosity of basal ice, and the exponent of the glacier flow law that relates stresses and strain rates (Kamb and Echelmeyer, 1986). Accordingly, we expect maximum SCLs in the interior of polar ice sheets ($\sim$4–10 ice thicknesses) and much shorter SCLs for temperate alpine glaciers ($\sim$1–3 ice thicknesses; Kamb and Echelmeyer, 1986). Early SCL approximations (e.g., Kamb and Echelmeyer, 1986) assumed equal longitudinal and shear effective viscosities, allowing SCL estimations from glacier geometry. In practice, most force balance analyses assume that the SCL can be approximated as a function of the ice thickness. However, methods relating SCL to ice thickness vary widely among analyses. For example, Kavanaugh and Cuffey (2009) estimated SCL scaling factors by relating the ice thickness to the length of the along-flow smoothing window that produces the best match between observed surface speeds and inferred basal speeds. In contrast, O'Neel et al. (2005) estimate the SCL as the length of the isotropic smoothing window required to minimize spurious (i.e., negative) values of basal drag while maintaining the characteristic out-of-phase relationship between the gravitational driving stress and longitudinal stress gradients.

In this paper we present a rigorous, automated method for estimating the SCL from high-resolution surface velocity and

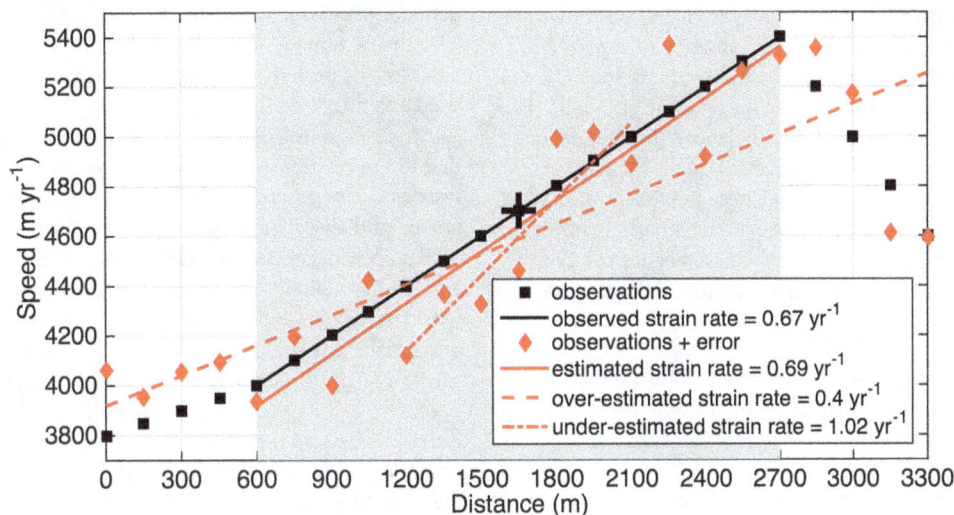

FIGURE 1 | A synthetic surface speed profile representing a fast-flowing tidewater glacier is shown in black and the speed profile plus random errors (± 300 m year^{-1}) are shown in red. The SCL is prescribed as 2100 m, which is reasonable for fast-flowing glaciers that are several hundred meters thick (Kamb and Echelmeyer, 1986). The gray shaded rectangle indicates the prescribed SCL, which defines the distance over which the strain rate should be calculated for the pixel marked by the black "+." The solid, dashed, and dashed-dotted lines display estimated strain rates (i.e., longitudinal speed gradients) for the true, over-estimated, and under-estimated SCLs, respectively.

elevation observations. Our empirical approach to estimate the SCL utilizes the depth-integrated force balance equations, and therefore requires that bed elevation estimates have similar spatial resolution and coverage as the surface observations. Additionally, ice flow must be dominated by basal sliding (i.e., plug flow conditions). These requirements have so far restricted the extraction of accurate SCL estimates and associated force balance applications to a small sample of marine-terminating glaciers with well-constrained bed elevations.. The recent development of mass-conserving bed elevation maps for Greenland (BedMachine Greenland; Morlighem et al., 2014), portions of Antarctica (e.g., Holt et al., 2006; Vaughan et al., 2006), and marine-terminating glaciers outside of the major ice sheets (e.g., McNabb et al., 2012) has, however, created the potential to perform high-resolution force balance analyses for a broad spectrum of glaciers. We focus here on data from Helheim Glacier, a well-studied outlet glacier in SE Greenland, to illustrate our method. The versatility of our empirical SCL estimation method is demonstrated through an additional application to Columbia Glacier, Alaska. Both glaciers have been the subject of extensive study over the last several decades and have observational records of comparable spatial and temporal resolution.

DATA AND METHODS

Our goal is to develop a method for estimating the SCL from the observational datasets required for both force balance analyses and numerical ice flow modeling. These datasets consist of maps of bed and surface elevation, as well as surface velocity, all with comparable spatial resolution. Although the Kamb and Echelmeyer (1986, Equation 32) theory provides the means to directly estimate the SCL from a comprehensive suite of observational data, the effective ice viscosity required to accurately estimate the SCL from theory is largely unknown for fast-flowing glaciers. Therefore, in the absence of *a priori* SCL estimates, we take a two-step approach to estimate the SCL: (1) solve the force balance equations across adjacent grid cells (i.e., compute the local, over-sampled force balance), then (2) estimate the SCL from the local, over-sampled stress fields themselves.

A brief overview of the force balance method is presented below, followed by a description of the SCL estimation approach. We refer the reader to van der Veen (2013, Section 11.2) for a more detailed description of the force balance method.

Force Balance Method

The gravitational force driving the down-slope flow of ice is balanced by resistive stress generated by frictional forces at the basal and lateral margins and by longitudinal and lateral stress gradients. Thus, where ice flow is dominated by basal sliding rather than internal deformation, the force balance over a vertical column of ice is calculated as

$$\tau_{dx} = \tau_{bx} - \frac{\partial}{\partial y}\left(HR_{xy}\right) - \frac{\partial}{\partial x}\left(HR_{xx}\right), \qquad (1)$$

$$\tau_{dy} = \tau_{by} - \frac{\partial}{\partial x}\left(HR_{xy}\right) - \frac{\partial}{\partial y}\left(HR_{yy}\right). \qquad (2)$$

Subscripts x and y indicate the along- and across-flow directions, τ_d is the gravitational driving stress, τ_b is the basal resistance, H is the ice thickness, and R is the tensile or compressional (subscripts xx and yy) and shear (subscript xy) resistive stresses. To solve (Equations 1, 2), gradients in surface elevation (i.e., surface slope), surface velocity (i.e., strain rates), and depth-integrated resistive stresses must be calculated. To minimize bias that can be introduced by random errors or by the truncation of the averaging window near the glacier margins, we take a hybrid approach of the force balance method and the control method (MacAyeal, 1992) and calculate gradients in elevation, speed, and resistive stresses using a linear regression across the averaging window (van der Veen, 2013), such as illustrated for the calculation of strain rates in **Figure 1**. The gravitational driving stress is calculated using an ice density of 917 kg m^{-3}. Depth-integrated resistive stresses are estimated from surface strain rates using Glen's flow law (Hooke, 2005, p. 66) with an exponent of $n = 3$. Basal drag is estimated as the residual that satisfies the force balance equalities.

Stress-Coupling Length Estimation

The stress coupling length in a spring under tension can be considered to be the wavelength of the displaced spring coils. To visualize this concept, imagine a spring, not under tension, anchored at both ends, resting on a high-friction surface. If you gently pull on one of the coils, only a few coils will be displaced because the force exerted through tension is rapidly dissipated with distance from the tensional source due to friction. As you increase the stress exerted on the spring (i.e., pull more forcefully), more coils begin to stretch and the length over which the coils are displaced increases. You are not directly pulling on the interior coils, but they unfurl because the force is distributed along a portion of the spring. The distance that the force exerted by your hand is transferred along the spring (i.e., SCL) is the distance over which the coils are displaced (i.e., the wavelength). The displacement of the coils themselves reveals the SCL. By analogy, the periodicity of oscillations in resistive stresses can be used as a proxy for the SCL in glaciers.

Stress coupling is more complicated for glaciers, however, due to the relationship between stress, strain rates, and effective viscosity. The differences in effective viscosity of ice undergoing tensile and shear stresses further complicate matters. As shown by Kamb and Echelmeyer (1986) and confirmed by Kavanaugh and Cuffey (2009), the SCL varies in the along-flow direction with the tensile and compressional (i.e., longitudinal) stresses and basal shear stress (i.e., basal drag). Following this precedent, we solve the force balance equations (Equations 1, 2) using a centered-differences approach and calculate the along-flow resistive stress as the sum of the local longitudinal stresses and basal drag. Stresses calculated in this manner are over-sampled, meaning the large-scale oscillations in longitudinal and basal stresses that reflect stress coupling will be superimposed on high-frequency stress anomalies introduced in part by random errors in the observational datasets. To ensure that potential spatial variations in the SCL are preserved, we extract the periodicity of the large-scale oscillations in stresses from multiple flow-following profiles (i.e., flowlines). The optimal number of profiles is somewhat subjective, but should be selected based on the glacier width and spatial resolution of the observational data. The number of profiles should also take into account considerable across-flow variations in thickness and/or rheology, such as those expected for complex glacier systems (i.e., glaciers with multiple tributaries), which may result in spatial variations in the SCL.

We use an automated, signal processing approach to estimate the SCL from the stress profiles. To assess whether this automated approach sufficiently resolves potential spatial variations in the SCL, we compare SCLs obtained using our automated approach to SCLs that are manually extracted from the stress profiles. For the automated estimates, we pre-process local resistive stress profiles for each observation date by normalizing them to vary about zero. The detrended profiles are tapered using a Hann (raised cosine) window that minimizes endpoint discontinuities to prevent spectral leakage. Resistive stress oscillations are then identified using periodograms constructed from the pre-processed profiles.

Assuming that the SCL does not vary considerably along each profile, the periodograms should return peaks in spectral power for the periods that correspond to the SCLs. In order to resolve the SCL in this manner, however, we must account for the exponential decrease in power with increased period length that is expected for spatially correlated data (i.e., a red noise distribution (Bartlett, 1955); power weighted toward lower frequencies). For each periodogram, we approximate the underlying red noise power spectrum using a linear regression of the logarithm of the periodogram (Vaughan, 2005) and calculate the residual power. We then estimate the SCL as the dominant periodicity (i.e., highest residual power) that falls within the analytically-derived minimum SCL of $1H$ (Bahr et al., 1994) and the theoretical maximum SCL of $10H$ for polar ice (Kamb and Echelmeyer, 1986), where H is the average ice thickness for the profile. Finally, we average stress profiles for as many time intervals as possible over a relatively quiescent period of dynamic change. Temporal averaging of the stress profiles has a similar effect on the SCL estimates as periodogram stacking because it minimizes observational errors, which increases the signal-to-noise ratio in the periodograms. The averaging also minimizes aliasing effects caused by short-term changes in ice flow and surface accumulation and/or ablation and reduces the effect of spatial data gaps, improving the accuracy of our automated SCL estimates.

Uncertainties in SCL Estimates

The surface speed and ice thickness observations used to estimate SCL benefit from higher spatial resolution and lower uncertainties than datasets that were available a decade ago. However, inversion of surface observations to estimate basal drag remains mathematically ill-posed (Bahr et al., 2014), leading to large uncertainties in the over-sampled stress estimates that we use to estimate the SCL. Several forms of observational errors exist, including: spatial and temporal variation in ice motion, and random errors associated with measuring ice motion as well as spatially systematic but transient errors (i.e., aliasing errors). Below we discuss each source of observational

uncertainty, emphasizing methods to minimize their impact on SCL estimates.

Ice motion and surface elevation data are subject to a diffusive smoothing algorithm (Perona and Malik, 1990), which increases spatial correlation between adjacent pixels and reduces random error while preserving the sharp gradients in ice speed observed near the lateral margins and the calving front (Martin, Pers. Commun., 2014). Remaining random errors in the observational data are further minimized through temporal averaging of the stress profiles. Diffusive smoothing of the velocity maps and temporal averaging of the stress profiles also minimizes the impacts of spatial and temporal aliasing errors that can be introduced due to variability in motion and ice thickness over tidal to seasonal time scales, as well as by in-filling of raster data gaps.

APPLICATIONS AND DISCUSSION

In practice, comprehensive determination of SCL using our empirical approach applied to contiguous flow-following profiles spanning entire glacierized catchments is computationally expensive. As a result, we choose to estimate the SCL for relatively sparse (~0.5–1 km separation) profile arrays then, following theory and convention, use these empirical SCL estimates to parameterize the SCL over the entire glacierized catchment as a function of ice thickness. The empirical scaling relationship between the SCL and thickness from our sparse profiles allows us to estimate the SCL for each pixel in the domain. We apply our SCL estimation approach and compute empirical thickness-scaling relationships for two well-studied fast-flowing tidewater glaciers: Helheim Glacier, SE Greenland and Columbia Glacier, Alaska, USA. Use of the standardized approach developed here will enable direct inter-comparisons of the force balance method at the two sites. In the examples presented below, we demonstrate the ability of our empirical SCL estimation approach to (1) resolve both intra- and inter-glacier variations in the SCL and (2) illustrate the sensitivity of SCL estimates and thickness-scaling parameterizations to observational uncertainties.

Observational Data

Surface elevations were extracted from stereo images collected by the Advanced Spaceborne Thermal Emission and Reflection Radiometer (ASTER) and WorldView-1 and -2 satellites. The 30 m-resolution ASTER digital elevation models (DEMs) are available from the NASA Earth Observing System Data and Information System (NASA EOSDIS; http://reverb.echo.nasa.gov). High-resolution (~1.5 m) WorldView DEMs were produced following the methods described in Shean et al. (2016). A total of 18 DEMs were compiled for each glacier (**Table 1**). The Helheim DEMs span the 2010–2014 period, with most (16/18) observations acquired from May-August of each year. Although DEMs of Columbia Glacier are also available over the entire 2010–2014 period, we restricted our DEM database to the 2012–2013 period when the glacier surface elevations remained fairly constant (i.e., within ± 10 m of the mean for the same period). Random errors

TABLE 1 | Dates of digital elevation models (DEMs) and velocities for Helheim Glacier, SE Greenland, and Columbia Glacier, Alaska, USA, that are used for our SCL estimates.

Helheim Glacier			Columbia Glacier		
DEM date	Velocity dates		DEM date	Velocity dates	
2010/06/01	2010/04/29	2010/06/23	2012/03/29	2012/03/14	2012/04/05
2010/06/24	2010/06/23	2010/07/04	2012/05/07	2012/04/10	2012/05/08
2010/07/08	2010/07/04	2010/07/15	2012/05/15	2012/05/08	2012/05/19
2011/03/19	2011/03/14	2011/04/27	2012/06/05	2012/05/19	2012/06/15
2011/06/15	2011/06/10	2011/06/21	2012/07/17	2012/07/13	2012/07/18
2011/06/28	2011/06/21	2011/07/02	2012/07/19	2012/07/18	2012/08/15
2012/05/13	2012/05/04	2012/05/15	2012/08/13	2012/07/18	2012/08/15
2012/06/24	2012/05/26	2012/06/24	2012/10/01	2012/09/06	2012/10/09
2012/07/08	2012/06/28	2012/07/09	2012/10/12	2012/10/09	2012/10/25
2013/04/23	2013/03/20	2013/05/14	2012/11/20	2012/11/11	2012/11/22
2013/06/16	2013/06/05	2013/06/21	2012/11/23	2012/11/22	2013/03/12
2014/05/09	2014/04/20	2014/05/12	2013/03/26	2013/03/12	2013/04/03
2014/06/01	2014/05/23	2014/06/03	2013/05/06	2013/04/25	2013/05/06
2014/06/11	2014/06/03	2014/06/14	2013/06/05	2013/05/06	2013/06/08
2014/06/28	2014/06/25	2014/07/06	2013/06/10	2013/06/08	2013/06/19
2014/07/02	2014/06/25	2014/07/06	2013/07/11	2013/06/19	2013/07/27
2014/07/03	2014/07/06	2014/07/06	2013/07/12	2013/06/19	2013/07/27
2014/07/31	2014/07/22	2014/08/08	2013/11/19	2013/11/14	2014/03/26

in DEM-derived surface elevations were estimated as ~7 m for the ASTER products (Stearns and Hamilton, 2007) and ~3 m for WorldView DEMs (Enderlin and Hamilton, 2014). Systematic biases in DEM-derived elevations were minimized through vertical co-registration of overlapping DEMs using exposed bedrock elevations (Nuth and Kääb, 2011). Horizontal uncertainties are comparable to the DEM pixel size (Stearns and Hamilton, 2007).

We calculated ice thickness as the difference between ice surface and bed elevations. Bed elevations for Helheim Glacier were taken from the 150 m-resolution BedMachine Greenland dataset (Morlighem et al., 2014; http://nsidc.org/data/docs/daac/icebridge/idbmg4/index.html), which is based on mass conservation and constrained by ice-penetrating radar observations. Uncertainties increase with distance from radar profiles, and were estimated as an average of ~50 m over the glacier's fast-flowing trunk (Morlighem et al., 2014). Columbia bed elevations were estimated using the same mass conservation algorithm and constrained using a subset of ice-penetrating radar observations from Rignot et al. (2013) that were manually checked for errors introduced by off-nadir reflections using a surface clutter prediction algorithm (Holt et al., 2006). Only bed echoes verified by this algorithm were retained for our study. Using these radar data, we found that the glacier is ~50–250 m thinner along its fast-flowing core and ~50–150 m thicker along its lateral margins than previously estimated (McNabb et al., 2012). As with Helheim, the uncertainties in the bed elevations increase with distance from the radar profiles, with an average vertical uncertainty of ~15 m. The filtered radar observations and bed elevation map

for Columbia Glacier can be accessed on the Maine DataVerse Network (MDVN; http://dataverse.acg.maine.edu/dvn/dv/eep).

Surface velocities for Helheim Glacier were extracted from the NASA Making Earth System Data Records for Use in Research Environments (MEaSUREs) product (NSIDC; http://nsidc.org/data/nsidc-0481/) produced using TerraSAR-X interferometric synthetic aperture radar (InSAR) measurements. The velocity product is posted at a 150 m spatial resolution. Columbia Glacier velocities were calculated using the same approach and have a spatial resolution of 100 m. Random errors were estimated from local variations in ice velocity and systematic errors were estimated as 3% of the speed (Joughin et al., 2010). Velocity dates are listed in **Table 1**.

To standardize the datasets and referencing systems, we down-sampled the co-registered DEMs from their native resolution and interpolated the bed elevation grids to the 150 m-resolution InSAR velocity grids using a linear distance-weighted approach similar to O'Neel et al. (2005). DEM data gaps were filled using an iterative nearest neighbor algorithm (Garcia, 2010). Velocities corresponding to each DEM were estimated by averaging the two InSAR velocity maps closest in time (typically within ~2 weeks) to the DEM acquisition date, which minimizes the impact of short-term (hourly to monthly) variability in ice dynamics. After determining SCLs, the force balance equations (Equations 1, 2) were solved using a temperature-dependent viscosity parameter of $B = 1.02 \times 10^8$ Pa s$^{1/3}$ corresponding to an average ice temperature of $-5°C$ for Helheim Glacier (Nick et al., 2009) and a viscosity parameter appropriate for temperate ice ($B = 7.47 \times 10^7$ Pa s$^{1/3}$) for Columbia Glacier (Nick et al., 2007).

Example Application: Helheim Glacier

The domain and input data for SCL estimates at Helheim Glacier in SE Greenland are shown in **Figure 2**. At Helheim, we extracted ice thickness and resistive stress along five flowlines spanning the main trunk and major tributaries (**Figure 2**, dashed lines). For **Figure 2** and for our SCL thickness-scaling parameterization, we use the median ice thickness from the 18 DEMs compiled for the 2010–2014 period. We use the median thickness, not the mean, to minimize biases in the derived thickness-scaling parameterization introduced by blunders in the DEMs.

Thickness and resistive stress profiles from the central flowline of Helheim Glacier are shown in **Figures 3A,B**, highlighting how temporal averaging preserves periodicity while reducing high-frequency noise in the stress profiles and resulting power spectra (**Figure 3C**). Temporal averaging also ensures that variability in the spatial extent of the datasets do not bias the periodogram-derived SCL estimates. The large variations in automated SCL estimates (**Figure 4A**) suggest that the SCL varies considerably within the glacier catchment. The presence of along-flow variations in the SCL is verified through manual inspection of the stress profiles and the derived periodograms. The distance between local resistive stress maxima are manually extracted from the stress profiles, as demonstrated in **Figure 3A**, and used as a proxy for along-flow changes in the SCL. Using this approach, we obtain manual SCL estimates that can be clustered into two SCLs: 2321 ± 43 m and 3524 ± 430 m. Manual inspection of the associated periodograms (**Figure 3C**) reveals primary and secondary significant peaks (i.e., positive residual power) at ~3800 and ~2000 m, respectively, that are in good agreement with the manual SCL estimates.

Although the along-flow variations in the SCL can be manually extracted from the resistive stress profiles, random errors in the observational data can obscure the local stress maxima, leading to subjective SCL estimates. To capture along-flow variations in the SCL, we instead use a moving window to segment the stress profiles and extract the SCL from each segment using our automated, periodogram-based approach. The segment length and sampling interval is chosen to minimize aliasing of SCL variability: the length of the moving window is slightly (10%) longer than $2 \times SCL_{mean}$, where SCL_{mean} is the mean SCL estimate for the profile, and there is 50% overlap between window segments. Given the theoretical and observed correlation between SCL and the ice thickness (**Figure 4A**), we estimate SCL_{mean} as the product of the mean profile thickness and the mean SCL:H ratio for the full-length time-averaged profiles (**Figure 4A**; $SCL = 4.2H$). SCL_{mean} values range from ~700 m to ~5.3 km, corresponding to segment lengths ranging from ~1.5 to ~11.7 km. The segmented automated approach yields an average SCL:H ratio for all segments of 3.7 (**Figure 4B**), which is in good agreement with the manually-derived average of 3.8. Profile segmentation also reduces variability in the SCL:H ratio and increases the strength of the correlation between the SCL and ice thickness: $\sigma = 2.5$ and $R = 0.56$ for the time-averaged profiles and $\sigma = 1.7$ and $R = 0.69$ for the segmented profiles, where σ is the standard deviation in the SCL:H ratio. Thus, in-line with theory and convention, we use the automated SCL estimates extracted from the segmented profiles to construct a thickness-scaling parameterization that allows us to extrapolate the SCL over the entire glacier domain.

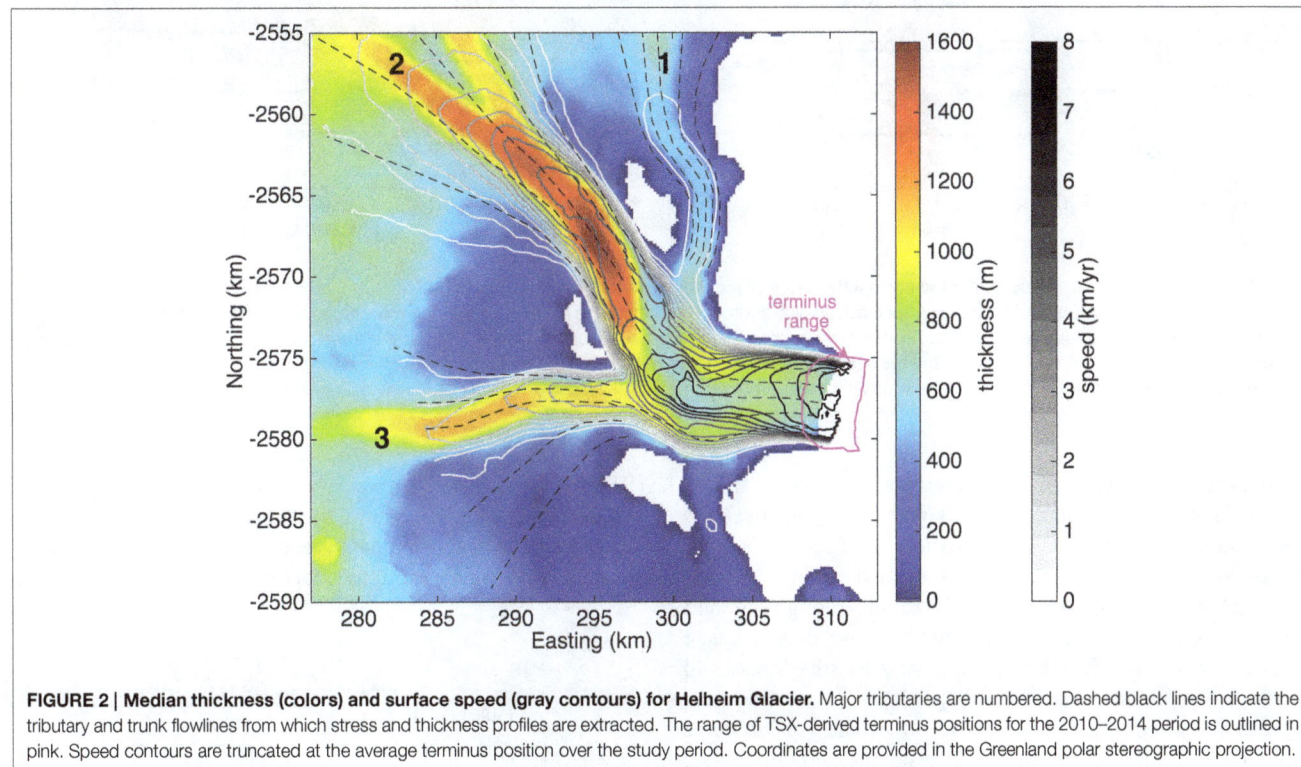

FIGURE 2 | Median thickness (colors) and surface speed (gray contours) for Helheim Glacier. Major tributaries are numbered. Dashed black lines indicate the tributary and trunk flowlines from which stress and thickness profiles are extracted. The range of TSX-derived terminus positions for the 2010–2014 period is outlined in pink. Speed contours are truncated at the average terminus position over the study period. Coordinates are provided in the Greenland polar stereographic projection.

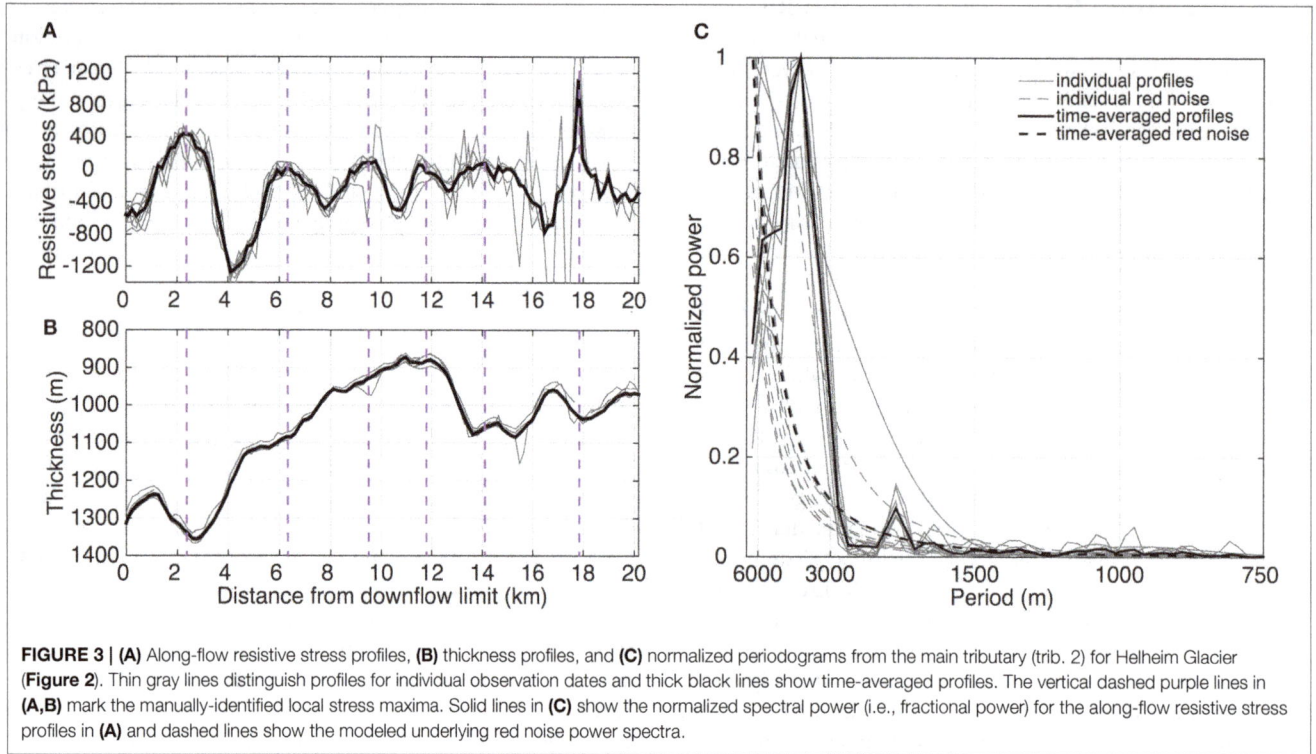

FIGURE 3 | **(A)** Along-flow resistive stress profiles, **(B)** thickness profiles, and **(C)** normalized periodograms from the main tributary (trib. 2) for Helheim Glacier (**Figure 2**). Thin gray lines distinguish profiles for individual observation dates and thick black lines show time-averaged profiles. The vertical dashed purple lines in (**A,B**) mark the manually-identified local stress maxima. Solid lines in (**C**) show the normalized spectral power (i.e., fractional power) for the along-flow resistive stress profiles in (**A**) and dashed lines show the modeled underlying red noise power spectra.

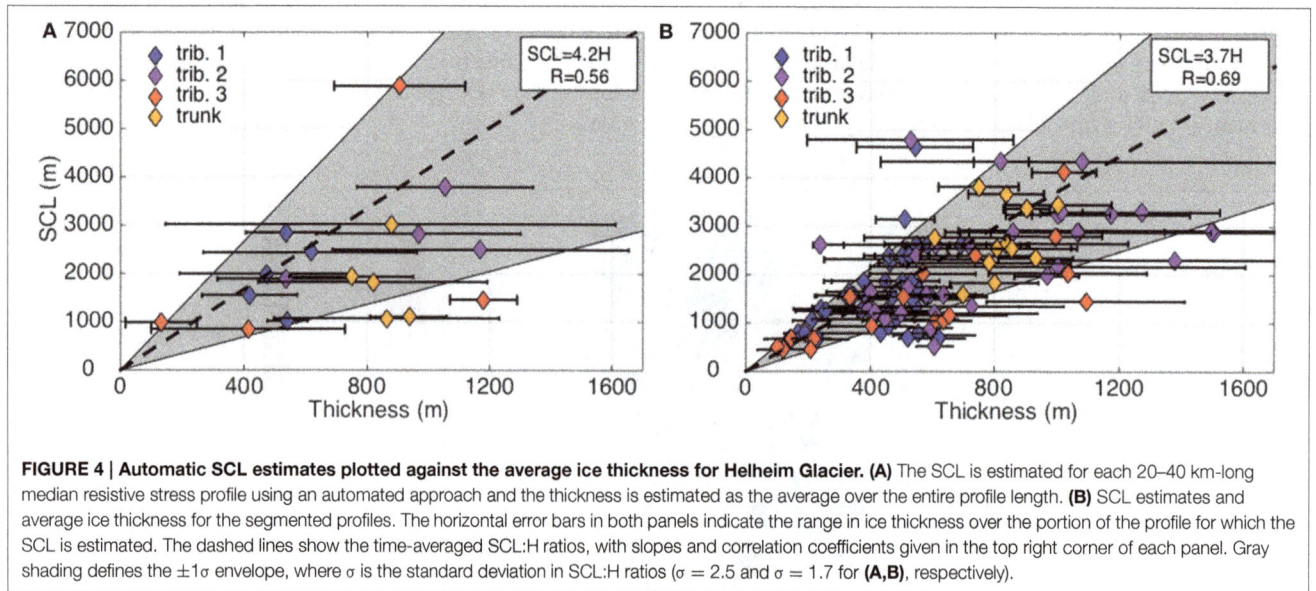

FIGURE 4 | Automatic SCL estimates plotted against the average ice thickness for Helheim Glacier. (A) The SCL is estimated for each 20–40 km-long median resistive stress profile using an automated approach and the thickness is estimated as the average over the entire profile length. **(B)** SCL estimates and average ice thickness for the segmented profiles. The horizontal error bars in both panels indicate the range in ice thickness over the portion of the profile for which the SCL is estimated. The dashed lines show the time-averaged SCL:H ratios, with slopes and correlation coefficients given in the top right corner of each panel. Gray shading defines the $\pm 1\sigma$ envelope, where σ is the standard deviation in SCL:H ratios ($\sigma = 2.5$ and $\sigma = 1.7$ for (**A,B**), respectively).

In order to analyze spatial and temporal variations in the force balance terms, Equations (1, 2) must be solved over the area spanning the along- and across-flow SCL. For each pixel, the stress-coupling area, or SCA, is defined as the centered rectangular area spanning the SCL in the along-flow and transverse directions. Following convention, we first calculate gradients in surface elevation, velocity, and resistive stresses in the x- and y- directions as the average slope of a least-squares regression spanning the SCA. These values allow us to solve Equations (1, 2), then rotate the stresses into flow-following coordinates. Estimated uncertainties and root mean square errors

(RMSEs) for each term are derived using a similar approach as van der Veen (2013, Section 3.6). We assume that the slope of the regression is approximately equal to the mean inter-pixel gradient and estimate random errors in the x and y directions as

$$
\begin{aligned}
\sigma_{\frac{\partial}{\partial \mathbf{x}}}(HR_{ij}) &= \sqrt{\sum_{k=1}^{n-1} \frac{\sigma^2_{HR_{ij}(k+1)} + \sigma^2_{HR_{ij}(k)}}{\partial \mathbf{x}^2}} \\
&\approx \frac{1}{|\partial \mathbf{x}|} \sqrt{\frac{2}{n-1} \sum_{k=1}^{n} \sigma^2_{HR_{ij}(k)}},
\end{aligned} \tag{3}
$$

where $i,j = x,y$ and n is the length of the SCL (in pixels). Errors are calculated using Equation (3), with $x = y$, for the y-direction. To estimate the total error, an additional 3% error is added to the depth-integrated strain rates to account for non-random uncertainties in the ice flow velocities. Force balance estimates obtained using our empirical SCL parameterization are demonstrated for Helheim Glacier in **Figure 5**. Following convention, the gravitational driving stress (**Figure 5A**) is positive along-flow and the resistive stresses are positive *against* flow (**Figures 5B–D**). Uncertainties associated with each force balance term are shown in **Figures 5E–H**. **Figures 5I–K** show the normalized RMSEs of the linear least-squares regressions (i.e., $RMSE/RMSE_{max}$), which provide a means to identify regions where under-estimation of the SCL would lead to large errors in force balance calculations: higher (lower) normalized RMSEs correspond to more (less) variability in local stresses over the SCA.

As expected, the gravitational driving stress is largest along the glacier's dominant tributary (tributary 2) where the ice thickness reaches a maximum of ~1600 m and the along-flow surface slope is ~0.05 m/m. Driving stress approaches zero along the main trunk where the surface slope approaches zero and

the ice nears flotation (De Juan et al., 2010; **Figure 5A**). The strong across-flow gradients in speed near the glacier's lateral margins lead to narrow zones where lateral stresses pull the slow-flowing margins in the direction of ice flow (**Figure 5B**, dark blue). Inward of the extensional shear margins, we observe wide bands of high lateral shear (>100 kPa). Alternating bands of longitudinal extension and compression out-of-phase with oscillations in the gravitational driving stress are evident for all three tributaries but are most pronounced at the confluence of the tributaries where the ice is advected over a bedrock or till ridge that rises ~200 m from the surrounding bed (**Figure 5C**) and the ice originating from the dominant tributary reaches its thickness minimum (**Figure 2**). Basal drag is generally highest near the lateral margins and increases inland from approximately zero at the down-flow limit (**Figure 5D**). A comparison of our basal drag estimates with the basal drag parameterization used to simulate the mid-2000s changes in Helheim dynamics (Nick et al., 2009) provides a means to assess the accuracy of our force balance results. We find that the difference between our width-averaged basal drag estimates and the basal drag parameterization in the 1-dimensional (i.e., width-averaged) ice flow model of Nick et al. (2009) is within the estimated uncertainty of ~20 kPa, indicating

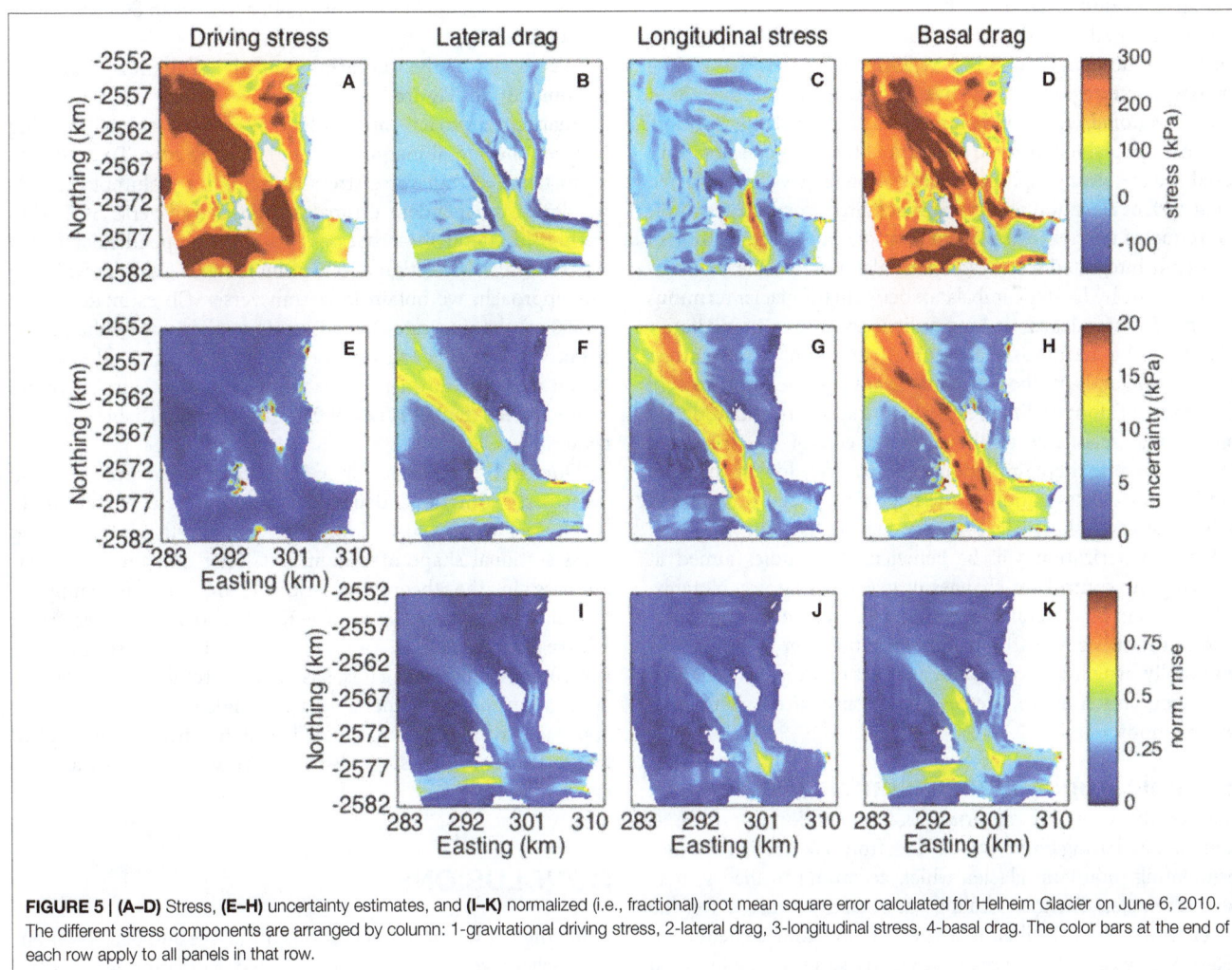

FIGURE 5 | (A–D) Stress, **(E–H)** uncertainty estimates, and **(I–K)** normalized (i.e., fractional) root mean square error calculated for Helheim Glacier on June 6, 2010. The different stress components are arranged by column: 1-gravitational driving stress, 2-lateral drag, 3-longitudinal stress, 4-basal drag. The color bars at the end of each row apply to all panels in that row.

that the force balance method yields reasonable estimates of basal drag for Helheim Glacier when calculated over the appropriate spatial scales.

Isolated patches of negative (>-50 kPa) basal drag are observed in several locations along the glacier trunk. Negative based drag (i.e., friction at the ice-bed interface pushing the glacier toward the terminus) is physically untenable, and can be attributed to observational errors or failure of the force balance method to realistically capture the physics of ice flow. Although, comparable in magnitude to the stress uncertainty introduced by observational errors, the spatial distribution of these patches of negative basal drag cannot be explained by random errors in observational data. Temporal aliasing introduced by the use of the unweighted average InSAR velocities to approximate surface velocities on the DEM acquisition dates may result in over- or under-estimation of basal drag. However, over the short time periods separating the repeat observations used here, it is unlikely that temporal variations in velocity can fully explain the presence of negative basal drag along the trunk. We instead attribute the region of negative basal drag spanning the trunk at easting ~301 km to failure in the depth-integrated force balance equations to capture the viscous and brittle deformation that likely occurs as the ice is advected through the relatively deep trough and over the ~200 m-tall bedrock ridge at the confluence of the tributaries. Within one SCL of the end of the glacier domain, prominent regions of negative basal drag emerge. These regions of negative basal drag are not due either to observational errors or failure in the force balance method, but can be explained by the truncation of the mass conservation-based bed elevation map. The bed elevation map was constructed using surface elevation observations acquired when the terminus was retracted relative to its present location, causing truncation of our depth-integrated stress estimates inland of the true terminus position. The hydrostatic imbalance between the glacier terminus and fjord water leads to longitudinal extension immediately inland of the terminus. Therefore, truncation of our stress estimates inland of the terminus leads to the omission of this zone of longitudinal extension and over-estimation of longitudinal resistance at the terminal end of the truncated domain. Despite these limitations, the relatively low uncertainty and high spatial resolution of the depth-integrated driving and resistive stresses shown in **Figure 5** suggest that our empirical SCL parameterization will be beneficial for studies aimed at resolving the controls of changes in glacier dynamics. Notably, although our SCL parameterization is developed using time-averaged local stresses, the parameterization can be applied to temporally-evolving force balance analyses to investigate changes in driving and resistive stresses in response to environmental perturbations.

Example Application: Columbia Glacier

The versatility of our empirical SCL estimation approach is demonstrated using observational data from a temperate marine-terminating mountain glacier, which according to theory, may have a smaller thickness-scaling ratio than estimated for the larger and colder Helheim Glacier (Kamb and Echelmeyer, 1986). We selected Columbia Glacier, Alaska over an additional

Greenland outlet glacier, for several reasons: it (1) is considerably smaller (~3 km-wide and <700 m-thick) and less viscous than Helheim Glacier, (2) has a detailed observational record of surface elevations and velocities and a bed elevation map of comparable spatial resolution to Helheim (**Figure 6**), and (3) has been the target of previous SCL and force balance analyses (e.g., Walters, 1989; van der Veen and Whillans, 1993; O'Neel et al., 2005) to which we can compare our analyses.

Given the sparse spatial coverage of ice-penetrating radar observations (Rignot et al., 2013), the spatial domain over which reliable thickness (and stress) estimates can be obtained is quite limited relative to the Helheim Glacier datasets. Thus, in order to ensure that our stress profiles are sufficiently long to accurately estimate the SCL using our automated approach (i.e., long periods are not truncated), we refrain from extracting separate stress profiles for the trunk and tributaries and instead extend the tributary profiles into the glacier trunk (**Figure 6**). For the full profiles, we obtain an average SCL of ~1400 m, which corresponds to a SCL:H ratio of 5.6 (**Figure 7A**). As with Helheim, we segment the resistive stress profiles in order to resolve potential spatial variations in SCL corresponding with along-flow changes in ice thickness. We find an average SCL of ~1150 m, a moderate strength relationship between variations in the SCL and ice thickness ($R = 0.54$), and a mean SCL:H ratio of 4.3 (**Figure 7B**).

We attribute the uniformity of the Columbia SCL that is apparent in **Figure 7** not to failure of our empirical SCL estimation approach but rather to strong stress coupling between the lateral margins and fast-flowing core. To assess the importance of transverse stress-coupling for Columbia's trunk, we derive independent estimates for the transverse SCL from profiles of lateral shear stresses, extracted perpendicular to the centerline flow direction at 600 m increments along-flow. Using this approach, we obtain long transverse SCL estimates (1610 ± 600 m) that indicate strong stress coupling across the glacier trunk. Thus, it is not surprising that the SCL extracted from the lateral margins (down-flow reaches of tributary 2 in **Figure 6**), where the ice thickness is <300 m, is indistinguishable from the centerline SCL.

Our SCL estimates for Columbia Glacier are in good agreement with the independent SCL estimate of 1200 m in (O'Neel et al., 2005). Taking into account the relatively flat cross-sectional shape of Columbia Glacier and the non-linear increase in the theoretical SCL:H ratio with flattening of glacial cross-sectional area (see Kamb and Echelmeyer, 1986; **Figure 5**), our SCL:H ratio of 4.3 ± 1.2 is also in good agreement with theory. Thus, based on the results of the example applications presented here, we are confident that our empirical approach for estimating the SCL can be successfully applied to glaciers encompassing a wide range of geometries and ice rheologies.

CONCLUSIONS

Using high-resolution observations of surface and bed elevations and surface velocities for two marine-terminating glaciers, we

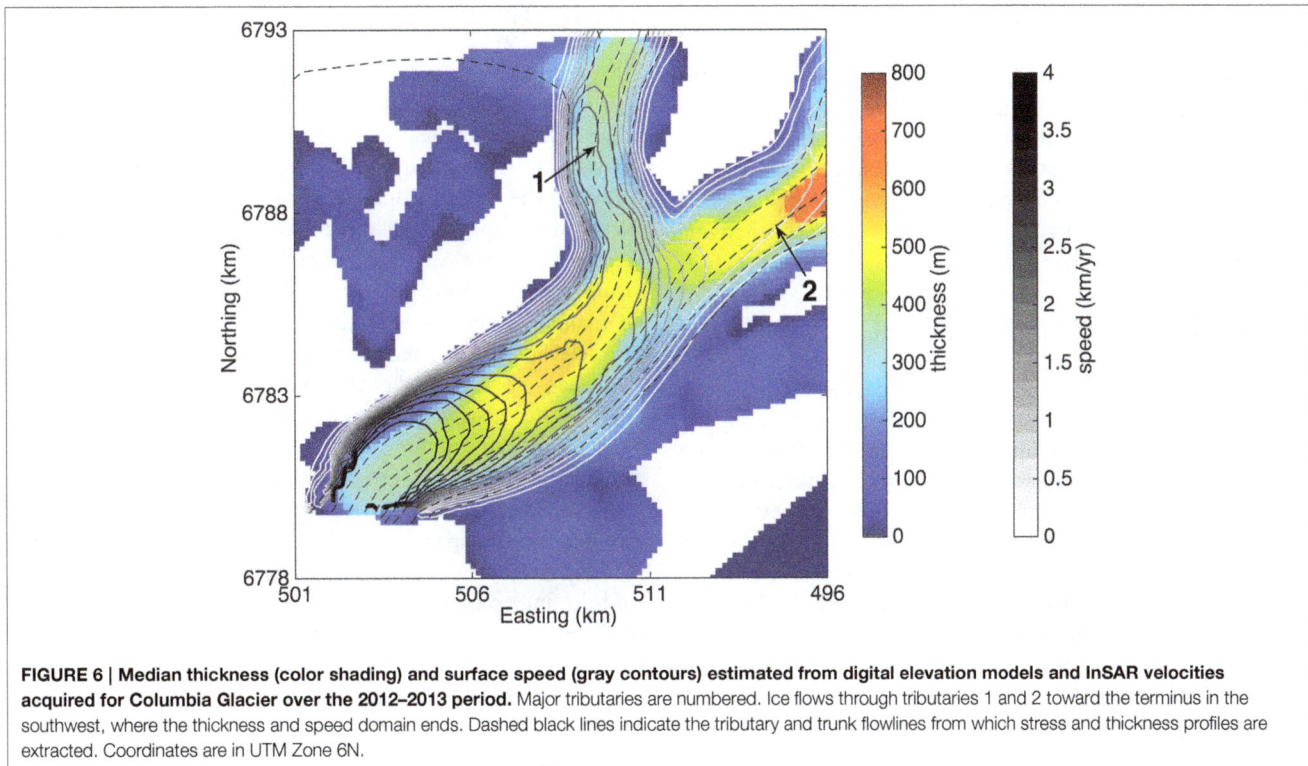

FIGURE 6 | Median thickness (color shading) and surface speed (gray contours) estimated from digital elevation models and InSAR velocities acquired for Columbia Glacier over the 2012–2013 period. Major tributaries are numbered. Ice flows through tributaries 1 and 2 toward the terminus in the southwest, where the thickness and speed domain ends. Dashed black lines indicate the tributary and trunk flowlines from which stress and thickness profiles are extracted. Coordinates are in UTM Zone 6N.

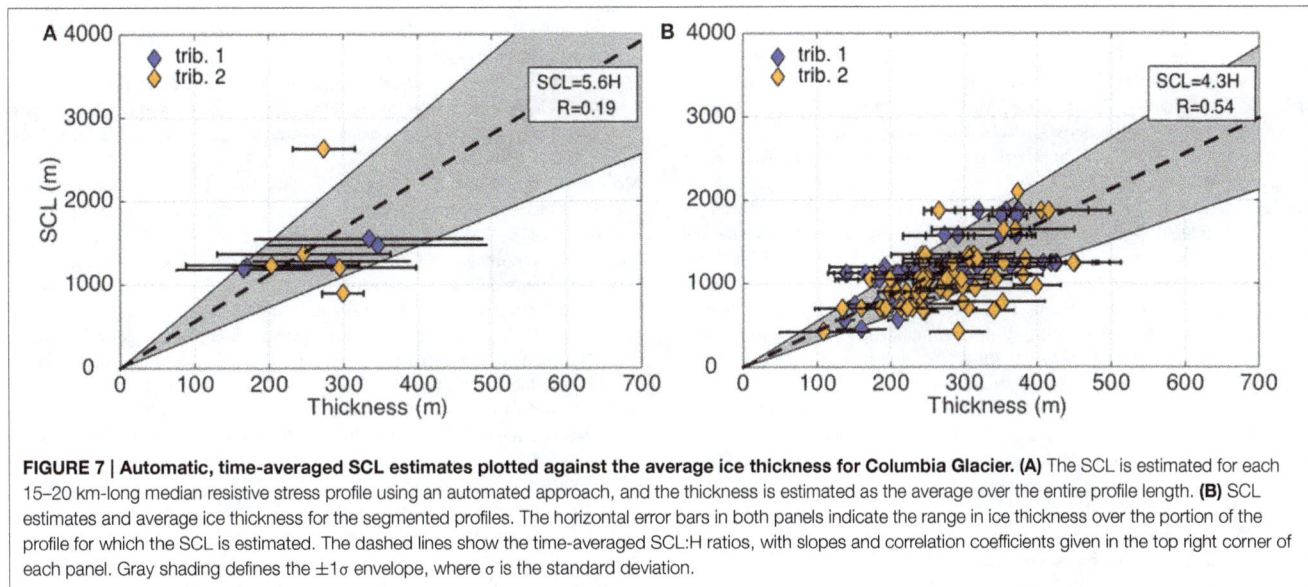

FIGURE 7 | Automatic, time-averaged SCL estimates plotted against the average ice thickness for Columbia Glacier. (A) The SCL is estimated for each 15–20 km-long median resistive stress profile using an automated approach, and the thickness is estimated as the average over the entire profile length. **(B)** SCL estimates and average ice thickness for the segmented profiles. The horizontal error bars in both panels indicate the range in ice thickness over the portion of the profile for which the SCL is estimated. The dashed lines show the time-averaged SCL:H ratios, with slopes and correlation coefficients given in the top right corner of each panel. Gray shading defines the ±1σ envelope, where σ is the standard deviation.

demonstrate an empirical approach for estimating the SCL that determines the spatial resolution of force balance analyses. We solve for the SCL as the dominant period of oscillations in the sum of longitudinal stresses and basal drag in the along-flow direction. In line with Kavanaugh and Cuffey (2009), we find that the SCL can be approximated as a linear function of the ice thickness for Helheim Glacier. Although the strength of the correlation is slightly weaker for Columbia Glacier, our empirical

SCL estimates are in good agreement with previous independent estimates and theory, indicating that the method can be applied to a range of glacier geometries and ice rheologies.

The use of a standardized method for estimating the SCL allows direct comparison of stress redistribution time series between glaciers, which is critical for improving the current understanding of glacier dynamics. Thus, we suggest that studies of the dynamic redistribution of driving and resistive stresses following a change in environmental forcing adopt a similar empirical approach to estimating the SCL. The code developed for our empirical approach is available on the MDVN (http://dataverse.acg.maine.edu/dvn/dv/eep). By using high-resolution datasets to simultaneously calculate the glacier force balance and the SCL, such as illustrated here, force balance analyses have the potential to improve the current understanding of glacier dynamics in a rapidly changing climate.

AUTHOR CONTRIBUTIONS

EE and SO conceived the initial project idea. EE developed the methodology, generated the digital elevation models, compiled the observational datasets, applied the method to the Helheim and Columbia observations, and wrote the manuscript. GH, SO, and TB provided feedback on the method and revised manuscript. JH revised bed picks from the Columbia Glacier radar dataset and MM used the revised picks to constrain the mass conservation-derived bed elevation map. JH and MM also contributed to final manuscript revisions.

ACKNOWLEDGMENTS

This work was supported by NASA award NNX14AH83G to EE, SO, and GH. WorldView images were distributed by the Polar Geospatial Center at the University of Minnesota (http://www.pgs.umn.edu/imagery/satellite) as part of an agreement between the US National Science Foundation and the US National Geospatial Intelligence Agency Commercial Imagery Program. DEMs were generated from WorldView images using supercomputing resources provided by the University of Maine Advanced Computing Group. The DEMs, Columbia bed elevation map and filtered radar profiles, and Matlab code used to extract the SCL estimates are archived on the Maine DataVerse Network (http://dataverse.acg.maine.edu/dvn/dv/eep). TSX terminus positions for Helheim Glacier were provided by Twila Moon. We thank David Bahr and Jason Amundson for providing feedback on the methods and Evan Burgess for his help with the initial stages of data processing. The findings and conclusions in this article are those of the author(s) and do not necessarily represent the views of any government agencies.

REFERENCES

Armstrong, W. H., Anderson, R. S., Allen, J., and Rajaram, H. (2016). Modeling the WorldView-derived season velocity evolution of Kennicott Glacier, Alaska. *J. Glaciol.* 62, 763–777. doi: 10.1017/jog.2016.66

Bahr, D. B., Pfeffer, W. T., and Kaser, G. (2014). Glacier volume estimation as an ill-posed inversion. *J. Glaciol.* 60, 922–934. doi: 10.3189/2014JoG14J062

Bahr, D. B., Pfeffer, W. T., and Meier, M. F. (1994). Theoretical limitations to englacial velocity calculations. *J. Glaciol.* 40, 509–518.

Bartlett, M. S. (1955). *An Introduction to Stochastic Processes.* Cambridge: Cambridge University Press.

De Juan, J., Elósegui, P., Nettles, M., Larsen, T. B., Davis, J. L., Hamilton, G. S., et al. (2010). Sudden increase in tidal response linked to calving and acceleration at a large Greenland outlet glacier. *Geophys. Res. Lett.* 37, L12501. doi: 10.1029/2010GL043289

Enderlin, E. M., and Hamilton, G. S. (2014). Estimates of iceberg submarine melting from high-resolution digital elevation models: application to Sermilik Fjord, East Greenland. *J. Glaciol.* 60, 1084–1092. doi: 10.3189/2014JoG14J085

Enderlin, E. M., Howat, I. M., Jeong, S., Noh, M. J., van Angelen, J. H., and van den Broeke, M. R. (2014). An improved mass budget for the Greenland ice sheet. *Geophys. Res. Lett.* 41, 866–872. doi: 10.1002/2013GL059010

Garcia, D. (2010). Robust smoothing of gridded data in one and higher dimensions with missing values. *Comput. Stat. Data Anal.* 54, 1167–1178. doi: 10.1016/j.csda.2009.09.020

Holt, J. W., Blankenship, D. D., Morse, D. L., Young, D. A., Peters, M. E., Kempf, S. D., et al. (2006). New boundary conditions for the West Antarctic Ice Sheet: subglacial topography of the Thwaites and Smith glacier catchments. *Geophys. Res. Lett.* 33, L09502. doi: 10.1029/2005GL025561

Hooke, R. L. (2005). *Principles of Glacier Mechanics, 2nd Edn.* Cambridge: Cambridge University Press.

Joughin, I., Smith, B. E., Howat, I. M., Scambos, T., and Moon, T. (2010). Greenland flow variability from ice-sheet-wide velocity mapping. *J. Glaciol.* 56, 415–430. doi: 10.3189/002214310792447734

Kamb, B., and Echelmeyer, K. A. (1986). Stress-gradient coupling in glacier flow: I. Longitudinal averaging of the influence of ice thickness and surface slope. *J. Glaciol.* 32, 267–284.

Kavanaugh, J. L., and Cuffey, K. M. (2009). Dynamics and mass balance of Taylor Glacier, Antarctica: 2. force balance and longitudinal coupling. *J. Geophys. Res.* 114, F04011. doi: 10.1029/2009jf001329

MacAyeal, D. R. (1992). The basal stress distribution of Ice Stream E, Antarctica, inferred by control methods. *J. Geophys. Res.* 97, 595–603. doi: 10.1029/91JB02454

Maxwell, D., Truffer, M., Avdonin, S., and Stuefer, M. (2008). An iterative scheme for determining glacier velocities and stresses. *J. Glaciol.* 54, 888–898. doi: 10.3189/002214308787779889

McNabb, R. W., Hock, R., O'Neel, S., Rasmussen, L. A., Ahn, Y., Braun, M., et al. (2012). Using surface velocities to calculate ice thickness and bed topography: a case study at Columbia Glacier, Alaska, USA. *J. Glaciol.* 58, 1151–1164. doi: 10.3189/2012JoG11J249

Morlighem, M., Rignot, E., Mouginot, J., Seroussi, H., and Larour, E. (2014). Deeply incised submarine glacial valleys beneath the Greenland ice sheet. *Nat. Geosci.* 7, 418–422. doi: 10.1038/ngeo2167

Nick, F. M., van der Veen, C. J., and Oerlemans, J. (2007). Controls on advance of tidewater glaciers: results from numerical modeling applied to Columbia Glacier. *J. Geophys. Res.* 112, F03S24. doi: 10.1029/2006JF000551

Nick, F. M., Vieli, A., Howat, I. M., and Joughin, I. (2009). Large-scale changes in Greenland outlet glacier dynamics triggered at the terminus. *Nat. Geosci.* 2, 110–114. doi: 10.1038/ngeo394

Nuth, C., and Kääb, A. (2011). Co-registration and bias corrections of satellite elevation data sets for quantifying glacier thickness change. *Cryosphere* 5, 271–290. doi: 10.5194/tc-5-271-2011

O'Neel, S., Pfeffer, W. T., Krimmel, R., and Meier, M. (2005). Evolving force balance at Columbia Glacier, Alaska, during its rapid retreat. *J. Geophys. Res.* 110, F03012. doi: 10.1029/2005JF000292

Perona, P., and Malik, J. (1990). Scale-space and edge detection using anisotropic diffusion. *IEEE Trans. Pattern Anal. Mach. Intell.* 12, 629–639. doi: 10.1109/34.56205

Rignot, E., Mouginot, J., Larsen, C. F., Gim, Y., and Kirchner, D. (2013). Low-frequency radar sounding of temperate ice masses in Southern Alaska. *Geophys. Res. Lett.* 40, 5399–5405. doi: 10.1002/2013GL057452

Sergienko, O. V., and Hindmarsh, R. C. A. (2013). Regular patterns in frictional resistance of ice-stream beds seen by surface data inversion. *Science* 342, 1086–1089. doi: 10.1126/science.1243903

Shean, D. E., Alexandrov, O., Moratto, Z. M., Smith, B. E., Joughin, I. R., Porter, C., et al. (2016). An automated, open-source pipeline for mass production of digital elevation models (DEMs) from very-high-resolution commercial stereo satellite imagery. *ISPRS J. Photogram. Remote Sens.* 116, 101–117. doi: 10.1016/j.isprsjprs.2016.03.012

Stearns, L. A., and Hamilton, G. S. (2007). Rapid volume loss from two East Greenland outlet glaciers quantified using repeat stereo satellite imagery. *Geophys. Res. Lett.* 34, L05503. doi: 10.1029/2006GL028982

van der Veen, C. J. (2013). *Fundamentals of Glacier Dynamics, 2nd Edn.* Boca Raton, FL: CRC Press.

van der Veen, C. J., Stearns, L. A., Johnson, J., and Csatho, B. (2014). Flow dynamics of Byrd Glacier, East Antarctica. *J. Glaciol.* 60, 1053–1064. doi: 10.3189/2014JoG14J052

van der Veen, C. J., and Whillans, I. M. (1989). Force budget. 1. Theory and numerical-methods. *J. Glaciol.* 35, 53–60. doi: 10.3189/0022143897937 01581

van der Veen, C. J., and Whillans, I. M. (1993). Location of mechanical controls on Columbia Glacier, Alaska, USA, prior to its rapid retreat. *Arctic Alpine Res.* 25, 99–105. doi: 10.2307/1551545

Vaughan, D. G., Corr, H. F. L., Ferraccioli, F., Frearson, N., O'Hare, A., Mach, D., et al. (2006). New boundary conditions for the West Antarctic ice sheet: subglacial topography beneath Pine Island Glacier. *Geophys. Res. Lett.* 33, L09501. doi: 10.1029/2005GL025588

Vaughan, S. (2005). A simple test for periodic signals in red noise. *Astron. Astrophys.* 431, 391–403. doi: 10.1051/0004-6361:20041453

Vieli, A., and Nick, F. M. (2011). Understanding and modelling rapid dynamic changes of tidewater outlet glaciers: issues and implications. *Surv. Geophys.* 32, 437–458. doi: 10.1007/s10712-011-9132-4

Walters, R. A. (1989). Small amplitude, short period variations in the speed of a tidewater glacier in south-central Alaska. *Ann. Glaciol.* 12, 187–191.

Whillans, I. M., Chen, Y. H., van der Veen, C. J., and Hughes, T. J. (1989). Force budget. 3. Application to 3-dimensional flow of Byrd Glacier, Antarctica. *J. Glaciol.* 35, 68–80. doi: 10.3189/002214389793701554

Conflict of Interest Statement: The authors declare that the research was conducted in the absence of any commercial or financial relationships that could be construed as a potential conflict of interest.

4

Scaling Precipitation Input to Spatially Distributed Hydrological Models by Measured Snow Distribution

Christian Vögeli[1]**, Michael Lehning*[1,2]*, Nander Wever*[1,2] *and Mathias Bavay*[1]

[1] *Research Unit Snow and Permafrost, WSL Institute for Snow and Avalanche Research SLF, Davos, Switzerland,* [2] *Civil and Environmental Engineering, CRYOS School of Architecture, École Polytechnique Fédérale de Lausanne, Lausanne, Switzerland*

Edited by:
Thomas Vikhamar Schuler,
University of Oslo, Norway

Reviewed by:
Ruzica Dadic,
Victoria University of Wellington,
New Zealand
Vincent Vionnet,
CNRM, CNRS and CEN, France

*Correspondence:
Christian Vögeli
christian.voegeli@slf.ch

Specialty section:
This article was submitted to
Cryospheric Sciences,
a section of the journal
Frontiers in Earth Science

Accurate knowledge on snow distribution in alpine terrain is crucial for various applications such as flood risk assessment, avalanche warning or managing water supply and hydro-power. To simulate the seasonal snow cover development in alpine terrain, the spatially distributed, physics-based model Alpine3D is suitable. The model is typically driven by spatial interpolations of observations from automatic weather stations (AWS), leading to errors in the spatial distribution of atmospheric forcing. With recent advances in remote sensing techniques, maps of snow depth can be acquired with high spatial resolution and accuracy. In this work, maps of the snow depth distribution, calculated from summer and winter digital surface models based on Airborne Digital Sensors (ADS), are used to scale precipitation input data, with the aim to improve the accuracy of simulation of the spatial distribution of snow with Alpine3D. A simple method to scale and redistribute precipitation is presented and the performance is analyzed. The scaling method is only applied if it is snowing. For rainfall the precipitation is distributed by interpolation, with a simple air temperature threshold used for the determination of the precipitation phase. It was found that the accuracy of spatial snow distribution could be improved significantly for the simulated domain. The standard deviation of absolute snow depth error is reduced up to a factor 3.4 to less than 20 cm. The mean absolute error in snow distribution was reduced when using representative input sources for the simulation domain. For inter-annual scaling, the model performance could also be improved, even when using a remote sensing dataset from a different winter. In conclusion, using remote sensing data to process precipitation input, complex processes such as preferential snow deposition and snow relocation due to wind or avalanches, can be substituted and modeling performance of spatial snow distribution is improved.

Keywords: mountain precipitation, spatial variability, Alpine3D, precipitation scaling, airborne digital sensor, snow depth, snow transport, preferential deposition

1. INTRODUCTION

The spatial distribution of snow depth can vary on scales of a few meters and from winter to winter. Different precipitation patterns, redistribution by wind, temperature and solar radiation strongly control snow distribution. The seasonal snow cover in alpine terrain accumulates and stores a large amount of water and has a major influence on the release and availability of water throughout the year. The water stored in snow is essential for hydro-power (Schaefli et al., 2007), water supply and recreational activities (Koenig and Abegg, 1997). Additionally, the development of flora and fauna in the alpine and lower elevated areas is influenced by the snow cover (Wipf et al., 2009) and the formation of avalanches is substantially influenced by the spatial and temporal distribution of the snow cover (Schweizer et al., 2003).

Using numerical models, snow distribution in alpine terrain can be simulated (Liston and Sturm, 1998; Gauer, 2001; Winstral et al., 2002; Vionnet et al., 2014). Simulations can be used to understand hydrological processes (Comola et al., 2015), for fore- and now-casting (Lehning et al., 1999; Bellaire et al., 2011) or the assessment of climate change (Bavay et al., 2009). The spatially distributed, physics-based model Alpine3D is suitable for simulations of snow in steep, alpine terrain (Lehning et al., 2006, 2008). It accounts for the most relevant processes leading to snow accumulation, ablation, metamorphism and the energy exchanges with soil, vegetation and the atmosphere. As simulations may become computationally expensive, not all processes can be explicitly accounted for in a simulation. Effects caused by wind, such as snow transport and preferential deposition, can be included (Lehning et al., 2008), but the computational demand increases dramatically. These effects are therefore often neglected, leading to errors in snow distribution. Also a detailed representation of the energy fluxes (Michlmayr et al., 2008) limits the applicability of large-scale simulations with high spatial resolution. Uncertainties in and lack of measurements of meteorological observations, which are usually used as input, contribute to errors in the simulation of snow distribution (Schölgl et al., 2016) and the measurements are often not representative for the vicinity of the automatic weather station (AWS) (Grünewald and Lehning, 2015).

Various simple methods to allocate precipitation to account for wind drift have been tested in previous studies. Most of them use topographic features such as elevation, slope, aspect and/or curvature (Schuler et al., 2007; Huss et al., 2008; Magnusson et al., 2014). More complex parameters are also calculated, e.g., topographic openness (Hanzer et al., 2016). The most successful topographic parameter is the Winstral redistribution parameter (Winstral et al., 2002, 2013), which has been shown to give good results even in fairly difficult conditions if at least the mean wind direction is known (Schirmer et al., 2011).

Recent advances in remote sensing techniques enable us to acquire information about the earth surface with high vertical accuracy and high spatial resolution and allow new quantitative insight in snow distribution (Trujillo et al., 2007; Deems et al., 2013). Terrestrial laser scanning (TLS) can be used to measure the snow surface continually (Grünewald et al., 2010; Deems et al., 2013) and with very high accuracy (less than 20 cm error on average Prokop et al., 2008). However, this method is labor intensive and limited to small areas only (Bühler et al., 2013). Airborne mounted scanning devices are well suited to gather information about the spatial distribution and thickness of the snow cover for large areas. Using an opto-electric scanner for large areas is cheaper than using laser scanning (Bühler et al., 2013). With an opto-electric scanner, stereo images are acquired and can be combined to a digital surface model (DSM) (Kresse

and Danko, 2012; Bühler et al., 2015). If snow free DSMs are subtracted from DSMs with snow covered areas, maps of spatial snow distribution can be generated. Such maps of snow depth, temporally close to the maximum seasonal snow depth, have been acquired in recent years for the region of Davos, Switzerland (Bühler et al., 2012, 2013). Such maps can be used for validation purposes of the performance of numerical models. They also offer the potential to be included as model input to improve the model performance (e.g., Revuelto et al., 2016). They used TLS data to adjust snow depths on dates where TLS scans were available in simulations throughout different winters. This method allows to correct the snow depth at time steps where TLS information is available and it was shown to improve the simulation of snow depth. However, the method either requires multiple TLS scans per winter, or it will only provide an accurate snow depth distribution on or directly after the TLS scan and during melting periods. During snow accumulation periods, TLS information is essential to improve simulation performance. Furthermore, snowpack layering and snow microstructure may not be well represented when simulated snow depth deviates from actual snow depth for extended periods of time.

In this work, a simple method of precipitation scaling using remote sensing datasets is presented. The method aims to achieve an accurate simulation of seasonal snow cover with one simulation run only. A characteristic of the model is that it distinguishes between different precipitation patterns depending on the precipitation phase. A comparison of the scaling methods with non-scaling methods and an iterative scaling approach is made to quantify the gain in accuracy based on the scaling. Problems and limitations of the methods are investigated and described.

2. METHODS

2.1. Site Description

To investigate the potential of precipitation scaling the area around the town Davos, Switzerland was chosen. For this area, several airborne digital sensor (ADS) remote sensing datasets of seasonal snow distribution are available (Bühler et al., 2015). This area has been the basis for numerous research projects and numerical simulations have been run in this area for various applications (e.g., Zappa et al., 2003; Bavay et al., 2009; Mott et al., 2011). The simulation domain covers an area of 21.5 × 21.5 km centered around the Dischma valley south-east of Davos (cf. **Figure 1**). The mean elevation of the domain is 2256 m.a.s.l. with a minimum elevation of 1255 m.a.s.l. and the highest peak at 3218 m.a.s.l. The lower areas, below ~1800 m.a.s.l., are mostly covered by forest (13.2% of the domain) and infrastructure and buildings of Davos (1.3% of the domain). The higher areas are mostly covered by alpine meadow (21.8% of the domain), changing into rocky and glacial areas in the highest elevations of the domain. The glacial areas cover 2.9% of the domain. The most dominant precipitation pattern is determined by the elevation gradient. Precipitation amounts at the Weissfluhjoch are about twice as high as in Davos (Wever, 2015).

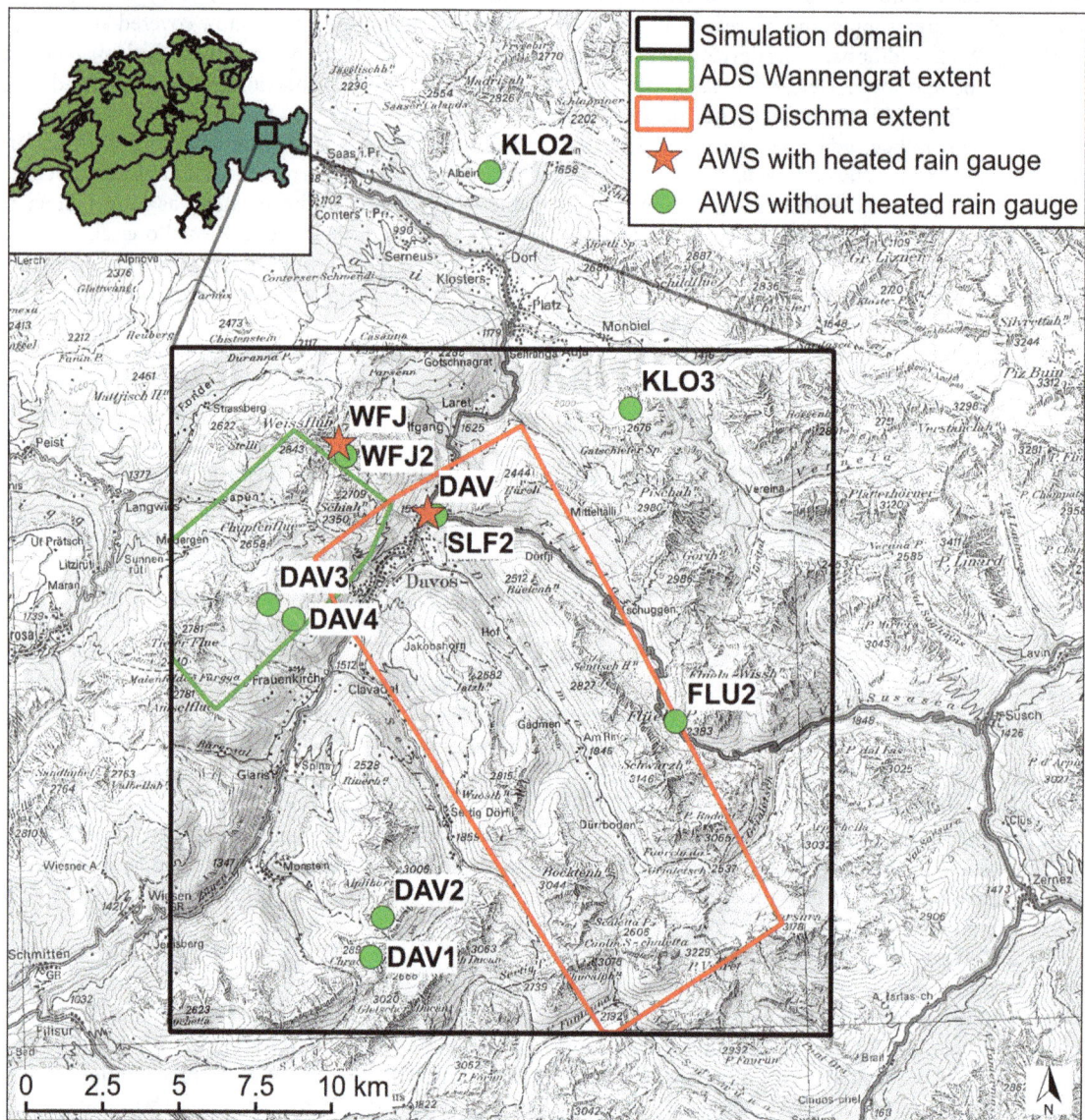

FIGURE 1 | Overview of the simulation domain in the area of Davos, Switzerland (black frame), the ADS (airborne digital sensor) Wannengrat extent (green frame) and the ADS Dischma extent (red frame). The AWSs used for Alpine3D input with (red stars) and without heated rain gauge (green dots). Maps reproduced by permission of swisstopo (JA100118).

Several meteorological stations are located within the simulation domain (cf. **Figure 1**). Most of them belong to the Intercantonal Measurement and Information System, a permanently operating network for avalanche warning (Lehning et al., 1999). These stations measure general meteorological parameters such as air, soil and snow temperatures, wind speed and direction, relative humidity and snow depth and reflected short-wave radiation but not precipitation. A meteorological station at the Weissfluhjoch (WFJ, 2540 m.a.s.l.) and one in Davos (DAV, 1596 m.a.s.l.), belonging to the SwissMetNet from MeteoSwiss (Heimo et al., 2007) are equipped with heated rain gauges, providing information about precipitation at different elevations. These stations are also equipped

with sensors to measure incoming short-wave radiation and incoming long-wave radiation. **Figure 1** gives an overview of the location and extent of the simulation domain and ADS extents, including the locations and code names of the AWS sites.

2.2. Remote Sensing Data

The maps of measured spatial snow depth distribution were generated using information acquired by the ADS technology (Bühler et al., 2015). The opto-electric line scanners ADS80 and ADS100 from Leica Geosystems were used to acquire summer and winter stereo images for different years over the Dischma and Wannengrat area near Davos (cf. **Figure 1**). The ADS dataset covering the Wannengrat area ($\sim 3.5 \times 7.5$ km) and the Dischma area ($\sim 7 \times 17$ km) are merged and handled as one dataset. The acquired images were processed to digital surface models (DSM) for summer and winter (Bühler et al., 2012). Subtracting the winter DSM from the summer DSM provides to maps of snow depth. The resolution of the final snow depth maps is 2 m with an average vertical accuracy of ± 30 cm (Bühler et al., 2015).

To use the ADS data for scaling, it was resampled to a 100 m resolution by averaging when more than 50% of the grid points contained data. If more than 50% of the grid points contained no data, also the resampled grid point was set to no data. Snow depth in areas covered with forest, scrub, buildings and water bodies can not be determined using ADS technology (Bühler et al., 2015) and are therefore masked out from the datasets. ADS datasets from 20 March 2012, 15 April 2013 and 17 April 2014 were used in this study. All snow depth maps were calculated using a summer DSM from 3 September 2013. **Figure 2** shows the ADS data for 20 March 2012 at 100 m resolution.

2.3. Numerical Modeling
2.3.1. Models

Alpine3D is an Open Source model of mountain surface processes with a special focus on snow cover. It has been designed for hydrological, meteorological and avalanche warning applications in steep, alpine terrain (Lehning et al., 2006). It consists of several modules simulating various processes: a module for snow transport (Lehning et al., 2008; Groot Zwaaftink et al., 2011), a module to simulate the radiation fields in complex terrain (Michlmayr et al., 2008; Helbig, 2009) or a module to provide runoff data in order to couple it with an hydrological routing scheme (Gallice et al., 2016). As the snow transport module requires the computation of 3D wind fields and comes at a very high computational cost, it has not been used for the simulations in this work. Similarly, the radiation fields have been treated by a simpler module that accounts for shading effects, atmospheric attenuation and a simplified representation of the radiation reflected by the surrounding terrain based on a view factors

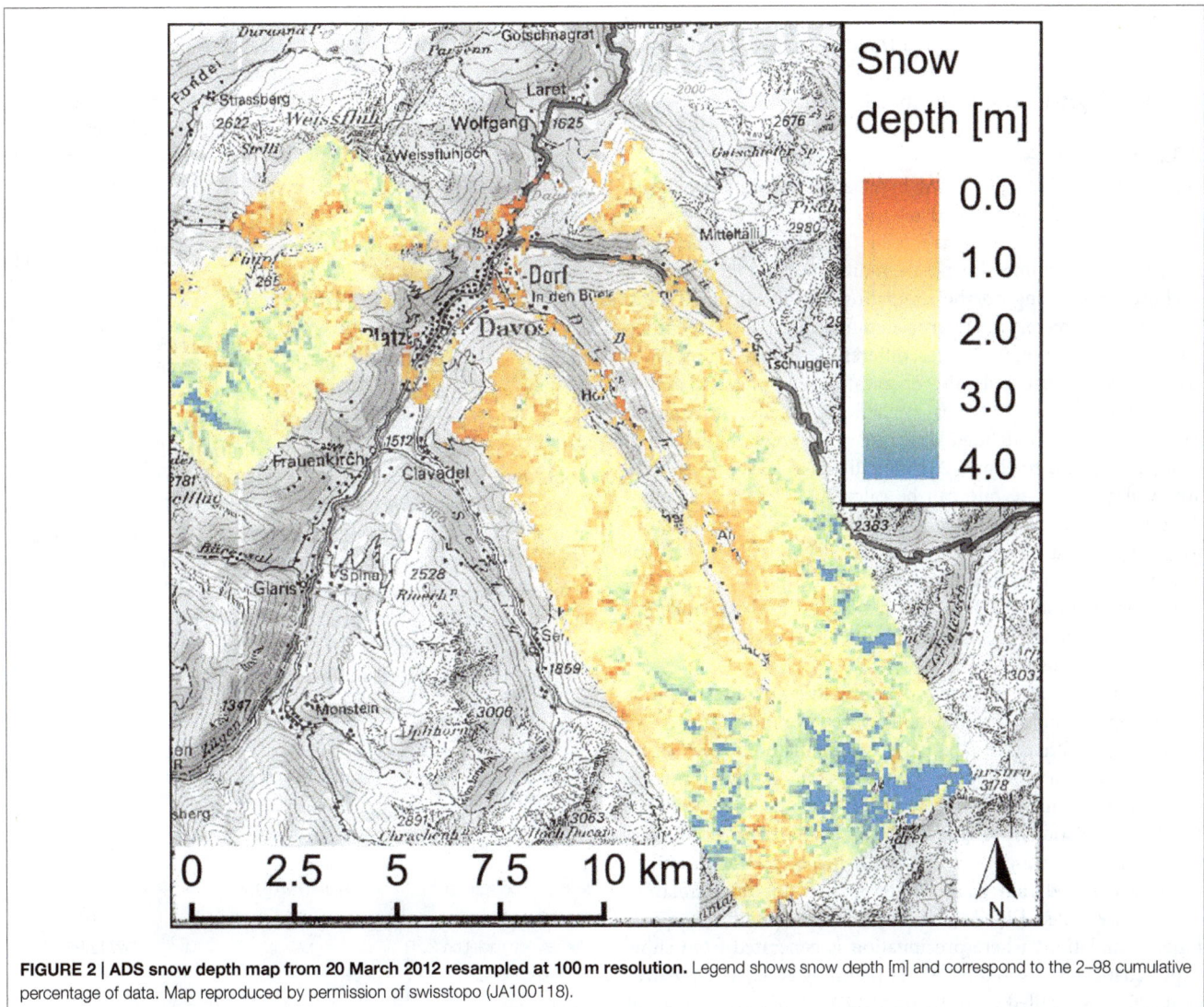

FIGURE 2 | ADS snow depth map from 20 March 2012 resampled at 100 m resolution. Legend shows snow depth [m] and correspond to the 2–98 cumulative percentage of data. Map reproduced by permission of swisstopo (JA100118).

approach (similarly to Anslow et al., 2008; Endrizzi et al., 2014).

At its core, Alpine3D is a distributed version of the SNOWPACK model (Bartelt and Lehning, 2002; Lehning et al., 2002a,b): 1D simulations are run for all grid points in the simulation domain and then processed by the other modules. The SNOWPACK model describes physical processes in and between soil, snow and vegetation and the atmospheric boundary layer with high accuracy (Wever et al., 2014, 2015). It can for example simulate the deposition and resublimation of surface hoar, the shortwave penetration into the snow cover and local phase changes. It uses a variable number of layers with a typical thickness of less than 2 cm and gives an overall reliable representation of the local snow mass balance if accurate forcing data is available.

The meteorological forcing data are taken from AWS. These raw data are filtered and preprocessed (more in Section 2.3.2) using the Open Source MeteoIO library (Bavay and Egger, 2014) before entering the core of the model. Since the MeteoIO library computes the distributed meteorological forcing, the precipitation scaling method presented here has been directly implemented into MeteoIO as another spatial interpolation algorithm. Alpine3D is used as a framework here to determine the potential of precipitation input scaling to improve the accuracy of spatial snow distribution in numerical models.

2.3.2. Input Data

For the simulations presented here, all input data are taken from AWSs and interpolated on the simulation domain. The two SwissMetNet stations are used for incoming short-wave radiation and incoming long-wave radiation. Precipitation input is chosen depending on the simulation and is explained later. The other parameters are interpolated from all AWS. Depending on the number of input sources, different interpolation methods are applied to distribute the observations on the domain. If none or only one source is available, a default or constant value (CST) or a predefined lapse rate (CST-LAPSE) are used to generate distributed input. Additionally, if at least two sources are available, a lapse rate can be calculated (LAPSE). If more than two sources are available, a combination of inverse distance weighting and a lapse rate (IDW-LAPSE) can be applied for interpolation (Bavay and Egger, 2014). For most input data except precipitation IDW-LAPSE is used for interpolation. For precipitation, generally only few inputs are available and IDW-LAPSE is rarely used.

In this study, we used two different interpolation methods for precipitation. For simulations with two rain gauges we interpolated with lapse rate only. For simulations with more than two precipitation inputs, we used a combination of inverse distance weighting and lapse rate (Bavay and Egger, 2014). For the simulations presented here two different precipitation input sources are used. Simulations using hourly observations from two heated rain gauges, interpolated on the simulation domain are called [PSUM], as they use precipitation sums as input. Simulations where precipitation is generated from snow depth changes measured by AWSs (Lehning et al., 1999; Wever et al., 2015) are called [HS]. Using SNOWPACK, a precipitation

rate corresponding to the observed change in snow depth is calculated, considering snow density and settling. In total 9 AWS are used to provide precipitation input. The [PSUM] and [HS] simulations are traditional, non-scaling methods used as reference to compare to the scaling methods and serve as the baseline precipitation field for the scaling methods. The interpolation methods for the different input sources are summarized in **Table 1**.

When using the heated rain gauges as precipitation input, the data from winter 2012/13 and 2013/14 are corrected for undercatch according to Goodison et al. (1998). Unusual meteorological circumstances in winter 2011/12 with strong winds and associated increased snow transport, led to a situation where undercatch correction was not necessary in order to accurately represent the snow depth at the WFJ and DAV measurement sites (Wever et al., 2015), and therefore, the undercatch correction was not applied for any simulations in this winter.

2.3.3. Direct Scaling Method

The direct scaling method is designed to scale precipitation input with one simulation run only and can be applied to any precipitation input. Here the [PSUM] and [HS] inputs are used to test direct scaling. First, precipitation is interpolated using LAPSE ([PSUM]) or IDW-LAPSE ([HS]) to provide the initial guess of the precipitation field over the total simulation domain. In a second step, the precipitation is redistributed using the snow depth information from the ADS data according to Equation (1):

$$P_{i,t} = \frac{P_{avg,t}}{HS_{avg}} \cdot HS_i \qquad (1)$$

$P_{i,t}$ is the amount of scaled precipitation distributed on a grid point i at time t. $P_{avg,t}$ is the spatial average interpolated precipitation at time t, HS_{avg} is the average snow depth of the ADS data for the same domain and HS_i is the ADS snow depth at grid point i. The redistribution of precipitation is done only for grid points where ADS information is available. $P_{avg,t}$ is calculated only for the area where ADS information is available to guarantee mass conservation. ADS information (e.g., HS_{avg}, HS_i) is always taken from one of the three ADS datasets available, depending on the simulated winter. This simple scaling assumes that the spatial precipitation correction factor is time-invariant. This scaling approach leads to a separation between the total precipitation input and scaling the distribution. The

---

TABLE 1 | Overview of the inputs and interpolation methods used in the simulations.

	[PSUM]	[HS]
Precipitation sources	Heated rain gauges	Snow depth changes
Precipitation inputs	2	9
Primary interpolation for P	LAPSE	IDW-LAPSE
Radiation inputs	2	2
Other inputs	9	9

total precipitation mass interpolated on the domain is strictly conserved while being redistributed and scaled onto the domain.

To account for the different precipitation patterns of solid and liquid precipitation, the precipitation is redistributed and scaled only when the air temperature (TA) on the grid cell is below a certain threshold value. Therefore, TA is interpolated over the simulation domain using IDW-LAPSE. If the air temperature at a grid point is above this threshold, only liquid precipitation is assumed and no scaling is applied, assuming the interpolation is already a good approximation for rainfall patterns. If the air temperature is below the threshold, a distribution according to Equation (1) is applied. The threshold for all simulations conducted in this study is set to 1.2°C. **Figure 3** schematically shows the structure of the scaling method. The simulations scaling [PSUM] input are called [PAT] (from PSUM-ADS-Threshold), the simulation scaling [HS] input are called [HAT] (from HS-ADS-Threshold).

2.3.4. Iterative Scaling Method

The results from an initial scaling may be improved considerably by an additional iteration step, which takes into account the deviation between simulated and measured snow depth after a first model run. The additional computational costs may be justified for applications were a precise snow depth distribution is important as well as simultaneously maintaining a correct spatial and timely simulation of individual snow fall events (e.g., snow avalanche prediction). Here, we also investigate the improvement in snow depth distribution by an iterative scaling approach [ALS2] (from Airborne Landscape Scans with 2 simulation runs), compared to direct scaling.

For [ALS2] simulations precipitation grids are calculated for each time step and fed into the model by assessing the precipitation on each grid point proportional to the ratio between the average snow depth in the ADS data at the elevation of a rain gauge (cf. Equation 2).

$$P_{i,t} = \frac{P_{DAV,t}}{HS_{1595}} \cdot HS_i \qquad (2)$$

$P_{i,t}$ is the amount of precipitation allocated to grid point i at time step t, $P_{DAV,t}$ is the measured amount of precipitation at time t at station Davos (DAV), HS_{1595} is the average snow depth of all grid points at 1595 m.a.s.l. ($\pm$ 10 m), which is the elevation of the station Davos and HS_i is the ADS snow depth on grid point i. It is assumed that the average snow depth at this elevation is represented by the precipitation measured at DAV.

From the resulting snow distribution at the date where the ADS dataset was acquired, a correction grid is calculated (cf. Equation 3). With this correction grid, all precipitation input grids from the first simulation are corrected (cf. Equation 4). The simulation is re-run using the corrected precipitation input grids.

$$f_{c,i} = \frac{HS_i}{HS_{ALS,i}} \qquad (3)$$

$$P_{2,i,t} = P_{i,t} \cdot f_{c,i} \qquad (4)$$

$P_{2,i,t}$ is the corrected precipitation for grid point i at time t. $P_{i,t}$ is the original precipitation field used in the first simulation run (cf. Equation 2), $f_{c,i}$ is the correction factor for grid point i, HS_i is the ADS snow depth and $HS_{ALS,i}$ is the snow depth resulting from the first run at grid point i. This iterative method requires substantially more effort and model adjustment because simulations have to be carried out twice.

2.4. Simulation Setups

For the two precipitation sources described above, non-scaling simulations ([PSUM] and [HS]) were run. For the same precipitation inputs used in [PSUM] and [HS], simulations were run using the direct scaling, which results in the experiments [PAT] and [HAT]. [PAT] is the scaled simulation of [PSUM] and [HAT] is the scaled simulation of [HS]. Additionally, simulations for the iterative scaling method [ALS2] were conducted. For all simulation setups only the precipitation input was varied. All other inputs remained unchanged and were equal for any simulation. Simulations were conducted for the winters 2011/12, 2012/13, and 2013/14. ADS data from March 2012, April 2013

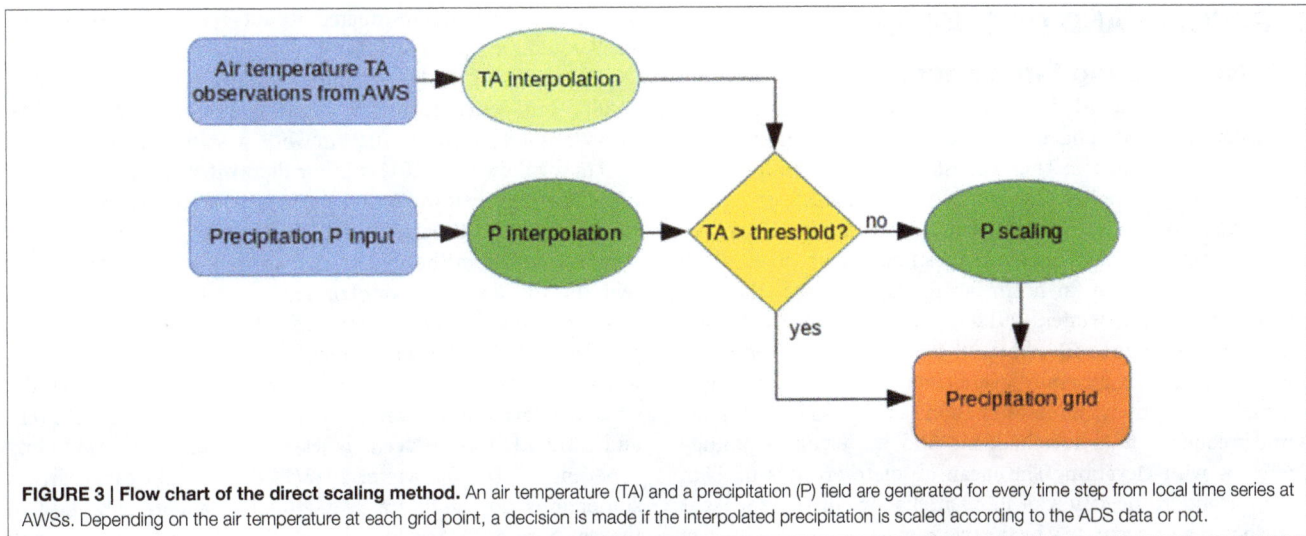

FIGURE 3 | Flow chart of the direct scaling method. An air temperature (TA) and a precipitation (P) field are generated for every time step from local time series at AWSs. Depending on the air temperature at each grid point, a decision is made if the interpolated precipitation is scaled according to the ADS data or not.

and 2014 were used for scaling. Simulations were run from August to July. To assess the potential of inter-annual scaling, the direct scaling simulations were run using all ADS datasets to scale all simulated winters e.g., winter 2011/12 was simulated using the corresponding ADS dataset 2012 for scaling, but also the datasets from 2013 and 2014. The same was done for winter 2012/13 and 2013/14. This inter-annual scaling can be considered as a cross-validation variant but is also of practical interest as it should show, how snow distribution acquired with quite some effort in a particular year can then be used in other years.

2.5. Data Analysis

The analysis of the simulation results was done for the dates of ADS data acquisitions. From simulated snow depths and ADS data, relative and absolute errors are calculated for each grid point and compared. The absolute errors are calculated according to Equation (5) and the relative errors according to Equation (6).

$$\Delta HS = HS_{sim} - HS_{ADS} \quad [m] \tag{5}$$

$$\delta HS = \frac{HS_{sim} - HS_{ADS}}{HS_{ADS} + 0.0001\,m} \quad [-] \tag{6}$$

ΔHS is the absolute error in snow depth, δHS is the relative error in snow depth, HS_{sim} is the snow depth simulated with Alpine3D at the same date where the ADS data is acquired and HS_{ADS} is the snow depth information from the 100 m ADS dataset. To avoid divisions by zero, 0.0001 m of snow depth where added to the ADS snow depth when calculating the relative errors. Additionally relative errors larger than 5 and smaller than −5 are classified as outliers and excluded from data analysis. At small snow depths, the relative error is highly sensitive and small absolute errors in snow depth can lead to very high relative errors. Nevertheless, the relative error is helpful to identify errors in snow depth simulations in complex terrain on the small scale, e.g., along ridges or depressions. Here we concentrate on the absolute error and maps of absolute errors of different simulations are shown. Maps of relative errors can be found in the Supplementary Material.

3. RESULTS AND DISCUSSION

3.1. Non-scaling Simulations

In non-scaling [PSUM] simulation for winter 2011/12 the local snow depth is over- and underestimated but maximum errors remain smaller than ±2 m. The strongest underestimations occurred in the highest elevations in the south-east of the simulation domain (cf. **Figures 4A,B**). This area is also most distant from the rain gauges. Overestimation occurred at all elevations and mostly in north facing slopes. Distributions of the errors for the winters 2012/13 and 2013/14 are similar (cf. **Table 2**). **Figure 4B** shows that small snow depths at low elevations are generally overestimated by the simulation. At higher elevations, underestimations occur almost with the same frequency as overestimations and the error magnitude increases with elevation. The mean absolute error is 0.044 m with a standard deviation of 0.569 m. The mean snow mass is distributed very accurately by the two rain gauges. Given the fact

that other studies (e.g., Grünewald and Lehning, 2011) found that two stations generally are not able to accurately represent elevation gradients in a mountain catchment, we consider it a coincidence that these two rain gauges are able to represent the mass distributed in this catchment accurately. A dependency of errors on the aspect of the slopes is present but not very strong. Small snow depths are generally found in south facing slopes whereas the largest snow depths are found in north facing slopes. Results for the winter 2012/13 are similar and summarized in **Table 2** and **Figure 5**. For winter 2013/14 the mean error was 0.174 m and the standard deviation was 0.399 m.

In [HS] simulations, the range of the absolute error is similar to the [PSUM] simulations. Most negative errors in snow depth are present in high alpine terrain and most positive errors occur at lower elevations (cf. **Figure 4C**). **Figure 4D** shows the deviation of snow depth from the ADS 2012 dataset for each grid point. It is clear that the range of snow depths of the simulation is limited between 1.5 and 2.5 m, whereas the ADS snow depths vary between 0.5 and more than 4 m. This indicates that an inadequate elevation gradient of precipitation has been calculated from the snow depth stations. Small snow depths at lower elevations are generally overestimated, high snow depths are generally underestimated. The mean absolute error is at -0.18 m with a standard deviation of 0.657 m. The smaller snow depths are generally simulated in south facing slopes whereas the largest snow depths are found in north facing slopes. Results for winter 2012/13 are similar to the ones from winter 2011/12. For winter 2013/14 the absolute mean error was 0.057 m only with a standard deviation of 0.512 m (cf. **Table 2**).

3.2. Direct scaling

The simulation using the [PAT] scaling approach represents the snow depth on average well (mean absolute error for winter 2011/12 = −0.01 m). Also the standard deviation of the error is rather low at 0.195 m. Small snow depths are generally slightly underestimated whereas large snow depths are slightly overestimated (cf. **Figures 4E,F**). An over- and underestimation depending on the slopes aspect is visible. The north facing slopes are generally overestimated whereas the snow depths in south facing slopes are underestimated. Results from other winters are similar and are shown in **Table 2**.

The results of the [HAT] simulation are similar to the [PAT] results, but generally underestimating the snow depths. The mean absolute error is -0.24 m with a standard deviation of 0.19 m for the winter 2011/12. For the winter 2012/13 the mean error was slightly lower and the standard deviation was reduced from 0.486 m to 0.162 m. For the winter 2013/14 the standard deviations were in the same range as for the other winters but with the lowest mean error of 0.018 m.

Comparing the non-scaling simulations to the direct scaling simulations with the same precipitation input (cf. **Figure 5A**), it can be stated that for any simulation using scaling the standard deviation is strongly reduced. For the winters 2011/12 and 2012/13 it is reduced to less than 0.23 m for all scaling experiments. For the winter 2013/14 for the rain gauge driven simulations no significant reduction in standard deviation of the absolute error results. This is mainly due to rather small

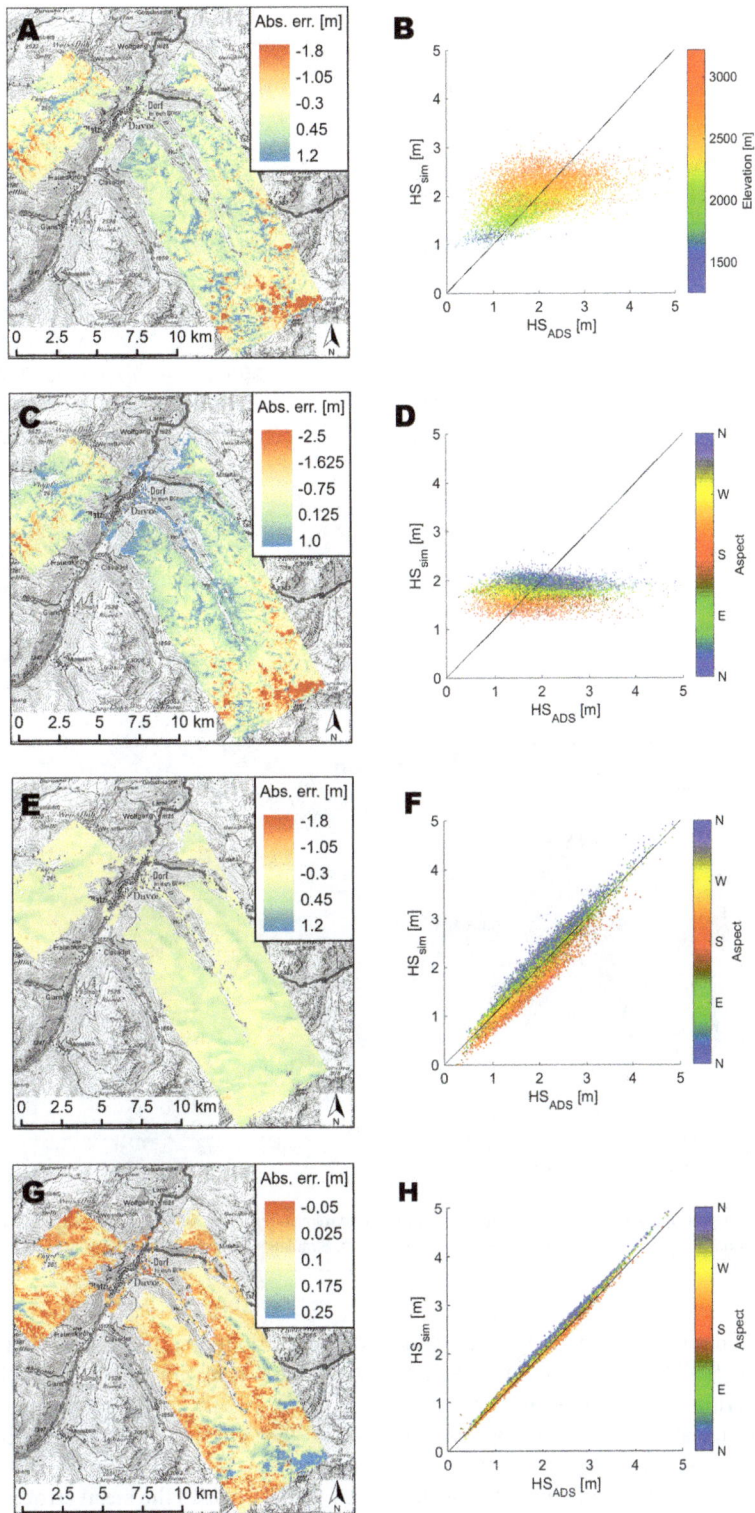

FIGURE 4 | (A,C,D,E,G) Maps of absolute errors in snow depth between simulation results and ADS dataset for 20 March 2012. (B,D,F,H) Scatter plots of simulated against ADS snow depth for 20 March 2012. The plots show [PSUM], [HS], [PAT], and [ALS2] simulation results from winter 2011/12. The range of the absolute errors show the 2–98 cumulative percentage of data for [PSUM], [HS], and [ALS2]. For [PAT] the data range is identical to the [PSUM] data range. Colors in scatter plots correspon to elevation in [PSUM] and aspect of the grid point for others. Maps reproduced by permission of swisstopo (JA100118).

TABLE 2 | Overview on the statistics of the relative (rel) and absolute (abs) errors [mean and standard deviation (std)] for all simulations conducted.

Simulation	Winter	ADS data	Mean rel. [–]	Std. rel. [–]	Mean abs. [m]	Std. abs. [m]
ALS2	2011/12	2012	0.033	0.037	0.064	0.06
ALS2	2012/13	2013	0.035	0.181	0.059	0.275
HAT	2013/14	2014	−0.171	0.432	0.018	0.253
HAT	2011/12	2012	−0.129	0.106	−0.244	0.192
HAT	2012/13	2013	−0.215	0.159	−0.292	0.162
HS	2011/12	–	0.022	0.418	−0.18	0.657
HS	2012/13	–	−0.036	0.443	−0.247	0.486
HS	2013/14	–	0.372	0.868	0.057	0.512
PAT	2011/12	2012	−0.011	0.115	−0.01	0.195
PAT	2013/14	2014	−0.055	0.498	0.179	0.387
PAT	2011/12	2013	0.008	0.277	0.006	0.46
PAT	2011/12	2014	−0.042	0.506	−0.02	0.893
PAT	2012/13	2012	−0.048	0.272	−0.083	0.319
PAT	2012/13	2013	−0.085	0.207	−0.07	0.225
PAT	2012/13	2014	−0.132	0.472	−0.078	0.654
PAT	2013/14	2012	0.224	0.642	0.136	0.352
PAT	2013/14	2013	0.17	0.555	0.151	0.304
PSUM	2011/12	–	0.118	0.381	0.044	0.569
PSUM	2012/13	–	0.073	0.41	−0.023	0.436
PSUM	2013/14	–	0.406	0.79	0.174	0.399

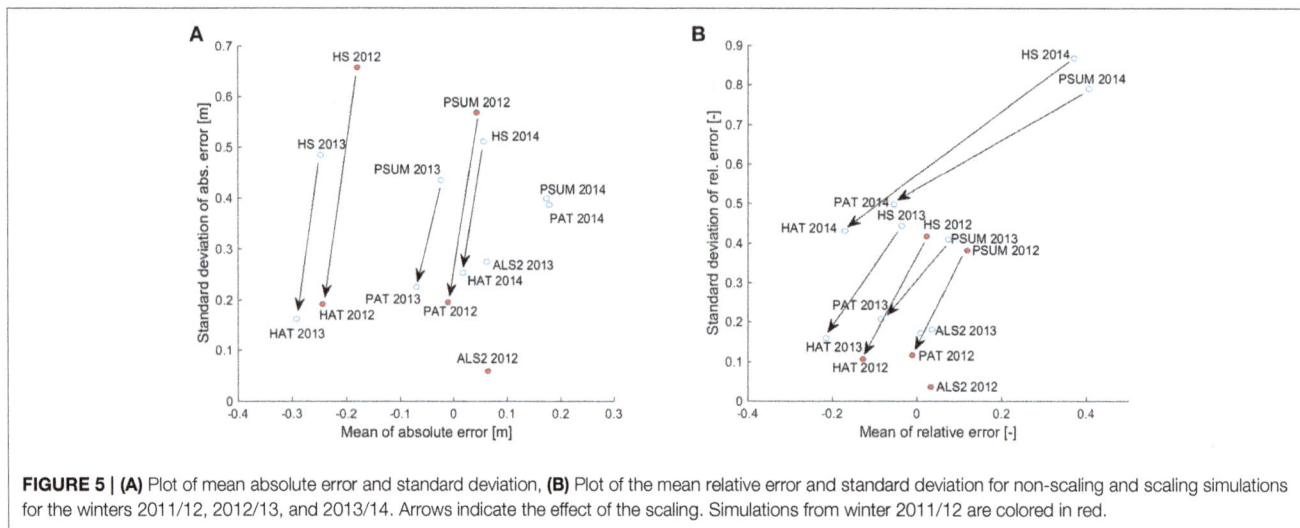

FIGURE 5 | (A) Plot of mean absolute error and standard deviation, **(B)** Plot of the mean relative error and standard deviation for non-scaling and scaling simulations for the winters 2011/12, 2012/13, and 2013/14. Arrows indicate the effect of the scaling. Simulations from winter 2011/12 are colored in red.

snow depths in this winter and an overestimation of the total precipitation in the [PSUM]/[PAT] simulations. This effect becomes visible when considering the change of relative errors (cf. **Figure 5B**). It can be seen that the standard deviation of the relative errors is significantly reduced. For all simulations, a reduction of total snow depth of about 10 cm is evident for the winters 2011/12 and 2012/13.

3.3. Iterative Scaling

Using the results from the first run to correct the precipitation field, the mean error and standard deviation are reduced remarkably in [ALS2] to a very good representation of snow

depth distribution (cf. **Figures 4G,H**). After the second, iterative simulation run, the snow depths in south facing slopes are generally slightly underestimated whereas in the north facing slopes they are slightly overestimated. The mean absolute error in winter 2011/12 is 0.064 m with a standard deviation of 0.06 m. For winter 2012/13 the mean error is similar with 0.059 m with a standard deviation of 0.12 m (cf. **Table 2**). For winter 2013/14 no simulation was run.

3.4. Discussion of Scaling Influence

The results of the simulations show that large variations in the results are obtained, depending on the precipitation input chosen and the applied scaling method. The non-scaling methods have the problem that the effective range of snow depths in the domain up to more than 4 m is not correctly represented. They typically occur as a result of local processes (e.g., snow transport) and/or geographical precipitation gradients that are not captured by the interpolation mechanisms.

As only two rain gauges are used in [PSUM] simulations, no IDW is applied and the interpolation cannot account for the climatological north-south gradient in the domain. Therefore, we assume that the interpolation using a lapse rate only is not able to represent the total range of the snow distribution for the simulation domain. As the average error in snow depth is rather small it can also be stated that the precipitation observations do represent the total precipitation well but the correct spatial allocation is missing. The accurate interpolation of total precipitation from two rain gauges also is coincidence since earlier investigations in the catchment (Grünewald and Lehning, 2011, 2015) have shown that individual stations are usually not representative in this type of terrain.

In the [HS] simulations the interpolation is done using IDW-LAPSE. Even though 9 precipitation inputs are interpolated, the range of snow depths was found to be not represented. We hypothesize that this is mainly due to the fact that the measured snow depths by AWSs show limited representativeness for the stations' vicinity (Grünewald and Lehning, 2015) and therefore also the calculated precipitation is not a sufficient input for the whole simulation domain. This shows that the precipitation input needs to be representative for a larger area than only the point where the observation is made. Also the stations are not located at very exposed locations and their data cannot account for the wind blown, shallow snowpack along ridges nor the drift-filled gullies in their vicinity, where generally no station is placed.

The [ALS2] method leads to the best results in terms of spread of the errors (standard deviation = 0.06 m) and a very small mean error of 0.064 m for 2011/12. This is mainly attributed to the fact that there is an iteration, in contrast to the other methods. This makes this method the most reliable one tested in this study, but requires two simulation runs to achieve good results. The small errors still evident in the [ALS2] result from the non-linear settling of the snow cover. A reduction of the precipitation by a factor of 2 does not lead to a reduction of the snow depth by a factor of 2, but a little less. With the adjustment of the precipitation, the settling caused by the snow covers own weight is influenced most. When snow mass is reduced, settling is disproportionately reduced due to missing mass and vice versa, leading to small errors in snow depth simulation. As the iterative approach is independent of direct precipitation measurements, the mean precipitation can be estimated with highest accuracy of all methods tested.

The direct scaling method redistributes the precipitation calculated by the base method [PSUM] or [HS]. Therefore, the influence of the scaling on the mean error is small and depends a lot on the precipitation input. The total precipitation mass allocated by the base method is not influenced and changes in the mean errors occur only due to the redistribution of the precipitation mass to other grid points, which are then influenced

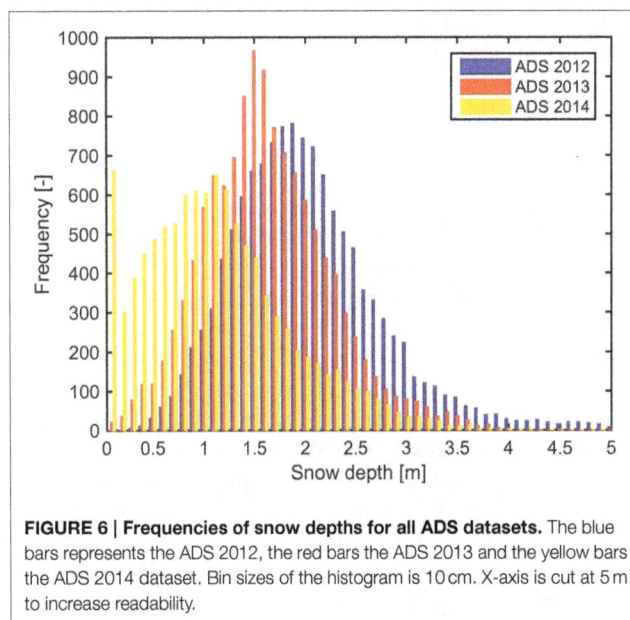

FIGURE 6 | Frequencies of snow depths for all ADS datasets. The blue bars represents the ADS 2012, the red bars the ADS 2013 and the yellow bars the ADS 2014 dataset. Bin sizes of the histogram is 10 cm. X-axis is cut at 5 m to increase readability.

differently by temperature, radiation, rain or metamorphism, which all influence settling and lead to a change in snow depth.

For all scaling runs, it is evident that for south facing slopes the snow depth is mostly underestimated whereas it is mostly overestimated in north facing ones. The influence of the aspect on the results will be discussed below.

Looking at the absolute error for the simulations [PSUM] and [PAT] the errors are strongly reduced. The mean error only varies slightly due to the mass conservation in the scaling method but the standard deviation is reduced by a factor of 2.9 from 0.569 m to 0.195 m, leading to errors less than ±1 m overall. The effect of the [HAT] method on the errors is comparable to the [PAT] method. The standard deviation of the absolute error is reduced by a factor of 3.4 to 0.192 m only. The mean error slightly shifts into a more negative area. This shift is mainly due to the redistribution of precipitation to other grid points leading to differences in settling as discussed above.

Figure 5A shows the improvement in spatial representation of the snow depths when using scaling methods. For any precipitation input, the standard deviation of the error is reduced up to a factor of 3.4 for the winter 2011/12 and a factor of 2 for the winter 2012/13. Only for the rain gauge driven simulations of the winter 2013/14 the effect of the scaling is rather small. This is due to the generally small snow depths in the ADS data of this winter (cf. **Figure 6**). When looking at the relative error (cf. **Figure 5B**), the standard deviation is reduced strongly also for the winter 2013/14. Still these values need to be looked at with caution. The relative error is very sensitive at small snow depths, where a little absolute error can produce a large relative error. This clearly shows that both, the relative and the absolute error need to be considered when investigating the model performance, especially when small snow depths are present frequently.

For the snow depth driven simulations for winter 2013/14, the effect of scaling is also clearly visible in the change of the absolute errors. We assume that not only the ADS data is crucial for good scaling results, but that also the total precipitation mass is important. If too much precipitation is allocated, the scaling will be less effective due to excessive supply of precipitation which leads to a smoothing of the snow cover. The smoothing stems from the fact that the proportional settling of deeper snow is larger than for thinner snow. If the correct or too little precipitation is allocated the scaling has a stronger effect on the correct redistribution. This can be seen when comparing the two simulations for winter 2013/14.

For all simulations using scaling, a negative drift of the mean error is evident. This is despite the fact that the scaling mechanism conserves the total precipitation mass allocated by the basic, non-scaling method (cf. Equation 1). The shift comes from the settling and melting information in the ADS datasets. The dataset represents a snow distribution toward the end of the winter but is applied to scale precipitation throughout the whole winter. The ADS data already accounts for settled snow depths and partial melting. If it is used for scaling precipitation throughout the whole winter and the model itself accounts for settling and melting on the snow cover too, these effects are overrepresented and lead to an overall reduction of snow volume. Also the local redistribution of precipitation due to scaling leads to different settling effects compared to the base method, leading to a decrease in overall snow depth. The average loss in snow depths is about 10 cm, independent of the precipitation input method.

For all the results presented here, the variation in the simulation setup is only in precipitation input and its interpolation and scaling. All other inputs are kept identical for any simulation. This allows us to best quantify the influence of variations in precipitation input and scaling. On the other hand, we did not quantify the influence and errors in simulation results due to other input quantities and their interpolation. An extended sensitivity analysis on various input quantities using Alpine3D in the Davos area has been conducted by Schölgl et al. (2016). The average vertical accuracy of the ADS data is estimated to ±30 cm (Bühler et al., 2015). For the simulations presented here, we had to assume the ADS data as being accurate to be able to scale precipitation and assess scaling performance. As the scaling methods presented here absolutely require ADS information, the quality of the simulation results highly depend on the available ADS data quality.

3.5. Inter-Annual scaling

If no ADS data is available to scale a specific winter, another dataset can be used. To assess the model performance for such a setting, the winters 2011/12, 2012/13, and 2013/14 have been simulated using [PAT] and were scaled with a different ADS dataset. The influence of a scaling with an ADS dataset from a different winter can be seen in **Figure 7**. The results from a non-scaling [PSUM] method are also shown for comparison.

Figure 7B shows the standard deviation of the absolute error in snow depth for the different simulations. The standard deviation is lowest for any simulation when it was scaled with the ADS dataset from the same year, except for winter 2013/14. Additionally, when being scaled with a different dataset, the standard deviation is lower compared to a non-scaling method,

except when using the ADS 2014 dataset to scale the winters 2011/12 and 2012/13.

Comparing the standard deviation for the winters 2011/12 and 2012/13 and the associated ADS datasets only, it can be stated that the scaling method leads to better results than the non-scaling method. This is due to the fact that the ADS snow distribution patterns in the winter 2011/12 and 2012/13 were similar and scaling with either dataset is possible (cf. **Figure 6**). Using the ADS 2014 dataset to scale other winters leads to worse results than using a non-scaling method. In March and April 2014 the air temperature reached maximum values of over 7°C at 2500 m.a.s.l. for several weeks, leading to intense melting of the snow cover below 2500 m.a.s.l.. This information is conserved in the ADS 2014 dataset which was acquired after this period. Therefore, scaling of other winters with this dataset leads to large errors in snow depth distribution with highest errors below 2500 m.a.s.l. in south facing slopes (not shown). Looking at **Figure 6**, it can be seen that in the ADS 2014 dataset a lot of grid points were already snow free. This influences the scaling strongly as no precipitation is allocated to these grid points. Therefore, the direct scaling mechanism presented here is not suitable to scale precipitation when grid points are snow free. For such cases, the scaling algorithm should be adapted or ADS data must be acquired earlier in the winter when these grid points are still snow covered.

The tests with inter-annual scaling show that the method has the potential to be used with a generalized ADS dataset to scale multiple winters. As only three ADS datasets were available, it was not possible to identify a generally valid snow distribution for the simulation domain. Nevertheless, results shown here indicate that ADS datasets 2012 and 2013 have the potential to be used as a general dataset to scale precipitation in the Davos area. The 2014 dataset was acquired too late and is not suitable for general snow scaling. Further studies using repeat ADS datasets within one winter are needed to find optimal scaling datasets and to assess the temporal consistency of ADS data within one winter and between winters.

3.6. Aspect-Dependency of Scaling Error

For all scaling simulations a relation between error in snow depth and the aspect of the grid point is evident. In north facing slopes the error in snow depth is generally smaller than in south facing slopes (cf. **Figure 8B**). Looking at non-scaling simulations such as [PSUM] (cf. **Figure 8A**), we find, however, only a small effect of aspect. For the iterative scaling approach [ALS2], this relation of error and aspect is also still evident (cf. **Figure 9**). However, compared to our other scaling simulations the amplitude of the error is smaller. We offer three possible explanations for these observations:

1. Using ADS data acquired toward the end of a winter, effects of settling and melting may be overestimated: The information about snow depth stored in the ADS datasets is based on the snow cover toward the end of the winter. Processes causing settling and melting throughout the winter are therefore represented by ADS data and applied on the distribution of precipitation. When the precipitation is distributed on the

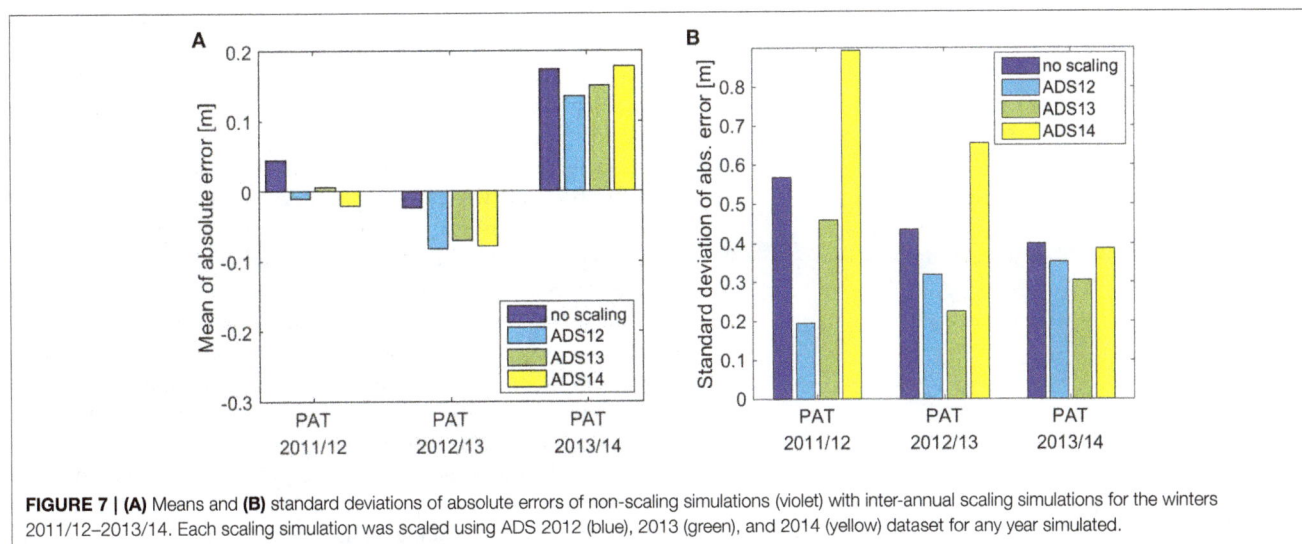

FIGURE 7 | (A) Means and (B) standard deviations of absolute errors of non-scaling simulations (violet) with inter-annual scaling simulations for the winters 2011/12–2013/14. Each scaling simulation was scaled using ADS 2012 (blue), 2013 (green), and 2014 (yellow) dataset for any year simulated.

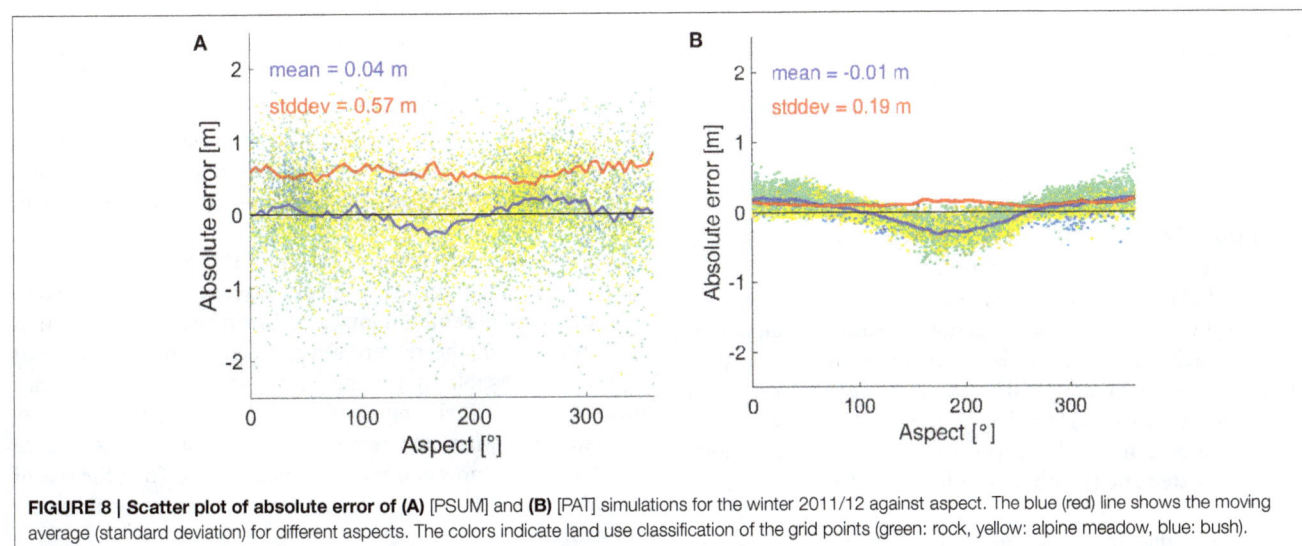

FIGURE 8 | Scatter plot of absolute error of (A) [PSUM] and (B) [PAT] simulations for the winter 2011/12 against aspect. The blue (red) line shows the moving average (standard deviation) for different aspects. The colors indicate land use classification of the grid points (green: rock, yellow: alpine meadow, blue: bush).

simulation domain, the model itself also calculates settling and melting. This leads to the problem that these effects are overestimated in the scaled simulations and lead to a strong underestimation of snow depth in slopes where settling is strong and melt has already occurred e.g., in south facing slopes. This can be confirmed when looking at the snow densities in the simulation domain. An analysis of settling and melting occurrence can be found in the Supplementary Material. It can be stated that the error is more dependent on the aspect than on the snow depth. We could show that high densities are more strongly related to thin snow covers in south facing slopes than to thick snow covers in north facing slopes. It is therefore assumed that the effect of radiation influences the snow density more than the gravitation effects due to a thick snow cover. It is therefore concluded that the aspect related errors are mainly caused by the snow settling due to radiation and temperature. Barely any melting had occurred until the date of ADS 2012 data acquisition (analysis

of SNOWPACK profiles; not shown). Therefore, we assume that melting effects are negligible compared to settling in this year.

2. The settling of the snow cover is non-linear. Therefore, scaling precipitation does not lead to the same scaling effect in the snow cover. When more precipitation is distributed on a grid point, more snow mass is generated there. The more mass, the more the snow cover is compacted by its own weight. This leads to a disproportional settling of the snow cover.

3. The radiation module used in Alpine3D may calculate too much radiation in south facing slopes: All simulations are run using the simple radiation module in Alpine3D, in which a clearness index is used to split the radiation into direct and diffuse radiation, which is known to sometimes produce unrealistic distributions (Lanini, 2010). If the splitting is overestimating the direct radiation, too much energy is distributed on a grid point, leading to disproportional settling.

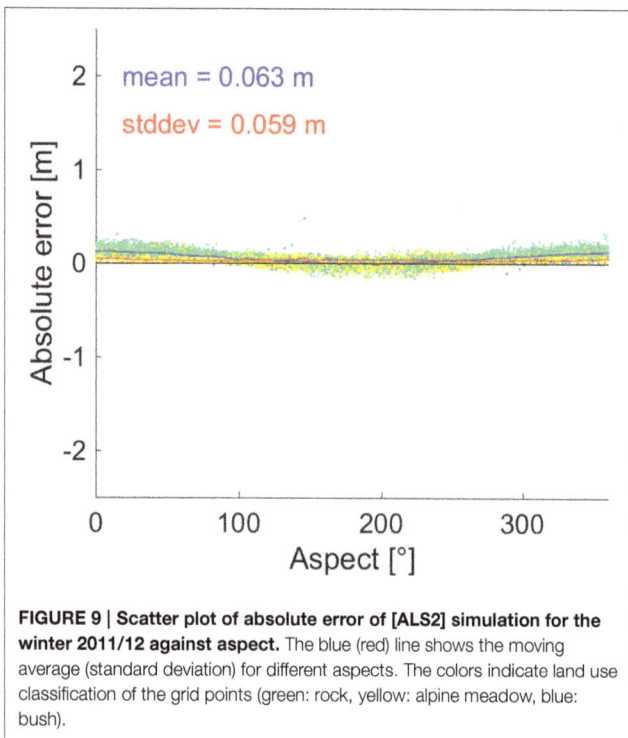

FIGURE 9 | Scatter plot of absolute error of [ALS2] simulation for the winter 2011/12 against aspect. The blue (red) line shows the moving average (standard deviation) for different aspects. The colors indicate land use classification of the grid points (green: rock, yellow: alpine meadow, blue: bush).

To estimate how much of the aspect related error is caused by the three effects described above, a comparison of [PAT] and [HAT] simulations with [ALS2] is suitable. As the iterative [ALS2] scaling is based on the correction of gridded precipitation data for each time step, it is independent of the settling information stored in the ADS datasets (1). Since this comparison shows that iterative scaling has a much smaller remaining error, it can be concluded that effect (1) dominates the aspect-dependent systematic error but that the other effects (2,3) are not negligible.

The error due to the scaling with ADS acquired late in the winter is larger than with an earlier acquisition. An option could be that ADS data is acquired earlier in the winter when less settling (and melting) has occurred. The error due to the radiation partitioning could be reduced by making the partitioning methods more robust. Even though the scaling leads to an aspect-dependent error as discussed, the mean error in distributed snow depths is reduced for most of the scaling methods and simulated winter.

4. CONCLUSIONS AND OUTLOOK

In this study, a simple precipitation scaling approach was developed and tested with the aim to improve the representation of the spatial distribution of snow in distributed numerical models, while maintaining low computational costs. This method has the potential to significantly improve spatial snow representation in distributed snow modeling without the need for spatially explicit modeling of wind transport (Lehning et al.,

2008; Vionnet et al., 2014), which either suffers from high computational demand, limited accuracy or both.

The performance of traditional simulation setups without scaling were tested and it was found that methods such as the [PSUM] or [HS] are able to generally represent the total snow mass in the simulation domain with good accuracy. However, these methods were found to not represent the total variability of the snow cover and often underestimate large snow depths at high elevations in particular whereas small snow depths are generally overestimated by the methods. Our proposed scaling approach reduced the standard deviation of the absolute error in spatial snow distribution significantly (up to a factor of 3.4) to less than 0.23 m. By choosing an appropriate base method to allocate the precipitation, the mean error of the snow cover could also be reduced to a few centimeters for the simulation domain. For these promising results, only one simulation run is required and one remote sensing dataset is sufficient for scaling. This shows that with the simple scaling method presented here, the overall performance in snow distribution modeling with Alpine3D or other snow models can be significantly improved. An iterative scaling approach [ALS2] is able to further compensate for errors generated in a first simulation run and the mean snow depth as well as the spatial variability are represented with very high accuracy. Drawbacks of iterative scaling are a high computational demand and the necessity to write, store and re-read large amounts of data.

Tests with inter-annual scaling showed that scaling approaches based on ADS measurements from a different year lead to good results as long as the relative snow distribution is represented in the remote sensing data used for scaling. It is therefore possible to use one remote sensing dataset to scale various winters (Lehning et al., 2011). scaling becomes less accurate only when the remote sensing data set has a large portion of little snow and snow free grid points. Then the risk of assigning too little or no precipitation to a grid point increases dramatically.

The performance of the scaling method depends on the quality of the remote sensing data. As this data fully determines how the precipitation is distributed and where a snow cover develops, its quality is critical in obtaining accurate snow depth distributions in the simulation. Errors in ADS data are difficult to identify, if the simulation output is compared to the remote sensing data. Therefore, it is important to use independent observations to verify the simulations results. In this respect, our inter-annual experiments can be seen as a successful cross-validation. To compare the simulation outputs with independent data, observations from AWSs are in principle a good choice. However, the station must be representative for its vicinity to allow a direct comparison, which is generally difficult for this type of terrain (Grünewald and Lehning, 2015). A very high grid resolution facilitates such a comparison. In this work it was not possible to find a suitable AWS to compare with the simulation results at the grid resolution of 100 m. Due to the extensive size of the simulation domain it was not feasible to further increase the simulation grid resolution within the context of this study.

In this study, it could be shown that the scaling method can substantially increase simulation performance regarding snow distribution although this method also brings along undesired effects leading to errors. The major drawback of the method is that it is sensitive to snow settling, melting and the date of ADS data acquisition. If remote sensing data are acquired from a snow cover where settling or melting has occurred, these effects influence the model and too little precipitation is allocated. This leads to an underestimation of the snow depth in certain areas. This not only decreases the performance of the spatial snow distribution simulation but can also have a major influence on the total snow volume in the simulation domain. To date, it was not possible to solve this issue, but it is assumed that with earlier snow depth acquisitions, the errors can further be reduced. Solving this issue is an important but also a challenging task in order to make more operational use of our scaling method in numerical snow modeling.

As an outlook on further work, a next step is to acquire more snow depth maps following individual storms and earlier in the winter. As briefly discussed above and shown in the Supplementary Material, a reasonable validation of time-resolved snow depth development at local weather stations is currently not possible. Based on temporal snow depth maps, the seasonal development can be studied in more detail and some of the errors discussed in our contribution can certainly be reduced. Along these lines, it is also interesting to decompose the weighting factor into a more regional simple trend, which takes larger-scale moisture gradients as well as the classical elevation into account. This base field can be applied to liquid and solid precipitation. The finer scale patterns, which can be seen as residuals from the larger-scale field, are then only be applied to snowfall. Finally, it should be studied, in how far spatial estimation of meteorological input other than precipitation amount can be improved to allow an optimized rain—snow distinction, which has a critical influence on our results.

AUTHOR CONTRIBUTIONS

All authors contributed to this work. This main paper was written by CV and is based on his master thesis. ML designed the study and supervised the working process. NW and MB also supervised the working process and gave assistance using Alpine3D, analysing results and implementing the scaling algorithm. All authors contributed written sections to this paper.

FUNDING

The paper is based on a master thesis by the first author and institutional funding from SLF Davos as well as EPFL is acknowledged. Furthermore, data and collaborators have partially been funded by the Swiss National Science Foundation (SNF) grants 200021_150146 and 200021E-160667.

ACKNOWLEDGMENTS

Leica Geosystems (Hexagon) supported us by flying the ADS camera free of charge. We thank Yves Bühler and Mauro Marty for the processing of the ADS data, Charles Fierz and Peter Molnar for the scientific inputs during the realization of the study and Marcia Phillips for proofreading the manuscript. We thank the two Referees and the Editor for their interest in our work and for the helpful comments that helped to improve the manuscript quality. Various institutions contributed with input data for model simulations such as MeteoSwiss, FOEN (Federal Office for the Environment) and the Canton of Grisons. The Alpine3D, MeteoIO and SNOWPACK source code is available under the (L)GPL version 3 at http://models.slf.ch.

REFERENCES

Anslow, F. S., Hostetler, S., Bidlake, W. R., and Clark, P. U. (2008). Distributed energy balance modeling of South Cascade Glacier, Washington and assessment of model uncertainty. *J. Geophys. Res.* 113:F02019. doi: 10.1029/2007JF000850

Bartelt, P., and Lehning, M. (2002). A physical SNOWPACK model for the Swiss avalanche warning: part I: numerical model. *Cold Regions Sci. Technol.* 35, 123–145. doi: 10.1016/S0165-232X(02)00074-5

Bavay, M., and Egger, T. (2014). MeteoIO 2.4.2: a preprocessing library for meteorological data. *Geosci. Model Dev.* 7, 3135–3151. doi: 10.5194/gmd-7-3135-2014

Bavay, M., Lehning, M., Jonas, T., and Löwe, H. (2009). Simulations of future snow cover and discharge in Alpine headwater catchments. *Hydrol. Process.* 23, 95–108. doi: 10.1002/hyp.7195

Bellaire, S., Jamieson, J. B., and Fierz, C. (2011). Forcing the snow-cover model SNOWPACK with forecasted weather data. *Cryosphere* 5, 1115–1125. doi: 10.5194/tc-5-1115-2011

Bühler, Y., Marty, M., Egli, L., Veitinger, J., Jonas, T., Thee, P., et al. (2015). Snow depth mapping in high-alpine catchments using digital photogrammetry. *Cryosphere* 9, 229–243. doi: 10.5194/tc-9-229-2015

Bühler, Y., Marty, M., and Ginzler, C. (2012). High resolution DEM generation in high-alpine terrain using airborne remote sensing techniques. *Trans. GIS* 16, 635–647. doi: 10.1111/j.1467-9671.2012.01331.x

Bühler, Y., Marty, M., and Ginzler, C. (2013). Grossflächige hochaufgelöste Schneehöhenkarten aus digitalen Stereoluftbildern. *Geomatik Schweiz Geoinformation und Landmanagement* 111, 508–510.

Comola, F., Schaefli, B., Ronco, P. D., Botter, G., Bavay, M., Rinaldo, A., et al. (2015). Scale-dependent effects of solar radiation patterns on the snow-dominated hydrologic response. *Geophys. Res. Lett.* 42:2015GL064075. doi: 10.1002/2015GL064075

Deems, J. S., Painter, T. H., and Finnegan, D. C. (2013). Lidar measurement of snow depth: a review. *J. Glaciol.* 59, 467–479. doi: 10.3189/2013JoG12J154

Endrizzi, S., Gruber, S., Dall'Amico, M., and Rigon, R. (2014). GEOtop 2.0: simulating the combined energy and water balance at and below the land surface accounting for soil freezing, snow cover and terrain effects. *Geosci. Model Dev.* 7, 2831–2857. doi: 10.5194/gmd-7-2831-2014

Gallice, A., Bavay, M., Brauchli, T., Comola, F., Lehning, M., and Huwald, H. (2016). StreamFlow 1.0: an extension to the spatially distributed snow model Alpine3D for hydrological modeling and deterministic stream temperature prediction. *Geosci. Model Dev. Discuss.* 167, 1–51. doi: 10.5194/gmd-2016-167

Gauer, P. (2001). Numerical modeling of blowing and drifting snow in Alpine terrain. *J. Glaciol.* 47, 97–110. doi: 10.3189/172756501781832476

Goodison, B., Louie, P., and Yang, D. (1998). *WMO Solid Precipitation Measurement Intercomparison–Final Report.* Technical Report 67, WMO/TD 872.

Grünewald, T., and Lehning, M. (2011). Altitudinal dependency of snow amounts in two small alpine catchments: can catchment-wide snow amounts be estimated via single snow or precipitation stations? *Ann. Glaciol.* 52, 153–158. doi: 10.3189/172756411797252248

Grünewald, T., and Lehning, M. (2015). Are flat-field snow depth measurements representative? A comparison of selected index sites with areal snow depth measurements at the small catchment scale. *Hydrol. Process.* 29, 1717–1728. doi: 10.1002/hyp.10295

Grünewald, T., Schirmer, M., Mott, R., and Lehning, M. (2010). Spatial and temporal variability of snow depth and ablation rates in a small mountain catchment. *Cryosphere* 4, 215–225. doi: 10.5194/tc-4-215-2010

Groot Zwaaftink, C. D., Löwe, H., Mott, R., Bavay, M., and Lehning, M. (2011). Drifting snow sublimation: a high-resolution 3-D model with temperature and moisture feedbacks. *J. Geophys. Res.* 116:D16107. doi: 10.1029/2011JD015754

Hanzer, F., Helfricht, K., Marke, T., and Strasser, U. (2016). Multilevel spatiotemporal validation of snow/ice mass balance and runoff modeling in glacierized catchments. *Cryosphere* 10, 1859–1881. doi: 10.5194/tc-10-1859-2016

Heimo, A., Calpini, B., Konzelmann, T., and Suter, S. (2007). *SwissMetNet: The New Automatic Meteorological Network of Switzerland.* Technical report, MeteoSwiss, Zurich.

Helbig, N. (2009). *Application of the Radiosity Approach to the Radiation Balance in Complex Terrain.* Dissertation, University of Zürich, Zürich.

Huss, M., Bauder, A., Funk, M., and Hock, R. (2008). Determination of the seasonal mass balance of four Alpine glaciers since 1865. *J. Geophys. Res.* 113:F01015. doi: 10.1029/2007JF000803

Koenig, U., and Abegg, B. (1997). Impacts of climate change on winter tourism in the Swiss alps. *J. Sustain. Tour.* 5, 46–58. doi: 10.1080/09669589708667275

Kresse, W., and Danko, D. M. (2012). *Springer Handbook of Geographic Information.* Dordrecht: Springer Science & Business Media.

Lanini, F. (2010). *Division of Global Radiation Into Direct Radiation and Diffuse Radiation.* Master thesis, University of Bern, Bern.

Lehning, M., Bartelt, P., Brown, B., and Fierz, C. (2002a). A physical SNOWPACK model for the Swiss avalanche warning: part III: meteorological forcing, thin layer formation and evaluation. *Cold Regions Sci. Technol.* 35, 169–184. doi: 10.1016/S0165-232X(02)00072-1

Lehning, M., Bartelt, P., Brown, B., Fierz, C., and Satyawali, P. (2002b). A physical SNOWPACK model for the Swiss avalanche warning: part II. Snow microstructure. *Cold Regions Sci. Technol.* 35, 147–167. doi: 10.1016/S0165-232X(02)00073-3

Lehning, M., Bartelt, P., Brown, B., Russi, T., Stöckli, U., and Zimmerli, M. (1999). Snowpack model calculations for avalanche warning based upon a new network of weather and snow stations. *Cold Regions Sci. Technol.* 30, 145–157. doi: 10.1016/S0165-232X(99)00022-1

Lehning, M., Grünewald, T., and Schirmer, M. (2011). Mountain snow distribution governed by an altitudinal gradient and terrain roughness. *Geophys. Res. Lett.* 38:L19504. doi: 10.1029/2011GL048927

Lehning, M., Löwe, H., Ryser, M., and Raderschall, N. (2008). Inhomogeneous precipitation distribution and snow transport in steep terrain. *Water Resour. Res.* 44:W07404. doi: 10.1029/2007WR006545

Lehning, M., Völksch, I., Gustafsson, D., Nguyen, T. A., Stähli, M., and Zappa, M. (2006). ALPINE3d: a detailed model of mountain surface processes and its application to snow hydrology. *Hydrol. Process.* 20, 2111–2128. doi: 10.1002/hyp.6204

Liston, G. E., and Sturm, M. (1998). A snow-transport model for complex terrain. *J. Glaciol.* 44, 498–516.

Magnusson, J., Gustafsson, D., Hüsler, F., and Jonas, T. (2014). Assimilation of point SWE data into a distributed snow cover model comparing two contrasting methods. *Water Resour. Res.* 50, 7816–7835. doi: 10.1002/2014WR015302

Michlmayr, G., Lehning, M., Koboltschnig, G., Holzmann, H., Zappa, M., Mott, R., et al. (2008). Application of the Alpine 3d model for glacier mass balance and glacier runoff studies at Goldbergkees, Austria. *Hydrol. Process.* 22, 3941–3949. doi: 10.1002/hyp.7102

Mott, R., Schirmer, M., and Lehning, M. (2011). Scaling properties of wind and snow depth distribution in an Alpine catchment. *J. Geophys. Res.* 116:D06106. doi: 10.1029/2010JD014886

Prokop, A., Schirmer, M., Rub, M., Lehning, M., and Stocker, M. (2008). A comparison of measurement methods: terrestrial laser scanning, tachymetry and snow probing for the determination of the spatial snow-depth distribution on slopes. *Ann. Glaciol.* 49, 210–216. doi: 10.3189/172756408787814726

Revuelto, J., Vionnet, V., López-Moreno, J.-I., Lafaysse, M., and Morin, S. (2016). Combining snowpack modeling and terrestrial laser scanner observations improves the simulation of small scale snow dynamics. *J. Hydrol.* 533, 291–307. doi: 10.1016/j.jhydrol.2015.12.015

Schaefli, B., Hingray, B., and Musy, A. (2007). Climate change and hydropower production in the Swiss Alps: quantification of potential impacts and related modelling uncertainties. *Hydrol. Earth Syst. Sci.* 11, 1191–1205. doi: 10.5194/hess-11-1191-2007

Schirmer, M., Wirz, V., Clifton, A., and Lehning, M. (2011). Persistence in intra-annual snow depth distribution: 1. Measurements and topographic control. *Water Resour. Res.* 47:W09516. doi: 10.1029/2010WR009426

Schöngl, S., Marty, C., Bavay, M., and Lehning, M. (2016). Sensitivity of alpine3d modeled snow cover to modifications in DEM resolution, station coverage and meteorological input quantities. *Environ. Model. Softw.* 83, 387–396. doi: 10.1016/j.envsoft.2016.02.017

Schuler, T., Loe, E., Taurisano, A., Eiken, T., Hagen, J., and Kohler, J. (2007). Calibrating a surface mass-balance model for Austfonna ice cap, Svalbard. *Ann. Glaciol.* 46, 241–248. doi: 10.3189/172756407782871783

Schweizer, J., Bruce Jamieson, J., and Schneebeli, M. (2003). Snow avalanche formation. *Rev. Geophys.* 41:1016. doi: 10.1029/2002RG000123

Trujillo, E., Ramírez, J. A., and Elder, K. J. (2007). Topographic, meteorologic, and canopy controls on the scaling characteristics of the spatial distribution of snow depth fields. *Water Resour. Res.* 43:W07409. doi: 10.1029/2006WR005317

Vionnet, V., Martin, E., Masson, V., Guyomarc'h, G., Naaim-Bouvet, F., Prokop, A., et al. (2014). Simulation of wind-induced snow transport and sublimation in alpine terrain using a fully coupled snowpack/atmosphere model. *Cryosphere* 8, 395–415. doi: 10.5194/tc-8-395-2014

Wever, N. (2015). *Liquid Water Flow in Snow and Hydrological Implications.* Dissertation, Ecole Polytechnique federal de Lausanne EPFL, Lausanne.

Wever, N., Fierz, C., Mitterer, C., Hirashima, H., and Lehning, M. (2014). Solving Richards Equation for snow improves snowpack meltwater runoff estimations in detailed multi-layer snowpack model. *Cryosphere* 8, 257–274. doi: 10.5194/tc-8-257-2014

Wever, N., Schmid, L., Heilig, A., Eisen, O., Fierz, C., and Lehning, M. (2015). Verification of the multi-layer SNOWPACK model with different water transport schemes. *Cryosphere* 9, 2271–2293. doi: 10.5194/tc-9-2271-2015

Winstral, A., Elder, K., and Davis, R. E. (2002). Spatial snow modeling of wind-redistributed snow using terrain-based parameters. *J. Hydrometeorol.* 3, 524–538. doi: 10.1175/1525-7541(2002)003<0524:SSMOWR>2.0.CO;2

Winstral, A., Marks, D., and Gurney, R. (2013). Simulating wind-affected snow accumulations at catchment to basin scales. *Adv. Water Resour.* 55, 64–79. doi: 10.1016/j.advwatres.2012.08.011

Wipf, S., Stoeckli, V., and Bebi, P. (2009). Winter climate change in alpine tundra: plant responses to changes in snow depth and snowmelt timing. *Climat. Change* 94, 105–121. doi: 10.1007/s10584-009-9546-x

Zappa, M., Pos, F., Strasser, U., Warmerdam, P., and Gurtz, J. (2003). Seasonal water balance of an alpine catchment as evaluated by different methods for spatially distributed snowmelt modelling. *Hydrol. Res.* 34, 179–202.

Conflict of Interest Statement: The authors declare that the research was conducted in the absence of any commercial or financial relationships that could be construed as a potential conflict of interest.

5

Exploring Google Earth Engine Platform for Big Data Processing: Classification of Multi-Temporal Satellite Imagery for Crop Mapping

Andrii Shelestov [1,2], Mykola Lavreniuk [1,2]*, Nataliia Kussul [1,2], Alexei Novikov [2] and Sergii Skakun [3,4]

[1] Department of Space Information Technologies and Systems, Space Research Institute (NASU-SSAU), Kyiv, Ukraine, [2] Department of Information Security, National Technical University of Ukraine "Igor Sikorsky Kyiv Polytechnic Institute," Kyiv, Ukraine, [3] Department of Geographical Sciences, University of Maryland, College Park, MD, USA, [4] NASA Goddard Space Flight Centre, Greenbelt, MD, USA

Edited by:
Monique Petitdidier,
Institut Pierre-Simon Laplace, France

Reviewed by:
Karine Zeitouni,
University of Versailles Saint-Quentin,
France
Télesphore Yao Brou,
University of La Réunion, France
Fabrizio Niro,
European Space Research Institute,
Italy

*Correspondence:
Mykola Lavreniuk
nick_93@ukr.net

Specialty section:
This article was submitted to
Environmental Informatics,
a section of the journal
Frontiers in Earth Science

Many applied problems arising in agricultural monitoring and food security require reliable crop maps at national or global scale. Large scale crop mapping requires processing and management of large amount of heterogeneous satellite imagery acquired by various sensors that consequently leads to a "Big Data" problem. The main objective of this study is to explore efficiency of using the Google Earth Engine (GEE) platform when classifying multi-temporal satellite imagery with potential to apply the platform for a larger scale (e.g., country level) and multiple sensors (e.g., Landsat-8 and Sentinel-2). In particular, multiple state-of-the-art classifiers available in the GEE platform are compared to produce a high resolution (30 m) crop classification map for a large territory ($\sim$28,100 km^2 and 1.0 M ha of cropland). Though this study does not involve large volumes of data, it does address efficiency of the GEE platform to effectively execute complex workflows of satellite data processing required with large scale applications such as crop mapping. The study discusses strengths and weaknesses of classifiers, assesses accuracies that can be achieved with different classifiers for the Ukrainian landscape, and compares them to the benchmark classifier using a neural network approach that was developed in our previous studies. The study is carried out for the Joint Experiment of Crop Assessment and Monitoring (JECAM) test site in Ukraine covering the Kyiv region (North of Ukraine) in 2013. We found that GEE provides very good performance in terms of enabling access to the remote sensing products through the cloud platform and providing pre-processing; however, in terms of classification accuracy, the neural network based approach outperformed support vector machine (SVM), decision tree and random forest classifiers available in GEE.

Keywords: Google Earth Engine, big data, classification, optical satellite imagery, land cover, land use, image processing

INTRODUCTION

Information on land cover/land use (LCLU) geographical distribution over large areas is extremely important for many environmental and monitoring tasks, including climate change, ecosystem dynamics analysis, food security, and others. Reliable crop maps can be used for more accurate agriculture statistics estimation (Gallego et al., 2010, 2012, 2014), stratification purposes (Boryan and Yang, 2013), better crop yield prediction (Kogan et al., 2013a,b; Kolotii et al., 2015), and drought risk assessment (Kussul et al., 2010, 2011; Skakun et al., 2016b). During the past decades, satellite imagery became the most promising data source for solving such important tasks as LCLU mapping. Yet, at present, there are no globally available satellite-derived crop specific maps at high-spatial resolution. Only coarse-resolution imagery (>250 m spatial resolution) has been utilized to derive global cropland extent (e.g., GlobCover, MODIS; Fritz et al., 2013). At present, a wide range of satellites provide objective, open and free high spatial resolution data on a regular basis. These new opportunities allow one to build high-resolution LCLU maps on a regular basis and to assess LCLU changes for large territories (Roy et al., 2014). With launches of Sentinel-1, Sentinel-2, Proba-V and Landsat-8 remote sensing satellites, there will be generated up to petabyte of raw (unprocessed) images per year. The increasing volume and variety of remote sensing data, dubbed as a "Big Data" problem, creates new challenges in handling datasets that require new approaches to extracting relevant information from remote sensing (RS) data from data science perspective (Kussul et al., 2015; Ma et al., 2015a,b). Generation of high resolution crop maps for large areas (>10,000 sq. km) using Earth observation data from space requires processing of large amount of satellite images acquired by various sensors. Images acquired at different dates during crop growth period are usually required to discriminate certain crop types. The following issues should be addressed while providing classification of multi-temporal satellite images for large areas: (i) non-uniformity of coverage of ground truth data and satellite scenes; (ii) seasonal differentiation of crop groups (e.g., winter and summer crops) and the need for incremental classification (to provide both in season and post season maps); (iii) the need to store, manage and seamlessly process large amount of data (big data issues).

The Google Earth Engine (GEE) provides a cloud platform to access and seamlessly process large amount of freely available satellite imagery, including those acquired by the Landsat-8 remote sensing satellite. The GEE also provides a set of the state-of-the-art classifiers for pixel-based classification that can be used for crop mapping. Though these methods are well-documented in the literature, there were no previous studies to compare all these methods for classification of multi-temporal satellite imagery for large scale crop mapping. Since the GEE platform does not include any neural network based models, we add a neural network based classifier (Skakun et al., 2007; Kussul et al., 2016; Lavreniuk et al., 2016; Skakun et al., 2016a) to the analysis to provide a more complete comparison. Hence, the paper aims to explore efficiency of using the GEE platform when classifying multi-temporal satellite imagery for crop mapping with potential to apply the platform for a larger scale (e.g., country level) and multiple sensors (e.g., Landsat-8 and Sentinel-2). Results are presented for a highly heterogeneous landscape with multiple cropping systems on the Joint Experiment of Crop Assessment and Monitoring (JECAM) test site in Ukraine with the area of more than 28,000 km^2.

STUDY AREA AND MATERIALS DESCRIPTION

The proposed study is carried out for the Joint Experiment for Crop Assessment and Monitoring (JECAM) test site in Ukraine. Agriculture is a major part of Ukrainian's economy accounting for 12% of the Ukrainian Gross Domestic Product (GDP). Globally, Ukraine was the largest sunflower producer (11.6 MT) and exporter, and the ninth largest wheat producer (22.2 MT) in the world in 2013, according to the U.S. Department of Agriculture (USDA) Foreign Agricultural Service (FAS) statistics.

The JECAM test site in Ukraine was established in 2011 as part of the collaborative activities within the GEOGLAM Community of Practice. The site covers the administrative region of Kyiv region with the geographic area of 28,100 km^2 and almost 1.0 M ha of cropland (**Figure 1**). Major crop types in the region include: winter wheat, maize, soybeans, vegetables, sunflower, barley, winter rapeseed, and sugar beet. The crop calendar is September-July for winter crops, and April-October for spring and summer crops (**Table 1**). Fields in Ukraine are quite large with size generally ranging up to 250 ha.

Ground surveys to collect data on crop types and other land cover classes were conducted in 2013 in Kyiv region (**Figure 2**). European Land Use and Cover Area frame Survey (LUCAS) nomenclature was used in this study as a basis for land cover/land use types. In total, 386 polygons were collected covering the area of 22,700 ha (**Table 2**). Data were collected along the roads following the JECAM adopted protocol (Waldner et al., 2016) using mobile devices with built-in GPS. All surveyed fields were randomly split into training set (50%) to train the classifiers and testing set (50%) for testing purposes. Fields were selected in such a way so there is no overlap between training and testing sets. All classification results, in particular overall accuracy (OA), user's (UA), and producer's (PA) accuracies are reported for testing set (Congalton, 1991; Congalton and Green, 2008). The input features were classified into one of the 13 classes.

Remote sensing images acquired by the Operational Land Imager (OLI) sensor aboard Landsat-8 satellite were used for crop mapping over the study region. Landsat-8/OLI acquires images in eight spectral bands (bands 1–7, 9) at 30 m spatial resolution and in panchromatic band 8 at 15 m resolution (Roy et al., 2014). Only bands 2 through 7 acquired at different time periods were used for crop classification maps. Bands 1 and 9 were not used due to strong atmospheric absorption. Three scenes with path/row coordinates 181/24, 181/25, and 181/26 covered the test site region. **Table 3** summarizes dates of image acquisitions and the fraction of missing values in images due to clouds and shadows.

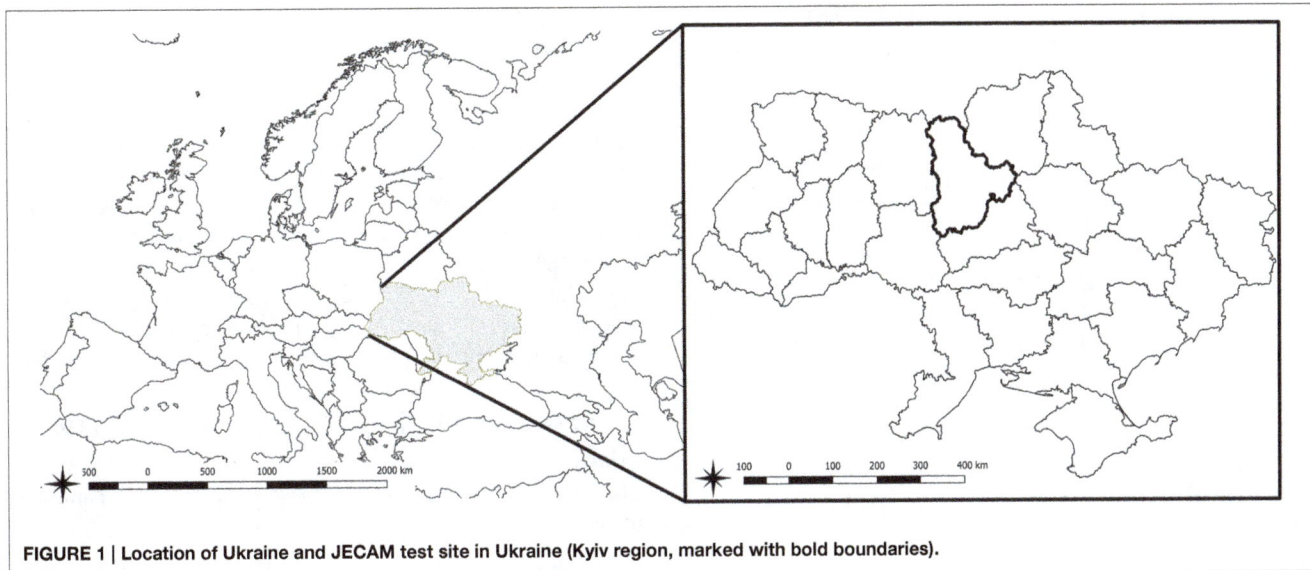

FIGURE 1 | Location of Ukraine and JECAM test site in Ukraine (Kyiv region, marked with bold boundaries).

TABLE 1 | Crop calendar with sowing and harvesting for Kyiv region.

N	Crop	Sowing	Harvesting
1	Winter wheat	End of September	Mid of July
2	Winter rapeseed	End of August	Mid of July
3	Maize	End of April	Mid of August—begin of September
4	Soybean	End of April—Begin of May	Mid of September—begin of October
5	Spring crops	Begin of April	Mid of July—August
6	Sugar beet	End of April—Mid of May	End of September—Mid of November
7	Sunflower	End of April—Begin of May	Begin of September

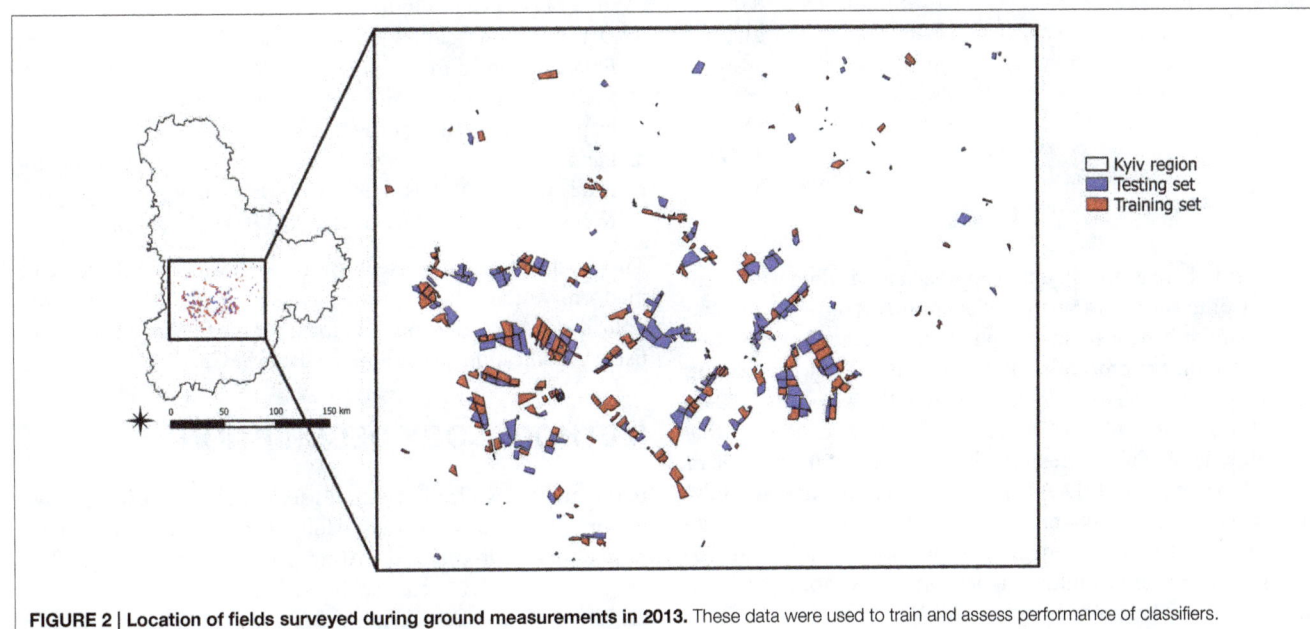

FIGURE 2 | Location of fields surveyed during ground measurements in 2013. These data were used to train and assess performance of classifiers.

The main issues that need to be addressed while dealing with satellite imagery for large areas (such as Kyiv region) are a non-regular coverage over the region and the presence of missing data due to the clouds and shadows. Therefore, before feeding satellite data to the classifiers, a pre-processing step should be performed to fill in missing values due to clouds and shadows.

TABLE 2 | Number of polygons and total area of crops and land cover types collected during the ground survey in 2013.

N	LUCAS class	Class	Polygons		Area	
			No.	%	ha	%
1	Axx	Artificial	6	1.6	23.0	0.1
2	B11	Winter wheat	51	13.2	3960.8	17.4
3	B32	Winter rapeseed	12	3.1	937.3	4.1
4	B12, B14	Spring crops	9	2.3	455.9	2.0
5	B16	Maize	87	22.5	7253.3	31.9
6	B22	Sugar beet	8	2.1	632.5	2.8
7	B31	Sunflower	30	7.8	2549.0	11.2
8	B33	Soybeans	60	15.5	3252.3	14.3
9	B19, B39, B40	Other cereals	32	8.3	1364.0	6.0
10	C10, B60	Forest	17	4.4	1014.3	4.5
11	E01, E02	Grassland	48	12.4	747.5	3.3
12	F00	Bare land	10	2.6	67.2	0.3
13	G01, G02	Water	16	4.1	448.3	2.0
		Total	386	100	22705.3	100

TABLE 3 | List of Landsat-8/OLI image acquisitions and estimate of missing values due to clouds and shadows.

Date (2013)	Path/ row	Missing values (%)	Date (2013)	Path/ row	Missing values (%)
April 16	181/24	4.77	June 19	181/24	62.58
	181/25	0.60		181/25	33.37
	181/26	0.02		181/26	26.43
May 02	181/24	0.01	July 05	181/24	35.30
	181/25	0.77		181/25	21.06
	181/26	4.38		181/26	15.69
May 18	181/24	9.06	August 06	181/24	24.35
	181/25	14.93		181/25	12.69
	181/26	14.32		181/26	40.86

At present, there is no a standard approach for dealing with these issues. Compositing is a very popular approach, however missing values can still happen in composite products (Yan and Roy, 2014). Another approach is to fill gaps using image processing techniques or ancillary data such as MODIS (Gao et al., 2006; Roy et al., 2008; Hilker et al., 2009). The main issue with this approach is spatial discrepancy between the 250 and 500 m MODIS products and the 30 m Landsat imagery. In this study, two types of approaches were explored: (i) compositing products available in GEE, so to benefit from products already available in GEE; (ii) restored multi-temporal images without involving ancillary data (Skakun and Basarab, 2014).

Composite Products Available at GEE

Different composites derived from Landsat-8 imagery and available in GEE were analyzed in the study. Landsat 8 8-Day Top-of-atmosphere (TOA) Reflectance Composites were used from GEE (**Figure 3**). As to the time of composition, 8 day composites were selected over 32 day composites to have a better temporal resolution. The reason for that is that 32 composites are composed based on the latest image, and this latest image can be of not the best quality. These Landsat-8 composites are made from Level L1T orthorectified scenes, using the computed TOA reflectance. The composites include all the scenes in each 8-day period beginning from the first day of the year and continuing to the 360th day of the year. The last composite of the year, beginning on day 361, will overlap the first composite of the following year by 3 days. All the images from each 8-day period are included in the composite, with the most recent pixel on top.

Landsat-8 Data Pre-Processing (Outside GEE)

The following pre-processing steps were applied for all Landsat-8 images:

1. Conversion of digital numbers (DNs) values to the TOA reflectance values using conversion coefficients in the metadata file (Roy et al., 2014).
2. To decrease an impact of atmosphere to the image quality conversion from the TOA reflectance to the surface reflectance (SR) has been done using the Simplified Model for Atmospheric Correction (SMAC; Rahman and Dedieu, 1994). The source code for the model was acquired from http://www.cesbio.ups-tlse.fr/multitemp/?p=2956. Parameters of the atmosphere to run the model (in particular, aerosol optical depth) were acquired from the Aeronet network's station in Kyiv (geographic coordinates +50.374N and +30.497E). The differences between TOA and SR satellite image is shown in **Figure 4**.
3. Detection of clouds and shadows were done using Fmask algorithm (Zhu and Woodcock, 2012). For this, a stand alone application was used from https://code.google.com/p/cfmask/.
4. Landsat-8 satellite images were reconstructed from missing pixel values (clouds and shadows) using self-organizing Kohonen maps (SOMs; Skakun and Basarab, 2014).

These pre-processing steps were performed outside GEE platform. After these products were generated they were uploaded in the GEE cloud platform for the further classification using classification algorithms available in GEE.

METHODOLOGY DESCRIPTION

In the study, classification of multi-temporal satellite imagery was performed at per-pixel basis. Multiple classification techniques were evaluated in the study. At first, all classification algorithms available in GEE were analyzed and used for classifying multi-temporal 8-days Landsat-8 TOA composites from GEE. Then, the best classification algorithms in terms of overall classification accuracy were compared with the neural network classifier that used multi-temporal SR values generated outside GEE. The presented approaches were compared in terms of classification accuracy at pixel level. GEE offers several

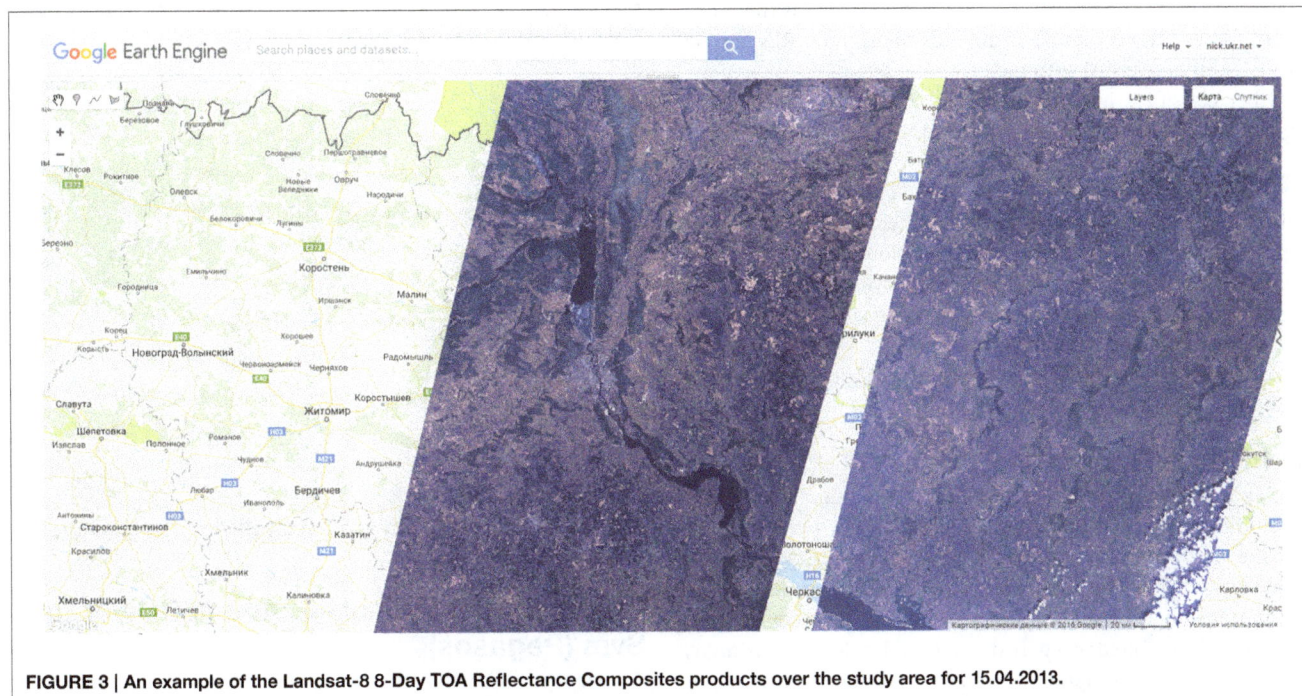

FIGURE 3 | An example of the Landsat-8 8-Day TOA Reflectance Composites products over the study area for 15.04.2013.

FIGURE 4 | **An example of TOA (A)** and SR **(B)** Landsat-8 image acquired on the 8th of August 2013. A true color composition of Landsat-8 bands 4-3-2 is shown.

classification algorithms among which are decision trees, random forests, support vector machine (SVM), and Naïve Bayes classifier.

Support Vector Machine (SVM)

SVM became popular in a recent decade for solving problems in classification, regression, and novelty detection. An important property of SVMs is that the determination of the model parameters corresponds to a convex optimization problem, and so any local solution is also a global optimum (Bishop, 2006). The SVM approaches classification problem through the concept of the margin, which is defined to be the smallest distance between the decision boundary and any of the samples. The decision boundary is chosen to be the one for which the margin is maximized. The margin is defined as the perpendicular distance between the decision boundary and the closest of the data points. Maximizing the margin leads to a particular choice of decision boundary. The location of this boundary is determined by a subset of the data points, known as support vectors.

Classification and Regression Tree (CART)

A decision tree (DT) classifier is built from a set of training data using the concept of information entropy. At each node of the tree, one attribute of the data that most effectively splits its set of samples into subsets enriched in one class or the other is selected. Its criterion is the normalized information gain that results from choosing an attribute for splitting the data. The attribute with the highest normalized information gain is chosen to make the decision. The algorithm then recurs on the smaller sublists. One disadvantage of the DT classifier is the considerable sensitivity to the training dataset, so that a small change to the training data can result in a very different set of subsets (Bishop, 2006).

Random Forests (RFs)

Since the major problem of the DT classifier is overfitting, RF overcomes it by constructing an ensemble of DTs (Breiman, 2001). More specifically, RF operates by constructing a multitude of DT at training time and outputting the class that is the mode of the classes (classification) of the individual trees. RFs correct for DT habit of overfitting to their training set.

Other classifiers included in the GEE platform that are less popular in the remote sensing community:

GMO Max Entropy

Multinomial logistic regression is generalization of linear regression using the softmax transformation function and the main task is to minimize an error function by taking the negative logarithm of the likelihood, which means cross-entropy. The main difference from other models and algorithms is the outcome score that could be considered as a probability value (Bishop, 2006; Haykin, 2008).

MultiClassPerceptron

This approach applied implementation of the linear perceptron to multiclass problems. The main idea of the perceptron is that the summing node of the neural model computes a linear combination of the inputs applied to its synapses, as well as incorporates an externally applied bias. The resulting sum, that is, the induced local field, is applied to a hard limiter. Accordingly, the neuron produces an output equal to 1 if the hard limiter input is positive, and -1 if it is negative (Haykin, 2008). Unfortunately, the perceptron is the simplest form of a neural network used for the classification of patterns said to be linearly separable.

Naïve Bayes

Bayes classifier is a simple probabilistic approach which is based on the Bayes theorem and assumption of independence between input features. Within the learning procedure, it minimizes the average risk of classification error. Main advantage of this classifier is that it requires a small number of training data to compute the decision surface. At the same time, its derivation is contingent on the assumption that the underlying distributions be Gaussian, which may limit its area of application (Haykin, 2008).

Intersection Kernel Passive Aggressive Method for Information Retrieval (IKPamir)

It is the specialized version of the SVM which represents a histogram intersection kernel SVMs (IKSVMs). The runtime complexity of the classifier is logarithmic in the number of support vectors as opposed to linear for the standard approach. It allows to put IKSVM classifiers in the same order of computational cost for evaluation as linear SVMs (Maji et al., 2008).

Winnow

The algorithm can be expressed as a linear-threshold algorithm that is similar to MultiClassPerceptron. However, while the perceptron uses an additive weight-update scheme, the Winnow classifier uses a multiplicative scheme. A primary advantage of this algorithm is that the number of mistakes that it makes is relatively little affected by the presence of large numbers of irrelevant attributes in the examples. The number of mistakes grows only logarithmically with the number of irrelevant

TABLE 4 | Overall classification accuracy achieved by GEE classifiers for TOA 8-day composites as an input.

Classifier	OA (%)
CART	75
GMO Max Entropy	72
Random Forest	68
MultiClassPerceptron	60
IKPamir	57
Winnow	49
FastNaiveBayes	32
Pegasos	–
VotingSvm	–
MarginSvm	–

attributes. Classifier is computationally efficient (both in time and space; Littlestone, 1988).

Primal Estimated sub-GrAdient SOlver for Svm (Pegasos)

A variant of SVM with simple and effective sub-GrAdient SOlver algorithm for approximately minimizing the objective function that has a fast rate of convergence results. At each iteration, a single training example is chosen at random and used to estimate a sub-gradient of the objective, and a step with pre-determined step-size is taken in the opposite direction. Solution is found in probability solely due to the randomization steps employed by the algorithm and not due to the data set. Therefore, the runtime does not depend on the number of training examples and thus Pegasos is especially suited for large datasets (Shalev-Shwartz et al., 2011).

Ensemble of Neural Networks

It should be noted that neural network (NN) classifiers are not available in GEE. Our proposed neural network approach based on committee of NNs, in particular Multi-Layer Perceptron (MLPs), is utilized to improve performance of individual classifiers. The MLP classifier has a hyperbolic tangent activation function for neurons in the hidden layer and logistic activation function in the output layer. Within training cross-entropy (CE) error function is minimized (Bishop, 2006)

$$E(\mathbf{w}) = -\ln p(\mathbf{T}|\mathbf{w}) = -\sum_{n=1}^{N} \sum_{k=1}^{K} t_{nk} \ln y_{nk} \tag{1}$$

where $E(\mathbf{w})$ is the CE error function that depends on the neurons' weight coefficients $\mathbf{w}$, $\mathbf{T}$ is the set of vectors of target outputs in the training set composed of N samples, K is the number of classes, t_{nk} and y_{nk} are the target and MLP outputs, respectively. In the target output for class k, all components of vector t_n are set to 0, except for the k-th component which is set to 1. The CE error $E(\mathbf{w})$ is minimized by means of the scaled conjugate gradient algorithm by varying weight coefficients $\mathbf{w}$ (Bishop, 2006).

A committee of MLPs was used to increase performance of individual classifiers. The committee was formed using MLPs

TABLE 5 | Classification results for CART, RF, and committee of NN for atmospherically corrected and restored Landsat-8 imagery.

		CART		RF		Committee of MLPs	
OA (%)		**76.9**		**69.9**		**84.7**	
		PA (%)	UA (%)	PA (%)	UA (%)	PA (%)	UA (%)
1	Artificial	87.1	33.2	80.8	54.9	64.8	94.9
2	Winter wheat	87.8	91.3	82.7	90.0	91.3	95.0
3	Winter rapeseed	88.1	92.8	67.2	86.5	97.7	92.9
4	Spring crops	10.4	6.5	13.2	14.2	39.2	45.2
5	Maize	75.3	92.7	73.3	83.9	85.9	89.7
6	Sugar beet	56.1	48.4	44.8	27.3	91.1	95.3
7	Sunflower	75.0	75.9	67.0	61.7	87.5	83.9
8	Soybeans	74.0	50.8	56.8	51.3	77.0	68.0
9	Other cereals	74.1	53.6	64.8	34.9	82.4	69.5
10	Forest	93.8	89.8	89.4	94.8	82.7	97.3
11	Grassland	60.1	77.2	52.7	61.6	72.8	93.1
12	Bare land	90.1	83.9	84.6	80.2	98.5	85.2
13	Water	98.5	99.6	99.3	99.8	98.1	99.8

Classification metrics are presented per pixel basis.

with different parameters trained on the same training data. Outputs from different MLPs were integrated using the technique of average committee. Under this technique the average class probability over classifiers is calculated, and the class with the highest average posterior probability for the given input sample is selected.

Ensemble based neural network model for crop classification was recently validated in the JECAM experiment within study areas in five different countries and agriculture conditions, and provided the best result compare to SVM, maximum likelihood, decision tree and logit regression (Waldner et al., 2016).

Therefore, this approach is used as a benchmark for assessing classification techniques available in GEE.

RESULTS

Input (Product) Selection

The first set of experiments was carried out to select the best input (TOA 8-day composites or restored values) and evaluating different classifiers available in GEE. **Table 4** shows the derived OA on polygons from a testing set using TOA 8-day composites as inputs. The best performance was achieved for CART at 75%. Somewhat surprisingly, an ensemble of DTs, i.e., RF, was outperformed by CART and yielded only 68%. Logistic regression (GMO Max Entropy) gave 72% accuracy. Linear classifiers, MultiClassPerceptron and Winnow, provided up to 60% accuracy, while variants of SVM achieve moderate accuracy of 57%. Unfortunately, it was unable to produce stable classification results for SVM classifiers which usually resulted into the Internal Server Error on invocation from Python.

Classifier Selection

One of the best GEE classifiers (CART and RF) on atmospherically corrected Landsat-8 imagery were compared to the committee of NN that was implemented outside GEE, in the Matlab environment using a Netlab toolbox (http://www.aston.ac.uk/eas/research/groups/ncrg/resources/netlab). **Table 5** summarizes classification metrics, in particular OA, PA, and UA for these three classification schemes. Committee of MLPs considerably outperformed DT-based classifiers: by +14.8% RF and by +7.8% DT.

DISCUSSION

Input Selection

The GEE platform offers powerful capabilities in handling large volumes of remote sensing imagery that can be used, for example, for classification purposes such as crop mapping for large territories. In order to deal with irregular observation and missing values due to clouds and shadows, a compositing approach was applied. Several compositing products were available. Also, using the rich JavaScript and/or Python APIs, it is possible to filling missing values using MODIS images using, for example, approaches developed by Roy et al. (2008), Hilker et al. (2009), and Gao et al. (2006). But these are not trivial procedures, are not available at hand and require a lot of efforts from the end-user to be implemented. It substantially reduced the quality of classification. In particular, the best OA achieved on composites from the GEE was 75%, while on atmospherically corrected and restored images the achieved accuracy was almost 77%. Therefore, while GEE offers several built-in composites, substantial efforts (including programming) are required from the end user to generate the required input data sets and to remove clouds/shadow effects.

Classification Algorithms

The GEE platforms provides a set of classification algorithms. The best results in the GEE were obtained for the DT-based

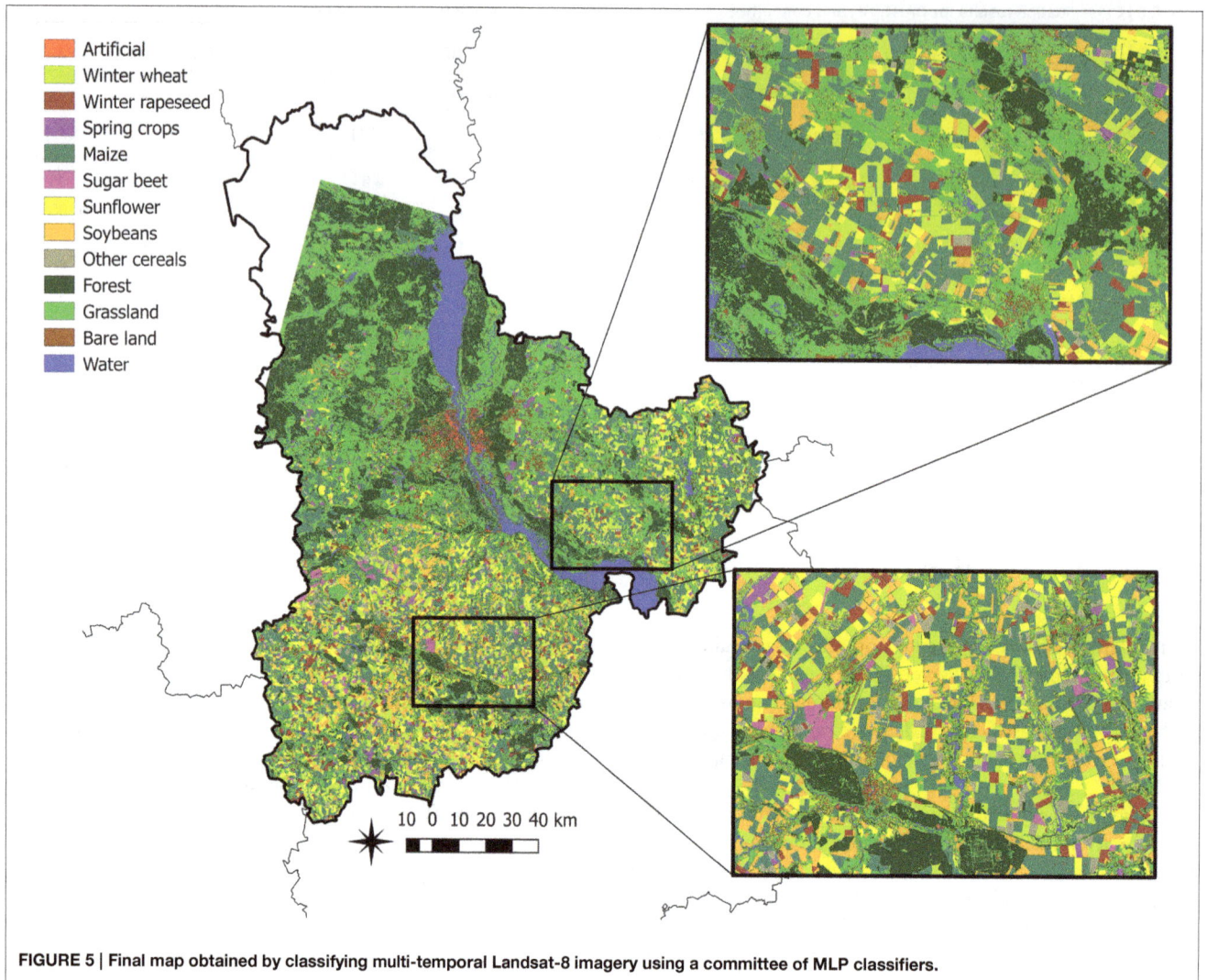

FIGURE 5 | Final map obtained by classifying multi-temporal Landsat-8 imagery using a committee of MLP classifiers.

classifiers, namely CART and RF. The best accuracy achieved on atmospherically corrected and restored Landsat-8 images was 76.9 and 69.9% for CART and RF, respectively. The classifier GMO Max Entropy shows better result than RF but there is no Python implementation (only Javascript). It is known that CART and RF tend to overfit and require fine-tuning. These accuracies were significantly lower than for the ensemble of MLPs that obtained 84.7% overall accuracy (**Figure 5**). Also, most SVM-based algorithms were not performing correctly within GEE, and therefore, performance of SVM was not possible to evaluate adequately.

Crop-Specific Classification Accuracies

Accuracy of 85% is usually considered as target accuracy for agriculture applications (McNairn et al., 2009). Among classifiers, considered in this study, the target accuracy was almost achieved only by the MLPs ensemble. As to the specific crops, this accuracy was achieved for the following crops:

- *winter wheat* (class 2, PA = 91.3%, UA = 95.0%): the major confusion was with other cereals (class 9).

- *winter rapeseed* (class 3, PA = 97.7%, UA = 92.9%): the major confusion was with other cereals (class 9).

- *maize* (class 5, PA = 85.9%, UA = 89.7%): the main confusion was with the soybeans (class 8) and partly with sunflower (class 7).

- *sugar beet* (class 6, PA = 91.1%, UA = 95.3%): the main confusion was with other cereals (class 9).

The crops that did not pass the 85% threshold for PA and UA:

- *spring crops* (class 4, PA = 39.2%, UA = 45.2%): classification using available set of satellite imagery failed to produce reasonable performance for spring crops. The main confusion of this class is with other cereals (class 9). Confusion with other cereals can be explained by almost identical vegetation cycle of spring barley and other cereals produced in the region, namely with rye and oats.

- *sunflower* (class 7, PA = 87.5%, UA = 83.9%): main confusion was with soybeans, maize, and other cereals.

- *soybeans* (class 8, PA = 77.0%, UA = 68.0%): this is the least discriminated summer crop with main confusion with maize.

DISCUSSION AND CONCLUSIONS

The activities within this paper were targeted on the comparison of pixel-based approaches to crop mapping in Ukraine, and exploring efficiency of the GEE cloud platform for large scale crop mapping with target to apply the platform for large areas involving multiple sensors and large volumes of data (at the order terabytes). Crop classification was performed by using multi-temporal Landsat-8 imagery over the JECAM tests site in Ukraine. Several inputs (products and composites) were evaluated by means of different classifiers available at GEE and our own approach. The classifiers included state-of-the-art techniques such as SVM, decision tree, random forest, and neural networks. Unlike other applications such as general land cover mapping or forest mapping, crop discrimination requires acquisition of multi-temporal profiles of crop growth dynamics. Therefore, images at multiple dates (or composites at several time intervals) need to be acquired or generated to discriminate particular crops. At the same, it causes the problem of missing data due to clouds and shadows when dealing with large territory, for example Kyiv oblast in the study. Therefore, it is very difficult to find an optimal solution in terms of temporal resolution and cloud-free composites, and missing data still occur. One way to handle this problem was to restore the missing values using self-organizing Kohonen maps. Another approach would be to use ancillary data such as MODIS to predict and fill in missing values in the Landsat time-series. But these are not trivial procedures, are not available at hand in GEE and require a lot of efforts from the end-user to be implemented. Therefore, 8-day TOA reflectance composites were used in GEE.

Better performance in terms of overall accuracy was shown for atmospherically corrected and restored Landsat images over the 8-day TOA reflectance composites from GEE. One of the explanations would that quality of restored data was better than TOA composites from GEE.

As to classification algorithms, ensemble of neural networks outperformed SVM, decision tree, and random forest classifiers. At some extent, it contradicts with recent studies where SMV and RF show better performance than neural networks. In our opinion, these are due: (i) in many cases, out-of-date neural networks techniques are used and not full potential of neural networks is explored in the remote sensing community; (ii) for SVM, DT and RF built in GEE parameters were used.

For the following agriculture classes an 85% threshold of producer's and user's accuracies was achieved: winter wheat, winter rapeseed, maize, and sugar beet. Such crops as sunflower, soybeans, and spring crops showed worse performance (below 85%).

In general, GEE provided very good performance in enabling access to remote sensing products through the cloud platform, and allowing seamless execution of complex workflow for satellite data processing. The user at once gets access to many satellite scenes or composites that are ready for processing. Although our particular study did not directly deal with large volumes of data, we still showed effectiveness of the GEE cloud platform to build large scale applications when accessing and processing satellite data with no reference to the volume. Indeed, the approaches addressed in the study can be applied at country level thanks to the capabilities offered in GEE. Another point addressed in the study was validity—in other words, what kind of accuracy can be expected for crop mapping with GEE when utilizing existing products and processing chains.

However, in our opinion, several improvements should be made to enable large scale crop mapping through GEE, especially within operational context:

- Provide at hand tools to deal with missing data due to clouds and shadows.
- To add neural networks classifiers (e.g., Tensorflow deep learning library) and allow optimization of existing classifiers especially SVM.

Future works should include:

- Large scale crop mapping (potentially at national scale) using optical and SAR imagery taking into account the coverage of Landsat-8, Proba-V, Sentinel-1, and Sentinel-2 imagery for the whole country (such as Ukraine).
- In the future we will implement parcel-based classification approach using GEE connected pixel method and some others, to improve a pixel-based crop classification map.

AUTHOR CONTRIBUTIONS

AS was the project manager. ML provided crop classification maps using ensemble of neural networks and conducted experiments in Google Earth Engine. NK and AN were scientific coordinators of the research and provided results interpretation. SS has prepared training and test sets and preprocessed Landsat-8 data for crop mapping.

ACKNOWLEDGMENTS

This research was conducted in the framework of the "Large scale crop mapping in Ukraine using SAR and optical data fusion" Google Earth Engine Research Award funded by the Google Inc.

REFERENCES

Bishop, C. M. (2006). *Pattern Recognition and Machine Learning*. New York, NY: Springer.

Boryan, C. G., and Yang, Z. (2013, July). "Deriving crop specific covariate data sets from multi-year NASS geospatial cropland data layers," in *2013 IEEE International Geoscience and Remote Sensing Symposium-IGARSS* (Melbourne, VIC), 4225–4228.

Breiman, L. (2001). Random forests. *Mach. Learn.* 45, 5–32. doi: 10.1023/A:1010933404324

Congalton, R. G. (1991). A review of assessing the accuracy of classifications of remotely sensed data. *Remote Sens. Environ.* 37, 35–46. doi: 10.1016/0034-4257(91)90048-B

Congalton, R. G., and Green, K. (2008). *Assessing the Accuracy of Remotely Sensed Data: Principles and Practices*. Boca Raton, FL: CRC Press.

Fritz, S., See, L., You, L., Justice, C., Becker-Reshef, I., Bydekerke, L., and Gilliams, S. (2013). The need for improved maps of global cropland. *Eos Trans. Am. Geophys. Union* 94, 31–32. doi: 10.1002/2013EO030006

Gallego, F. J., Kussul, N., Skakun, S., Kravchenko, O., Shelestov, A., and Kussul, O. (2014). Efficiency assessment of using satellite data for crop area estimation in Ukraine. *Int. J. Appl. Earth Observ. Geoinform.* 29, 22–30. doi: 10.1016/j.jag.2013.12.013

Gallego, J., Carfagna, E., and Baruth, B. (2010). "Accuracy, objectivity and efficiency of remote sensing for agricultural statistics," in *Agricultural Survey Methods*, eds R. Benedetti, M. Bee, G. Espa, and F. Piersimoni (Chichester, UK: John Wiley & Sons, Ltd.). doi: 10.1002/9780470665480.ch12

Gallego, J., Kravchenko, A. N., Kussul, N. N., Skakun, S. V., Shelestov, A. Y., and Grypych, Y. A. (2012). Efficiency assessment of different approaches to crop classification based on satellite and ground observations. *J. Autom. Inf. Sci.* 44, 67–80. doi: 10.1615/JAutomatInfScien.v44.i5.70

Gao, F., Masek, J., Schwaller, M., and Hall, F. (2006). On the blending of the Landsat and MODIS surface reflectance: predicting daily Landsat surface reflectance. *IEEE Trans. Geosci. Remote Sens.* 44, 2207–2218. doi: 10.1109/TGRS.2006.872081

Haykin, S. (2008). *Neural Networks and Learning Machines, 3rd Edn.* Upper Saddle River: NJ: Prentice Hall.

Hilker, T., Wulder, M. A., Coops, N. C., Seitz, N., White, J. C., Gao, F., et al. (2009). Generation of dense time series synthetic Landsat data through data blending with MODIS using a spatial and temporal adaptive reflectance fusion model. *Remote Sens. Environ.* 113, 1988–1999. doi: 10.1016/j.rse.2009.05.011

Kogan, F., Kussul, N., Adamenko, T., Skakun, S., Kravchenko, O., Kryvobok, O., et al. (2013a). Winter wheat yield forecasting in Ukraine based on Earth observation, meteorological data and biophysical models. *Int. J. Appl. Earth Observ. Geoinform.* 23, 192–203. doi: 10.1016/j.jag.2013.01.002

Kogan, F., Kussul, N. N., Adamenko, T. I., Skakun, S. V., Kravchenko, A. N., Krivobok, A. A., et al. (2013b). Winter wheat yield forecasting: A comparative analysis of results of regression and biophysical models. *J. Autom. Inf. Sci.* 45, 68–81. doi: 10.1615/JAutomatInfScien.v45.i6.70

Kolotii, A., Kussul, N., Shelestov, A., Skakun, S., Yailymov, B., Basarab, R., and Ostapenko, V. (2015). Comparison of biophysical and satellite predictors for wheat yield forecasting in Ukraine. *Int. Arch. Photogramm. Remote Sens. Spat. Inf. Sci.* 40, 39. doi: 10.5194/isprsarchives-XL-7-W3-39-2015

Kussul, N., Lemoine, G., Gallego, F. J., Skakun, S. V., Lavreniuk, M., and Shelestov, A. Y. (2016). Parcel-based crop classification in ukraine using landsat-8 data and sentinel-1A data. *IEEE J. Select. Top. Appl. Earth Observ. Rem. Sens.* 9, 2500–2508. doi: 10.1109/JSTARS.2016.2560141

Kussul, N. N., Sokolov, B. V., Zyelyk, Y. I., Zelentsov, V. A., Skakun, S. V., and Shelestov, A. Y. (2010). Disaster risk assessment based on heterogeneous geospatial information. *J. Autom. Inf. Sci.* 42, 32–45. doi: 10.1615/JAutomatInfScien.v42.i12.40

Kussul, N., Shelestov, A., Basarab, R., Skakun, S., Kussul, O., and Lavrenyuk, M. (2015). "Geospatial Intelligence and Data Fusion Techniques for Sustainable Development Problems," in *ICTERI* (Lviv), 196–203.

Kussul, N., Shelestov, A., and Skakun, S. (2011). "Grid technologies for satellite data processing and management within international disaster monitoring projects," in *Grid and Cloud Database Management*, eds S. Fiore and G. Aloisio (Berlin; Heidelberg: Springer), 279–305.

Lavreniuk, M. S., Skakun, S. V., Shelestov, A. J., Yalimov, B. Y., Yanchevskii, S. L., Yaschuk, D. J., et al. (2016). Large-scale classification of land cover using retrospective satellite data. *Cybern. Syst. Anal.* 52, 127–138. doi: 10.1007/s10559-016-9807-4

Littlestone, N. (1988). Learning quickly when irrelevant attributes abound: a new linear-threshold algorithm. *Mach. Learn.* 2, 285–318. doi: 10.1007/BF00116827

Ma, Y., Wang, L., Liu, P., and Ranjan, R. (2015a). Towards building a data-intensive index for big data computing–a case study of Remote Sensing data processing. *Inf. Sci.* 319, 171–188. doi: 10.1016/j.ins.2014.10.006

Ma, Y., Wu, H., Wang, L., Huang, B., Ranjan, R., Zomaya, A., et al. (2015b). Remote sensing big data computing: challenges and opportunities. *Future Generation Comput. Syst.* 51, 47–60. doi: 10.1016/j.future.2014.10.029

Maji, S., Berg, A. C., and Malik, J. (2008). "Classification using intersection kernel support vector machines is efficient," in *IEEE Conference on Computer Vision and Pattern Recognition* (Anchorage, AK), 1–8.

McNairn, H., Champagne, C., Shang, J., Holmstrom, D. A., and Reichert, G. (2009). Integration of optical and Synthetic Aperture Radar (SAR) imagery for delivering operational annual crop inventories. *ISPRS J. Photogramm. Remote Sens.* 64, 434–449. doi: 10.1016/j.isprsjprs.2008.07.006

Rahman, H., and Dedieu, G. (1994). SMAC: a simplified method for the atmospheric correction of satellite measurements in the solar spectrum. *Remote Sens.* 15, 123–143. doi: 10.1080/01431169408954055

Roy, D. P., Ju, J., Lewis, P., Schaaf, C., Gao, F., Hansen, M., et al. (2008). Multi-temporal MODIS–Landsat data fusion for relative radiometric normalization, gap filling, and prediction of Landsat data. *Remote Sens. Environ.* 112, 3112–3130. doi: 10.1016/j.rse.2008.03.009

Roy, D. P., Wulder, M. A., Loveland, T. R., Woodcock, C. E., Allen, R. G., Anderson, M. C., et al. (2014). Landsat-8: science and product vision for terrestrial global change research. *Remote Sens. Environ.* 145, 154–172. doi: 10.1016/j.rse.2014.02.001

Shalev-Shwartz, S., Singer, Y., Srebro, N., and Cotter, A. (2011). Pegasos: primal estimated sub-gradient solver for svm. *Math. Programming* 127, 3–30. doi: 10.1007/s10107-010-0420-4

Skakun, S., Kussul, N., Shelestov, A., and Kussul, O. (2016b). The use of satellite data for agriculture drought risk quantification in Ukraine. *Geomatics Nat. Hazards Risk*, 7, 901–917. doi: 10.1080/19475705.2015.1016555

Skakun, S., Kussul, N., Shelestov, A. Y., Lavreniuk, M., and Kussul, O. (2016a). Efficiency assessment of multitemporal C-Band Radarsat-2 intensity and Landsat-8 surface reflectance satellite imagery for crop classification in Ukraine. *IEEE J. Select. Top. Appl. Earth Observ. Remote Sens.* 9, 3712–3719. doi: 10.1109/JSTARS.2015.2454297

Skakun, S. V., and Basarab, R. M. (2014). Reconstruction of missing data in time-series of optical satellite images using self-organizing Kohonen maps. *J. Autom. Inf. Sci.* 46, 19–26. doi: 10.1615/JAutomatInfScien.v46.i12.30

Skakun, S. V., Nasuro, E. V., Lavrenyuk, A. N., and Kussul, O. M. (2007). Analysis of applicability of neural networks for classification of satellite data. *J. Autom. Inf. Sci.* 39, 37–50. doi: 10.1615/JAutomatInfScien.v39.i3.40

Waldner, F., De Abelleyra, D., Verón, S. R., Zhang, M., Wu, B., Plotnikov, D., et al. (2016). Towards a set of agrosystem-specific cropland mapping methods to address the global cropland diversity. *Int. J. Remote Sens.* 37, 3196–3231. doi: 10.1080/01431161.2016.1194545

Yan, L., and Roy, D. P. (2014). Automated crop field extraction from multi-temporal Web Enabled Landsat Data. *Remote Sens. Environ.* 144, 42–64. doi: 10.1016/j.rse.2014.01.006

Zhu, Z., and Woodcock, C. E. (2012). Object-based cloud and cloud shadow detection in Landsat imagery. *Remote Sens. Environ.* 118, 83–94. doi: 10.1016/j.rse.2011.10.028

Conflict of Interest Statement: The authors declare that the research was conducted in the absence of any commercial or financial relationships that could be construed as a potential conflict of interest.

6

The Influence of Reactive Oxygen Species on Local Redox Conditions in Oxygenated Natural Waters

Andrew L. Rose *

School of Environment, Science and Engineering, Southern Cross University, Lismore, NSW, Australia

Redox conditions in natural waters are a fundamental control on biogeochemical processes and ultimately many ecosystem functions. While the dioxygen/water redox couple controls redox thermodynamics in oxygenated aquatic environments on geological timescales, it is kinetically inert in the extracellular environment on the much shorter timescales on which many biogeochemical processes occur. Instead, electron transfer processes on these timescales are primarily mediated by a relatively small group of trace metals and stable radicals, including the reactive oxygen species superoxide. Such processes are of critical biogeochemical importance because many of these chemical species are scarce nutrients, but may also be toxic at high concentrations. Furthermore, their bioavailability and potentially toxicity is typically strongly influenced by their redox state. In this paper, I examine to what extent redox conditions in oxygenated natural waters are expected to be reflected in the redox states of labile redox-active compounds that readily exchange electrons with the dioxygen/superoxide redox couple, and potentially with each other. Additionally, I present the hypothesis that the relative importance of the dioxygen/superoxide and superoxide/hydrogen peroxide redox couples exerts a governing control on local redox conditions in oxygenated natural waters on biogeochemically important timescales. Given the recent discovery of widespread extracellular superoxide production by a diverse range of organisms, this suggests the existence of a fundamental mechanism for organisms to tightly regulate local redox conditions in their extracellular environment in oxygenated natural waters.

Keywords: superoxide, oxygen, reactive oxygen species, labile redox-active compounds, redox conditions, natural waters

Edited by:
Leanne C. Powers,
Skidaway Institute of Oceanography,
USA

Reviewed by:
Julian Blasco,
Spanish National Research Council,
Spain
Paul R. Erickson,
ETH Zurich, Switzerland

*Correspondence:
Andrew L. Rose
andrew.rose@scu.edu.au

Specialty section:
This article was submitted to
Marine Biogeochemistry,
a section of the journal
Frontiers in Earth Science

INTRODUCTION

Electron transfer and transport in natural waters is a fundamental control on chemical conditions, biological processes, and ecosystem structure. Oxidation and reduction (redox) reactions involve the transfer of electrons, and associated chemical energy, from one chemical species to another. In most natural surface waters, chemical energy is originally captured from solar energy by photosynthesis, and then redistributed by chemical and physical transport processes. This chemical energy is the substrate for life on Earth. Chemical redox reactions combined with physical transport of redox active species therefore dictate, at a fundamental level, the functioning of biogeochemical systems on Earth from cellular to ecosystem scales.

In most natural surface waters, oxygen controls redox conditions on geological timescales. The Great Oxidation Event approximately 2.4 billion years ago saw a massive shift in redox conditions on Earth resulting from photosynthetic release of O_2 into the atmosphere, following the evolution of cyanobacteria (Sverjensky and Lee, 2010). However, the O_2/H_2O redox couple is extremely kinetically inert because the dioxygen molecule possesses an unusual electronic structure, with its lowest energetic state (ground state) possessing two unpaired electrons with parallel spins, i.e., its ground state is a triplet state biradical (Sawyer, 1991). Quantum mechanical restrictions (the Pauli exclusion principle) dictate that, under typical environmental conditions, triplet state dioxygen can react only extremely slowly with molecules possessing singlet state electronic configurations (e.g., most organic compounds), but much more readily with molecules possessing unpaired electrons (e.g., free radicals and transition metal ions; Fridovich, 1998).

Usually, there are relatively few free radical species present in oxygenated natural waters at sufficiently high concentration to facilitate the thermal, abiotic reduction of O_2 at substantial rates. In natural waters, free radicals typically include only a select few trace metals, certain organic radicals, and a few other trace species such as nitric oxide (Goldstein and Czapski, 1995). Enzymatic reactions (respiration) inside living cells catalyze reduction of oxygen to water, and in aquatic environments supporting sufficient biomass, this can be the major pathway for oxygen consumption, even resulting in complete hypoxia when respiration rates are high. Conversely, photosynthetic organisms can enzymatically catalyse the sunlight-driven oxidation of water by CO_2 (or other compounds). In the absence of biological catalysis, sunlight can also mediate abiotic photochemical oxygen consumption (Amon and Benner, 1996; Andrews et al., 2000). Overall, however, rates of oxygen consumption and production are balanced over long timescales, and O_2 and H_2O coexist at high concentrations in the majority of natural surface waters, defying thermodynamic expectations.

The electronic structure of O_2 also dictates that the transfer of four electrons required to reduce O_2 to H_2O or oxidize H_2O to O_2 must occur via single electron transfer steps (Sawyer, 1991). The sequential reduction of O_2 to H_2O under near neutral pH conditions results first in the formation of superoxide (O_2^-) then hydrogen peroxide (H_2O_2), and finally hydroxyl ($HO^\bullet$), which collectively belong to a group known as reactive oxygen species (ROS). Despite the overall thermodynamic impetus, reduction of O_2 to O_2^- does not guarantee the complete reduction of O_2 to H_2O as the electronic structure of O_2^- again favors reaction only with molecules possessing unpaired electrons (Sawyer, 1991). In addition, under conditions typical of natural waters O_2^- is relatively reducing; i.e., O_2^- is relatively easily oxidized back to O_2, while its reduction to H_2O_2 is unfavorable due to the thermodynamic instability of the immediate product of the reaction (Sawyer, 1991). In contrast, H_2O_2 and $HO^\bullet$ are powerful oxidants such that once formed, they will ultimately be almost exclusively further reduced to H_2O. Thus, reduction of O_2^- to H_2O_2 represents a major barrier to the overall reduction of O_2 to H_2O, and whether O_2^- is being primarily oxidized to O_2 or primarily reduced to H_2O_2 in a particular environment has the potential to exert a major influence on local redox chemistry (Rose et al., 2010).

In this paper, I explore the influence of ROS on local redox conditions in oxygenated natural waters on environmentally relevant timescales and discuss their significance to understanding the redox biogeochemistry of natural waters on a fundamental level, with a particular emphasis on the role of O_2^-. Furthermore, while acknowledging the importance of biological processes, the focus of the paper is on non-enzymatic reactions in the extracellular environment, which is the dominant pathway for redox cycling of a substantial number of redox active elements (Borch et al., 2009). On the basis of the theory presented, I also present the hypothesis that that the relative importance of the dioxygen/superoxide and superoxide/hydrogen peroxide redox couples exerts a governing control on local redox conditions in oxygenated natural waters on biogeochemically important timescales. While the approach taken can be applied to any natural waters, the scope of this paper is restricted primarily to carbonate-buffered, oxygen saturated waters with a nominal pH of 8.1 and dissolved O_2 concentration of 250 μM.

REDOX THERMODYNAMICS OF OXYGENATED NATURAL WATERS

The redox thermodynamics of oxygenated natural waters are dominated by the chemistry of oxygen, which is present in both its most common oxidized state O_2 and its most common reduced state H_2O; the latter, of course, also provides the solvent in which aqueous reactions occur. For a chemical species to be thermodynamically able to exist in aqueous solution in a particular redox state, the redox potential of the half reaction that defines its oxidation or reduction to another redox state must lie between the redox potentials for oxidation of H_2O to O_2, or reduction of H^+ to H_2 (Fraústo Da Silva and Williams, 2001). Under standard conditions (unit activity of all reactants and products at 298 K), the standard redox potentials (E^0) for oxidation of H_2O to O_2 and reduction of H^+ to H_2 are +1270 and 0 mV, respectively, when defined with respect to the standard hydrogen electrode. However, the redox potential E varies as defined by the Nernst equation:

$$E = E^0 - \frac{59.16}{n} \log \frac{\prod \{Ox\}}{\prod \{Red\}} \qquad (1)$$

where n is the number of electrons transferred in the half-reaction, $\{Ox\}$ refers to the activity of the oxidized species, $\{Red\}$ refers to the activity of the reduced species, E and E^0 are in mV, and the coefficient of 59.16 mV applies at 25°C.

Thus at pH 8.1, which is typical for carbonate buffered natural surface waters and where $\{H^+\} = 10^{-8.1}$ M, the redox potentials for oxidation of H_2O to O_2 and reduction of H^+ to H_2 are instead +790 and −480 mV, respectively, due to the presence of protons in the respective reduction half reactions. While these values represent the ultimate thermodynamic constraints on stability in aqueous solution, in practice the former is conditional on a mechanism existing to catalyze the four electron oxidation of H_2O to O_2. As previously discussed, in the absence of catalysis the oxidation of H_2O to O_2 must proceed initially via the one-electron oxidation of H_2O to $HO^\bullet$. This step is much less thermodynamically favorable than the overall oxidation, with a redox potential of $+2060$ mV under standard conditions at pH 8.1. In most oxygenated natural waters, photosynthesis is the only significant process that can facilitate multi-electron transfer steps during oxidation of H_2O to O_2. Thus, in the extracellular environment, couples with redox potentials for one-electron transfers lying between -480 and $+2060$ mV are practically stable with respect to their aqueous environment at pH 8.1.

Figure 1 illustrates the thermodynamic stability at pH 8.1 of various redox states of the major redox active elements C, N and S, along with several biogeochemically important trace metals, with regard to the oxidation or reduction of water. Stability in aqueous solution at pH 8.1 is shown with respect to both the theoretical limit imposed by H_2O/O_2 couple (hereafter referred to as absolute stability) and the practical limit imposed by the $H_2O/HO^\bullet$ couple (hereafter referred as conditional stability). It is clear that a limited number of redox states for both major elements and trace metals are permissible under these conditions.

Stability in aqueous solution with respect to the H_2/H^+, H_2O/O_2, or $H_2O/HO^\bullet$ couples does not mean that a particular species is stable against any redox transformation, however. Most notably, the presence of O_2 should drive other redox active elements toward the most oxidized thermodynamically stable species. However, for the major redox active elements C, N, and S (Figure 1A), transformations between many of the various stable redox states are not possible via one-electron steps, as the intermediate species are thermodynamically unstable in aqueous solution. Thus, highly reduced species such as CH_4, NH_4^+, and HS^- are unable to be oxidized by O_2 directly without catalysis. In contrast, transformation among most of the permissible redox states for the trace metals shown in Figure 1B is able to occur via one-electron transfers, enabling direct oxidation or reduction by O_2, H_2O, and/or ROS.

Of course, this reasoning applies only to direct reactions between these various redox couples and O_2, H_2O, or ROS: reactions between two redox species can occur independently of O_2, H_2O, or ROS, and in many such reactions multi-electron transfers are possible (and indeed favored), as detailed by Luther (2010). Furthermore, one-electron transfers between C, N or S species and O_2, H_2O, or ROS can be facilitated via trace metals, as has been shown to facilitate sulfide oxidation by O_2 in natural waters (Luther et al., 2011). Nonetheless, the role of such reactions in oxygenated waters is limited by the fact that where such alternate pathways to oxidation exist, accumulation of reduced compounds to relatively high concentrations is unlikely to occur.

While this approach is useful to constrain the redox thermodynamics of oxygenated natural waters on a broad level, a more realistic analysis must also account for the complex speciation of many elements. In particular, trace metals potentially exist as a range of different solid phases or in complexes with a vast array inorganic and organic ligands. The redox potentials of solid phases or complex species can be very different to those for the simple aquated species; for example, the redox potential for reduction of Fe(III) to Fe(II) at pH 7 ranges

from -750 mV when complexed with the strong Fe(III) binding ligand enterobactin to 1150 mV when complexed with the strong Fe(II) binding ligand phenanthroline (Pierre et al., 2002). The difference in thermodynamic behavior of the "average" speciation of dissolved inorganic Fe(III) in seawater, denoted as Fe(III)', and amorphous $Fe(OH)_3$ solid is illustrated in Figure 1.

A second limitation with the analysis presented in Figure 1 is that standard conditions are defined by unit activities of all species (except, in this case, H^+). As seen in the Nernst equation (Equation 1), the redox potential varies with the ratio of the oxidized and reduced species in the couple such that the slope of the lines in Figure 1 will change by 59 meV for each order of magnitude difference between the activities of the oxidized and reduced species in the couple. Thus, while a particular process may be thermodynamically favorable (or unfavorable) under standard conditions, this may not necessarily be the case under the conditions actually encountered in a particular natural water body where the concentrations of the oxidized and reduced species may be orders of magnitude apart. The effect of using typical concentrations of O_2 and ROS found in many natural surface waters rather than standard conditions for thermodynamic analysis is illustrated in Figure 2. It is evident that accounting for actual concentrations of these species, rather than using unit activities, brings the redox potentials of all one-electron steps in the reduction of O_2 to H_2O much closer to the overall redox potential for the four-electron process. As a result, some species that were seen to be conditionally unstable (with respect to the H_2O/O_2 couple) but practically stable (with respect to $H_2O/HO^\bullet$) under standard conditions (Figure 1) are practically unstable under the conditions shown in Figure 2, for example N_2O and Co^{3+}. However, the latter case applies to the inorganic redox couple, whereas in marine waters Co is known to exist predominantly in organically complexed form (Vraspir and Butler, 2009), again illustrating the importance of accounting for complex speciation.

TIMESCALES OF REDOX PROCESSES IN OXYGENATED NATURAL WATERS

Thermodynamic analysis is based on the requirement that sufficient time has passed for equilibrium to have been reached. While redox equilibrium is likely to be reached on geological timescales, it is unlikely to occur on timescales on the order of seconds to days, because environmental conditions in natural waters typically vary significantly over these timescales due to processes such as diel variations in solar radiation, hydrologic fluctuations (e.g., inputs of rainwater or variations in streamflow), and biological activity. Furthermore, chemistry imposes substantial constraints on the kinetics of some redox reactions, which allows thermodynamically incompatible redox species to coexist at potentially high concentrations, as exemplified by the coexistence of O_2 and H_2O in oxygenated natural waters.

Redox active compounds in natural waters vary considerably in terms of their reactivity, and hence longevity. At one end of this scale are relatively kinetically inert compounds that

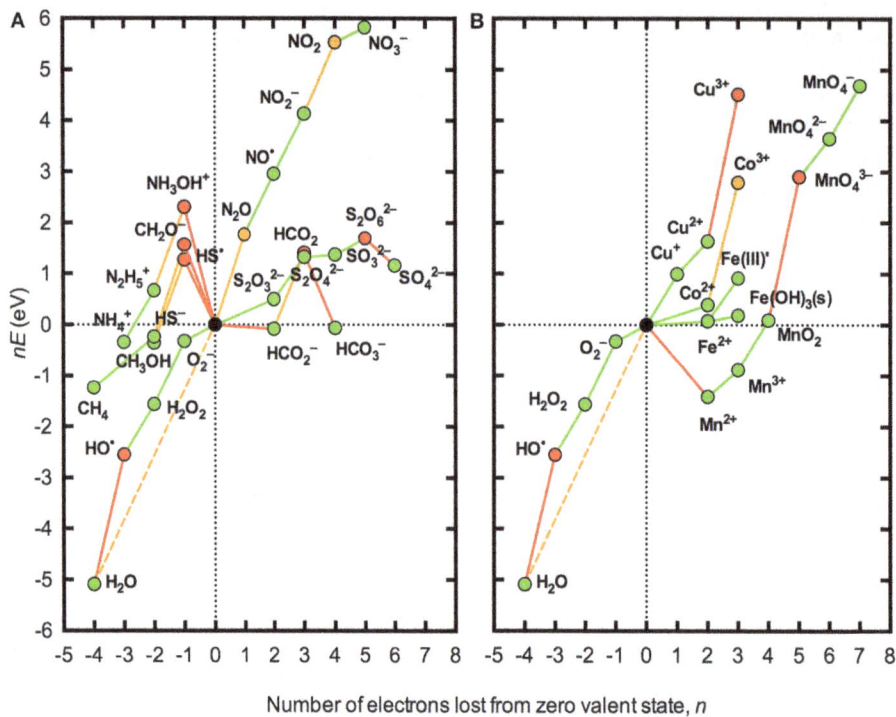

FIGURE 1 | Frost diagrams for some common redox active elements in natural waters under standard conditions at pH 8.1. (A) Representative species for the major elements C, N, and S. **(B)** Trace metals Co, Cu, Fe, and Mn. All redox potentials (E) are defined with reference to the standard hydrogen electrode at pH 8.1 ($Pt,H_2|a_{H^+} = 10^{-8.1}$). The compounds shown represent the dominant acid-base species at pH 8.1, allowing for hydrolysis of trace metals. The zero valent states for the elements shown are a representative carbohydrate "CH_2O" equivalent to 1/6 of a glucose molecule (Morel and Hering, 1993), N_2 (g), S_8 (s), Co^0 (s), Cu^0 (s), Fe^0 (s), Mn^0 (s), and O_2 (aq). Where the necessary thermodynamic data were available, one-electron transfer steps are shown. Red lines indicate redox couples that are unstable with respect to one-electron transfer processes involving the $H_2O/HO^\bullet$ couple (lines with positive slope) or the H_2/H^+ couple (lines with negative slope); orange lines indicate redox couples that are unstable with respect to multi-electron transfer processes involving the H_2O/O_2 couple; and green lines indicate redox couples that are thermodynamically permissible in aqueous solution at pH 8.1. The dashed orange line represents the redox potential associated with the four electron transfer for complete reduction of O_2 to H_2O. Red symbols represent species that are thermodynamically unable to exist due to instability with respect to one-electron transfer processes involving the $H_2O/HO^\bullet$ couple (lines with positive slope) or the H_2/H^+ couple; orange symbols represent species that are thermodynamically able to exist in the absence of processes to catalyze multi-electron transfer processes involving the H_2O/O_2 couple; and green symbols represent species that are thermodynamically able to coexist with both the H_2/H^+ and H_2O/O_2 couples. Thermodynamic relationships used to construct the diagrams are listed in the Appendix in Supplementary Material. The format of the Frost diagrams was adapted from Figures 1.7 and 1.8 in Fraústo Da Silva and Williams (2001).

typically undergo redox reactions on timescales of greater than a day, and can therefore potentially persist at relatively high concentrations in oxygenated natural waters even in thermodynamically unfavorable redox states. Many of the commonly occurring compounds of C, N, and S in aquatic environments behave in this way. The ability of such compounds to persist in thermodynamically unstable redox states at relatively high concentrations provides a critical store of chemical free energy in the environment that can be harnessed by organisms for biological use if the kinetic inertness of such compounds can be overcome. "Activation" of thermodynamically unstable but kinetically inert redox active compounds typically occurs through photochemical catalysis or reaction with free radicals (e.g., via enzymes containing trace metal or organic radical centers), however such reactions are often spatially and/or temporally compartmentalized, for example during daylight hours or inside cells. In the extracellular environment, "activation" tends to be a relatively slow process due to the relatively low abundance of

potential catalysts such as free (non-organically complexed) trace metals in many natural waters. Therefore, these kinetically inert compounds usually have a limited ability to mediate electron transfer and transport processes in the extracellular environment in oxygenated natural waters on timescales of less than a day.

At the other end of the scale are compounds that undergo redox reactions on timescales of less than a second, and which thus typically exist only at sub-picomolar concentrations in most natural waters. The short lifetimes and hence low concentrations of these transient species mean that they are unable to diffuse far from the site of their production. Thus, while potentially important in specific microenvironments, their low concentrations and extreme localisation means that such highly reactive species are typically unable to influence redox conditions in natural waters on a large spatial scale. Between these two extremes exist redox active compounds that have half-lives (or turnover times) of between a second and a day. This range has particular significance because it corresponds

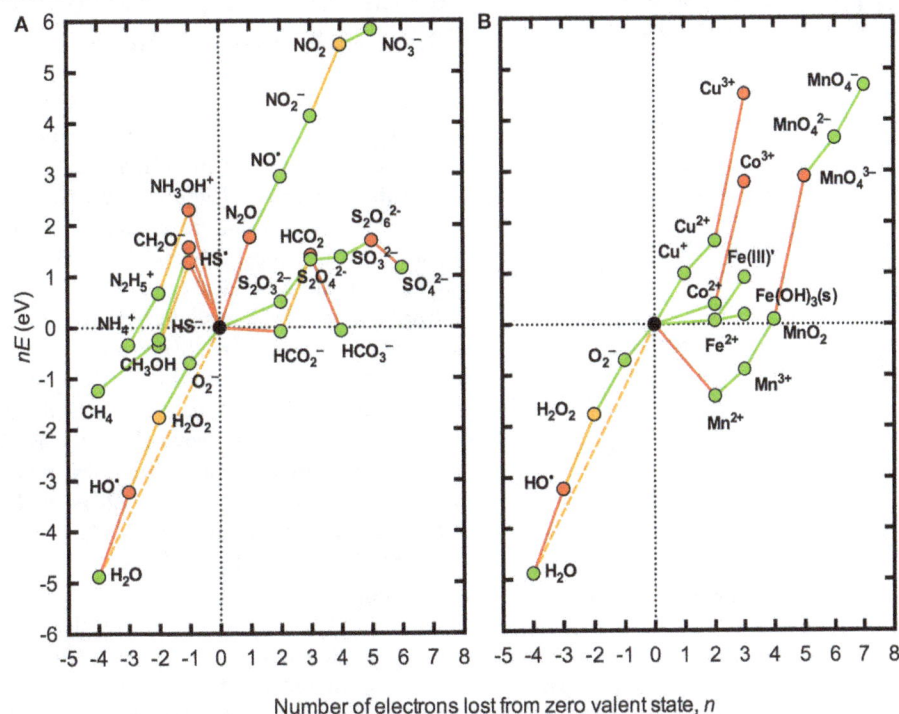

FIGURE 2 | **Frost diagrams for some common redox active elements in natural waters at pH 8.1 accounting for typical activities of oxygen redox species in natural waters. (A)** Representative species for the major elements C, N, and S. **(B)** Trace metals Co, Cu, Fe, and Mn. Typical activities were taken as $[O_2]$ = 250 μM, $[O_2^-]$ = 100 pM (Rose et al., 2008; Hansard et al., 2010), $[H_2O_2]$ = 100 nM (Yuan and Shiller, 2005; Mostofa and Sakugawa, 2009; Vermilyea et al., 2010; Rusak et al., 2011) and $[HO^\bullet]$ = 1 fM (Zhou and Mopper, 1990). These values are based primarily on measurements from seawater, since this is where most measurements of ROS concentrations in natural surface waters have been made. All other conditions and the lines and symbols used are the same as in **Figure 1**.

to the timescale of many critical biological and environmental processes. Furthermore, compounds with lifetimes in this range are typically sufficiently long-lived to persist at biologically significant concentrations. This longevity also enables such redox active compounds to facilitate physical transport of electrons on biologically significant spatial scales of at least a few millimeters.

A second critical property of certain redox active compounds is their ability to undergo reversible redox cycling. In the extracellular environment in oxygenated natural waters, this is only possible with the existence of two or more thermodynamically stable redox states (with respect to the H_2/H^+ and $H_2O/HO^\bullet$ redox couples) that can be accessed by one-electron transfers. As can be seen for the selection of elements shown in **Figure 2**, such compounds are relatively uncommon (although the selection shown in **Figure 2** is not exhaustive). Compounds which possess this ability to undergo reversible redox cycling and for which this cycling typically occurs with a turnover time between a second and a day are hereafter referred to as labile redox active compounds (LRACs). LRACs possess a unique combination of properties: sufficient reactivity to accept and donate electrons on biologically important timescales but sufficient longevity to transport these labile electrons over biologically important spatial scales, and the ability to do so repeatedly. This confers an ability for LRACs to potentially act as "electron shuttles" in oxygenated natural

waters, and thereby influence overall local redox conditions in a biologically significant way.

In the extracellular environment in oxygenated natural waters, relatively few compounds satisfy the definition of LRACs. Under conditions typical of natural waters, such compounds are primarily redox active trace metals, organic moieties and potentially a few trace non-metal compounds that can exist in two or more redox states that can be accessed via one electron transfers. **Table 1** lists a range of trace metals that may meet the definition of LRACs in natural waters on the basis of their ability to undergo reversible redox cycling. However, the ability to undergo redox cycling in this manner does not necessarily mean that these compounds do in fact function as LRACs, since the kinetics of redox cycling in many cases remains relatively poorly understood. Furthermore, the list in **Table 1** is not exhaustive. In particular, several other elements (mostly metals and metalloids) are well known to exist in natural waters in multiple redox states separated by two or more electron transfers (e.g., As and Se), however there is a lack of information on the kinetics and mechanisms of redox transformations for many of these species that is needed to determine if such transformations may be able to occur via one-electron transfers.

Other LRACs in natural waters may include certain reduced sulfur species such as thiols and metal sulfide clusters, which may

TABLE 1 | Trace metals that can potentially be interconverted between multiple redox states through one electron transfers in oxygenated natural waters.

Element	Possible stable redox states that are accessible by one-electron transfers	"Typical" concentration (nM)		References
		Marine waters	River waters	
Fe	+2, +3	0.5	720	Mason, 2013
Cu	+1, +2	4	24	Mason, 2013
Mn	+2, +3, +4	0.3	145	Mason, 2013
Co	+2, +3	0.02	3.4	Mason, 2013
Mo	+4, +5, +6	110	5.2	Wang, 2012; Mason, 2013
V	+2, +3, +4	40	variable	Huang et al., 2015

be found at measurable concentrations in oxic waters (Bowles et al., 2003; Luther and Rickard, 2005; Rickard and Luther, 2006); organic moieties such as semiquinones, which appear to be ubiquitous components of humic type natural organic compounds (Aeschbacher et al., 2010); and the ROS O_2^-. In the case of the O_2^- anion, resonance stabilization of the unpaired electron results in a so-called "three electron bond" (Neuman, 1934; Pauling, 1979) that renders the molecule to be very selective in its reactions in aqueous solution (Sawyer, 1991). Thus, the O_2/O_2^- couple undergoes reversible electron transfer on a timescale characteristic of LRACs, but only with the relatively small set of compounds that satisfy the strict requirement to possess unpaired electrons.

The ROS H_2O_2 also has a typical lifetime in oxygenated waters on the order of a few days (Cooper and Zepp, 1990; Yuan and Shiller, 2005; Vermilyea et al., 2010), which is close to the timescale characteristic of LRACs. However, one electron transfers involving H_2O_2 are not readily reversible. The reduction of H_2O_2 to $HO^{\bullet}$ is essentially irreversible owing to the strongly oxidizing nature of $HO^{\bullet}$, while oxidation of H_2O_2 to O_2^- appears very unusual in natural waters. Consequently, H_2O_2 behaves predominantly as a relatively long-lived trace oxidant rather than a LRAC, with limited ability to facilitate "electron shuttling," unlike O_2^-.

REDOX CYCLING OF LABILE REDOX ACTIVE SPECIES IN OXYGENATED NATURAL WATERS

LRACs collectively act as carriers for a pool of labile electrons that can effectively maintain locally reducing conditions relative to the overall redox potential set by the O_2/H_2O system in oxygenated natural waters. The size of this pool of electrons, and hence the intensity of the reducing conditions, will be determined by the steady-state balance of electrons entering and leaving the pool. As these electrons can potentially be exchanged on timescales of seconds to days, the redox speciation of various LRACs would be expected to reflect the prevailing redox conditions within this pool on these timescales.

Sources of Electrons to the Pool of Labile Redox Active Compounds

The input of electrons into the LRAC pool in oxygenated natural waters is primarily driven through the slow oxidation of kinetically inert reduced compounds and/or physical transport of reduced LRACs from more reducing environments such as sediments or rainwater (**Figure 3**). As kinetically inert reduced compounds are resistant to oxidation otherwise, this release of electrons into the LRAC pool is achieved mostly through biological or photochemical catalysis.

The precise mechanisms by which each of these processes occur are, in general, relatively poorly understood. There is now strong evidence for biological input of electrons into the extracellular LRAC pool in a wide range of marine and freshwater aquatic environments, as demonstrated by extensive measurements of biologically mediated extracellular O_2^- production (Diaz et al., 2013; Zhang et al., 2016). However, it remains unclear whether O_2 is reduced to O_2^- directly by cell surface oxidoreductase enzymes, or whether reductases first reduce other LRACs that subsequently react with dioxygen to yield O_2^-. It may also be possible that some reductase enzymes are able to reduce multiple LRACs, as has been shown for the ferric reductase system of the yeast *Saccharomyces cerevisiae* (Lesuisse et al., 1996; Kosman, 2003), or that some organisms release soluble reduced LRACs such as small quinone molecules directly, as has been shown for the bacterium *Escherichia coli* (Korshunov and Imlay, 2006). In the case of non-photosynthetic organisms, the electrons must presumably originate from respiratory processes, however there is evidence that in at least some photosynthetic organisms, the electrons are primarily derived from photosynthesis (Marshall et al., 2002).

Mechanisms of abiotic photochemical processes in sunlit surface waters are also not fully resolved. Natural organic matter is thought to undergo photochemical oxidation via photoexcitation and subsequent electron transfer processes, potentially involving quinone, phenol and or other similar moieties within the heterogeneous molecular structure (Zhang et al., 2012; Sharpless and Blough, 2014). Simpler organic molecules, such as siderophores containing carboxylic acid moieties, may undergo direct photodecarboxylation reactions, also potentially liberating electrons (Barbeau et al., 2003). In addition, complex formation between organic compounds and metals such as iron and copper can lead to formation of charge transfer complexes that increase the efficiency of photochemical processes, and potentially result in direct formation of reduced LRACs through photochemical ligand-to-metal charge transfer (Jones et al., 1985; Barbeau, 2006).

Cycling of Electrons among the Pool of Labile Redox Active Compounds

The defining characteristic of LRACs is their ability to undergo reversible redox cycling on relatively short timescales, while persisting at biologically significant concentrations. While the kinetics of electron transfer between many specific redox couples in natural waters have been well studied, knowledge of electron transfer between multiple LRAC species simultaneously is poor.

FIGURE 3 | Sources, sinks and cycling of electrons in the LRAC pool. The major sources of electrons are: **(A)** Photosynthetic oxidation of water inside cells, with subsequent donation of electrons to LRACs in the extracellular environment through membrane associated oxidoreductase enzymes, or excretion of reduced LRACs through the cell membrane; **(B)** Respiratory oxidation of organic carbon or other non-labile redox active compounds (NRACs) inside cells, with subsequent donation of electrons to LRACs in the extracellular environment through membrane associated oxidoreductase enzymes, or excretion of reduced LRACs through the cell membrane; **(C)** Respiratory oxidation of other reduced NRACs in reducing environments such as sediments, with subsequent physical transport of reduced LRACs into overlying oxygenated surface waters; **(D)** Abiotic, photochemical oxidation of reduced NRACs in sunlit surface waters; and **(E)** Abiotic, photochemical oxidation of water or reduced NRACs in the atmosphere, with subsequent deposition of reduced LRACs into oxygenated surface waters through dust or rain. **(F)** Once in the LRAC pool, electrons can potentially be readily exchanged between a range of different LRAC redox couples, including the O_2/O_2^- couple. **(G)** The primary sink of electrons in the pool is through oxidation of O_2^- by reduced LRACs, resulting in the essentially irreversible formation of H_2O_2 and, ultimately, water.

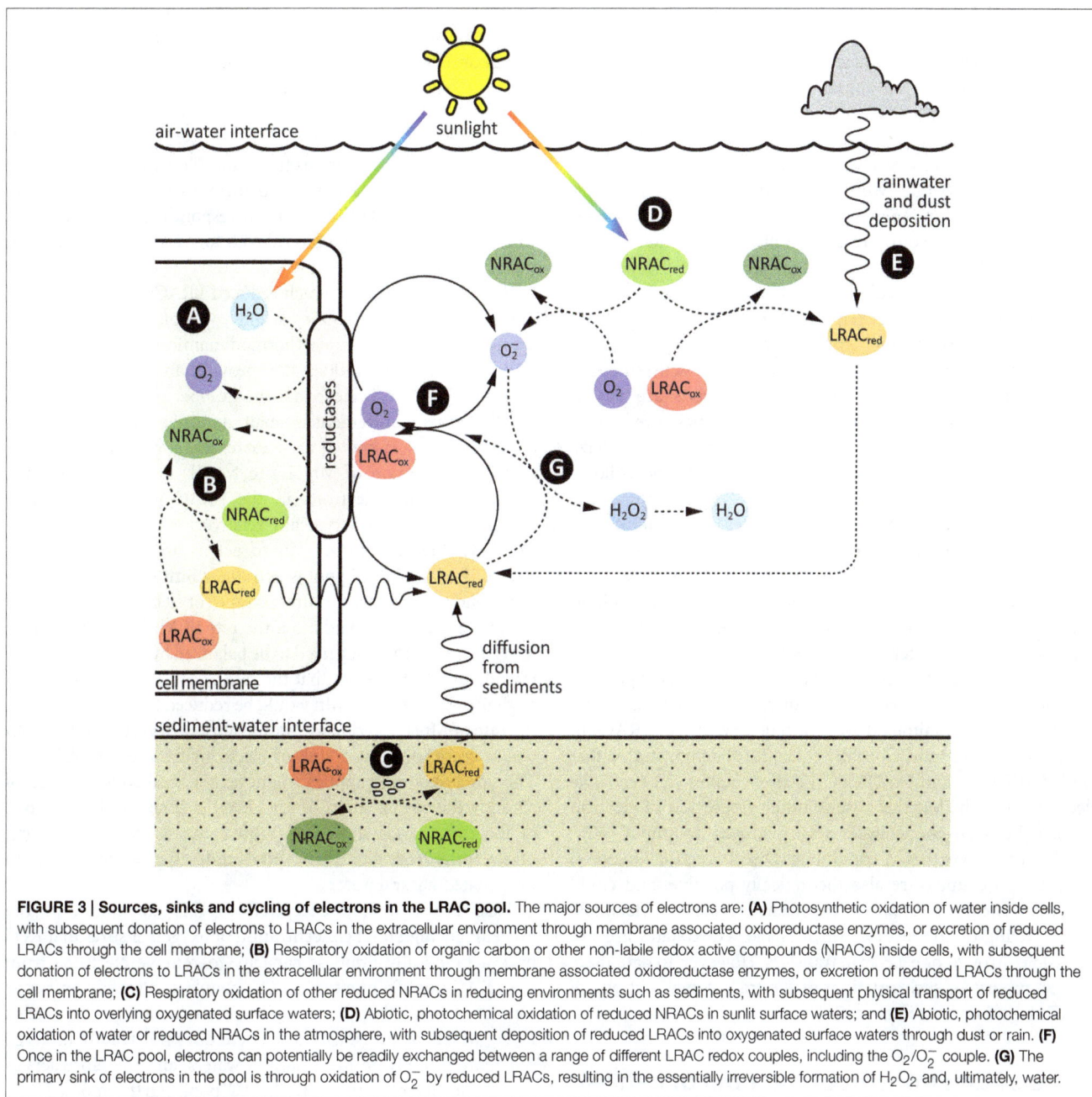

Because O_2 is present at high concentrations in all oxygenated natural waters while most other LRACs are present at trace concentrations, transfer of electrons to O_2 by reduced LRACs likely represents a major pathway for electron transfer among the LRAC pool. The oxygenation of reduced metal LRACs including Fe(II), Mn(II), V(II), Cu(I), and Co(II) is thought to occur via the Haber-Weiss mechanism (Haber and Weiss, 1934):

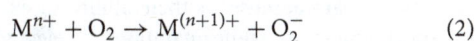

$$M^{n+} + O_2 \rightarrow M^{(n+1)+} + O_2^- \qquad (2)$$

where M^{n+} represents the reduced form of the metal and $M^{(n+1)+}$ its oxidized form.

The kinetics of this reaction vary by many orders of magnitude between different metals, and also between different complex species of the same metal. For example, the second order rate constant is around $4\ M^{-1}.s^{-1}$ for oxygenation of inorganic Fe(II) in 0.7 M NaCl and 2 mM bicarbonate at pH 8.1 (Santana-Casiano et al., 2005) and around $2\ M^{-1}.s^{-1}$ for inorganic Cu(I) under similar conditions (González-Dávila et al., 2009), but $<10^{-4}\ M^{-1}.s^{-1}$ for Mn(II) in seawater at pH 8.1 (Morgan, 2005). Similarly, rate constants for the oxygenation kinetics of Fe(II) in seawater at pH 8.1 when complexed by natural organic ligands extracted from coastal soils ranged from 2 to $1000\ M^{-1}.s^{-1}$ (Rose and Waite, 2003). Thus, the rate at which electrons are cycled between LRACs and O_2 can vary substantially depending on the LRAC involved.

The oxygenation of reduced trace metals is expected to produce O_2^-, however the initial product of the reaction may involve coordination of O_2^- in the inner sphere of the metal. Oxygenation of V^{2+}, Co^{2+}, and Fe^{2+} appears to proceed via an outer-sphere mechanism, resulting in production of non-coordinated O_2^-, while Mn^{2+} and hydrolysed Fe(II) species appear to react via an inner-sphere process, such that O_2^- would initially be coordinated to the metal center (Rosso and Morgan, 2002). While it is likely that O_2^- is ultimately released from inner sphere complexes, the details of this process are not well known.

As previously stated, O_2^- can function as a mild one-electron reductant under conditions typical of natural waters (Sawyer, 1991). Thus, many of the oxygenation reactions discussed above are reversible, with the net effect that electrons are passed between O_2, O_2^-, and various other LRACs. Kinetics of reduction by O_2^- under conditions relevant to natural surface waters have been documented for a range of Fe(III) complexes and colloids (Rush and Bielski, 1985; Rose and Waite, 2005; Fujii et al., 2006), Cu(II) (Zafiriou et al., 1998; Voelker et al., 2000), and Mn(III) (Hansard et al., 2011). One electron reduction of various complexes of Co(III), Ni(III), and Mo(V) by O_2^- has also been observed in aqueous solution (Afanas'ev, 1989), although in some cases under conditions that are not necessarily relevant to oxygenated natural waters around neutral pH.

Due to the ubiquitousness of the O_2/O_2^- couple in oxygenated natural waters, the mechanisms discussed above are thus expected to be ubiquitous drivers of redox cycling of LRACs in these systems. Recently, González et al. (2016) demonstrated that the Fe(II)/Fe(III) and Cu(I)/Cu(II) redox couples can exchange electrons directly, but that a substantial proportion of electron transfer between the couples was mediated by the O_2/O_2^- couple under oxygenated conditions. Direct reactions between other LRAC couples are also theoretically possible, and could potentially include multiple electron transfers that could facilitate redox cycling of couples for which stable redox states are not accessible through one electron transfers. However, to date there are few reports of such reactions at the trace concentrations of LRACs found in most in natural waters.

Sinks of Electrons from the Pool of Labile Redox Active Compounds

Because oxidation of kinetically inert reduced compounds is largely irreversible in the extracellular environment, electrons must leave the LRAC pool via other pathways. The major potential sinks for electrons in the LRAC pool are therefore expected to be physical transport of reduced LRACs out of the local environment, scavenging of reduced LRACs through non-redox processes, or redox processes yielding kinetically non-labile products for which reduced LRACs are precursors.

Mechanisms for loss due to physical transport potentially include biological uptake of reduced LRACs such as Fe(II), or transport into more oxidizing environments. In the latter case electrons will be lost from the local LRAC pool, but will still remain associated with the overall LRAC pool until removed by one of the other processes discussed. Scavenging of reduced LRACs through non-redox processes can potentially occur by complexation (and hence stabilization of the reduced LRAC against reoxidation), radical addition reactions, or precipitation of reduced LRACs into non-reactive particulate forms. However, these processes are likely to be relatively minor in many oxygenated natural waters, as the reduced forms of many LRACs (especially trace metals) tend to form weaker complexes and remain more soluble than their corresponding oxidized forms, while scavenging through radical addition reactions is not known to occur at significant rates.

The final pathway in which reduced LRACs react to produce kinetically non-labile species is only possible where the LRAC species can exist in multiple, thermodynamically accessible redox states. While **Table 1** does not provide an exhaustive list of possible LRACs, it is clear that the number of potential LRACs satisfying this requirement is small. As discussed earlier, abiotic oxidation Mn^{2+} is known to be extremely slow (Morgan, 2005), such that reduction of Mn(III) to Mn(II) represents one such possible sink for electrons in the LRAC pool. However, as Mn concentrations are typically relatively low in natural waters (**Table 1**), the capacity of Mn to act as an ongoing sink for electrons in the LRAC pool is extremely limited—under steady-state conditions with a continuous input of electrons to the LRAC pool through NRAC oxidation, the rate of Mn(III) reduction to Mn(II) would ultimately need to be balanced by the rate of Mn(II) reoxidation. In the event that the rate of Mn(II) reoxidation was negligibly slow, then all Mn would be reduced to Mn(II) such that the rate of Mn(III) reduction would approach zero, and thus Mn would cease to act as a sink of electrons from the LRAC pool. The same reasoning applies whenever the total concentration of all allowable redox states of an LRAC is small, such that none of the compounds shown in **Table 1** can act as a major sink of electrons from the LRAC pool under typical conditions in oxygenated natural waters.

ROS are a unique exception to this constraint, as the exceedingly high total concentration of oxygen ensures that ROS have a practically limitless capacity to remove electrons from the LRAC pool. While the O_2/O_2^- couple can readily exchange electrons with other LRACs, further reduction of O_2^- yields H_2O_2 which, as previously discussed, does not readily undergo reoxidation to O_2^- in most natural waters, and therefore acts as a sink for electrons from the LRAC pool. Furthermore, H_2O_2 can undergo further irreversible reduction through reaction with reduced LRACs to generate $HO^\bullet$, which can undergo one further irreversible reduction step to water.

IMPORTANCE OF ROS FOR LOCAL REDOX CHEMISTRY IN OXYGENATED NATURAL WATERS

At least in part because of their ability to exchange electrons on short timescales while maintaining selectivity in reactions, LRACs are integral components of many redox-active enzymes in biological systems (Fraústo Da Silva and Williams, 2001). Thus, the bioavailability of LRACs is a critical influence on ecosystem scale structure and function in natural waters, given that acquisition of LRACs from the extracellular environment is critical for cells to synthesize and maintain these enzymes. For many LRACs, especially trace metals, there is a fine balance between scarcity (sometimes to the point of growth limitation) under conditions of low bioavailability and toxicity under conditions of high bioavailability. This is directly related to redox conditions, which typically strongly influence the bioavailability of trace metals and potentially other LRACs (Fraústo Da Silva and Williams, 2001; Borch et al., 2009).

The ability of O_2^- to influence the bioavailability of Fe in oxygenated waters by mediating its redox cycling has already been demonstrated under a range of scenarios (Kustka et al., 2005; Rose et al., 2005; Garg et al., 2007; Rose, 2012). Biologically produced extracellular O_2^- has also been shown to drive redox cycling of Mn in fungi (Hansel et al., 2012) and bacteria (Learman et al., 2011) under oxygenated conditions. However, from the discussion above it is evident that O_2^- may be expected to play a more general role in controlling the loss of electrons from the LRAC pool, and therefore in controlling local redox conditions. While functioning as a reductant within the O_2/O_2^- couple, O_2^- plays a major role in facilitating redox cycling among the LRAC pool, however once reduced to H_2O_2, it facilitates irreversible loss of electrons from the LRAC pool. Strictly speaking, O_2^- itself is unable to directly accept electrons in aqueous solution, however it exists in equilibrium with its conjugate acid $HOO^\bullet$, which is a powerful oxidant. Although O_2^- is dominant at pH 8.1, given that the pK_a for the acid-base pair is 4.8 (Bielski et al., 1985), the existence of even small concentrations of $HOO^\bullet$ in equilibrium with O_2^- implies that loss of electrons from the LRAC pool via this pathway is unavoidable. However, provided that the rate of loss is relatively low, then ongoing inputs of electrons through the processes described earlier can potentially maintain substantial steady-state concentrations of reduced LRACs.

Assuming that reduction of O_2^- to H_2O_2 is the primary pathway for loss of electrons from the LRAC pool, the rate of loss, P_{loss}, will be given by:

$$P_{loss} = [HOO^\bullet] \sum k_{ox} [LRAC_{red}]$$
$$= \alpha_{HOO^\bullet}[O_2^-]_T \sum k_{ox} \alpha_{red}[LRAC]_T \qquad (3)$$

where $\alpha_{HOO^\bullet}$ represents the fraction of O_2^- that is present in the protonated $HOO^\bullet$ form, k_{ox} represents the rate constant for oxidation of a particular reduced LRAC by $HOO^\bullet$, α_{red} represents the fraction of a particular LRAC that is in the reduced state, and $[LRAC]_T$ represents the total concentration of that LRAC.

The three major assumptions underlying the development of Equation (3) are that: (i) reduction of O_2^- to H_2O_2 represents the major pathway for loss of electrons from the LRAC pool; (ii) the LRAC pool rapidly reaches redox (pseudo)equilibrium when subjected to a perturbation; and (iii) the species comprising the LRAC pool can be accurately defined. The first assumption is essentially based on the principle that O_2 is the primary oxidant of reduced LRACs; under conditions where other oxidants may be present at significant concentrations, Equation (3) will not be

valid and the hypothesis presented is unlikely to be correct. As discussed in Section Sinks of Electrons from the Pool of Labile Redox Active Compounds, other processes that might remove electrons from the LRAC pool are unlikely to be significant in many oxygenated natural waters. The second assumption is based primarily on the notion that the O_2/O_2^- couple can mediate relatively rapid electron transfer between various LRAC couples. While this is relatively well known for some LRAC couples, as discussed in Section Cycling of Electrons among the Pool of Labile Redox Active Compounds the kinetics of this process can potentially vary by several orders of magnitude for different LRAC couples. The third assumption is based on the principle that kinetically inert redox active compounds and highly reactive transient species can be neatly and precisely distinguished from LRACs, and that their contribution to electron transfer within the LRAC pool can be ignored. This is conceptually difficult, since in reality the timescales of redox processes involving different couples can vary almost continuously, and the distinction between LRAC and non-LRAC couples is somewhat arbitrary. Furthermore, even if such a distinction can be accurately made, the kinetics of redox processes involving different couples is strongly influenced by complex speciation and pH, which can vary enormously between different types of natural waters, as discussed in Section Cycling of Electrons among the Pool of Labile Redox Active Compounds.

While it would seem reasonable to suggest that these assumptions might be valid at least in certain aquatic environments, it is clear that a substantial amount of additional information is required to establish under what conditions, if any, this will be the case. In particular, site-specific information on the kinetics of redox transformations between the full range of redox couples present, including accounting for complex speciation, is required to test the underlying assumptions of the theory presented and the validity of the resulting Equation (3) for any particular natural water body. Explicitly testing the hypothesis that the relative importance of the O_2/O_2^- and O_2^-/H_2O_2 redox couples exerts a governing control on local redox conditions will require such information to be gathered from a wide variety of natural waters to determine the range of aquatic environments in which this hypothesis might hold.

Should this theory prove to hold in a particular aquatic system, the consequences for redox biogeochemistry are profound. From Equation (3), it follows that under steady-state conditions (where the rate of input of electrons into the LRAC pool must equal the rate of loss):

1. Maintaining a higher steady-state O_2^- concentration requires a higher rate of input of electrons into the LRAC pool for a given set of k_{ox} and $[LRAC]_T$. However this dependence is not linear, as increasing the steady-state O_2^- concentration will also increase values of α_{red}, assuming that the redox state of the O_2/O_2^- couple reflects the redox state of the entire LRAC pool. In other words, a relatively low rate of electron input is needed to maintain slightly reducing local redox conditions within the LRAC pool (relative to the overall redox potential set by the O_2/H_2O system), but greatly increasing rates are needed to maintain more strongly reducing conditions.

2. Systems containing higher total LRAC concentrations will require higher rates of input of electrons into the LRAC pool to maintain a given steady-state O_2^- concentration. In other words, a relatively low rate of electron input is needed to maintain slightly reducing local redox conditions in systems containing low total LRAC concentrations compared with systems containing high total LRAC concentrations.

By adjusting rates of extracellular electron transport, through direct production of O_2^- or via other reduced LRACs, organisms could therefore have the ability to tightly regulate local redox conditions, at least within a limited window of redox potentials, in the extracellular environment on the critical biogeochemical timescales of seconds to days. Given the increasing evidence for the ubiquity of extracellular O_2^- production by aquatic microorganisms (Diaz et al., 2013; Zhang et al., 2016), this suggests that, at least in some oxygenated natural waters, ROS (and in particular O_2^-) may play a critical role in mediating overall local redox conditions, with major implications for the biogeochemistry of these systems.

AUTHOR CONTRIBUTIONS

All work for this manuscript was conducted by AR.

REFERENCES

Aeschbacher, M., Sander, M., and Schwarzenbach, R. P. (2010). Novel electrochemical approach to assess the redox properties of humic substances. *Environ. Sci. Technol.* 44, 87–93. doi: 10.1021/es902627p

Afanas'ev, I. B. (1989). *Superoxide Ion Chemistry and Biological Implications*. Boca Raton, FL: CRC Press.

Amon, R. M. W., and Benner, R. (1996). Photochemical and microbial consumption of dissolved organic carbon and dissolved oxygen in the Amazon River system. *Geochim. Cosmochim. Acta* 60, 1783–1792. doi: 10.1016/0016-7037(96)00055-5

Andrews, S. S., Caron, S., and Zafiriou, O. C. (2000). Photochemical oxygen consumption in marine waters: A major sink for colored dissolved organic matter? *Limnol. Oceanogr.* 45, 267–277. doi: 10.4319/lo.2000.45.2.0267

Barbeau, K. A. (2006). Photochemistry of organic iron(III) complexing ligands in oceanic systems. *Photochem. Photobiol.* 82, 1505–1516. doi: 10.1111/j.1751-1097.2006.tb09806.x

Barbeau, K., Rue, E. L., Trick, C. G., Bruland, K. W., and Butler, A. (2003). Photochemical reactivity of siderophores produced by marine heterotrophic bacteria and cyanobacteria based on characteristic Fe(III) binding groups. *Limnol. Oceanogr.* 48, 1069–1078. doi: 10.4319/lo.2003.48.3.1069

Bielski, B. H. J., Cabelli, D. E., Arudi, R. L., and Ross, A. B. (1985). Reactivity of HO_2/O_2^- radicals in aqueous solution. *J. Phys. Chem. Ref. Data* 14, 1041–1100. doi: 10.1063/1.555739

Borch, T., Kretzschmar, R., Kappler, A., Cappellen, P. V., Ginder-Vogel, M., Voegelin, A., et al. (2009). Biogeochemical redox processes and their impact on contaminant dynamics. *Environ. Sci. Technol.* 44, 15–23. doi: 10.1021/es9026248

Bowles, K. C., Ernste, M. J., and Kramer, J. R. (2003). Trace sulfide determination in oxic freshwaters. *Anal. Chim. Acta* 477, 113–124. doi: 10.1016/S0003-2670(02)01370-3

Bratsch, S. G. (1989). Standard electrode potentials and temperature coefficients in water at 298.15 K. *J. Phys. Chem. Ref. Data* 18, 1–21. doi: 10.1063/1.555839

Cooper, W. J., and Zepp, R. G. (1990). Hydrogen peroxide decay in waters with suspended soils: evidence for biologically mediated processes. *Can. J. Fish. Aquat. Sci.* 47, 888–893. doi: 10.1139/f90-102

Diaz, J. M., Hansel, C. M., Voelker, B. M., Mendes, C. M., Andeer, P. F., and Zhang, T. (2013). Widespread production of extracellular superoxide by heterotrophic bacteria. *Science* 340, 1223–1226. doi: 10.1126/science.1237331

Fraústo Da Silva, J. R. R., and Williams, R. J. P. (2001). *The Biological Chemistry of the Elements: The Inorganic Chemistry of Life*. Oxford: Oxford University Press.

Fridovich, I. (1998). Oxygen toxicity: a radical explanation. *J. Exp. Biol.* 201, 1203–1209.

Fujii, M., Rose, A. L., Waite, T. D., and Omura, T. (2006). Superoxide-mediated dissolution of amorphous ferric oxyhydroxide in seawater. *Environ. Sci. Technol.* 40, 880–887. doi: 10.1021/es051622t

Garg, S., Rose, A. L., Godrant, A., and Waite, T. D. (2007). Iron uptake by the ichthyotoxic *Chattonella marina* (Raphidophyceae): impact of superoxide generation. *J. Phycol.* 43, 978–991. doi: 10.1111/j.1529-8817.2007.00394.x

Goldstein, S., and Czapski, G. (1995). Kinetics of nitric oxide autoxidation in aqueous solution in the absence and presence of various reductants. The nature of the oxidizing intermediates. *J. Am. Chem. Soc.* 117, 12078–12084.

González, A. G., Pérez-Almeida, N., Magdalena Santana-Casiano, J., Millero, F. J., and González-Dávila, M. (2016). Redox interactions of Fe and Cu in seawater. *Mar. Chem.* 179, 12–22. doi: 10.1016/j.marchem.2016.01.004

González-Dávila, M., Santana-Casiano, J. M., González, A. G., Pèrez, N., and Millero, F. J. (2009). Oxidation of copper(I) in seawater at nanomolar levels. *Mar. Chem.* 115, 118–124. doi: 10.1016/j.marchem.2009.07.004

Haber, F., and Weiss, J. (1934). The catalytic decomposition of hydrogen peroxide by iron salts. *Proc. R. Soc. Lond. A Mathe. Phys. Sci.* 147, 332–351. doi: 10.1098/rspa.1934.0221

Hansard, S. P., Easter, H. D., and Voelker, B. M. (2011). Rapid reaction of nanomolar Mn(II) with superoxide radical in seawater and simulated freshwater. *Environ. Sci. Technol.* 45, 2811–2817. doi: 10.1021/es104014s

Hansard, S. P., Vermilyea, A. W., and Voelker, B. M. (2010). Measurements of superoxide radical concentration and decay kinetics in the Gulf of Alaska. *Deep Sea Res. I Oceanogr. Res. Papers* 57, 1111–1119. doi: 10.1016/j.dsr.2010.05.007

Hansel, C. M., Zeiner, C. A., Santelli, C. M., and Webb, S. M. (2012). Mn(II) oxidation by an ascomycete fungus is linked to superoxide production during asexual reproduction. *Proc. Natl. Acad. Sci. U.S.A.* 109, 12621–12625. doi: 10.1073/pnas.1203885109

Huang, J.-H., Huang, F., Evans, L., and Glasauer, S. (2015). Vanadium: Global (bio)geochemistry. *Chem. Geol.* 417, 68–89. doi: 10.1016/j.chemgeo.2015.09.019

Jones, G. J., Waite, T. D., and Smith, J. D. (1985). Light-dependent reduction of copper(II) and its effect on cell-mediated thiol-dependent superoxide production. *Biochem. Biophys. Res. Commun.* 128, 1031–1036. doi: 10.1016/0006-291X(85)90151-2

Korshunov, S., and Imlay, J. A. (2006). Detection and quantification of superoxide formed within the periplasm of *Escherichia coli*. *J. Bacteriol.* 188, 6326–6334. doi: 10.1128/JB.00554-06

Kosman, D. J. (2003). Molecular mechanisms of iron uptake in fungi. *Mol. Microbiol.* 47, 1185–1197. doi: 10.1046/j.1365-2958.2003.03368.x

Kustka, A., Shaked, Y., Milligan, A., King, D. W., and Morel, F. M. M. (2005). Extracellular production of superoxide by marine diatoms: contrasting effects on iron redox chemistry and bioavailability. *Limnol. Oceanogr.* 50, 1172–1180. doi: 10.4319/lo.2005.50.4.1172

Learman, D. R., Voelker, B. M., Vazquez-Rodriguez, A. I., and Hansel, C. M. (2011). Formation of manganese oxides by bacterially generated superoxide. *Nat. Geosci.* 4, 95–98. doi: 10.1038/ngeo1055

Lesuisse, E., Casteras-Simon, M., and Labbe, P. (1996). Evidence for the *Saccharomyces cerevisiae* ferrireductase system being a multicomponent electron transport chain. *J. Biol. Chem.* 271, 13578–13583. doi: 10.1074/jbc.271.23.13578

Luther, G. W. III. (2010). The role of one- and two-electron transfer reactions in forming thermodynamically unstable intermediates as barriers in multi-electron redox reactions. *Aquat. Geochem.* 16, 395–420. doi: 10.1007/s10498-009-9082-3

Luther, G. W. III., Findlay, A. J., Macdonald, D. J., Owings, S. M., Hanson, T. E., Beinart, R. A., et al. (2011). Thermodynamics and kinetics of sulfide oxidation by oxygen: a look at inorganically controlled reactions and biologically mediated processes in the environment. *Front. Microbiol.* 2:62. doi: 10.3389/fmicb.2011.00062

Luther, G. W. III., and Rickard, D. T. (2005). Metal sulfide cluster complexes and their biogeochemical importance in the environment. *J. Nanopart. Res.* 7, 389–407. doi: 10.1007/s11051-005-4272-4

Marshall, J.-A., Hovenden, M., Oda, T., and Hallegraeff, G. M. (2002). Photosynthesis does influence superoxide production in the ichthyotoxic alga *Chattonella marina* (Raphidophyceae). *J. Plankton Res.* 24, 1231–1236. doi: 10.1093/plankt/24.11.1231

Mason, R. P. (2013). *Trace Metals in Aquatic Systems.* Chichester: Wiley-Blackwell.

Morel, F. M. M., and Hering, J. G. (1993). *Principles and Applications of Aquatic Chemistry.* New York, NY: Wiley.

Morgan, J. J. (2005). Kinetics of reaction between O_2 and Mn(II) species in aqueous solutions. *Geochim. Cosmochim. Acta* 69, 35–48. doi: 10.1016/j.gca.2004.06.013

Mostofa, K. M. G., and Sakugawa, H. (2009). Spatial and temporal variations and factors controlling the concentrations of hydrogen peroxide and organic peroxides in rivers. *Environ. Chem.* 6, 524–534. doi: 10.1071/EN09070

Neuman, E. W. (1934). Potassium superoxide and the three-electron bond. *J. Chem. Phys.* 2, 31–33. doi: 10.1063/1.1749353

Pauling, L. (1979). The discovery of the superoxide radical. *Trends Biochem. Sci.* 4, N270–N271. doi: 10.1016/0968-0004(79)90203-2

Pierre, J. L., Fontecave, M., and Crichton, R. R. (2002). Chemistry for an essential biological process: the reduction of ferric iron. *BioMetals* 15, 341–346. doi: 10.1023/A:1020259021641

Rickard, D., and Luther, G. W. III. (2006). Metal sulfide complexes and clusters. *Rev. Mineral. Geochem.* 61, 421–504. doi: 10.2138/rmg.2006.61.8

Rose, A. L. (2012). The influence of extracellular superoxide on iron redox chemistry and bioavailability to aquatic microorganisms. *Front. Microbiol.* 3:124. doi: 10.3389/fmicb.2012.00124

Rose, A. L., Godrant, A., Furnas, M., and Waite, T. D. (2010). Dynamics of nonphotochemical superoxide production in the Great Barrier Reef lagoon. *Limnol. Oceanogr.* 55, 1521–1536. doi: 10.4319/lo.2010.55.4.1521

Rose, A. L., Salmon, T. P., Lukondeh, T., Neilan, B. A., and Waite, T. D. (2005). Use of superoxide as an electron shuttle for iron acquisition by the marine cyanobacterium *Lyngbya majuscula. Environ. Sci. Technol.* 39, 3708–3715. doi: 10.1021/es048766c

Rose, A. L., and Waite, T. D. (2003). Effect of dissolved natural organic matter on the kinetics of ferrous iron oxygenation in seawater. *Environ. Sci. Technol.* 37, 4877–4886. doi: 10.1021/es034152g

Rose, A. L., and Waite, T. D. (2005). Reduction of organically complexed ferric iron by superoxide in a simulated natural water. *Environ. Sci. Technol.* 39, 2645–2650. doi: 10.1021/es048765k

Rose, A. L., Webb, E. A., Waite, T. D., and Moffett, J. W. (2008). Measurement and implications of nonphotochemically generated superoxide in the equatorial Pacific Ocean. *Environ. Sci. Technol.* 42, 2387–2393. doi: 10.1021/es7024609

Rosso, K. M., and Morgan, J. J. (2002). Outer-sphere electron transfer kinetics of metal ion oxidation by molecular oxygen. *Geochim. Cosmochim. Acta* 66, 4223–4233. doi: 10.1016/S0016-7037(02)01040-2

Rusak, S. A., Peake, B. M., Richard, L. E., Nodder, S. D., and Cooper, W. J. (2011). Distributions of hydrogen peroxide and superoxide in seawater east of New Zealand. *Mar. Chem.* 127, 155–169. doi: 10.1016/j.marchem.2011.08.005

Rush, J. D., and Bielski, B. H. J. (1985). Pulse radiolytic studies of the reactions of HO_2/O_2^- with Fe(II)/Fe(III) ions. The reactivity of HO_2/O_2^- with ferric ions and its implication on the occurrence of the Haber-Weiss reaction. *J. Phys. Chem.* 89, 5062–5066. doi: 10.1021/j100269a035

Santana-Casiano, J. M., González-Dávila, M., and Millero, F. J. (2005). Oxidation of nanomolar levels of Fe(II) with oxygen in natural waters. *Environ. Sci. Technol.* 39, 2073–2079. doi: 10.1021/es049748y

Sawyer, D. T. (1991). *Oxygen Chemistry.* New York, NY: Oxford University Press.

Sharpless, C. M., and Blough, N. V. (2014). The importance of charge-transfer interactions in determining chromophoric dissolved organic matter (CDOM) optical and photochemical properties. *Environ. Sci. Proc. Impacts* 16, 654–671. doi: 10.1039/c3em00573a

Sverjensky, D. A., and Lee, N. (2010). The Great Oxidation Event and mineral diversification. *Elements* 6, 31–36. doi: 10.2113/gselements.6.1.31

Vermilyea, A. W., Hansard, S. P., and Voelker, B. M. (2010). Dark production of hydrogen peroxide in the Gulf of Alaska. *Limnol. Oceanogr.* 55, 580–588. doi: 10.4319/lo.2009.55.2.0580

Voelker, B. M., Sedlak, D. L., and Zafiriou, O. C. (2000). Chemistry of superoxide radical in seawater: reactions with organic Cu complexes. *Environ. Sci. Technol.* 34, 1036–1042. doi: 10.1021/es990545x

Vraspir, J. M., and Butler, A. (2009). Chemistry of marine ligands and siderophores. *Ann. Rev. Mar. Sci.* 1, 43–63. doi: 10.1146/annurev.marine.010908.163712

Wang, D. (2012). Redox chemistry of molybdenum in natural waters and its involvement in biological evolution. *Front. Microbiol.* 3:427. doi: 10.3389/fmicb.2012.00427

Wardman, P. (1989). Reduction potentials of one-electron couples involving free radicals in aqueous solution. *J. Phys. Chem. Ref. Data* 18, 1637–1755. doi: 10.1063/1.555843

Yuan, J., and Shiller, A. M. (2005). Distribution of hydrogen peroxide in the northwest Pacific Ocean. *Geochem. Geophys. Geosys.* 6:Q09M02. doi: 10.1029/2004GC000908

Zafiriou, O. C., Voelker, B. M., and Sedlak, D. L. (1998). Chemistry of the superoxide radical (O_2^-) in seawater: reactions with inorganic copper complexes. *J. Phys. Chem. A* 102, 5693–5700. doi: 10.1021/jp980709g

Zhang, T., Hansel, C. M., Voelker, B. M., and Lamborg, C. H. (2016). Extensive dark biological production of reactive oxygen species in brackish and freshwater ponds. *Environ. Sci. Technol.* 50, 2983–2993. doi: 10.1021/acs.est.5b03906

Zhang, Y., Del Vecchio, R., and Blough, N. V. (2012). Investigating the mechanism of hydrogen peroxide photoproduction by humic substances. *Environ. Sci. Technol.* 46, 11836–11843. doi: 10.1021/es3029582

Zhou, X., and Mopper, K. (1990). Determination of photochemically produced hydroxyl radicals in seawater and freshwater. *Mar. Chem.* 30, 71–88. doi: 10.1016/0304-4203(90)90062-H

Conflict of Interest Statement: The author declares that the research was conducted in the absence of any commercial or financial relationships that could be construed as a potential conflict of interest.

7

A Plastic Network Approach to Model Calving Glacier Advance and Retreat

Lizz Ultee * and *Jeremy N. Bassis*

Department of Climate and Space, University of Michigan, Ann Arbor, MI, USA

Calving glaciers contribute substantially to sea level rise, but they are challenging to represent in models. Fine resolution is required for continental-scale models to accurately resolve calving dynamics, and in many cases glacier geometry is too complicated to be adequately reflected by more simplified models. Flowline models are able to resolve flow along the main branch of a glacier, but many of those in current use either ignore tributaries entirely or parameterize their effect using a measure of "equivalent width." Here we present a simple method to simulate terminus advance and retreat for an interacting network of glacier branches, based on a model extending Nye's (1953) perfect plastic flow approximation to calving glaciers. We apply the method to case studies of four marine-terminating glaciers: Jakobshavn Isbræ and Helheim Glacier of Greenland, and Columbia and Hubbard Glaciers of Alaska. Given bed topography and upstream elevation history, our method reproduces observed patterns of terminus advance and retreat in all cases, as well as centerline profiles for all branches.

Keywords: tidewater glacier dynamics, Jakobshavn Isbræ, Columbia Glacier, Helheim Glacier, Hubbard Glacier, plastic approximation, flowline, network

Edited by:
Timothy C. Bartholomaus,
University of Idaho, USA

Reviewed by:
Christine F. Dow,
University of Waterloo, Canada
Andreas Vieli,
University of Zurich, Switzerland

*Correspondence:
Lizz Ultee
ehultee@umich.edu

Specialty section:
This article was submitted to
Cryospheric Sciences,
a section of the journal
Frontiers in Earth Science

1. INTRODUCTION

The global sea level rise contribution from land ice is large and growing (Rignot et al., 2011; van den Broeke et al., 2016). Much of the increase in the land ice contribution to sea level rise comes from the dynamic response of marine-terminating glaciers, including those draining the Greenland and Antarctic Ice Sheets (Church et al., 2013; Straneo et al., 2013; Vaughan et al., 2013). Where glaciers drain the large ice sheets, there is often a transition from laterally unconfined ice (width $\gtrsim 10^2$ km) to flow through narrow fjords (width ~ 10 km or smaller). For example, most of the 199 widest Greenland outlet glaciers studied by Murray et al. (2015) do not exceed 3 km in width, and many smaller glaciers less than one kilometer in width are excluded from consideration. Where tidewater glaciers drain alpine ice fields, as in coastal Alaska, the ice flux through the terminus may come from tens or hundreds of tributary branches upstream. The largest glaciers with the greatest potential contribution to sea level rise also have the most tributary branches—more than 400 in the case of Hubbard Glacier, Alaska (Kienholz et al., 2015).

The relatively small scale and large number of outlet glaciers and tributaries present a challenge to effective ice sheet modeling: the fine resolution required to capture small-glacier dynamics is prohibitively expensive to apply in full-Stokes models of entire ice sheets. Adaptive mesh refinement (AMR) has been applied in some models, e.g., BISICLES (Cornford et al., 2013), as a first step toward addressing this challenge. However, the finest-resolution cells (~ 500 m) in AMR grids are generally those nearest the grounding line, potentially leaving upstream tributaries of width ~ 1 km and smaller unresolved. By contrast, simplified centerline models applied to individual glaciers may use grid spacing of < 10 m over the entire glacier length, but such models often ignore tributary branches entirely.

In Ultee and Bassis (2016), we extended the perfect plastic approximation of Nye (1951, 1952, 1953) to a centerline model of tidewater glaciers that self-consistently predicts terminus advance and retreat forced with upstream elevation change. We now generalize this model to account for the intersection of networks of tributaries. Our plastic model is well-suited to this problem because the condition for intersections is straightforward: ice thickness H must match at intersection points between branches. Here, we describe the model's application to intersecting glacier networks through four case studies, including glaciers from both Alaska and Greenland. We also discuss two alternative forcing methods that could bring glacier-wide mass balance—and its changes due to climate—directly into the model.

As illustrative cases, we choose four large, well-studied glaciers: Jakobshavn Isbræ and Helheim Glacier, Greenland, and Columbia and Hubbard Glaciers, Alaska. These glaciers were chosen based on their diversity of behavior and because of the (relatively) abundant data available. For example, geometry of these glaciers varies from Jakobshavn's single major channel to Hubbard's more than 400 mountain tributaries. Moreover, in recent decades Columbia and Jakobshavn have been undergoing sustained retreat (Krimmel, 2001; Joughin et al., 2004, 2012, 2014; O'Neel et al., 2005; McNabb et al., 2012), Hubbard has been advancing (Trabant et al., 2003; Ritchie et al., 2008) and Helheim has experienced both advance and retreat episodes (Howat et al., 2005, 2010; Murray et al., 2010), providing tests of our model's ability to resolve calving dynamics in a range of environments. Further, with the case studies selected here we go beyond the work of Ultee and Bassis (2016) to show that the model can reproduce patterns of retreat and advance in both Greenland and Alaska and that the model can be used to analyze glaciers with more than one branch.

2. METHODS

Our method extends the perfect plastic approximation of Nye (1951, 1952, 1953) to calving glaciers with a network of tributaries. The perfect plastic approximation corresponds to the assumption that glacier ice is perched at a yield strength. Applying the approximation to a glacier centerline as in Ultee and Bassis (2016), we obtain an equation for the surface elevation profile along the centerline:

$$|h - b| \frac{\partial h}{\partial x} = \frac{\tau_b}{\rho_i g}, \tag{1}$$

where h is glacier surface elevation, b bed elevation, $\rho_i = 920 \text{ kg m}^{-3}$ the density of glacier ice, $g = 9.81 \text{ m s}^{-2}$ the acceleration due to gravity, and τ_b the basal yield strength.

At the calving front, we require that tractions balance across the ice-water interface, i.e.,

$$\int_{b(x)}^{h(x)} \sigma_{xx} \, dz = \int_{b(x)}^{0} -\rho_w g z \, dz, \tag{2}$$

with $\rho_w = 1020 \text{ kg m}^{-3}$ the density of sea water. We also require that the ice at the terminus is at the yield strength of

ice, τ_y, and integrating as in Ultee and Bassis (2016) provides a corresponding condition on the ice thickness at the terminus:

$$H_{\text{terminus}} = 2 \frac{\tau_y}{\rho_i g} + \sqrt{\frac{\rho_w D^2}{\rho_i} + 2 \frac{\tau_y}{\rho_i g}} \tag{3}$$

where

$$D = \begin{cases} |b(x)| & \text{for } b(x) < 0 \\ 0 & \text{for } b(x) \geq 0 \end{cases} \tag{4}$$

represents water depth. For a given terminus position, our model uses Equations (1) and (3) to construct the surface elevation profile along the glacier centerline with a self-consistent terminus position.

A general plastic model admits two yield strengths: that of the bed, τ_b, and that of the ice, here called τ_y. Generally speaking, $\tau_b \leq \tau_y$. When the glacier substrate is weaker than the glacier ice, stress at the calving front (Equation 3) is limited by the yield strength of ice, and basal stress (Equation 1) is limited by the lower yield strength of the bed. The especially simple solution we present here assumes $\tau_b = \tau_y$, which may not be realistic but has shown promising results when applied to Columbia Glacier (Ultee and Bassis, 2016). τ_y thus becomes our single adjustable parameter.

This simple model may be run with a constant yield strength, or with a Coulomb yield criterion:

$$\tau_y = \tau_0 + \mu N, \quad \text{where } N = (\rho_i g H - \rho_w g D) \tag{5}$$

with μ a constant cohesion coefficient, $H = h - b$ the ice thickness, other terms as above, and τ_0 replacing τ_y as the directly adjustable yield parameter. N in Equation (5) represents an effective pressure at the glacier bed, such that basal substrates below sea level can be assumed saturated and promoting faster flow (through a weaker bed and/or more deformation in the adjacent layer of ice). Basal water pressure beneath real glaciers is affected by hydrological factors such as variable meltwater flux and evolution of the basal drainage network, but accounting for such effects is beyond the scope of this simple model. In all four case studies presented here, we keep $\mu = 0.01$ constant to avoid excessive "tuning." This value was chosen to reflect a Mohr-Coulomb condition with low friction angle—as would be found for soft marine sediments near tidewater glacier termini—and consistent with the laboratory estimates of Cohen et al. (2005).

To apply our one-dimensional centerline model to a network of glacier tributaries, we define the network of centerlines, identify an appropriate value of the yield strength τ_y for the glacier in question, and simulate upstream thinning/thickening over time.

2.1. Define Network Centerlines

Each centerline comprises a sequence of point coordinates. For relatively simple tributary networks, we manually select centerlines from maps of glacier thickness and bed topography. We save the lists of point coordinates and note which are points of intersection between branches. For general cases, it is more practical to extract point coordinates from automatically-generated sets of centerlines such as those created by Machguth and Huss (2014) or Kienholz et al. (2014). With coordinate sequence in hand, we generate a set of evenly-spaced points from the (common) terminus to the head of each glacier branch and we re-express the points of each line in terms of arc length. Expressing the lines in terms of arc length allows a continuous representation of input quantities along each line, allowing model resolution to be adjusted.

2.2. Find Best-Fit Yield Strength

Our model rheology depends on a single adjustable parameter: the yield strength τ_y. Note once again that, though a general plastic model might include separate yield strengths for the glacier bed and glacier ice, τ_b and τ_y, respectively, we assume here that $\tau_b = \tau_y$.

Where observed surface elevation profiles are available, we can find the best-fit value for yield strength τ_y (or τ_0 of Equation 5) by minimizing model mismatch with observation. We minimize model misfit only along the main branch, where observations are likely to be higher-quality, testing a range of values of τ_y. We quantify the mismatch with observation using the coefficient of variation of the root-mean-square error, CV_{RMS}. We take the value of τ_y corresponding to minimum CV_{RMS} to be the best-fit yield strength, and we apply the same best-fit τ_y to all branches of a given glacier.

For glaciers lacking observations of surface elevation, such as those that have retreated significantly from a less-well-documented initial state, a best-guess value of τ_y may be used instead. Laboratory and field observations indicate that τ_y should be approximately 50–300 kPa (O'Neel et al., 2005; Cuffey and Paterson, 2010), with lower values for soft, flat beds and higher values for hard, steep beds.

2.3. Simulate Advance/Retreat over Time

Using our best-fit or best-guess yield strength, we generate a reference profile along each glacier branch corresponding to an initial terminus position. For the plastic model, terminus retreat corresponds to upstream thinning and advance to thickening. We select an upstream point along the reference profile, ideally a point with good observations available for multiple years, where we will apply changes in ice thickness. We first simulate main branch thinning/thickening as described in Ultee and Bassis (2016), finding the terminus position satisfying Equation 3 for each new glacier profile. Because all tributaries share a common terminus, we use the terminus position identified by the main branch profile and step the plastic model upstream into the tributaries from there. Thus, we generate a consistent set of branch profiles, which by definition agree at the points of intersection, for each time step.

2.4. Numerical Considerations

We discretize Equation 1 with an Euler forward step. The model is fast and can be run with small grid spacing; in the following case studies, grid spacing ranges from 1–5 m to ensure negligible numerical error. Our results are robust and insensitive even to order-of-magnitude increases in grid spacing. We enforce our boundary condition of matching ice thickness at tributary branching points in the model initialization, stepping the model upstream into the tributaries from a common trunk where ice thickness agrees by definition. For more details of our numerical implementation, see Ultee and Bassis (2016).

3. CASE STUDIES

To demonstrate the application of the network method, we now turn to four case studies, arranged in order of increasing complexity. The case studies include calving glaciers in both Alaska and Greenland, with networks ranging from the very simple single branch of Jakobshavn Isbræ to the highly complex Hubbard Glacier network of more than 400 branches (of which we treat eight). Each case illustrates a different capability of the model: matching average retreat based on relatively sparse data, reproducing more detailed retreat including a network split, simulating retreat and readvance without adjusting model settings, and matching a well-studied terminus advance with a forcing fit to observations. In the final case study, we illustrate the application of the model to automatically generated networks of centerlines, dramatically improving the scalability of this approach for large and complex networks of calving glaciers.

3.1. Jakobshavn Isbræ, Greenland—1 Centerline

Jakobshavn Isbræ is the fastest-flowing glacier currently known, with summer flow speeds near the terminus reaching 16 km a^{-1} (Joughin et al., 2014). It drains approximately 7% of the Greenland Ice Sheet by volume (Joughin et al., 2004; Csatho et al., 2008) and is a substantial contributor to global sea level rise: nearly 1 mm between 2000 and 2011 (Joughin et al., 2014; Howat et al., 2011). A CReSIS level 3 data product (CReSIS, 2016b) provides gridded ice thickness, surface elevation, and ice-bottom elevation for Jakobshavn Isbræ. The CReSIS product is a composite of data collected between 2006 and 2014, with ice bottom reconstructed by kriging of radar lines. Because Jakobshavn lost its floating ice tongue before the period of the CReSIS product (2006–2014), we assume that the terminus is grounded and that ice-bottom elevation corresponds to bed topography everywhere. Three intersecting troughs are visible in the bed topography, but we consider only the 60 kilometers closest to the terminus, common to all three branches. Thus, we begin our investigation with the simplest possible network geometry: a single centerline. **Figure 1** shows the selected centerline on a map of Jakobshavn bed topography.

According to the CReSIS data, our centerline terminates (reaches the end of available data) in ice more than 1400 m thick, grounded more than 1100 m below sea level. If this were the true terminus, it would imply an ice cliff 314 m above the water line—much larger than observations suggest (Joughin et al., 2004, 2014; Csatho et al., 2008). For such a cliff to be stable, the yield strength of ice would have to be more than

3.5 MPa, which is an order of magnitude larger than the values we would expect from observation and our previous work (see O'Neel et al., 2005; Bassis and Walker, 2012; Bassis and Jacobs, 2013; Ultee and Bassis, 2016). Thus, we suspect that the seaward end of our centerline does not represent a true "terminus" with ice-water interface, making it unsuitable for our optimization procedure initialized with the balance thickness (Equation 3). We instead initialize with the observed terminal thickness and proceed with optimization upstream, finding the best fit with $\tau_y = 355$ kPa. **Figure 2** compares the centerline profile generated by our plastic model with the observed centerline profile from CReSIS data. The plastic model fit to observation is good, with $CV_{RMS} = 3.3\%$.

In recent years, Jakobshavn has thinned and retreated substantially. Using an average thinning rate based on Csatho et al. (2008), we investigate whether the plastic model produces appropriate retreat. **Figure 3** shows the simulated change in terminus position with 30 m a^{-1} thinning applied 15 km upstream from the original terminus (reference point marked in blue in **Figure 2**). After a few years of thinning without change in terminus position, retreat begins in the third year of the simulation and accelerates to just over 1 km a^{-1} in years 5–15. Over the entire 15-year period, average retreat rate is 763 m a^{-1}.

Our simulation of the retreat is simplistic, but not unreasonable. Because we initialize from a dataset compositing observations between 2006 and 2014, the onset date of our retreat could be anytime in that range, and comparison with observed retreat is inexact. However, we note that between 1991 and 2006, Csatho et al. (2008) report a 15-year average retreat rate of 830 m a^{-1}, comparable to the 15-year average retreat rate of 763 m a^{-1} found in our simulation. Further, for the 5-year period 2001–2006, closer to the probable time of our simulated retreat, the average retreat rate reported by Csatho et al. (2008) is 2.23 km a^{-1}, on the order of our retreat rate of 1.1 km a^{-1} in the years after onset of retreat in the simulation. Our results are of similar magnitude to those reported in Joughin et al. (2004, 2012, 2014) for the early 2000's as well. Though we do not capture the strong seasonal cycling observed at Jakobshavn's terminus (Joughin et al., 2014) under our constant annual upstream thinning, our

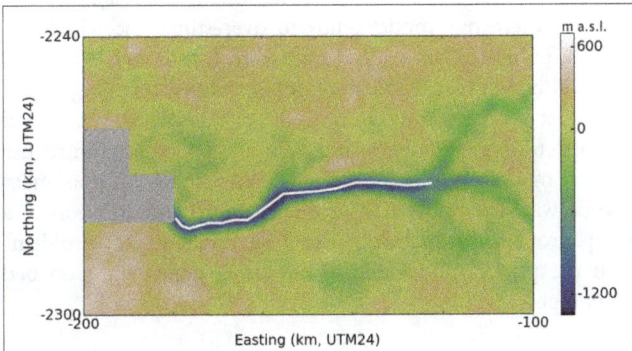

FIGURE 1 | CReSIS 2006–2014 composite bed topography of Jakobshavn Isbræ, with hand-selected centerline (white) for terminal 60 km. Note the dark gray area of no data immediately in front of the terminus.

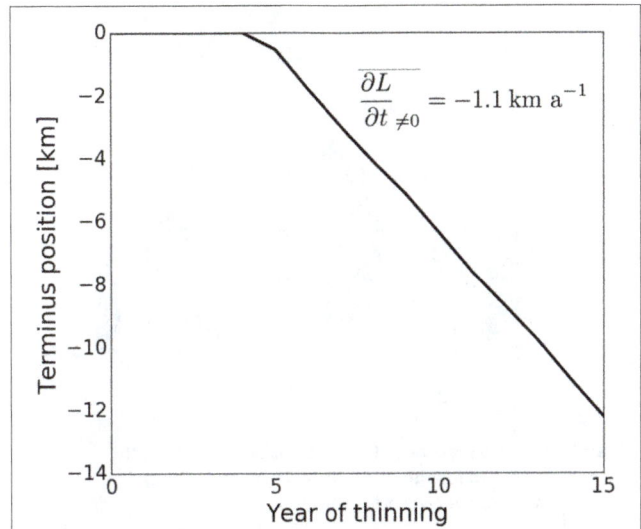

FIGURE 3 | Jakobshavn Isbræ terminus retreat under 15 years of 30 m a^{-1} upstream thinning. Average retreat rate $\overline{\frac{\partial L}{\partial t}}$ is listed for the years after onset of retreat; compare with rates of 0.83–2.23 km a^{-1} reported by Csatho et al. (2008) for a similar time period.

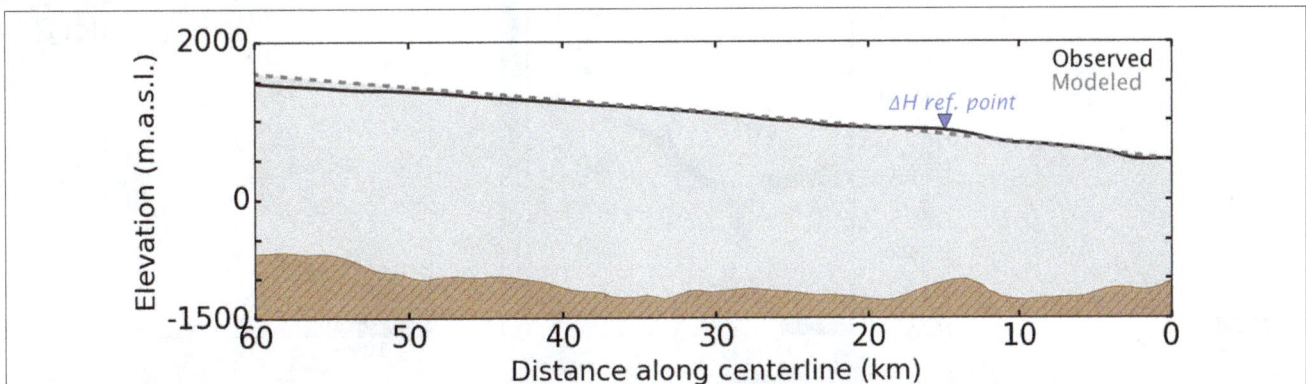

FIGURE 2 | Centerline profiles of Jakobshavn Isbræ: CReSIS composite of surface observations 2006–2014 (black curve) and plastic model with $\tau_y = 355$ kPa (gray dashed curve). Blue marker indicates the upstream reference point where forcing was applied to simulate retreat (see **Figure 3**).

results are satisfactory for our primary interest in annual- to decadal-scale change.

3.2. Columbia Glacier, Alaska—3 Centerlines

Alaska's Columbia Glacier is among the most well-documented cases of tidewater glacier retreat. The glacier has been monitored by the United States Geologic Survey (USGS) since the early twentieth century, and over the past 35 years it has undergone rapid terminus retreat and upglacier thinning. Datasets available

from McNabb et al. (2012) and Krimmel (2001) document the state of Columbia Glacier before and during its recent retreat. McNabb et al. (2012) offers topography and ice thickness reconstructed for 1957 and 2007 from observations, using a mass-conservation algorithm based on Glen's flow law; Krimmel (2001) gives more frequent flightline observations of the glacier surface elevation.

Figure 4 shows the structure of our hand-selected tributary network over Columbia Glacier bed topography. A Coulomb yield criterion with $\tau_0 = 130$ kPa gives the best model fit to observation. Each panel of **Figure 5** is an aerial view of the same Columbia Glacier network, illustrating different aspects of the model results. **Figure 5A** shows glacier surface elevation for plastic steady-state profiles generated with the 1957 terminus position; **Figure 5B** shows ice thickness for the same profiles; **Figure 5C** compares the modeled profiles to observation (1957 USGS topographic map, as presented in McNabb et al., 2012). The maximum error is 58% overestimation of ice thickness on parts of the east branch and 44% underestimation of ice thickness on the upper reaches of the main branch, though the percent error in modeled ice thickness is highly spatially variable, as seen in **Figure 5C**. Overall, the plastic model fit to observed surface elevation as measured by CV_{RMS} is good along the main and west branches (CV_{RMS} = 6.2% and 9.5%, respectively) but weaker along the east branch (19.9%). The east branch, where our model tends to overestimate ice thickness by 50% or more, has seldom been directly studied (Krimmel, 2001), although recent radar-sounding and bed-mapping efforts (Rignot et al., 2013; Enderlin et al., 2016) have better constrained the bed topography there and can be applied in future use of our model. We note also that the stretch of the main branch with the poorest fit to observation is the location of a sharp drop in the bedrock, where we might expect problems with the plastic model due to its strong dependence on bed topography.

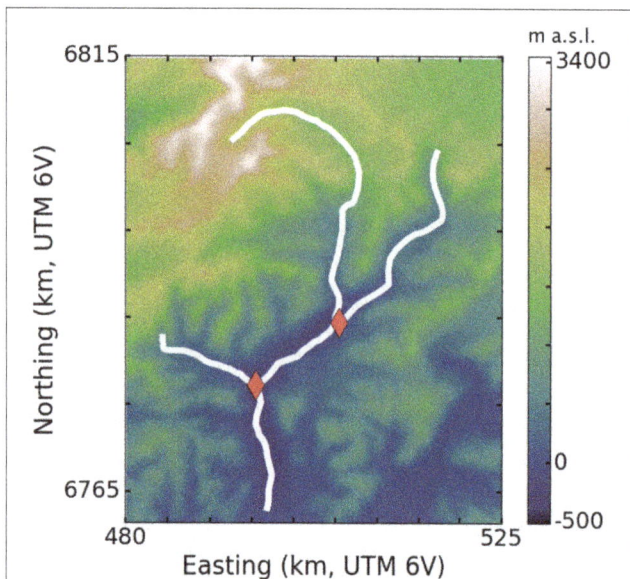

FIGURE 4 | Three branches of Columbia Glacier (white) with reconstructed bed topography from McNabb et al. (2012). Red diamonds mark intersections of the branches.

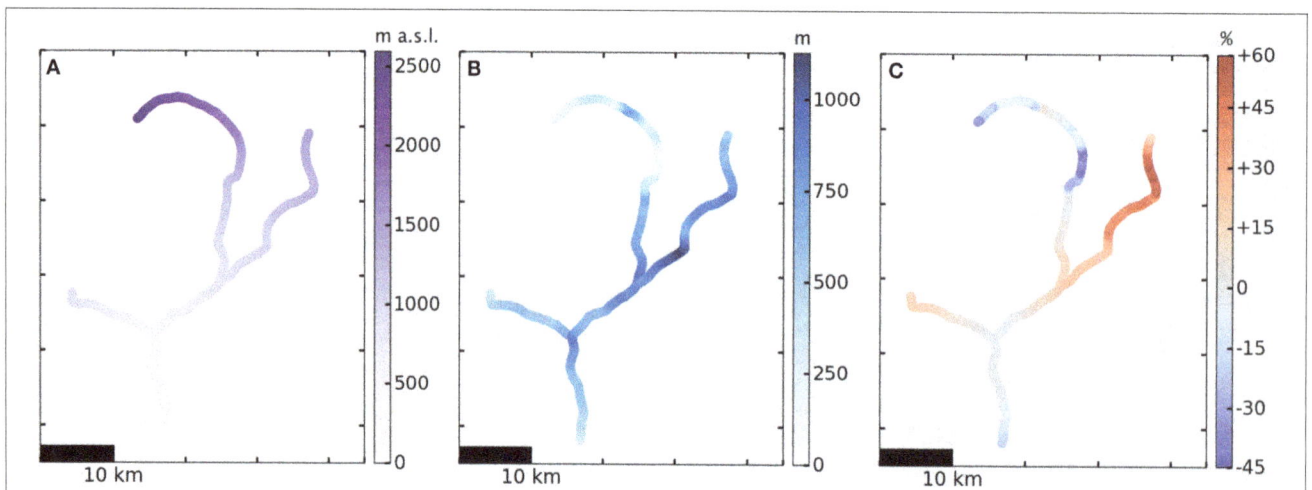

FIGURE 5 | Plastic model with $\tau_y = 130$ kPa $+ \mu N$ applied to a network of three major tributaries of Columbia Glacier, aerial view, 1957. (A) Glacier surface elevation, (B) Ice thickness, (C) Percent error model-observation (McNabb et al., 2012). A positive percent error (red colors in C) corresponds to model overestimation of observed ice thickness, and a negative percent error (blue colors) corresponds to underestimation.

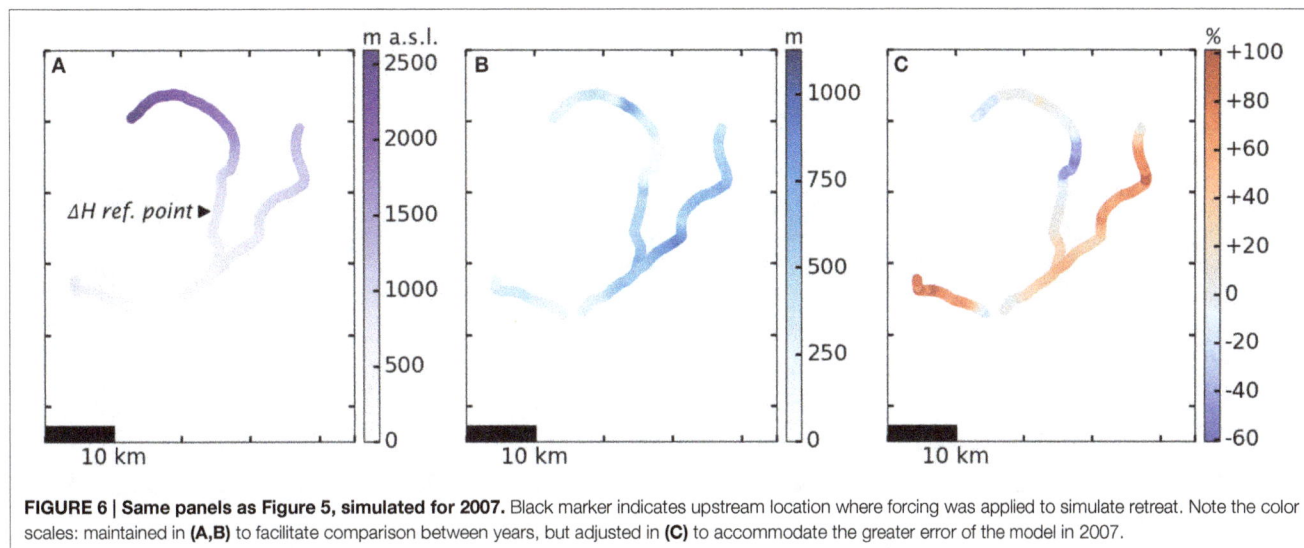

FIGURE 6 | Same panels as Figure 5, simulated for 2007. Black marker indicates upstream location where forcing was applied to simulate retreat. Note the color scales: maintained in **(A,B)** to facilitate comparison between years, but adjusted in **(C)** to accommodate the greater error of the model in 2007.

We explore the retreat patterns of Columbia Glacier in detail using a single-branch model in Ultee and Bassis (2016). Here, we apply constant upstream thinning of 8 m a^{-1} for 25 a, 1982–2007, and show the simulated 2007 state of Columbia Glacier in **Figure 6**. The color schemes in the two figures have been kept the same (with the exception of the percent error plotted in **Figure 6C**) to allow for direct comparison. We note that the most pronounced changes occur in the lower reaches of the glacier, including a striking terminus retreat of more than 20 km. Very little change is visible in the upper reaches of the main branch, which agrees with observations of Columbia's dynamics above and below a "hinge point" (an icefall) approximately 40 km upstream from the 1957 terminus (McNabb et al., 2012).

During the dramatic retreat between 1957 and 2007, the terminus of Columbia Glacier reached the intersection point of two branches and eventually split into two calving fronts, shedding icebergs from each branch. That split is visible in **Figure 6**. Once the terminus retreated upstream of the intersection point, it was necessary to run the model separately on the two smaller networks comprising the single-branch network to the west and the two-branch network to the east. Our model is able to handle this separation of tributaries with little difficulty.

3.3. Helheim Glacier, Greenland—5 Centerlines

Helheim Glacier is another of Greenland's largest outlet glaciers, located along its southeast coast. In recent years, Helheim was observed to thin, retreat, and accelerate, then slow its retreat and begin readvancing (Stearns and Hamilton, 2007; Howat et al., 2010, 2005; Murray et al., 2015). A CReSIS level 3 data product (CReSIS, 2016a) provides gridded ice thickness, surface elevation, and ice bottom elevation for Helheim Glacier. The product we are using is a composite of data collected between 2006 and 2014, with the ice bottom reconstructed by kriging of radar lines.

Figure 7 shows the ice bottom elevation underlying five branches of Helheim Glacier, with hand-selected centerlines for

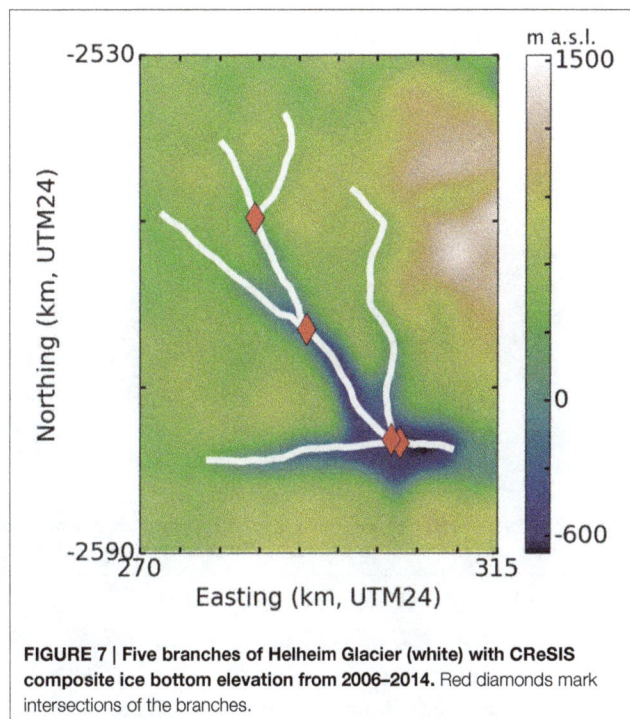

FIGURE 7 | Five branches of Helheim Glacier (white) with CReSIS composite ice bottom elevation from 2006–2014. Red diamonds mark intersections of the branches.

the branches marked in white. We optimize for the yield strength and find that the Coulomb yield criterion with $\tau_0 = 245$ kPa provides the best fit to observation.

Figure 8, analogous to **Figure 5**, shows the plastic model applied to the network of five branches of Helheim. The maximum errors are 47.5% overestimation of ice thickness on parts of the easternmost branch and 5.8% underestimation of ice thickness on upstream portions of the two central branches. For four of the five branches studied, the plastic model fit to the CReSIS composite is very good ($CV_{RMS} \leq 7.9\%$). For the easternmost branch, the fit is weaker, with $CV_{RMS} = 23.8\%$. This

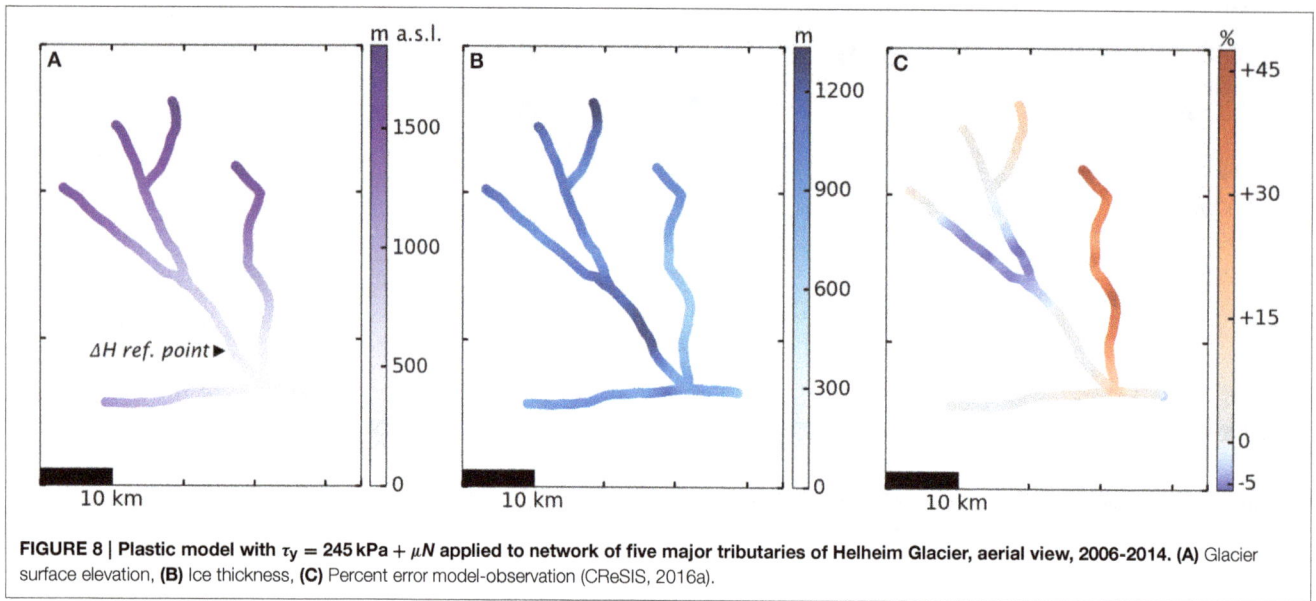

FIGURE 8 | Plastic model with $\tau_y = 245$ kPa $+ \mu N$ applied to network of five major tributaries of Helheim Glacier, aerial view, 2006-2014. (A) Glacier surface elevation, **(B)** Ice thickness, **(C)** Percent error model-observation (CReSIS, 2016a).

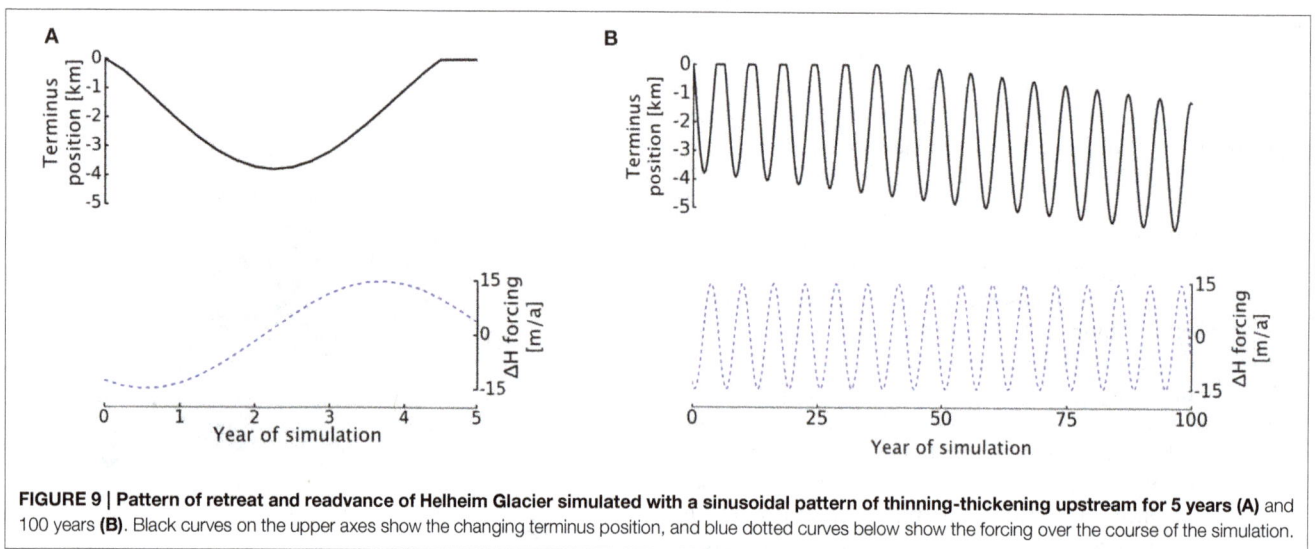

FIGURE 9 | Pattern of retreat and readvance of Helheim Glacier simulated with a sinusoidal pattern of thinning-thickening upstream for 5 years (A) and 100 years **(B)**. Black curves on the upper axes show the changing terminus position, and blue dotted curves below show the forcing over the course of the simulation.

is likely because the best-fit Coulomb yield strength, $\tau_0 = 245$ kPa, was optimized for the main branch and is too high for the easternmost branch. Running the optimization procedure for each branch individually, we find that the best-fit Coulomb yield strength is between 220 and 250 kPa for all other branches, but only 140 kPa for that easternmost branch. It is possible that ice-dynamical factors unique to that branch, e.g., a different balance of basal/wall drag, affect the optimal yield strength.

Given Helheim's recent pattern of retreat and readvance, we now investigate whether the plastic model can produce retreat and readvance within a single simulation. We fit a sinusoidal function to the upstream surface elevation changes reported in Murray et al. (2015) and run the simulation for 100 years. **Figure 9A** shows the retreat and readvance of the terminus in the first 5 years of simulation, with **Figure 9B** showing the continued oscillation over the 100-year period. Note that

positive cumulative changes are not possible, because we have no information about the bed topography for terminus positions more advanced than the initial state.

Under the time-varying upstream forcing, our model shows Helheim's terminus undergoing multi-year cycles of retreat and readvance over a longer-term retreat signal. Data limitations preclude a direct comparison between our simulation and observed patterns of advance and retreat, but we note that the simulated retreat is of the correct order of magnitude. For 2 years, the terminus retreats at an average rate of 1.65 km a^{-1} before slowing its retreat and readvancing, consistent with what was observed in the early 2000s (Howat et al., 2005; Murray et al., 2015). The 100-year simulation results also demonstrate that the model can capture short-term oscillations over a longer-term retreat, which is promising for eventual treatment of seasonal cycling.

3.4. Hubbard Glacier, Alaska—8 Auto-Selected Centerlines

Finally, we explore the performance of the plastic network method for a glacier with more complicated geometry. Hubbard Glacier is Alaska's largest tidewater glacier by area, covering 2450 km^2. Unlike many other tidewater glaciers in the region, and apparently independent of climate forcing, it thickened and advanced throughout the 20th century (Arendt et al., 2002; Trabant et al., 2003; Stearns et al., 2015). The advance has been well-documented due to the glacial outburst flood hazard created when the advancing glacier terminus closes Russell Fjord, which has happened twice in the past 30 years (Ritchie et al., 2008).

Huss and Hock (2015) have reconstructed ice thickness for Alaska glaciers following the method of Huss and Farinotti (2012). We estimate bed topography for Hubbard Glacier by subtracting that reconstructed ice thickness from a digital elevation model of the surface provided by Kienholz et al. (2015). The automated algorithm described by Kienholz et al. (2014) identifies 459 tributary centerlines on Hubbard Glacier, from which we choose the 8 that exceed 20 km in length. **Figure 10** shows the bed topography underlying the network of eight branches, with automatically-selected centerlines of the branches in white and intersection points marked in red. Following the optimization procedure described above, we find the best model fit with $\tau_y = 200$ kPa.

Figure 11 shows the plastic model applied to all eight branches. Note that we have restricted the colorbar of **Figure 11C** to saturate at 100% error. Overall, model error on Hubbard Glacier is comparable to other cases, with mean error of +52% and median of +27%. The maximum underestimation of ice thickness is 52%, also comparable to other cases. However, the upper reaches of several branches show high error, with five points reaching 1000% error and an anomalously high 4237% overestimation of ice thickness found at one point in the main branch. The reconstruction method of Huss and Farinotti (2012) constrains ice thickness to be 0 at the ice divide , which likely accounts for such large increases in model error upstream (%$_{\text{error}}$ = $100(H_{\text{model}} - H_{\text{recon}})/H_{\text{recon}}$). Indeed, the point with the highest percent error corresponds to a 747 m overestimation of ice thickness, which is only about 70% of the maximum observed ice thickness for Hubbard Glacier. We expect the plastic model to be more applicable close to the terminus—where stresses are higher and the ice is flowing faster—and the increasing error far upstream agrees with this expectation.

Observations of Hubbard Glacier are sparse above the lowest 15 km (Trabant et al., 2003) and the highest model error coincides with poorly-constrained areas of the glacier. Further, our optimization procedure shows that optimizing for τ_y over the lower 30 km of Hubbard Glacier provides the best, least sensitive fit to observation. For these reasons, we focus our attention on the better-constrained downstream portion of the glacier.

Figure 12 shows the plastic model results on the lower 30 km of Hubbard Glacier. The percent error of the model with respect to the reconstruction of Huss and Hock (2015) in

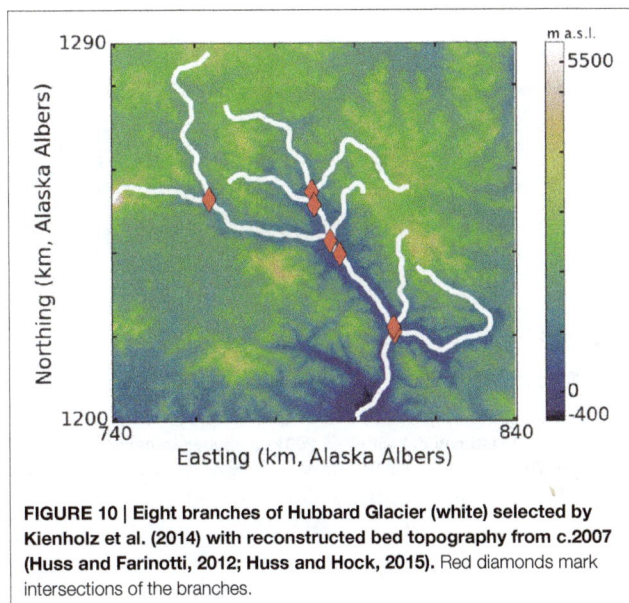

FIGURE 10 | Eight branches of Hubbard Glacier (white) selected by Kienholz et al. (2014) with reconstructed bed topography from c.2007 (Huss and Farinotti, 2012; Huss and Hock, 2015). Red diamonds mark intersections of the branches.

this region is quite low, comparable to percent error on other glaciers. The maximum errors are 27.7% overestimation and 22.3% underestimation of ice thickness, a marked improvement on the error seen on the full eight-branch network.

3.4.1. Advance 1948–present

Previous case studies, as well as our earlier work on Columbia Glacier, documented that our plastic model reproduces observed patterns of retreat (Ultee and Bassis, 2016) as well as rudimentary multi-year cycles of advance and retreat. Now, using observations of the lowest 10 km of Hubbard Glacier from 1948 to the present, we investigate how well the model can reproduce observed tidewater glacier advance. Several published studies (Trabant et al., 2003; U. S. Geological Survey, 2003; McNabb and Hock, 2014) and a US Geological Survey dataset documenting the observed advance provide a useful basis for evaluation.

We choose a reference point 6 km upstream, where the successive longitudinal profiles of Trabant et al. (2003) show a total of 70 m of thickening between 1948 and 2000. The rate of thickening was not constant, but for simplicity we impose a constant rate equal to the 1948–2000 mean— approximately 1.3 m a^{-1} at the reference point. We run the model for every year 1948-2000 with constant annual thickening, and find the terminus advance plotted in **Figure 13**. Mean terminus advance rate over the entire period is 29.7 m a^{-1}, and total advance 1948–2000 is 1.55 km. For comparison, black diamonds in **Figure 13** show the advance reported by the USGS. The modeled total advance of 1.54 km 1948–2000 agrees with the total advance of 1.4 km reported by Trabant et al. (2003), as well as the 1.75 km advance reported for 1948–2012 by McNabb and Hock (2014). The mean advance rate of 29.7 m a^{-1} agrees with the 28 m a^{-1} rate reported by Trabant et al. (2003) and the 23 − 36 m a^{-1} reported by Stearns et al. (2015).

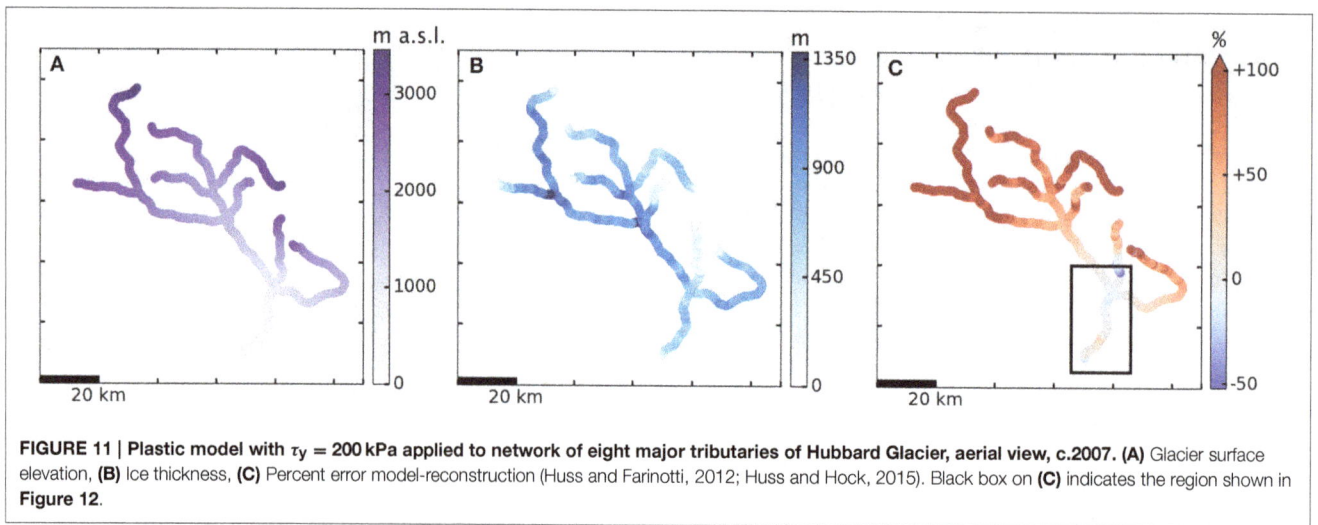

FIGURE 11 | Plastic model with $\tau_y = 200$ kPa applied to network of eight major tributaries of Hubbard Glacier, aerial view, c.2007. (A) Glacier surface elevation, **(B)** Ice thickness, **(C)** Percent error model-reconstruction (Huss and Farinotti, 2012; Huss and Hock, 2015). Black box on **(C)** indicates the region shown in **Figure 12**.

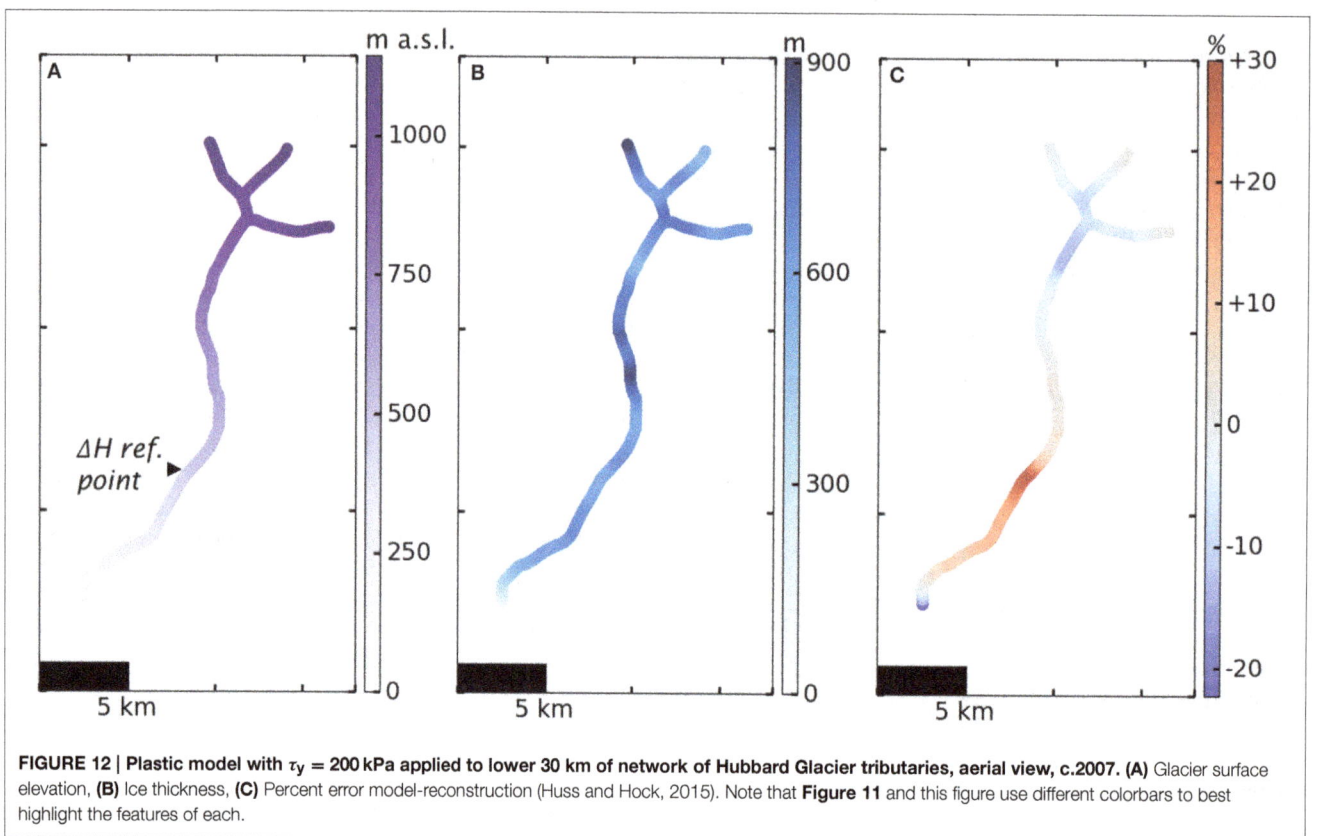

FIGURE 12 | Plastic model with $\tau_y = 200$ kPa applied to lower 30 km of network of Hubbard Glacier tributaries, aerial view, c.2007. (A) Glacier surface elevation, **(B)** Ice thickness, **(C)** Percent error model-reconstruction (Huss and Hock, 2015). Note that **Figure 11** and this figure use different colorbars to best highlight the features of each.

Finally, we explore what can be gained by running the model with a more realistic forcing, deriving the upstream thickening from a fit to observations rather than the 52-year mean. We run the model with time-variable upstream thickening fit to observations and find the terminus advance shown by the dashed curve in **Figure 13**. The variable forcing produces more realistic advance, but only marginally so. Mean terminus advance rate over the entire period is 28.0 m a^{-1}, very similar to the mean advance rate found using a constant forcing. We conclude that using a constant average forcing is an acceptable simplification.

For both forcings, our model shows advance rate decreasing over time, while observations show an increasing rate of advance punctuated by one slow period c.1972–1984 (Stearns et al., 2015). The construction of a moraine shoal through sediment buildup has been shown to be fundamentally important for tidewater glacier advance (Meier and Post, 1987; Oerlemans and Nick, 2006; Goff et al., 2012); we suspect that the lack of sediment transport in our plastic model prevents it from maintaining or accelerating rapid advance as a real glacier could.

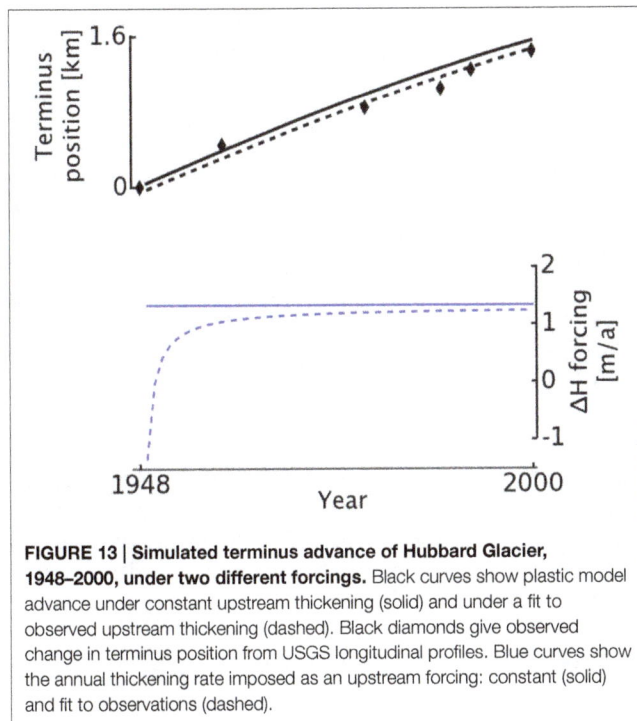

FIGURE 13 | Simulated terminus advance of Hubbard Glacier, 1948–2000, under two different forcings. Black curves show plastic model advance under constant upstream thickening (solid) and under a fit to observed upstream thickening (dashed). Black diamonds give observed change in terminus position from USGS longitudinal profiles. Blue curves show the annual thickening rate imposed as an upstream forcing: constant (solid) and fit to observations (dashed).

4. DISCUSSION

The yield strength τ_y, as the sole adjustable parameter of our model, merits careful examination. We note that there is not one single value of τ_y that works equally well for all observed glaciers; rather, there is variation in the best-fit values. The two Alaska glaciers studied are best fit by lower yield strengths (130–200 kPa), while the Greenland glaciers are best fit by higher yield strengths (245–355 kPa). We hypothesized that the greater physical sophistication of the Coulomb yield criterion would better match observed glacier profiles, but the results of our case studies are inconsistent on this point. For example, CV_{RMS} between modeled and observed main branch profiles is minimized with the Coulomb criterion for Columbia and Helheim Glaciers, but with the constant-yield-strength criterion for Jakobshavn Isbræ and Hubbard Glacier. We suggest that some of the diversity of yield-strength regimes found in our case studies is a reflection of glacier physics, and some is a side effect of the plastic model's simplicity. For example, the two Alaskan glaciers are temperate and receive much more precipitation than the Greenland outlet glaciers, which may contribute to the lower yield strengths of the former. If the model could be adjusted for ice-dynamical factors such as fjord width, temperature, presence of proglacial mélange or ice tongue, etc., we might expect to adjust for those factors with their own parameters and find a more consistent best-fit yield criterion across all observed glaciers. With τ_y as the only parameter of the model, however, it is reasonable that a diverse set of glaciers with unique characteristics would have similar diversity in their best-fit yield strengths. This phenomenon is also recognizable in the inter-branch yield strength disparity found for Helheim Glacier. Nevertheless, we note that the range of τ_y in our case studies

is well within the range of laboratory and field observations (roughly 75–500 kPa).

For networks of two or more branches, we must also consider the inter-branch variability in best-fit τ_y. On Helheim Glacier, that variability is especially noteworthy, with more than 100 kPa difference in the best-fit Coulomb yield strength τ_0 between the small north-easternmost branch and the larger, deeper main branch. In the case studies presented here, we used a single value of τ_y (constant yield strength) or τ_0 (Coulomb yield condition) for all branches of a given glacier network, but future model development could include varying the yield strength by branch. To avoid discontinuities in surface slope, it would be necessary to vary the yield strength smoothly across the junction between branches rather than allowing a step change at the branch point.

Though our plastic network model fits observed profiles of tidewater glaciers remarkably well given its simplicity, some clear model weaknesses remain. For example, the simplest version of the model, making use of a single constant yield strength τ_y, cannot reproduce retreat that occurs while the glacier is thickening, nor advance that occurs while the glacier is thinning. In observed glaciers, upstream thinning usually corresponds to terminus retreat, and thickening to advance, but not always. For example, about 10% of the Alaska glaciers sampled by Arendt et al. (2002) exhibited thickening-retreat or thinning-advance. Further, the model response to forcing is instantaneous. Driving the model upstream produces instant change in terminus position, and driving the model at the terminus produces instant change in the upstream elevation profile. Therefore, we remain unable to comment on the causal relationship between retreat and thinning (see also Ultee and Bassis, 2016). More fundamentally, lack of a sediment transport mechanism hampers our model's simulation of realistic tidewater glacier advance, as illustrated by the Hubbard Glacier case study. Another obstacle to wide implementation of our model is the input data required: bed topography and bathymetry, which are not globally available. We have treated case studies of Columbia and Hubbard Glaciers, which are among the best-constrained Alaska tidewater glaciers. While extending the model to other Alaska tidewater glaciers seems a natural next step, further study is hampered by limited observational constraints on glacier bed topography.

4.1. Introducing Climate Forcing

The highly simplified model implementation presented here relies on manual input of an upstream thickening or thinning rate for each glacier. Manual input may be constrained by observational data, but it is a poor substitute for true climate forcing. Further, the technique is not scalable to the hundreds of tidewater glaciers worldwide, limiting the utility of the model. Two options exist to remedy this situation: introducing climate forcing directly into the plastic model, for example by adding functionality to treat changes in surface mass balance, or coupling upstream with a more sophisticated model that handles climate forcing itself. We discuss the first tactic in Bassis and Ultee (in review) and focus our attention here on the second tactic.

Coupling with a more complex glacier dynamics model (e.g., Marzeion et al., 2012) may be possible, but will require judicious choice of coupling location x_{couple}. We need only require that ice thickness remains continuous at the junction, but the best x_{couple} may depend on the glacier and the choice of upstream model. Without the benefit of coupling experiments to guide us, we suggest that the most consistent choice of x_{couple} will match not only ice thickness, but basal shear stress at the junction point. The plastic model assumes that basal stress over the entire glacier is exactly the yield strength of ice, τ_y, which may vary spatially according to the Coulomb condition of Equation (5) (see also Ultee and Bassis, 2016). In theory, basal stress of a tidewater glacier flowing down a valley without large pinning points should increase downstream throughout the accumulation area, which should be reflected in the glacier models with which our model could couple. The appropriate location to induce coupling, x_{couple}, is the point along the centerline of the main glacier branch where the basal stress according to the upstream model matches our model's yield strength of ice to within a certain tolerance. Large, abrupt features such as upstream ice falls—like the one on Columbia Glacier's main branch—may affect selection of x_{couple}. Careful initial experiments will help in designing a general method of avoiding such problems.

At x_{couple}, we require that the ice thickness of the plastic glacier, $H_{couple} \equiv H(x_{couple})$, match that found by the upstream model. As the upstream model evolves in time, the plastic glacier downstream will respond to changes in H_{couple}. In this way, the plastic model can respond to changes in climate forcing applied upstream. We note that the climate forcing relevant to tidewater glaciers includes not only atmosphere forcing, but ocean forcing as well. However, coupling our model to an ocean model at the calving front would be considerably more complicated to implement, and we have not explored that possibility.

5. CONCLUSION

We have presented a model that balances the simplicity of one-dimensional flowline modeling with the power of explicitly representing intersections between glacier branches. Our case studies indicate that the model can produce surface elevation profiles of multiple branches with low error, as well as realistically simulate the advance and retreat of Alaska

and Greenland tidewater glaciers. Though our method gains important validation from detailed application to individual glaciers, its true appeal is in scaling up to large networks of tidewater glaciers, such as those draining the Greenland Ice Sheet. Simulation of those larger networks can be enhanced by explicit inclusion of surface mass balance—as described in Bassis and Ultee (2017, submitted manuscript)—or coupling to a more sophisticated model upstream.

Our initial case studies, performed without high-quality local climate data or coupling to a sophisticated ice dynamics model, give reason for optimism. In particular, the cases of Columbia Glacier (see Ultee and Bassis, 2016) and Hubbard Glacier demonstrate that a simple constant upstream forcing can reproduce terminus advance and retreat with surprising accuracy. More precise forcing based on observations of the glacier or the local climate does offer marginal improvement in model performance, as demonstrated for Hubbard Glacier above, but it is not essential. Thus, our model offers useful projections despite the inherent uncertainty in models of future climate.

AUTHOR CONTRIBUTIONS

LU wrote the model, ran simulations, and wrote most of the manuscript. JNB provided direction in model development and collaborated in writing the manuscript. Both authors contributed to designing the project.

FUNDING

This work was supported by National Science Foundation grant ANT-114085 and the National Oceanic and Atmospheric Administration, Climate Process Team: Iceberg Calving grant NA13OAR4310096.

ACKNOWLEDGMENTS

We thank Christian Kienholz, Matthias Huss, and Shad O'Neel for providing essential data.

REFERENCES

Arendt, A. A., Echelmeyer, K. A., Harrison, W. D., Lingle, C. S., and Valentine, V. B. (2002). Rapid wastage of Alaska glaciers and their contribution to rising sea level. *Science* 297, 382–386. doi: 10.1126/science.1072497

Bassis, J. N., and Jacobs, S. (2013). Diverse calving patterns linked to glacier geometry. *Nat. Geosci.* 6, 833–836. doi: 10.1038/ngeo1887

Bassis, J. N., and Walker, C. C. (2012). Upper and lower limits on the stability of calving glaciers from the yield strength envelope of ice. *Proc. R. Soc. Lond. A Math. Phys. Eng. Sci.* 468, 913–931. doi: 10.1098/rspa.2011.0422

Church, J. A., Clark, P. U., Cazenave, J. M., Jevrejeva, S., Levermann, A., Merrifield, M. A., et al. (2013). "Sea level change," in *Climate Change 2013: The Physical Science Basis*, eds T. F. Stocker, D. Qin, G. K. Plattner, M. Tignor, S. K. Allen, J. Boschung, A. Nauels, Y. Xia, V. Bex, and P. M. Midgley (Cambridge, UK; New York, NY: Cambridge University Press), 1137–1216.

Cohen, D., Iverson, N. R., Hooyer, T. S., Fischer, U. H., Jackson, M., and Moore, P. L. (2005). Debris-bed friction of hard-bedded glaciers. *J. Geophys. Res. Earth Surf.* 110:F02007. doi: 10.1029/2004JF000228

Cornford, S. L., Martin, D. F., Graves, D. T., Ranken, D. F., Brocq, A. M. L., Gladstone, R. M., et al. (2013). Adaptive mesh, finite volume modeling of marine ice sheets. *J. Comput. Phys.* 232, 529–549. doi: 10.1016/j.jcp.2012.08.037

CReSIS (2016a). *CReSIS L3 Radar Depth Sounder Data: Helheim Glacier*. Available online at: http://data.cresis.ku.edu/

CReSIS (2016b). *CReSIS L3 Radar Depth Sounder Data: Jakobshavn Isbræ*. Available online at: http://data.cresis.ku.edu/

Csatho, B., Schenk, T., van der Veen, C., and Krabill, W. B. (2008). Intermittent thinning of Jakobshavn Isbræ, West Greenland, since the Little Ice Age. *J. Glaciol.* 54, 131–144. doi: 10.3189/002214308784409035

Cuffey, K., and Paterson, W. (2010). *The Physics of Glaciers, 4th Edn.* Burlington; Kidlington: Elsevier Science.

Enderlin, E. M., Hamilton, G. S., O'Neel, S., Bartholomaus, T. C., Morlighem, M., and Holt, J. W. (2016). An empirical approach for estimating stress-coupling lengths for marine-terminating glaciers. *Front. Earth Sci.* 4:104. doi: 10.3389/feart.2016.00104

Goff, J. A., Lawson, D. E., Willems, B. A., Davis, M., and Gulick, S. P. S. (2012). Morainal bank progradation and sediment accumulation in disenchantment bay, alaska: response to advancing hubbard glacier. *J. Geophys. Res. Earth Surf.* 117:F02031. doi: 10.1029/2011JF002312

Howat, I. M., Ahn, Y., Joughin, I., van den Broeke, M. R., Lenaerts, J. T. M., and Smith, B. (2011). Mass balance of Greenland's three largest outlet glaciers, 2000–2010. *Geophys. Res. Lett.* 38:L12501. doi: 10.1029/2011GL047565

Howat, I. M., Box, J. E., Ahn, Y., Herrington, A., and McFadden, E. M. (2010). Seasonal variability in the dynamics of marine-terminating outlet glaciers in Greenland. *J. Glaciol.* 56, 601–613. doi: 10.3189/002214310793146232

Howat, I. M., Joughin, I., Tulaczyk, S., and Gogineni, S. (2005). Rapid retreat and acceleration of Helheim Glacier, east Greenland. *Geophys. Res. Lett.* 32:L22502. doi: 10.1029/2005GL024737

Huss, M., and Farinotti, D. (2012). Distributed ice thickness and volume of all glaciers around the globe. *J. Geophys. Res. Earth Surf.* 117:F04010. doi: 10.1029/2012JF002523

Huss, M., and Hock, R. (2015). A new model for global glacier change and sea-level rise. *Front. Earth Sci.* 3:54. doi: 10.3389/feart.2015.00054

Joughin, I., Abdalati, W., and Fahnestock, M. (2004). Large fluctuations in speed on Greenland's Jakobshavn Isbrae glacier. *Nature* 432, 608–610. doi: 10.1038/nature03130

Joughin, I., Smith, B. E., Howat, I. M., Floricioiu, D., Alley, R. B., Truffer, M., et al. (2012). Seasonal to decadal scale variations in the surface velocity of Jakobshavn Isbræ, Greenland: observation and model-based analysis. *J. Geophys. Res. Earth Surf.* 117:F02030. doi: 10.1029/2011JF002110

Joughin, I., Smith, B. E., Shean, D. E., and Floricioiu, D. (2014). Brief communication: further summer speedup of Jakobshavn Isbræ. *Cryosphere* 8, 209–214. doi: 10.5194/tc-8-209-2014

Kienholz, C., Herreid, S., Rich, J. L., Arendt, A. A., Hock, R., and Burgess, E. W. (2015). Derivation and analysis of a complete modern-date glacier inventory for Alaska and northwest Canada. *J. Glaciol.* 61, 403–420. doi: 10.3189/2015JoG14J230

Kienholz, C., Rich, J. L., Arendt, A. A., and Hock, R. (2014). A new method for deriving glacier centerlines applied to glaciers in Alaska and northwest Canada. *Cryosphere* 8, 503–519. doi: 10.5194/tc-8-503-2014

Krimmel, R. M. (2001). *Photogrammetric Data Set, 1957–2000, and Bathymetric Measurements for Columbia Glacier, Alaska*. Reston, VA: US Geological Survey.

Machguth, H., and Huss, M. (2014). The length of the world's glaciers—a new approach for the global calculation of center lines. *Cryosphere* 8, 1741–1755. doi: 10.5194/tc-8-1741-2014

Marzeion, B., Jarosch, A. H., and Hofer, M. (2012). Past and future sea-level change from the surface mass balance of glaciers. *Cryosphere* 6, 1295–1322. doi: 10.5194/tc-6-1295-2012

McNabb, R., Hock, R., O'Neel, S., Rasmussen, L., Ahn, Y., Braun, M., et al. (2012). Using surface velocities to calculate ice thickness and bed topography: a case study at Columbia Glacier, Alaska, USA. *J. Glaciol.* 58, 1151–1164. doi: 10.3189/2012JoG11J249

McNabb, R. W., and Hock, R. (2014). Alaska tidewater glacier terminus positions, 1948–2012. *J. Geophys. Res. Earth Surf.* 119, 153–167. doi: 10.1002/2013JF002915

Meier, M. F., and Post, A. (1987). Fast tidewater glaciers. *J. Geophys. Res. Solid Earth* 92, 9051–9058. doi: 10.1029/JB092iB09p09051

Murray, T., Scharrer, K., James, T. D., Dye, S. R., Hanna, E., Booth, A. D., et al. (2010). Ocean regulation hypothesis for glacier dynamics in southeast greenland and implications for ice sheet mass changes. *J. Geophys. Res. Earth Surf.* 115:F03026. doi: 10.1029/2009JF001522

Murray, T., Scharrer, K., Selmes, N., Booth, A. D., James, T. D., Bevan, S. L., et al. (2015). Extensive retreat of Greenland tidewater glaciers, 2000–2010. *Arctic Antarctic Alpine Res.* 47, 427–447. doi: 10.1657/AAAR0014-049

Nye, J. F. (1951). The flow of glaciers and ice-sheets as a problem in plasticity. *Proc. R. Soc. Lond. A Math. Phys. Eng. Sci.* 207, 554–572. doi: 10.1098/rspa.1951.0140

Nye, J. F. (1952). The mechanics of glacier flow. *J. Glaciol.* 2, 82–93. doi: 10.1017/S0022143000033967

Nye, J. F. (1953). The flow law of ice from measurements in glacier tunnels, laboratory experiments and the Jungfraufirn borehole experiment. *Proc. R. Soc. Lond. A Math. Phys. Eng. Sci.* 219, 477–489. doi: 10.1098/rspa.1953.0161

Oerlemans, J., and Nick, F. (2006). Modelling the advance–retreat cycle of a tidewater glacier with simple sediment dynamics. *Glob. Planet. Change* 50, 148–160. doi: 10.1016/j.gloplacha.2005.12.002

O'Neel, S., Pfeffer, W. T., Krimmel, R., and Meier, M. (2005). Evolving force balance at Columbia Glacier, Alaska, during its rapid retreat. *J. Geophys. Res. Earth Surf.* 110:F03012. doi: 10.1029/2005JF000292

Rignot, E., Mouginot, J., Larsen, C. F., Gim, Y., and Kirchner, D. (2013). Low-frequency radar sounding of temperate ice masses in southern Alaska. *Geophys. Res. Lett.* 40, 5399–5405. doi: 10.1002/2013GL057452

Rignot, E., Velicogna, I., van den Broeke, M. R., Monaghan, A., and Lenaerts, J. T. M. (2011). Acceleration of the contribution of the greenland and antarctic ice sheets to sea level rise. *Geophys. Res. Lett.* 38:L05503. doi: 10.1029/2011GL046583

Ritchie, J. B., Lingle, C. S., Motyka, R. J., and Truffer, M. (2008). Seasonal fluctuations in the advance of a tidewater glacier and potential causes: Hubbard Glacier, Alaska, USA. *J. Glaciol.* 54, 401–411. doi: 10.3189/002214308785836977

Stearns, L. A., and Hamilton, G. S. (2007). Rapid volume loss from two East Greenland outlet glaciers quantified using repeat stereo satellite imagery. *Geophys. Res. Lett.* 34:L05503. doi: 10.1029/2006GL028982

Stearns, L. A., Hamilton, G. S., vander Veen, C. J., Finnegan, D. C., O'Neel, S., Scheick, J. B., et al. (2015). Glaciological and marine geological controls on terminus dynamics of Hubbard Glacier, southeast Alaska. *J. Geophys. Res. Earth Surf.* 120, 1065–1081. doi: 10.1002/2014JF003341

Straneo, F., Heimbach, P., Sergienko, O., Hamilton, G., Catania, G., Griffies, S., et al. (2013). Challenges to understanding the dynamic response of Greenland's marine terminating glaciers to oceanic and atmospheric forcing. *Bull. Amer. Meteorol. Soc.* 94, 1131–1144. doi: 10.1175/BAMS-D-12-00100.1

Trabant, D. C., Krimmel, R. M., Echelmeyer, K. A., Zirnheld, S. L., and Elsberg, D. H. (2003). The slow advance of a calving glacier: Hubbard Glacier, Alaska, U.S.A. *Ann. Glaciol.* 36, 45–50. doi: 10.3189/172756403781816400

Ultee, L., and Bassis, J. N. (2016). The future is Nye: an extension of the perfect plastic approximation to tidewater glaciers. *J. Glaciol.* 62, 1143–1152. doi: 10.1017/jog.2016.108

U. S. Geological Survey (2003). *Hubbard Glacier, Alaska: Growing and Advancing in Spite of Global Climate Change and the 1986 and 2002 Russell Lake Outburst Floods*. USGS Fact Sheet 001-03.

van den Broeke, M. R., Enderlin, E. M., Howat, I. M., Kuipers Munneke, P., Noël, B. P. Y., van de Berg, W. J., et al. (2016). On the recent contribution of the greenland ice sheet to sea level change. *Cryosphere* 10, 1933–1946. doi: 10.5194/tc-10-1933-2016

Vaughan, D. G., Comiso, J., Allison, I., Carrasco, J., Kaser, G., Kwok, R., et al. (2013). "Observations: cryosphere," in *Climate Change 2013: The Physical Science Basis*, eds T. F. Stocker, D. Qin, G. K. Plattner, M. Tignor, S. K. Allen, J. Boschung, A. Nauels, Y. Xia, V. Bex, and P. M. Midgley (Cambridge, UK; New York, NY: Cambridge University Press), 317–382.

Conflict of Interest Statement: The authors declare that the research was conducted in the absence of any commercial or financial relationships that could be construed as a potential conflict of interest.

8

New Advances in Dial-Lidar-Based Remote Sensing of the Volcanic CO_2 Flux

Alessandro Aiuppa[1,2*], Luca Fiorani[3], Simone Santoro[1,4], Stefano Parracino[4,5], Roberto D'Aleo[1], Marco Liuzzo[2], Giovanni Maio[4,6†] and Marcello Nuvoli[3]

[1] Dipartimento DiSTeM, Università di Palermo, Palermo, Italy, [2] Istituto Nazionale di Geofisica e Vulcanologia, Palermo, Italy, [3] Fusion and Technology for Nuclear Safety and Security Department, ENEA, Frascati, Italy, [4] ENEA Guest, Frascati, Italy, [5] Department of Industrial Engineering, University of Rome "Tor Vergata", Rome, Italy, [6] Vitrociset SpA, Roma, Italy

Edited by:
John Stix,
McGill University, Canada

Reviewed by:
Jacob B. Lowenstern,
United States Geological Survey, USA
Marie Edmonds,
University of Cambridge, UK

*Correspondence:
Alessandro Aiuppa
alessandro.aiuppa@unipa.it

†Present Address:
Giovanni Maio,
ARES Consortium, Rome, Italy

Specialty section:
This article was submitted to
Volcanology,
a section of the journal
Frontiers in Earth Science

We report here on the results of a proof-of-concept study aimed at remotely sensing the volcanic CO_2 flux using a Differential Adsorption lidar (DIAL-lidar). The observations we report on were conducted in June 2014 on Stromboli volcano, where our lidar (LIght Detection And Ranging) was used to scan the volcanic plume at $\sim$3 km distance from the summit vents. The obtained results prove that a remotely operating lidar can resolve a volcanic CO_2 signal of a few tens of ppm (in excess to background air) over km-long optical paths. We combine these results with independent estimates of plume transport speed (from processing of UV Camera images) to derive volcanic CO_2 flux time-series of $\approx$16–33 min temporal resolution. Our lidar-based CO_2 fluxes range from 1.8 ± 0.5 to 32.1 ± 8.0 kg/s, and constrain the daily averaged CO_2 emissions from Stromboli at 8.3 ± 2.1 to 18.1 ± 4.5 kg/s (or 718–1565 tons/day). These inferred fluxes fall within the range of earlier observations at Stromboli. They also agree well with contemporaneous CO_2 flux determinations (8.4–20.1 kg/s) obtained using a standard approach that combines Multi-GAS-based in-plume readings of the CO_2/SO_2 ratio ($\approx$8) with UV-camera sensed SO_2 fluxes (1.5–3.4 kg/s). We conclude that DIAL-lidars offer new prospects for safer (remote) instrumental observations of the volcanic CO_2 flux.

Keywords: volcanic CO_2, DIAL-lidar, Stromboli, remote sensing, CO_2 flux

INTRODUCTION

A major step forward in ground-based volcano monitoring has recently arisen from the advent of modern instrumental techniques and networks for volcanic gas observations (Galle et al., 2010; Oppenheimer et al., 2014; Saccorotti et al., 2014; Fischer and Chiodini, 2015). Such technical advances provide improved temporal resolution relative to traditional direct sampling techniques (Symonds et al., 1994; Giggenbach, 1996). As longer-term volcanic gas records increase in number and quality, full empirical evidence is finally emerging for increased CO_2 flux emissions prior to eruption of mafic to intermediate volcanoes (Aiuppa, 2015). Precursory plume CO_2 flux increases have been now detected at several volcanoes, including Etna (Aiuppa et al., 2008; Patanè et al., 2013), Kilauea (Poland et al., 2012), Redoubt (Werner et al., 2013), Turrialba (de Moor et al., 2016a), and Poas (de Moor et al., 2016b).

At Stromboli (in Italy), however, CO_2 flux observations have been particularly valuable for interpreting, and eventually predicting, the volcano's behavior (Aiuppa et al., 2010a, 2011). On

Stromboli, the "regular" mild strombolian activity is occasionally interrupted by larger-scale vulcanian-style explosions, locally referred as "*major explosions*" or (in the most extreme events) "*paroxysms*" (Rosi et al., 2006, 2013; Andronico and Pistolesi, 2010; Pistolesi et al., 2011; Pioli et al., 2014). These explosions, although short-lived (tens of seconds to a few minutes), represent a real hazard for local populations, tourists and volcanologists, since they produce fallout of coarse pyroclastic materials over wide dispersal areas (Rosi et al., 2013). In addition, such events are not anticipated by any detectable anomaly in the geophysical or volcanological record, perhaps because they originate deep in the crustal roots of the volcano's plumbing system (Bertagnini et al., 2003; Métrich et al., 2005, 2010; Allard, 2010). Observational evidence suggests, however, that "*major explosions*" (Aiuppa et al., 2011) and "*paroxysms*" (Aiuppa et al., 2010a) are both systematically preceded by days/weeks of anomalous CO_2-rich gas leakage from Stromboli's deep (8–10 km) magma storage zone (Aiuppa et al., 2010b). CO_2 flux emissions from the open-vent crater plume have become, therefore, a unique monitoring tool for volcanic hazard assessment and mitigation on the volcano.

On Stromboli, as at other volcanoes, the volcanic gas CO_2 flux is calculated from a combination of co-measured SO_2 fluxes and plume CO_2/SO_2 ratios (Burton et al., 2013; Aiuppa, 2015). While the SO_2 flux can remotely be sensed by UV spectroscopy (Oppenheimer, 2010; Oppenheimer et al., 2011), measuring the CO_2/SO_2 ratio requires *in-situ* direct sampling and/or measurements via Multi-GAS (Aiuppa et al., 2010a) or Fourier Transform Infra-Red Spectrometry (La Spina et al., 2013)

in the vicinity of hazardous active vents. As such, implementation of novel techniques for the remote observation of the volcanic CO_2 flux, from more distal (and safer) locations, remains highly desirable.

New prospects for ground-based *remote* detection of the volcanic CO_2 flux have recently become available from the advent of a new lidar (Light Detection and Ranging) using the DIAL (Differential Absorption lidar) technique (Aiuppa et al., 2015; Fiorani et al., 2015, 2016). DIAL-lidars (Weitkamp, 2005; Fiorani, 2007) use backscattering of artificial light (laser) from atmospheric back-scatterers and/or from the volcanic plume itself, and are therefore potentially ideal for remote volcanic CO_2 detection (Fiorani et al., 2013; Queißer et al., 2015, 2016). In previous work, we demonstrated the ability of our lidar to remotely resolve the volcanic CO_2 flux from a relatively proximal measuring site (<200 m from the source vents) (Aiuppa et al., 2015). Here, we extend this work by reporting on a successful CO_2 flux detection at Stromboli over a far longer optical path (~3 km distance from the vents). Results of this proof-of-concept experiment confirm lidars as promising tools for remote monitoring of the volcanic CO_2 flux (Aiuppa et al., 2015; Queißer et al., 2016).

MATERIALS AND METHODS

The Bridge Lidar

Our measurements on Stromboli (**Figure 1**) were obtained using the same DIAL-lidar described in Aiuppa et al. (2015) and Fiorani et al. (2015, 2016), and realized within the context of the FP7-

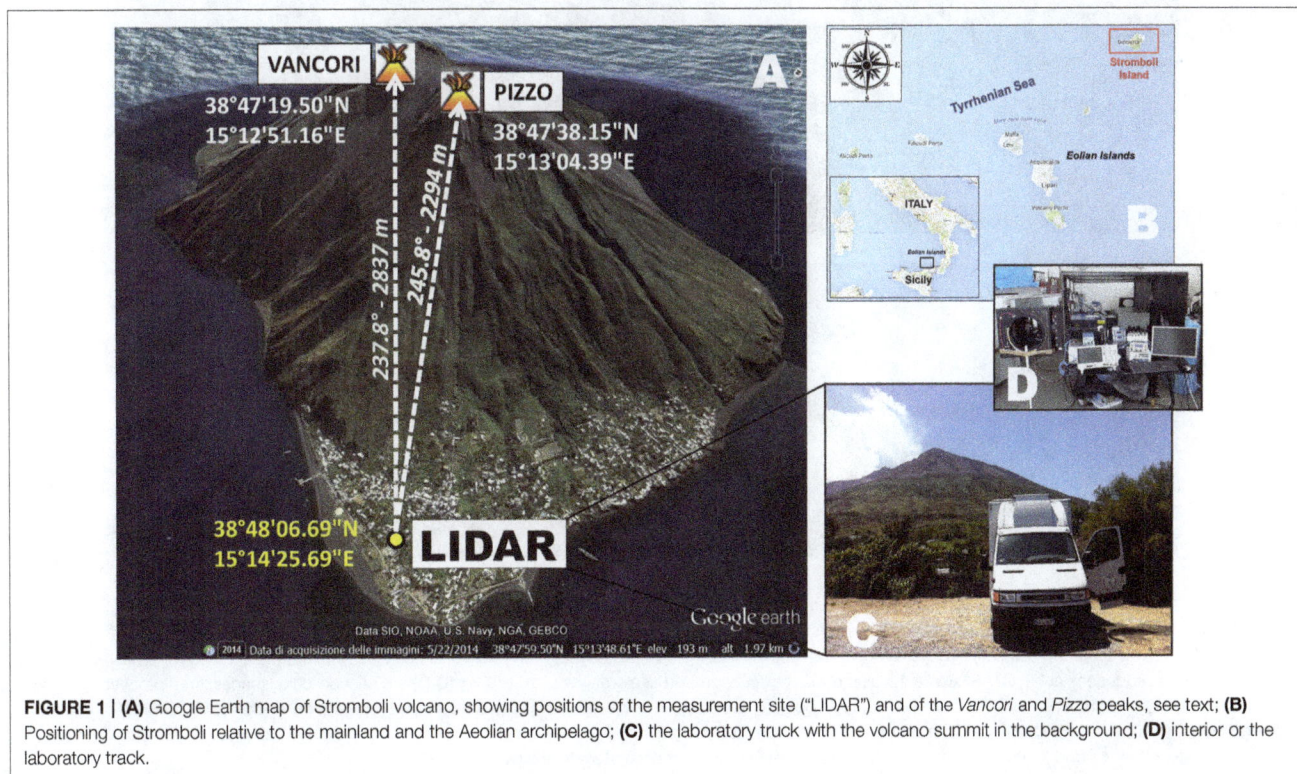

FIGURE 1 | (A) Google Earth map of Stromboli volcano, showing positions of the measurement site ("LIDAR") and of the *Vancori* and *Pizzo* peaks, see text; **(B)** Positioning of Stromboli relative to the mainland and the Aeolian archipelago; **(C)** the laboratory truck with the volcano summit in the background; **(D)** interior or the laboratory track.

ERC project *Bridge* (www.bridge.unipa.it). Only key information is reported here, and the reader is referred to previous studies for a detailed description of the instrument. In brief, the Bridge lidar (**Figure 1D**) uses a complex transmitter that integrates (i) an injection seeded Nd:YAG laser with (ii) a double grating dye laser. This transmitter is used to generate laser radiation at ~2010 nm, a region of the electromagnetic spectrum absorbed by atmospheric CO_2, while showing minimal cross-sensitivity to H_2O (Fiorani et al., 2013). At the ON and OFF wavelengths selected for this experiment, the differential cross section of CO_2 is five orders of magnitude larger than that of H_2O (Rothman et al., 2013). Considering a CO_2 mixing ratio of 400 ppm, and with the upper and lower ranges of H_2O mixing ratios used in atmospheric models (Berk et al., 2014), i.e., from 2.59% (tropics, sea level) to 0.141% (high latitude, winter, sea level), the respective CO_2 absorption is 3 and 5 orders of magnitude larger than that of H_2O. The 2.59% H_2O mixing ratio is not far from the saturated water vapor pressure at standard atmospheric conditions. We conclude that, even in a condensing volcanic plume, H_2O absorption is negligible compared to that of CO_2.

A piezo-electric element is used to sequentially switch the wavelength of the transferred laser beam, from λ_{ON} (2009.537 mm: maximum CO_2 absorption) to λ_{OFF} (2008.484 nm: no CO_2 absorption), at 10 Hz repetition rate. These closely spaced pairs of laser beams are sequentially transmitted into the atmosphere, where they are eventually scattered back by atmospheric back-scatterers (aerosols, water droplets, particles) in either the volcanic plume or the background atmosphere. During their atmospheric propagation, the laser beams are also reflected by any obstacle encountered along the optical path, e.g., in our specific case, the *Pizzo* and *Vancori* walls/rims in front of or behind the volcanic plume (see **Figures 1**, **2**). The returned signal is captured by the lidar receiver (a Newtonian telescope, diameter: 310 mm), and then detected and amplified by an InGaAs PIN photodiode module, directly connected with the analog-to-digital converter (ADC).

Field Operations

During our experiment, the DIAL-lidar operated from a small laboratory truck (**Figure 1C**), positioned in a fixed position at the base of the volcano in the *Scari* area, ~2–2.5 km from the degassing vents on the volcano's summit (**Figures 1**, **2**). The lidar operated during June 24–29, 2015, including an initial instrumental setup phase. Stable weather conditions

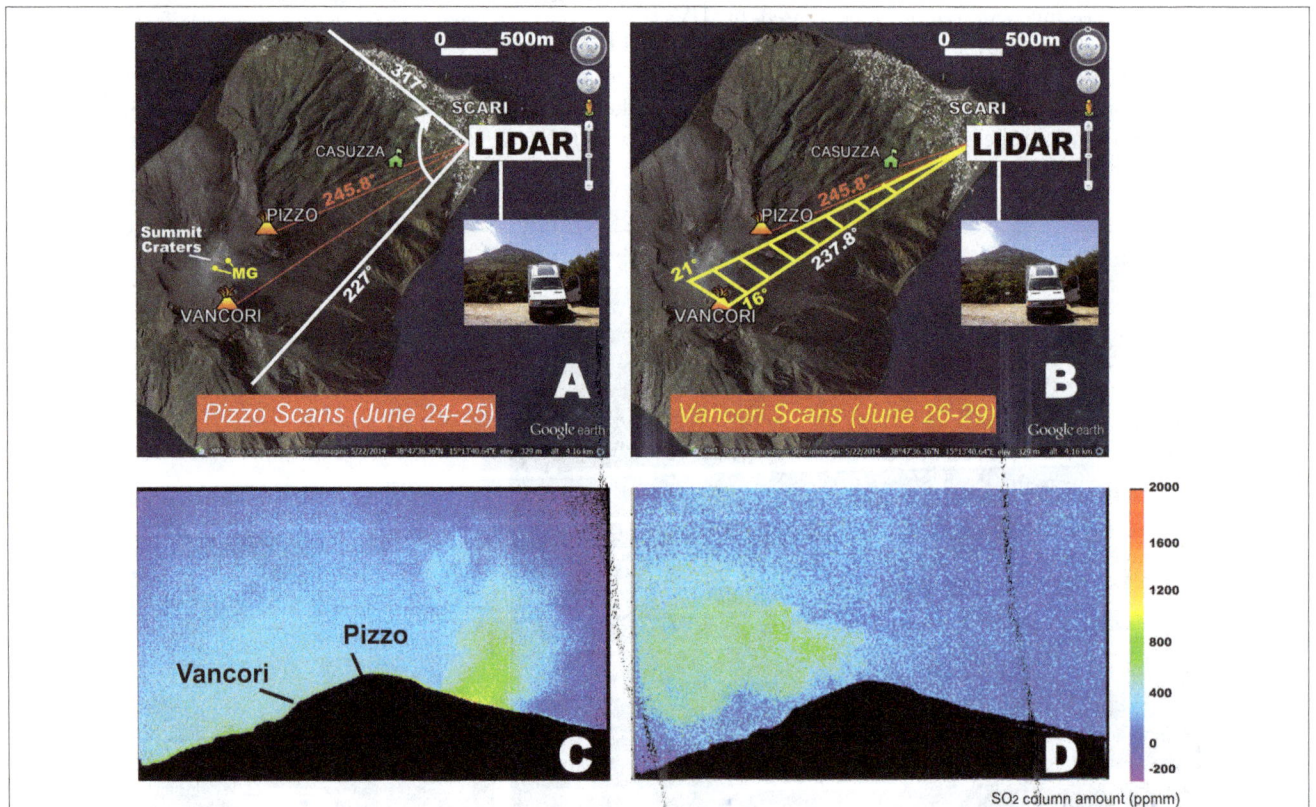

FIGURE 2 | Maps illustrating geometries of (A) *Pizzo scans* and **(B)** *Vancori scans*. In the horizontal *Pizzo scan* in **(A)**, the Field Of View (FOV) of the lidar was sequentially rotated (at constant elevation) at heading angles ranging from 227° to 317° (the *Pizzo* morphological peak was intercepted at ~245.8°). In **(B)**, the heading angle was kept constant at 237.8°, while the plume was vertically profiled at elevations of 16 to 21°. **(C,D)** are pseudo-color images, from processing of UV camera data, showing distribution of SO_2 column amounts (in ppm m, see scale). The locations of *Pizzo* and *Vancori* are indicated in **(C)**. During June 24–25, the UV camera images (see example in **C**) identified the plume as a nearly vertically rising band of peak SO_2 column amount, north of the *Pizzo* area; on June 26–29, the plume was instead transported south-southeast of *Pizzo* by the prevailing north-northwesterly winds (see image **D**).

(temperature, 22–26 °C; no rainfall) persisted during the entire measurement window.

During operations, two large motorized elliptical mirrors (major axis: 450 mm) simultaneously aimed the laser beams and the telescope, allowing the laser beam of the lidar to scan the volcanic plume either horizontally (**Figure 2A**) or vertically (**Figure 2B**). In particular, during June 24–25, the volcanic plume was mainly dispersed northwards by gentle southerly winds. From our *Scari* observation point (**Figure 2**), the plume was seen to rise nearly vertically north of the *Pizzo* area (**Figure 2C**). The Line Of Sight (LOS) of the lidar was therefore pointed north of *Pizzo* and the horizontal scan mode was preferred (heading angles: 227–317°; **Figure 2A**). Vertical scans above the *Pizzo* area were also performed. For simplicity, we refer below to these June 24–25 measurements as the *Pizzo scans* (**Figure 2A**).

On June 26–29, the plume was instead transported south-southeast by the prevailing north-northwesterly winds (**Figure 2D**). Vertical scans were therefore preferred that were operated at constant heading angle (237.8°) and at elevation angles from 16 to 21° (**Figure 2B**). In such conditions, the *Pizzo* and *Vancori* peaks were intercepted at elevation angles of 16.98° and 17.78°, respectively, and the volcanic plume was in all cases encountered in the 2300–2700 m range. We hereafter refer to these scans as the *Vancori scans* (**Figure 2B**).

During each profile, 100 lidar returns, 50 at λ_{ON}, 50 at λ_{OFF}, and interlaced (OFF after ON, OFF again and so on), were emitted at a 10 Hz rate, then co-added and averaged to increase the signal-to-noise ratio, reducing the signal sampling frequency to 0.1 Hz (temporal resolution of 10 s). The spatial resolution was about 5 m (corresponding to the rise time of the detector module due to its bandwidth). Plume scans, both horizontal and vertical, were retrieved combining about 50 profiles in <10 min. Typically, 10 scans at different elevations were repeated, obtaining a three-dimensional tomography of the volcanic plume.

A cell filled with standard CO_2 gas was periodically used during operations, for check of wavelength accuracy, repeatability and stability. In brief, our calibration procedure involved measuring-by photoacoustic spectroscopy—the absorption of the CO_2 gas cell as a function of wavelength. This calibration, limited to a small interval near the predicted λ_{ON}, allowed identifying the wavelength at which cell absorption is maximum. The laser system was finally forced to transmit at this radiation. The CO_2 absorption cross-section used in our calculations was based on HITRAN data (Rothman et al., 2013).

UV Camera

Concurrently with our lidar observations, a dual-UV camera system (Kantzas et al., 2010; Tamburello et al., 2012; Burton et al., 2014) was used to monitor the temporal variations of the SO_2 flux and plume transport speed. A fully autonomous system, similar to that used in other recent work (D'Aleo et al., 2016), was mounted on the roof of the laboratory truck and operated every day from 6 am to 4 pm (Local Time). The UV camera system acquired sequential images of the plume at ∼0.5 Hz using two JAI CM 140 GR cameras. Both cameras

had 10-bit digitization and 1392 × 1040 pixels, using an Uka Optics UV lens with a ∼37° field of view. Distinct band-pass filters, centerd at either 310 nm (where SO_2 absorbs) or 330 nm (no SO_2 absorption), were mounted on the back on the lenses of the two cameras. Each set of co-acquired images from the two UV cameras was processed using the methodology of Kantzas et al. (2010) and integrated into the Vulcamera software (Tamburello et al., 2011, 2012), to calculate an absorbance for each camera pixel. Absorbance was converted into an SO_2 column amount from readings of a co-exposed Ocean optics USB2000+ UV Spectrometer, as outlined in Lübcke et al. (2013). Cameras and spectrometer were both controlled by a mini-pc Jetway. To calculate SO_2 flux time-series, we used Vulcamera to derive temporal records of SO_2 integrated column amounts (ICAs) along a plume cross-section, perpendicular to the plume transport direction. The obtained ICA time-series were then combined with high-temporal resolution (∼1 Hz) records of plume transport speed. This latter was derived using an Optical Flow sub-routine using the Lukas/Kanade algorithm (Bruhn et al., 2005; Peters et al., 2015), integrated in Vulcamera. In our specific case, the Lucas-Kanade method was used to track movements of gas fronts (e.g., gas-rich and/or ash-free portions of the plume, having well distinct absorbance features) in consecutive UV camera frames, which allowed us quantifying plume transport speed at 0.5 Hz. We tested performance of this method by using artificial images with known particle velocities, and obtained errors in estimated velocities of <5%. **Table 1** lists daily means (±1 SD: standard deviation) of both SO_2 fluxes and plume transport speed (V_p) during our observational period.

RESULTS

Characteristics of the DIAL-Lidar Signal

According to lidar theory (Fiorani, 2007), the optical power returned to the lidar receiver at any time t is produced by back-scattering of the laser beam by an atmospheric layer at distance R (range) from the source, where $R = ct/2$ and c is the speed of light. As such, the lidar offers range-resolved information on atmospheric structure and properties (aerosols, particles and gas molecules) along the laser beam, in the form of an intensity (I) vs. range plot (**Figure 3**).

Upon its atmospheric propagation, the beam intensity decreases approximately (a) exponentially, due to atmospheric extinction, according to the Lambert-Beer law; and (b) as $1/R^2$, because the solid angle subtended by the receiver is A/R^2, where A is the telescope's effective area. The two processes are superimposed. As such, in order to better observe the atmospheric back-scattering, a "range corrected signal, S" is commonly used, being given by: $S = \ln(I\ R^2)$ (see below). Since the system works in DIAL mode, each intensity profile is in fact acquired at two distinct wavelengths, λ_{ON}–absorbed by CO_2–and λ_{OFF}–not absorbed by CO_2 (**Figure 3**). The two wavelengths are so close that atmospheric behavior, except from CO_2 absorption, is practically identical. The measured intensity contrast between the co-emitted λ_{ON} and λ_{OFF} signals allow

TABLE 1 | Results of volcanic gas plume observations at Stromboli volcano, 24–29 June 2015.

Date	Plume speed (m/s)		SO_2 flux (kg/s)		CO_2/SO_2 (Molar ratio)		CO_2 flux[a] (kg/s)		CO_2 flux[b] (kg/s)	
Method	Mean	(1 SD)	Mean	(1 SD)	Mean	(1 SD)	Mean	(Uncertainty)[c]	Mean	(Uncertainty)[d]
	UV camera		UV camera		Multi-GAS		Multi-GAS + UV camera		DIAL-LIDAR	
24/6/15	5.5	0.3	1.5	1.2	8	1.2	8.4	8.0	8.3	2.1
25/6/15	4.8	0.3	n.d.	n.d.	n.d.	n.d.	n.d.	n.d.	18.1	4.5
26/6/15	5.2	0.3	2.5	0.8	7.9	1.2	13.4	6.3	16.5	4.1
27/6/15	6.1	0.3	3.4	2.1	8.6	1.3	20.1	15.3	13.5	3.4
28/6/15	6.1	0.4	2.7	2.3	8.9	1.3	16.7	16.6	8.6	2.2
29/6/15	5.5	0.4	n.d.	n.d.	n.d.	n.d.	n.d.	n.d.	9.2	2.3

Plume speed and SO_2 flux are obtained by processing UV camera images. For both parameters the daily average and its standard deviation (SD) are quoted (the latter is taken as representative of uncertainty). The plume volcanic gas CO_2/SO_2 ratios are derived from in-situ Multi-GAS observations taken on the volcano's summit; each quoted ratio is the average (+1 SD) over a 30-min observational period, from 16 to 16:30 Local Time. No successful Multi-GAS plume detection was obtained in other daily observational windows (04-04:30; 10-10:30-22-22:30). Two independent estimates of the CO_2 flux are reported, based on either.
[a]Multiplying the SO_2 flux by the CO_2/SO_2 ratio, or
[b]Processing of DIAL-LIDAR results. Uncertainties in the derived CO_2 fluxes are from either
[c]Error propagation on SO_2 fluxes and CO_2/SO_2 ratios (taken as 1 SD), or
[d]Estimated at ±25% (see appendix).

range-resolved CO_2 concentrations in the volcanic plume to be obtained.

An example of a lidar-based atmospheric profile, obtained at Stromboli during a typical *Vancori scan*, is illustrated in **Figure 3**. As described above, the lidar registered one such profile every 10 s, since 100 lidar returns acquired at 10 Hz were co-added and averaged to increase the signal-to-noise ratio. Each of the atmospheric profiles (e.g., **Figure 3**) acquired during the *Vancori scan* contains the following characteristic features:

1) at R ~0, a first strong intensity peak is recorded for both λ_{ON} and λ_{OFF} (**Figure 3A**); this peak, which we refer to as $I_{0,ON}$ and $I_{0,OFF}$, is due to scattering inside the laboratory truck of some photons of the transmitted laser pulse. This peak yields the pulse transmission zero-time, and its intensity is proportional to the transmitted energy (used for signal normalization),

2) for R between 0 and ~500 m, a weak signal is observed that is returned from atmospheric back-scatterers encountered by the laser beam along the optical path (**Figure 3A**); this signal, as explained before, attenuates with distance and vanishes at R ~500 m,

3) a $I_{P,ON}$ and $I_{P,OFF}$ peak at R ~1900 m (**Figures 3A,B**); this is produced by reflection of the lidar beam by the southeastern margin of the *Pizzo* morphological peak (see **Figure 2**),

4) a series of weak but resolvable peaks observed in the range interval 2300–2700 m (**Figure 3B**); in these peaks, the λ_{ON} signal appears strongly attenuated relative to the co-acquired λ_{OFF} signal, a fact due to laser absorption by CO_2 molecules in the volcanic plume,

5) a $I_{V,ON}$ and $I_{V,OFF}$ peak at R ~2800 m, which is produced by reflection of the laser beam by the *Vancori* peak (**Figure 3B**).

Atmospheric profiles obtained during the *Pizzo scans* of June 24–25 share similar characteristics, except that the *Pizzo* morphological peak is intercepted by the lidar beam at R ~2300 m, and the plume is encountered either before or after the *Pizzo* (**Figure 4**). The *Vancori* peak was obviously not encountered.

Data Processing and Calculation of CO_2 Concentrations

We processed each acquired atmospheric profile using a Matlab analysis routine, with the aim of calculating the CO_2 concentrations in the atmospheric background and in the volcanic plume. The data processing routine consists of the following steps, all based on the Lambert-Beer law relation:

a) Initially, the CO_2 concentration in the natural background atmosphere, C_0, is calculated as:

$$\ln\left[\frac{I_{P,OFF}/I_{0,OFF}}{I_{P,ON}/I_{0,ON}}\right] = 2\,\Delta\sigma\,C_0\,R_P \qquad (1)$$

where $I_{P,ON}$ ($I_{P,OFF}$) stands for intensity of the ON (OFF) lidar signal [(3) in **Figure 3A**] caused by reflection of the laser beam off the surface of the *Pizzo* wall ($R_P = 2294$ m); $I_{0,ON}$ ($I_{0,OFF}$) is the intensity of the ON (OFF) lidar peak caused by laboratory scattering of the laser pulse [(1) in **Figure 3A**]; and $\Delta\sigma$ is the CO_2 differential absorption cross-section.

b) Secondly, ΔC, the average excess CO_2 concentration in the volcanic plume cross-section between *Pizzo* and *Vancori* [(3,5) in **Figure 3B**], is derived from:

$$\ln\left[\frac{I_{V,OFF}/I_{P,OFF}}{I_{V,ON}/I_{P,ON}}\right] = 2\,\Delta\sigma\,(C_0 + \Delta C)\,(R_V - R_P) \qquad (2)$$

Where $I_{V,ON}$ ($I_{V,OFF}$) is the peak intensity of the ON (OFF) lidar signal caused by reflection of the laser beam off the surface of the *Vancori* rock wall (at $R_V = 2837$ m).

c) Thirdly, $C_{CO2,i}$, the excess CO_2 concentration corresponding to each i-th ADC channel of the lidar profile (**Figure 3C**) is calculated from:

$$C_{CO2,i} = k\,S_i \qquad (3)$$

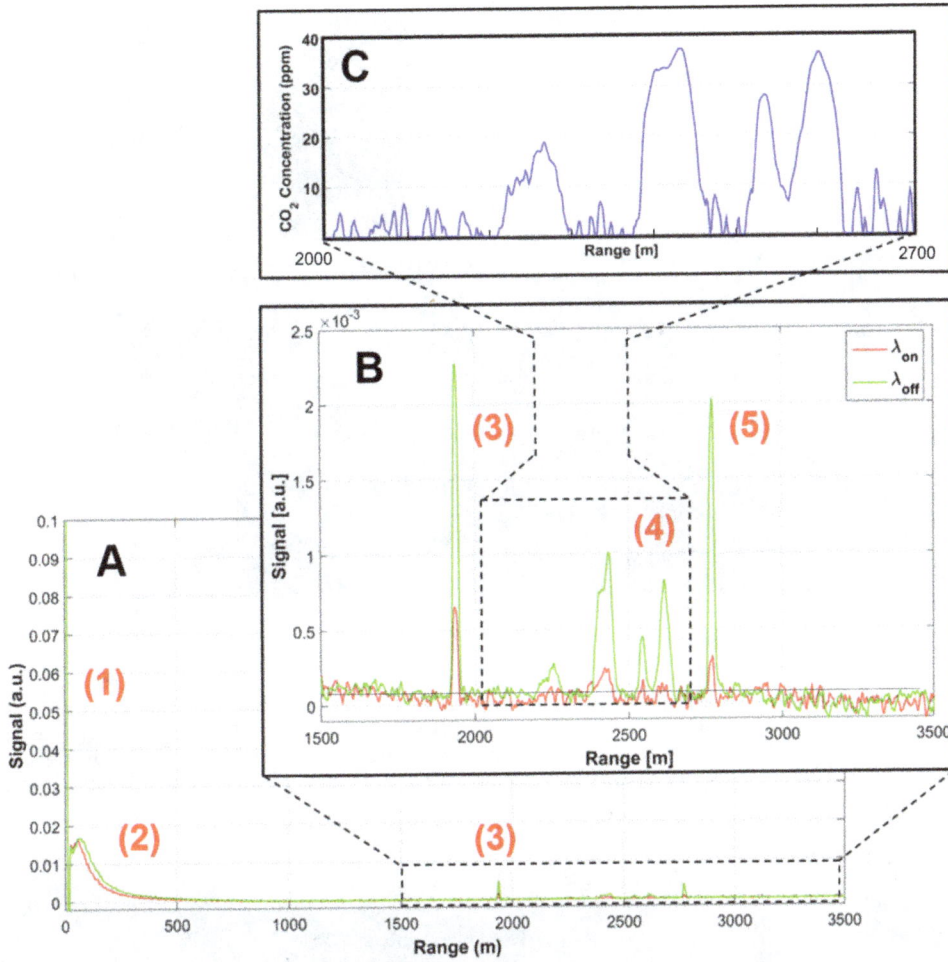

FIGURE 3 | (A) Example of a lidar-based atmospheric profile, obtained at Stromboli during a typical *Vancori scan*, in the form of a range (distance) vs. signal intensity (arbitrary units, a.u.) plot. Peak (1) yields the pulse transmission zero-time (scattering, inside the laboratory truck, of some photons of the transmitted laser pulse); peak (2) is the returned signal from atmospheric back-scatterers along the laser optical path; peak (3) is the returned signal produced by reflection of the lidar beam by southeastern margin of the *Pizzo* morphological peak. **(B)** is a detail of **(A)**, for ranges between 1500 and 3500 m. In this panel, peak (3) is as in **(A)**; the series of peaks observed in the range interval 2300–2700 m (4) are due to back-scattering of the laser beam from the volcanic plume; peak (5) is produced by reflection of the laser beam by the *Vancori* peak; **(C)** a profile of in-plume excess CO2 concentrations, in the 2000–2700 m range interval, calculated from processing of the lidar signal in **(B)**. See text for the procedure used.

$$k = \frac{\Delta C \ (R_V - R_P)}{\Delta R \sum_i S_i} \qquad (4)$$

$$S_i = \ln \left(I_{i,OFF} \ R_i^2 \right) \qquad (5)$$

where ΔR is the range interval corresponding to each ADC channel, and $I_{i,OFF}$ and R_i are the OFF lidar signal and the range of the i-th ADC channel (the OFF signal has been chosen because its signal-to-noise ratio is higher). **Figure 3C** shows an example of in-plume excess CO2 concentration profile, obtained by applying the procedure above to the lidar profile of **Figure 3B** (in the 2100–2700 m range interval, where the volcanic plume was detected).

In-Plume CO2 Concentration Maps

A series of CO2 concentration profiles (one every 10 s), all similar to those shown in **Figure 3C**, were obtained as the volcanic plume was sequentially scanned by our DIAL-lidar, either horizontally or vertically, during the *Pizzo/Vancori scans*. By interpolating all CO2 concentration profiles obtained during a single scan, we obtained sequences of CO2 concentration maps, examples of which are shown in **Figures 4, 5**. Since a full scan of the plume was completed in ~1000–2000 s, each map is in fact obtained from the combination of ~50 to ~100 atmospheric profiles.

The maps illustrate the 2D distribution of CO2 concentrations as a function of azimuth angle [°] (X axis) and range [m] (Y axis) for horizontal scans (**Figure 4**); or as a function of range [m] (X axis), and elevation angle [°] (Y axis) for the vertical scans (**Figure 5**). In both plots, the color scales (from blue to red) illustrate the level of CO2 concentrations (in [ppm]) in the investigated space.

Figures 4, 5 demonstrate the ability of our DIAL-lidar to resolve in-plume volcanic CO2 from the atmospheric

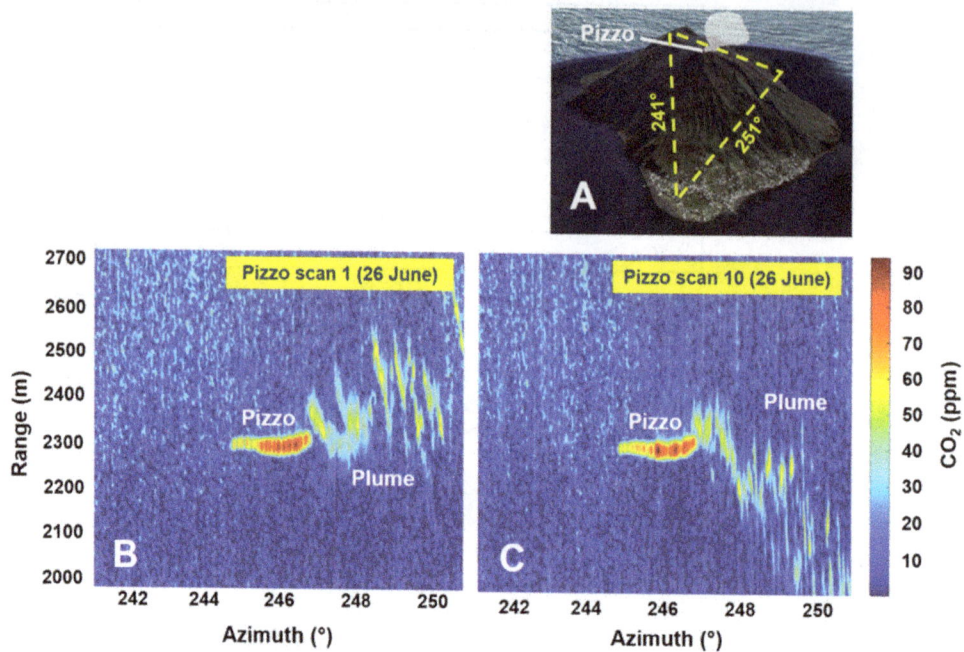

FIGURE 4 | Example of two CO$_2$ concentration maps (B,C) obtained during a *Pizzo* horizontal *scan* on June 26. Geometry of the scans and location of the plume are schematically shown in **(A)**. The maps show the distribution of CO$_2$ concentrations in the lidar's Field Of View (FOV), as a function of heading angle and range. Each map was obtained by interpolation of all CO$_2$ concentration profiles (e.g., same as 3A), obtained during a given *Pizzo scan*. In the maps, the red colored horizontal bands identify the margin of the *Pizzo* peak (heading angle: 244–245°), while the volcanic plume is the band of peak CO$_2$ concentration (up to 60 ppm) areas at heading angles of 245–250°.

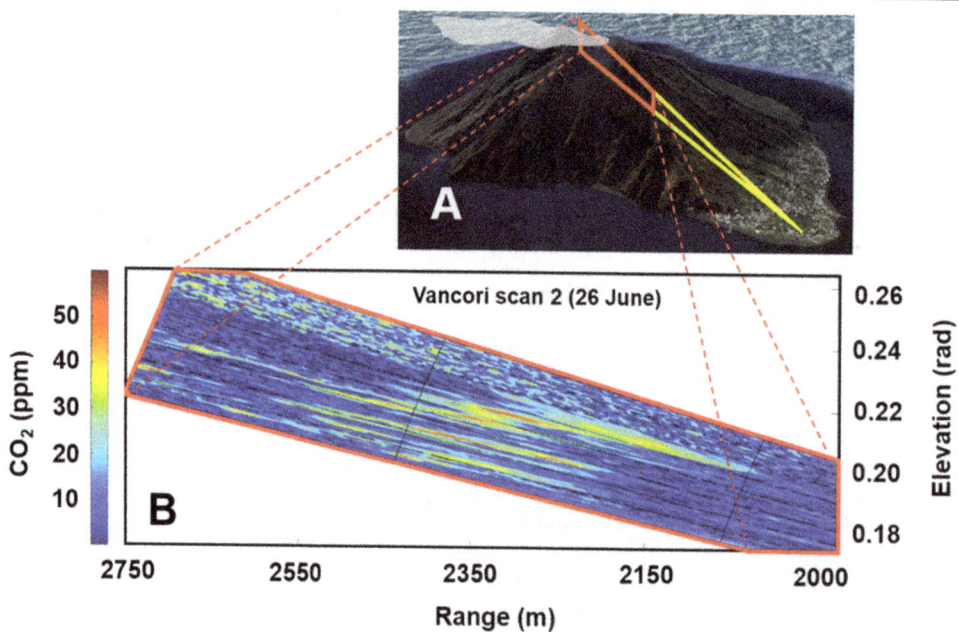

FIGURE 5 | Example of a CO$_2$ concentration map (B) obtained during a *Vancori* vertical *scan* on June 26. Geometry of the scan and location of the plume is schematically shown in **(A)**. The map shows the distribution of CO$_2$ concentrations in the lidar's FVO, as a function of range and elevation, and was obtained by interpolation of all CO$_2$ concentration profiles (e.g., same as 3A) obtained during a given *Vancori scan*. In the map, the volcanic plume corresponds to the cluster of high CO$_2$ concentrations (up to 60 ppmv) in the range interval 2200–2500 m and 0.2–0.22 rad (17–19.5°) elevations. The blue colored areas in the 2000–2000 m range correspond to near ambient (<10 ppm above background) CO$_2$ concentrations.

background CO_2 (blue colors). In the CO_2 distribution maps, clusters of peak CO_2 concentration areas (marked by red, orange and yellow colors) identify the geometry of the plume. The lidar-based plume locations are consistent with visual and UV observations of volcanic plume dispersion (**Figure 2**). In the *Pizzo* horizontal scans, the plume was intercepted north of the *Pizzo* peak (heading angle: 244–245°), and is identified in the maps of **Figure 4** as a cluster of peak CO_2 concentrations (up to 60 ppm above ambient air) at heading angles of 245–250°. The plume was detected over a relative wide range interval ($R = 2000$–2400 m), relative to the *Pizzo* peak (R $\sim$2300 m). This is consistent with the slightly variable plume transport directions during our June 24–25 observation period that dispersed the plume either toward (**Figure 4B**) or away from (**Figure 4A**) the lidar observation point ($R = 0$). A few *Pizzo* vertical scans (not shown) confined the vertical extension of the plume to a diagonal band, extending from $R = 2300$ m and elevation $\sim$19° (the *Pizzo* area) to R $\sim$2700 m and $\sim$20° elevation. **Figure 5** is an example of a CO_2 distribution map obtained during a vertical *Vancori* scan. The map exhibits a clear volcanic plume signal, as marked by a cluster of high CO_2 concentrations (up to 60 ppm) in the range interval 2200–2500 m and 17–19.5° elevation. CO_2 remained at background air levels for range distances <2000 and >2800 m.

CO2 Flux

The CO_2 concentration maps served as basis for calculating the CO_2 flux. To this aim, and by analogy with previous work (Aiuppa et al., 2015), we integrated the background-corrected (excess) CO_2 concentrations over the entire plume cross-sectional area covered by each scan, and multiplied this integrated column amount by the plume transport speed. Mathematically, the CO_2 flux (Φ_{CO_2}, in kg·s^{-1}) was obtained from:

$$\Phi_{CO_2} = v_P \cdot \frac{PM_{CO_2}}{10^3 N_A} \cdot N_{molCO2-total} \qquad (6)$$

where v_P is the plume transport speed (in [m/s]) obtained from processing of UV camera images (**Table 1**); $N_{molCO2-total}$ is the total-plume CO_2 molecular density (expressed in molecules m^{-1}); and PM_{CO2} and N_A are, respectively, the CO_2 molecular weight and Avogadro's constant. The term $N_{molCO2-total}$ was obtained by integrating the effective average excess CO_2 concentrations ($\overline{C_{exc,i}}$ *[ppm]*) over the entire plume cross section, according to:

$$N_{molCO2-total} = N_h \cdot 10^{-6} \cdot \sum_i \overline{C_{exc,i}} \cdot A_i \qquad (7)$$

where: N_h is the atmospheric number density (molecules m^{-3}) at the crater's summit height, the term 10^{-6} converts $\overline{C_{exc,i}}$ into a dimensionless quantity, and A_i represents the i-th effective plume area, given by:

$$A_i = l_i \cdot \Delta R \qquad (8)$$

where ΔR is the spatial resolution of the lidar (1.5 *m*) and l_i is the *i-th* arc of circumference (**Figure 6B**):

$$l_i = R_i \cdot \theta \qquad (9)$$

In relation (9), R_i is the *i-th* distance vector (in meters) and θ is the angular resolution of the system expressed in radians (ranging from 0.04° π/180 = 1.75 10^{-4} rad to 0.1° π/180 = 0.00175 rad) (**Figure 6B**).

Our obtained CO_2 fluxes, shown in **Figure 6A**, range from 1.8 to 32.1 kg/s. The lidar-based CO_2 flux time-series (**Figure 6A**) has maximum temporal resolution of 16–33 min (the time required to complete a full scan of the plume for our instrumental configuration). Temporal gaps in the dataset are caused by decreases in the signal-to-noise ratio (SNR) that prevent us from accurately detecting a clear CO_2 excess. These SNR decreases are likely caused by reduction of the backscattering coefficient of the probed air parcel, reflecting temporal variations in condensation extent of the volcanic gas plume. Visual (and UV camera) observations confirmed that the plume was variably condensed during our measurement interval, possibly due to slight changes in atmospheric conditions.

We evaluate the overall uncertainty in our derived CO_2 fluxes at ± 25% at 1s (see appendix).

DISCUSSION

The scarcity of volcanic CO_2 flux data in the geological literature (see Burton et al., 2013 for a recent review) is a direct consequence of the technical challenges in resolving the volcanic CO_2 signal from the large atmospheric background ($\approx$400 ppmv). In contrast to SO_2, which is present at the part per billion level in the background atmosphere, allowing the volcanic flux to be routinely measured from ground and space using UV spectroscopy (Oppenheimer, 2010), remote sensing of volcanic CO_2 has only been achieved during eruptions of mafic volcanoes. In such circumstances, magma/hot rocks can effectively be used as a light source for ground-based Fourier Transform Infra-Red (FTIR) spectrometers (Allard et al., 2005; Burton et al., 2007; Oppenheimer and Kyle, 2008). In contrast, measurement of the far more common "passive" CO_2 emissions from quiescent volcanoes has required access to hazardous volcano's summit craters for direct sampling of fumaroles (Fischer and Chiodini, 2015) or *in-situ* measurement of plumes via either Multi-GAS instruments (Aiuppa, 2015) or active-FTIR (Burton et al., 2000; La Spina et al., 2013; Conde et al., 2014).

A major breakthrough has recently arisen from the possible application of lidars to remote volcanic CO_2 sensing (Fiorani et al., 2013, 2015, 2016; Aiuppa et al., 2015; Queißer et al., 2015, 2016). Aiuppa et al. (2015) were the first to report on a DIAL-lidar-based remote measurement of the volcanic CO_2 flux at Campi Flegrei volcano, but their observations were limited to short (<200 m) measurement distances. Here, we have extended this earlier work to demonstrate that DIAL-lidars can successfully detect volcanic CO_2 at tens of ppmv above the atmospheric background over optical paths up to $\approx$3 km (**Figures 4, 5**). Similar results have recently been obtained at Campi Flegrei volcano by Queißer et al. (2016), suggesting that lidar may soon become an important operational tool in volcanic-gas research.

Our results constrain the CO_2 flux at Stromboli during June 24–29, 2015 (**Figure 6A**). Averaging all successful results during

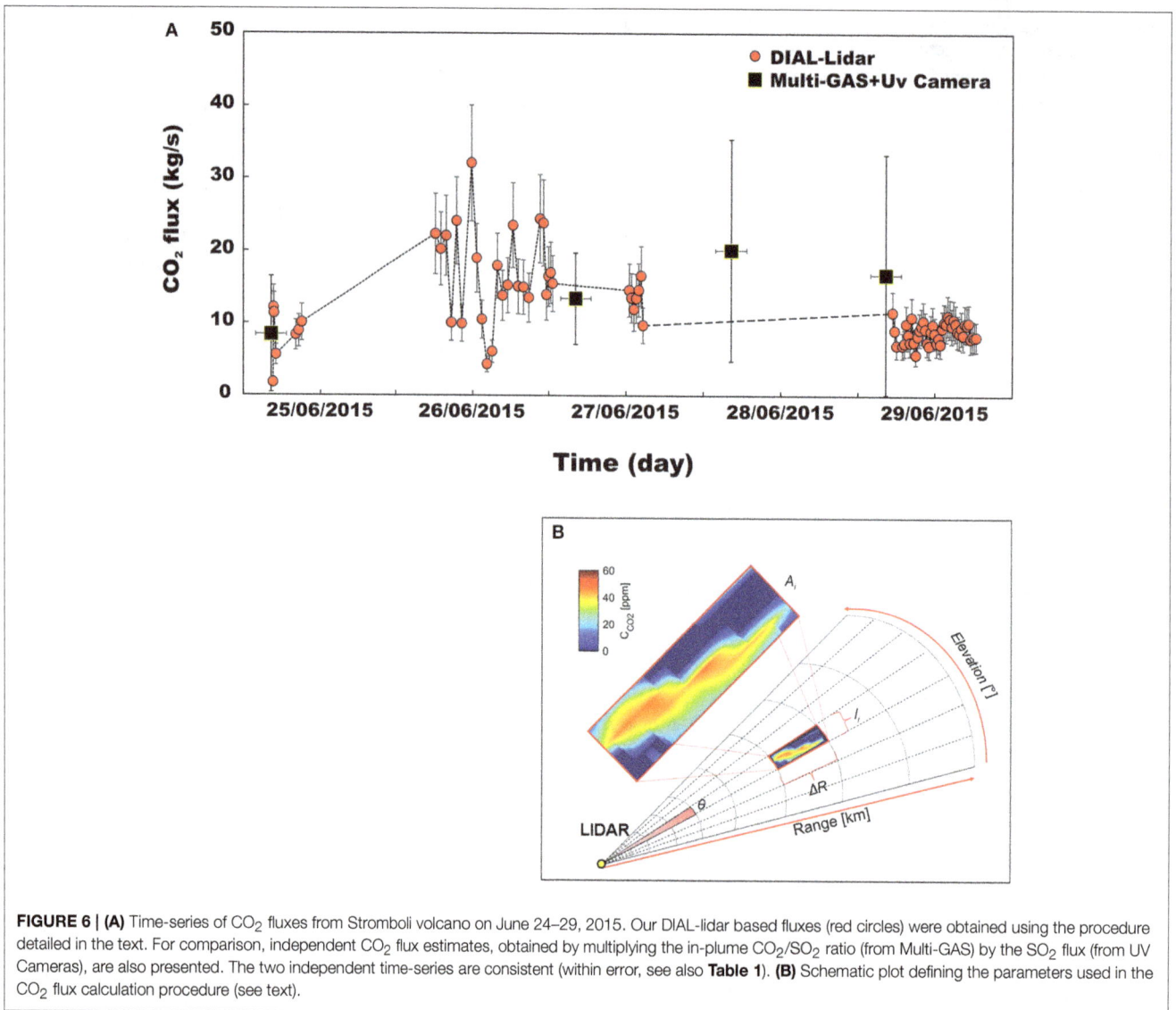

FIGURE 6 | (A) Time-series of CO_2 fluxes from Stromboli volcano on June 24–29, 2015. Our DIAL-lidar based fluxes (red circles) were obtained using the procedure detailed in the text. For comparison, independent CO_2 flux estimates, obtained by multiplying the in-plume CO_2/SO_2 ratio (from Multi-GAS) by the SO_2 flux (from UV Cameras), are also presented. The two independent time-series are consistent (within error, see also **Table 1**). **(B)** Schematic plot defining the parameters used in the CO_2 flux calculation procedure (see text).

each measurement day, we obtain daily averages of the CO_2 flux between 8.3 ± 2.1 (June 24) and 18.1 ± 4.5 (June 25) kg/s, which correspond to cumulative daily outputs of 718 and 1565 tons, respectively. These results fall well in the range of previous CO_2 measurements on Stromboli. Aiuppa et al. (2010a, 2011) found that the CO_2 flux exhibits large temporal oscillations on Stromboli, from as low as 60 tons/day to as high as 11,000 tons/day, the highest values being observed in the days prior to *paroxysmal* and/or *major* explosions. The time-averaged CO_2 flux from Stromboli has been evaluated at 550 tons/day (Aiuppa et al., 2011) and at 1040–1200 tons/day (Allard, 2010). Our lidar-based CO_2 flux for the entire (June 24–29) measurement period is reasonably close, averaging at 1050 ± 250 tons/day (mean of 80 individual measurements).

Figure 6A offers further confirmation to the robustness of our results. In the figure, we compare our lidar-based CO_2 fluxes with independent estimates, in which the CO_2 flux was derived by multiplying the CO_2/SO_2 ratio of the plume by the SO_2 flux. This latter approach has been used at volcanoes for years (Aiuppa, 2015), and at Stromboli involves use of two fully automated Multi-GAS instruments, operating on the volcano's summit to measure the in-plume CO_2/SO_2 ratio (**Figure 2A**; Aiuppa et al., 2009, 2010a,b; Calvari et al., 2014). This is combined with SO_2 fluxes, delivered from either the FLAMES network of scanning UV spectrometers (Burton et al., 2009) or from UV camera observations (Tamburello et al., 2012), to obtain the CO_2 flux.

Problems with this Multi-GAS + SO_2 flux approach include issues of different temporal resolutions and poor temporal alignment of the two time-series. Successful Multi-GAS measurements of plume composition on Stromboli (Aiuppa et al., 2009, 2010b) are restricted to periods when the volcanic plume is dispersed by the local wind field into the *Pizzo* area, where the instruments are deployed (see **Figure 2A**). In addition, the Multi-GAS cannot operate continuously, but only during four equally spaced measurement cycles per day, each being 30 min long (Aiuppa et al., 2009). As such, the temporal resolution of the

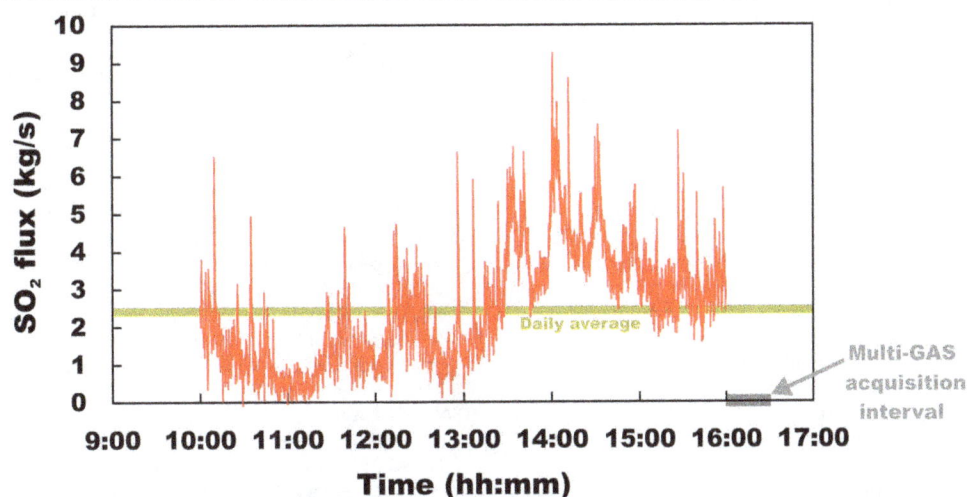

FIGURE 7 | Temporal record of the volcanic SO$_2$ flux from Stromboli on June 26 th, 2015, as derived from our UV camera observations. The figure exemplifies misalignment between Multi-GAS and SO$_2$ flux time-series; plume CO$_2$/SO$_2$ ratios on June 26th were successfully measured only during the 1600–1630 local time Multi-GAS acquisition period, immediately after the end of the SO$_2$ flux acquisition window (0900–1600 local time). Poor temporal alignment is a flaw in the technique of estimating the CO$_2$ flux through a combination of Multi-GAS and UV camera records.

CO$_2$/SO$_2$ ratio time-series is 6 h at best. In contrast, the temporal resolution of UV spectrometers/cameras is higher, from ~10 to 20 min (Burton et al., 2009) to 0.5 s (Tamburello et al., 2012), but observations are intrinsically limited to daylight hours and to good meteorological conditions (no clouds).

Figure 7 exemplifies the issue related to misalignment between Multi-GAS and UV observations. In the June 26th example, the only successful Multi-GAS acquisition period (from 1600 to 1630 h local time) clearly did not overlap with the SO$_2$ flux acquisition window (0900 to 1600 h local time). To overcome this problem, the common practice is to average out available Multi-GAS and UV spectroscopy data to obtain *daily means* of the CO$_2$ flux (Aiuppa et al., 2010a). Owing to the large inter-daily variability of SO$_2$ flux (e.g., **Figure 7**), however, large uncertainties are associated with these derived CO$_2$ fluxes (see **Table 1**, and errors bars in **Figure 6A**).

In spite of the issues above, we find overall consistency between the lidar-based and the traditional (Multi-GAS + UV spectroscopy-based) CO$_2$ fluxes (**Figure 6A**). This provides mutual validation for both quantification approaches. Our lidar-based CO$_2$ flux time-series (**Figure 6A**) are manifestly more continuous and of better temporal resolution (16–33 min). In addition, the lidar as with other remote sensing techniques is intrinsically safer. We caution, however, that further development is required before the lidar can become an operative tool for volcano monitoring. Improvements will need to occur in portability (the prototype weighs ~1100 kg), and reduced power requirement (6.5 kW) and costs (300 kUS $). In addition, the current measurement protocol is complex and thus requires great familiarity with the technique. Efforts are now being made to make the lidar more simple, user-friendly and fully automated, including development of an on-line remote control system and of a self-checking routine of the laser's wavelength settings.

Electro-optics and laser/lidar private manufacturers need to be directly involved to transition the prototype into a more widely accessible, commercial instrument.

CONCLUSIONS

Our proof-of-concept study demonstrates the ability of DIAL-lidars to remotely (≈3 km distance) measure the volcanic CO$_2$ flux. Our reported lidar-based CO$_2$ fluxes at Stromboli volcano (1.8 ± 0.5 to 32.1 ± 8.0 kg/s) are in the same range as those obtained using standard techniques that require in-situ observations and are intrinsically more risky for operators. Our results, with those of Queißer et al. (2016), open new prospects for the use of lidars for instrumental remote monitoring of volcanic CO$_2$ flux. Further work is warranted in order to standardize and widen potential applications of Lasers in volcanic gas studies.

AUTHOR CONTRIBUTIONS

AA, LF, and SS conceived the idea. AA, LF, SS, GM, and MN conducted the lidar/UV camera experiment. SP and RD processed the data, with help from AA, LF, and SS. ML provided the Multi-GAS results. AA drafted the manuscript with help from all co-authors.

FUNDING

The research leading to these results has received funding from the European Research Council under the European Union's Seventh Framework Program (FP7/2007/2013)/ERC grant agreement n 305377 (PI, Aiuppa), and from the DECADE-DCO research initiative.

ACKNOWLEDGMENTS

Società Enel Produzione spa is kindly acknowledged for logistical support and for kindly providing access to "Centrale ENEL di Stromboli" during the field campaign. The authors wish to thank the Comune di Lipari, Circoscrizione di Stromboli, for authorizing use of the lidar on Stromboli.

REFERENCES

Aiuppa, A. (2015). "Volcanic gas monitoring," in *Volcanism and Global Environmental Change*, eds A. Schmidt, K. E. Fristad, and L. T. Elkins-Tanton (Cambridge: Cambridge University Press), 81–96.

Aiuppa, A., Bertagnini, A., Métrich, N., Moretti, R., Di Muro, A., Liuzzo, M., et al. (2010b). A model of degassing for Stromboli volcano, earth planet. *Sci. Lett.* 295, 195–204. doi: 10.1016/j.epsl.2010.03.040

Aiuppa, A., Burton, M., Allard, P., Caltabiano, T., Giudice, G., Gurrieri, S., et al. (2011). First observational evidence for the CO_2-driven origin of Stromboli's major explosions. *Solid Earth* 2, 135–142. doi: 10.5194/se-2-135-2011

Aiuppa, A., Burton, M., Caltabiano, T., Giudice, G., Gurrieri, S., Liuzzo, M., et al. (2010a). Unusually large magmatic CO_2 gas emissions prior to a basaltic paroxysm. *Geophys. Res. Lett.* 37, L17303. doi: 10.1029/2010GL043837

Aiuppa, A., Federico, C., Giudice, G., Giuffrida, G., Guida, R., Gurrieri, S., et al. (2009). The 2007 eruption of Stromboli volcano: insights from real-time measurement of the volcanic gas plume CO_2/SO_2 ratio. *J. Volcanol. Geotherm. Res.* 182, 221–230. doi: 10.1016/j.jvolgeores.2008.09.013

Aiuppa, A., Fiorani, L., Santoro, S., Parracino, S., Nuvoli, M., Chiodini, G., et al. (2015). New ground-based lidar enables volcanic CO_2 flux measurements. *Sci. Rep.* 5:13614. doi: 10.1038/srep13614

Aiuppa, A., Giudice, G., Gurrieri, S., Liuzzo, M., Burton, M., Caltabiano, T., et al. (2008). Total volatile flux from Mount Etna. *Geophys. Res. Lett.* 35, L24302. doi: 10.1029/2008gl035871

Allard, P. (2010). A CO_2-rich gas trigger of explosive paroxysms at Stromboli basaltic volcano, Italy. *J. Volcanol. Geoth. Res.* 189, 363–374. doi: 10.1016/j.jvolgeores.2009.11.018

Allard, P., Burton, M. R., and Mure, F. (2005). Spectroscopic evidence for a lava fountain driven by previously accumulated magmatic gas. *Nature* 433, 407–410. doi: 10.1038/nature03246

Andronico, D., and Pistolesi, M. (2010). The November 2009 paroxysmal explosions at Stromboli. *J. Volcanol. Geotherm. Res.* 196, 120–125. doi: 10.1016/j.jvolgeores.2010.06.005

Berk, A., Conforti, P., Kennett, R., Perkins, T., Hawes, F., and van den Bosch, J. (2014). "MODTRAN6: a major upgrade of the MODTRAN radiative transfer code," in *Proceedings SPIE 9088, Algorithms and Technologies for Multispectral, Hyperspectral, and Ultraspectral Imagery XX, 90880H* (Baltimore, MD) (Accessed June 13, 2014).

Bertagnini, A., Métrich, N., Landi, P., and Rosi, M. (2003). Stromboli an open window on the deep feeding system of a steady state volcano. *J. Geophys. Res.* 108, 2336. doi: 10.1029/2002JB002146

Bruhn, A., Weickert, J., and Schnörr, C. (2005). Lucas/kanade meets horn/schunck: combining local and global optic flow methods international. *J. Comput. Vis.* 61, 211–231. doi: 10.1023/B:VISI.0000045324.43199.43

Burton, M., Allard, P., Murè, F., and La Spina, A. (2007). Depth of slug-driven strombolian explosive activity. *Science* 317, 227–230. doi: 10.1126/science.1141900

Burton, M. R., Caltabiano, T., Murè, F., and Randazzo, D. (2009). SO_2 flux from Stromboli during the 2007 eruption: results from the FLAME network and traverse measurements. *J. Volcanol. Geotherm. Res.* 182, 214–220. doi: 10.1016/j.jvolgeores.2008.11.025

Burton, M. R., Oppenheimer, C., Horrocks, L. A., and Francis, P. W. (2000). Remote sensing of CO_2 and H_2O emission rates from Masaya Volcano, Nicaragua. *Geology* 28, 915–918. doi: 10.1130/0091-7613(2000)28<915:RSOCAH>2.0.CO;2

Burton, M. R., Prata, F., and Platt, U. (2014). Volcanological applications of SO_2 cameras. *J. Volcanol. Geotherm. Res.* 300, 2–6. doi: 10.1016/j.jvolgeores.2014.09.008

Burton, M. R., Sawyer, G. M., and Granieri, D. (2013). Deep carbon emissions from volcanoes. *Rev. Mineral. Geochem.* 75, 323–354. doi: 10.2138/rmg.2013.75.11

Calvari, S., Bonaccorso, A., Madonia, P., Neri, M., Liuzzo, M., Salerno, G. G., et al. (2014). Major eruptive style changes induced by structural modifications of a shallow conduit system: the 2007-2012 Stromboli case. *Bull. Volcanol.* 76, 1–15. doi: 10.1007/s00445-014-0841-7

Conde, V., Robidoux, P., Avard, G., Galle, B., Aiuppa, A., Muñoz, A., et al. (2014). Measurements of SO_2 and CO_2 by combining DOAS, Multi-GAS and FTIR: study cases from Turrialba and Telica volcanoes. *Int. J. Earth Sci.* 103, 8, 2335–2347. doi: 10.1007/s00531-014-1040-7

D'Aleo, R., Bitetto, M., Delle Donne, D., Tamburello, G., Battaglia, A., Coltelli, M., et al. (2016). Spatially resolved SO_2 flux emissions from Mt Etna. *Geophys. Res. Lett.* 43, 7511–7519. doi: 10.1002/2016GL069938

de Moor, J. M., Aiuppa, A., Avard, G., Wehrmann, H., Dunbar, N., Muller, C., et al. (2016a). Turmoil at Turrialba Volcano (Costa Rica): degassing and eruptive processes inferred from high-frequency gas monitoring. *J. Geophys. Res. Solid Earth* 121, 5761–5775. doi: 10.1002/2016JB013150

de Moor, J. M., Aiuppa, A., Pacheco, J., Avard, G., Kern, C., Liuzzo, M., et al. (2016b). Short-period volcanic gas precursors to phreatic eruptions: insights from Poás Volcano, Costa Rica. *Earth Planet. Sci. Lett.* 442, 218–227. doi: 10.1016/j.epsl.2016.02.056

Fiorani, L. (2007). "Environmental monitoring by laser radar," in *Lasers and Electro-optics Research at the Cutting Edge*, ed S. B. Larkin (Hauppauge, NY: Nova Science Publishers), 119–171.

Fiorani, L., Santoro, S., Parracino, S., Nuvoli, M., Minopoli, C., Aiuppa, A., et al. (2015). Volcanic CO_2 detection with a DFM/OPA-based lidar. *Opt. Lett.* 40, 1034–1036. doi: 10.1364/OL.40.001034

Fiorani, L., and Durieux, E. (2001). Comparison among error calculations in differential absorption lidar measurements. *Opt. Laser Technol.* 3, 371–377. doi: 10.1016/S0030-3992(01)00041-X

Fiorani, L., Saleh, W. R., Burton, M., Puiu, A., and Queißer, M. (2013). Spectroscopic considerations on DIAL measurement of carbon dioxide in volcanic emissions. *J. Optoelectron. Adv. Mater.* 15, 317–325.

Fiorani, L., Santoro, S., Parracino, S., Maio, G., Nuvoli, M., and Aiuppa, A. (2016). Early detection of volcanic hazard by lidar measurement of carbon dioxide. *Nat. Hazards* 83, 21. doi: 10.1007/s11069-016-2209-0

Fischer, T. P., and Chiodini, G. (2015). "Volcanic, magmatic and hydrothermal gas discharges," in *Encyclopaedia of Volcanoes, 2nd Edn*, eds H. Sigurdsson, B. Houghton, S. McNutt, H. Rymer, and J. Stix (London: Academic Press; Elsevier), 779–797.

Galle, B., Johansson, M., Rivera, C., Zhang, Y., Lehmann, T., Platt, U., et al. (2010). Network for Observation of Volcanic and Atmospheric Change (NOVAC)-A global network for volcanic gas monitoring: network layout and instrument description. *J. Geophys. Res.* 115:D05304. doi: 10.1029/2009JD011823

Giggenbach, W. F. (1996). "Chemical composition of volcanic gases," in *Monitoring and Mitigation of Volcanic Hazards*, ed R. Scarpa and R. J. Tilling (Heidelberg: Springer), 221–256.

Kantzas, E. P., McGonigle, A. J. S., Tamburello, G., Aiuppa, A., and Bryant, R. G. (2010). Protocols for UV camera volcanic SO_2 measurements. *J. Volcanol. Geotherm. Res.* 94, 55–60. doi: 10.1016/j.jvolgeores.2010.05.003

La Spina, A., Burton, M. R., Harig, R., Mure, F., Rusch, P., Jordan, M., et al. (2013). New insights into volcanic processes at Stromboli from Cerberus, a remote-controlled open-path FTIR scanner system. *J. Volcanol. Geotherm. Res.* 249, 66–76. doi: 10.1016/j.jvolgeores.2012.09.004

Lübcke, P., Bobrowski, N., Illing, S., Kern, C., Alvarez Nieves, J. M., Vogel, L., et al. (2013). On the absolute calibration of SO_2 cameras. *Atmos. Meas. Tech.* 6, 677–696. doi: 10.5194/amt-6-677-2013

Métrich, N., Bertagnini, A., and Di Muro, A. (2010). Conditions of magma storage, degassing and ascent at Stromboli: new insights into the volcano plumbing system with inferences on the eruptive dynamics. *J. Petrol.* 51, 603–626. doi: 10.1093/petrology/egp083

Métrich, N., Bertagnini, A., Landi, P., Rosi, M., and Belhadj, O. (2005). Triggering mechanism at the origin of paroxysms at Stromboli (Aeolian archipelago, Italy): the 5 April 2003 eruption. *Geophys. Res. Lett.* 32:L103056. doi: 10.1029/2004GL022257

Oppenheimer, C. (2010). Ultraviolet sensing of volcanic sulfur emissions, *Elements* 6, 87–92. doi: 10.2113/gselements.6.2.87

Oppenheimer, C., Fischer, T. P., and Scaillet, B. (2014). "Volcanic Degassing: Process and Impact," in *Treatise on Geochemistry, The Crust, 4*, eds H. D. Holland and K. K. Turekian (Amsterdam: Elsevier), 111–179.

Oppenheimer, C., and Kyle, P. R. (2008). Probing the magma plumbing of Erebus volcano, Antarctica, by open-path FTIR spectroscopy of gas emissions. *J. Volcanol. Geotherm. Res.* 177, 743–754. doi: 10.1016/j.jvolgeores.2007.08.022

Oppenheimer, C., Scaillet, B., and Martin, R. S. (2011). Sulfur degassing from volcanoes: source conditions, surveillance, plume chemistry and earth system impacts. in Sulfur in magmas and melts: its importance for natural and technical processes. *Rev. Miner.* 73, 363–422. doi: 10.2138/rmg.2011.73.13

Patanè, D., Aiuppa, A., Aloisi, M., Behncke, B., Cannata, A., Coltelli, M., et al. (2013). Insights into magma and fluid transfer at Mount Etna by a multiparametric approach: a model of the events leading to the 2011 eruptive cycle. *J. Geophys. Res. Solid Earth* 118, 3519–3539. doi: 10.1002/jgrb.50248

Peters, N., Hoffmann, A., Barnie, T., Herzog, M., and Oppenheimer, C. (2015). Use of motion estimation algorithms for improved flux measurements using SO2 cameras. *J. Volcanol. Geotherm. Res.* 300, 58–69. doi: 10.1016/j.jvolgeores.2014.08.031

Pioli, L., Pistolesi, M., and Rosi, M. (2014). Transient explosions at open-vent volcanoes: the case of Stromboli (Italy). *Geology* 42, 863–866. doi: 10.1130/G35844.1

Pistolesi, M., Donne, D. D., Pioli, L., Rosi, M., and Ripepe, M. (2011). The 15 March 2007 explosive crisis at Stromboli volcano, Italy: Assessing physical parameters through a multidisciplinary approach. *J. Geophys. Res. Solid Earth* 116, B12206. doi: 10.1029/2011JB008527

Poland, M. P., Miklius, A., Sutton, J. A., and Thornber, C. R. (2012). A mantle-driven surge in magma supply to Kīlauea Volcano during 2003-2007. *Nat. Geosci.* 5, 295–300. doi: 10.1038/ngeo1426

Queißer, M., Burton, M., and Fiorani, L. (2015). Differential absorption lidar for volcanic CO_2 sensing tested in an unstable atmosphere. *Opt. Express* 23, 6634–6644. doi: 10.1364/OE.23.006634

Queißer, M., Granieri, D., and Burton, M. (2016). A new frontier in CO_2 flux measurements using a highly portable DIAL laser system. *Sci. Rep.* 6:33834. doi: 10.1038/srep33834

Rosi, M., Bertagnini, A., Harris, A. J. L., Pioli, L., Pistolesi, M., and Ripepe, M. (2006). A case history of paroxysmal explosion at Stromboli: timing and dynamics of the April 5, 2003 event. *Earth Planet. Sci. Lett.* 243, 594–606. doi: 10.1016/j.epsl.2006.01.035

Rosi, M., Pistolesi, M., Bertagnini, A., Landi, P., Pompilio, M., and Di Roberto, A. (2013). Stromboli volcano, Aeolian Islands (Italy): present eruptive activity and hazards. *Geol. Soc. Memoir* 37, 473–490. doi: 10.1144/M37.14

Rothman, L. S., Gordon, I. E., Babikov, Y., Barbe, A., Chris Benner, D., Bernath, P. F., et al. (2013). The HITRAN2012 molecular spectroscopic database. *J. Quant. Spectrosc. Radiat. Transf.* 130, 4–50. doi: 10.1016/j.jqsrt.2013.07.002

Saccorotti, G., Iguchi, M., and Aiuppa, A. (2014). "In situ Volcano monitoring: present and future," in *Volcanic Hazards, Risks and Disasters*, ed P. Papale (Amsterdam: Elsevier), 169–202.

Symonds, R. B., Rose, W. I., Bluth, G. J. S., and Gerlach, T. M. (1994). Volcanic-gas studies: methods, results and applications. *Rev. Mineral. Geochem.* 30, 1–66.

Tamburello, G., Aiuppa, A., Kantzas, E. P., McGonigle, A. J. S., and Ripepe, M. (2012). Passive vs. active degassing modes at an open-vent volcano (Stromboli, Italy). *Earth Planet. Sci. Lett.* 359, 106–116. doi: 10.1016/j.epsl.2012.09.050

Tamburello, G., Kantzas, E. P., McGonigle, A. J. S., and Aiuppa, A. (2011). Vulcamera: a program for measuring volcanic SO_2 using UV cameras. *Ann. Geophys.* 54, 2. doi: 10.4401/ag-5181

Weitkamp, C. (ed.). (2005). "Lidar: range-resolved optical remote sensing of the atmosphere," in *Springer Series in Optical Sciences, Vol. 102.* (New York, NY: Springer).

Werner, C., Kelly, P. J., Doukas, M., Lopez, T., Pfeffer, M., McGimsey, R., et al. (2013). Degassing of CO_2, SO_2, and H_2S associated with the 2009 eruption of Redoubt Volcano, Alaska. *J. Volcanol. Geotherm. Res.* 259, 270–284. doi: 10.1016/j.jvolgeores.2012.04.012

Conflict of Interest Statement: The authors declare that the research was conducted in the absence of any commercial or financial relationships that could be construed as a potential conflict of interest.

APPENDIX–UNCERTAINTY AND ERROR ANALYSIS

Our lidar-based CO_2 fluxes are affected by the following error sources:

i. systematic error in CO_2 concentration measurement,
ii. statistical error in CO_2 concentration measurement,
iii. error in plume transport speed,
iv. error in identifying the integration area.

  i. *Systematic error of the CO_2 concentration measurement*- It is well known that the DIAL-lidar systematic error is dominated by imprecision in wavelength setting (Fiorani et al., 2015), leading to inaccuracy in differential absorption cross section and thus in gas concentration. To minimize this error, we implemented a photo-acoustic cell filled with pure CO_2 at atmospheric pressure and temperature, close to the laser exit, in order to control the transmitted wavelength before each atmospheric measurement. This procedure allows us to set the ON/OFF wavelengths with better accuracy than the laser linewidth (Fiorani et al., 2016). Assuming that the error in the wavelength setting is ± 0.02 cm^{-1} (half laser linewidth), in the wavelength region used in this study, the systematic error of the CO_2 concentration measurement is 5.5%.

  ii. *Statistical error of the CO_2 concentration measurement* - The statistical error has been calculated by standard error propagation techniques from the standard deviation of the lidar signal at each ADC channel. As discussed in Fiorani and Durieux (2001), the statistical error of the lidar signal increases with range. As a consequence, the uncertainty associated with the derived CO_2 concentrations also increases with range. In the distance range between *Pizzo* and *Vancori*, representing a mean measurement range, and at typical atmospheric and plume conditions encountered during this study, the statistical error of the CO_2 concentration measurement was about 2%. The statistical error exceed 5% at 4 km (well beyond our measurement range).

  iii. *Error in plume transport speed* - The standard deviation and the average value of the wind speed have been calculated for each measurement session, and the corresponding relative error was evaluated (by error propagation technique) at 3%.

  iv *Error in identifying the integration area* - The integration area in which an excess CO_2 concentration is actually present is probably the most difficult parameter to retrieve accurately, and therefore represents the main error source in our calculated volcanic CO_2 fluxes. The following procedure has been followed. For each CO_2 concentration map (e.g., **Figure 5**), we initially measured: 1) A_{15}, the area where the excess CO_2 concentration was larger than 15 ppm; and 2) A_{25} the area where the excess CO_2 concentration was larger than 25 ppm. Then, the average between A_{15} and A_{25} was taken as the best-estimated area, and their semi-difference as the error ($\sim$25%). The above thresholds have been chosen because below 15 ppm noise becomes significant, while above 25 ppm the plume area is reduced to its core. Use of a 15 ppm threshold likely underestimates the area (and thus the flux) of the order of magnitude of the measurement error, i.e., 10–20%.

Assuming that each error source is statistically independent, we can quadratically sum all the errors and obtain a cumulative error of $\sim$ 25% (dominated by the area error).

Microbial and Biogeochemical Dynamics in Glacier Forefields Are Sensitive to Century-Scale Climate and Anthropogenic Change

James A. Bradley [1,2]*, Alexandre M. Anesio [2] and Sandra Arndt [2,3]

[1] Department of Earth Sciences, University of Southern California, Los Angeles, CA, USA, [2] School of Geographical Sciences, University of Bristol, Bristol, UK, [3] Department of Earth and Environmental Sciences, Université Libre de Bruxelles, Brussels, Belgium

Edited by:
Samuel Abiven,
University of Zurich, Switzerland

Reviewed by:
Alejandro Mateos-Rivera,
University of Bergen, Norway
Anja Miltner,
Das Helmholtz-Zentrum für
Umweltforschung, Germany

*Correspondence:
James A. Bradley
jbradley8365@gmail.com

Specialty section:
This article was submitted to
Biogeoscience,
a section of the journal
Frontiers in Earth Science

The recent retreat of glaciers and ice sheets as a result of global warming exposes forefield soils that are rapidly colonized by microbes. These ecosystems are dominant in high-latitude carbon and nutrient cycles as microbial activity drives biogeochemical transformations within these newly exposed soils. Despite this, little is known about the response of these emerging ecosystems and associated biogeochemical cycles to projected changes in environmental factors due to human impacts. Here, we applied the model SHIMMER to quantitatively explore the sensitivity of biogeochemical dynamics in the forefield of Midtre Lovénbreen, Svalbard, to future changes in climate and anthropogenic forcings including soil temperature, snow cover, and nutrient and organic substrate deposition. Model results indicated that the rapid warming of the Arctic, as well as an increased deposition of organic carbon and nutrients, may impact primary microbial colonizers in Arctic soils. Warming and increased snow-free conditions resulted in enhanced bacterial production and an accumulation of biomass that was sustained throughout 200 years of soil development. Nitrogen deposition stimulated growth during the first 50 years of soil development following exposure. Increased deposition of organic carbon sustained higher rates of bacterial production and heterotrophic respiration leading to decreases in net ecosystem production and thus net CO_2 efflux from soils. Pioneer microbial communities were particularly susceptible to future changes. All future climate simulations encouraged a switch from allochthonously-dominated young soils (<40 years) to microbially-dominated older soils, due to enhanced heterotrophic degradation of organic matter. Critically, this drove remineralisation and increased nutrient availability. Overall, we show that human activity, especially the burning of fossil fuels and the enhanced deposition of nitrogen and organic carbon, has the potential to considerably affect the biogeochemical development of recently exposed Arctic soils in the present day and for centuries into the future. These effects must be acknowledged when attempting to make accurate predictions of the future fate of Arctic soils that are exposed over large expanses of presently ice-covered regions.

Keywords: SHIMMER, glacier forefield, microbial dynamics, Arctic soils, climate change

INTRODUCTION

Over the past century, the Arctic mean surface air temperature has increased at a rate twice as fast as the global mean and is predicted to warm a further > 4°C by 2,100 (Stocker et al., 2013). The observed warming has been accompanied by alarming changes in ice cover, increasing length of glacier melt seasons, and changing precipitation patterns and hydrology (Macdonald et al., 2005). These changes to the physical environment exert cascading effects on Arctic carbon and nutrient dynamics with potentially important, yet underexplored implications for both Arctic ecosystems (Kirchman et al., 2009), as well as global climate (Serreze et al., 2000; Screen and Simmonds, 2010; Stocker et al., 2013). Specifically, ecosystems in Polar regions are thought to be among the most vulnerable to global climate change, due to the adaptation of microbial processes to extreme environmental conditions (Vincent, 2010), and the vulnerability of Polar regions to tipping points (Lenton, 2012). However, there is high uncertainty in the future ecosystem response to predicted Arctic warming (Ciais et al., 2013). Warming of Arctic soils may increase soil respiration and thus CO_2 fluxes to the atmosphere, thereby contributing to a positive feedback effect (Billings, 1987; Oechel et al., 1993; Goulden et al., 1998). On the other hand, there is also evidence that Arctic ecosystems in particular may acclimatize to warming over decadal timescales (Oechel et al., 2000). Assessing the potential implications of accelerated climate change in Polar regions on regional and global biogeochemical cycles and the climate system requires a better quantitative understanding of Arctic ecosystems, as well as predictions of their response to ongoing climate change.

Glacier forefields are one of the most intriguing ecosystems in the Arctic as they act as pioneer sites for ecosystem development and soil formation. The retreat of glaciers since the end of the Little Ice Age has led to the emergence of terrestrial landscapes that were previously locked underneath ice (Paul et al., 2011). Glacier forefields act as bioreactors and intermediates between the icy biome (Anesio and Laybourn-Parry, 2012) and downstream ecosystems (Fountain et al., 2004, 2008; Kastovska et al., 2005; Mindl et al., 2007). The physical, geochemical and biological development of exposed soils following glacier retreat have been studied using chronosequence approaches (Bradley et al., 2014 and references therein). Decades of empirical research in glacier forefields has shown that microbes support enhanced weathering rates, the development of complex community structures, the colonization of plants, as well as the physical process of soil formation (Schulz et al., 2013; Bradley et al., 2014). More recently, data derived from field campaigns in the Alps, the Canadian Arctic, and Svalbard, as well as laboratory experiments, were integrated with numerical modeling using the Soil biogeocHemIcal Model for Microbial Ecosystem Response (SHIMMER) to explore microbial dynamics and nutrient fluxes along soil chronosequences (Bradley et al., 2015, 2016b). This integrated model-data approach revealed that autotrophic and heterotrophic biomass has likely accumulated in the forefield of Midtre Lovénbreen glacier (Svalbard) over the last century. Low measured microbial growth efficiency had a potentially important role for nutrient accumulation by

enhancing the degradation of organic matter (Bradley et al., 2016b). In addition, simulation results emphasized that microbial communities play a key role in fixing and recycling carbon and nutrients. Furthermore, results indicated that both allochthonous carbon inputs, as well as microbial necromass, are important in sustaining a pool of organic material in older soils, that feeds heterotrophic bacteria.

There are relatively few studies that have explored the effect of environmental factors, such as temperature, snow cover and nutrient supply on microbial dynamics in developing forefield soils. These mostly empirical investigations have correlated microclimatic environments and weather-related seasonal variations with distinct patterns of microbial diversity (Lazzaro et al., 2015) and demonstrated that allochthonous nutrient deposition stimulates microbial activity (Brankatschk et al., 2011; Goransson et al., 2011). Whilst these experiments and observations provide important insights, they often explore only the short-term (seasonal) response to changes in one single environmental factor for a spatially discreet geographical location and thus do not allow for assessment of the mid-term (decadal) to long-term (century) response of the system to holistic environmental perturbations. As such, little is known about the influence of ongoing and projected Arctic environmental change on microbial dynamics and the resulting implications for carbon and nutrients imported and exported from these environments.

In this respect, a mathematical model, which is constructed on the basis of current mechanistic knowledge and tested against experimental datasets, is a powerful means to assess the potential, long-term response of microbial and biogeochemical dynamics to projected climate changes, using a scenario-based approach (Bradley et al., 2016a). Here, we used the novel numerical model SHIMMER (Bradley et al., 2015, 2016b) to quantitatively predict the response of the microbial community as well as the implications for carbon and nutrient transformations and fluxes in the forefield of the Midtre Lovénbreen glacier (Svalbard) over climate-relevant time periods (decades to centuries). To assess the induced changes, simulation results of (1) a baseline scenario that has previously been calibrated (Bradley et al., 2016b) are compared to five different anthropogenic change scenarios: (2) increased soil temperature, (3) increased soil temperature and earlier spring snow melt, (4) increased deposition of reactive nitrogen, (5) increased deposition of organic carbon, and (6) a combination of all of these factors. Finally, regional implications of projected changes are discussed on the basis of model results.

METHODS

Study Site

We focused the investigation on the forefield of Midtre Lovénbreen, an Arctic polythermal valley glacier on the Kongsfjorden, Western Svalbard (78°55′N, 12°10′E). This glacier forefield has previously been the subject of an integrated and comprehensive field, laboratory and modeling study (Bradley et al., 2016b). In the framework of this study, determination of model sensitivity helped identify critical model parameters, that were then constrained by laboratory measurement of bacterial growth rates, growth efficiencies and temperature response for this specific habitat, and helped refine model predictions. The Midtre Lovénbreen catchment is roughly 5 km East of Ny-Ålesund, where several long-term monitoring programs and weather stations provide contextual information and forcing data. Midtre Lovénbreen has experienced negative mass balance throughout much of the Twentieth Century. Since the end of the Little Ice Age maximum (in the 1900s) in Svalbard, the glacier snout has retreated considerably in response to a warming of mean annual temperatures (Lefauconnier et al., 1999). This retreat continues to the present day (Fleming et al., 1997; Moreau et al., 2008). The regional climate is predominantly influenced by the North Atlantic Current, resulting in a maritime Polar climate that is uncharacteristically mild for the region. The Midtre Lovénbreen catchment falls within the tundra zone in Svalbard, amid the areas that are the richest in fauna and flora. The glacier forefield zone, however, is extremely sparsely vegetated, mostly consisting of lichens, *Dryas octopetala* and *Saxifraga oppositifolia* in micro-local habitats in soil that has been exposed for roughly 100 years (Moreau et al., 2008). Nearby cliffs are heavily populated with nesting birds, and the tundra is frequented by mammals including the Svalbard reindeer, the Arctic fox and the polar bear. Snow-cover persists over winter (typically from October until June), containing biologically significant concentrations of nutrients and organic material (Larose et al., 2013a,b).

The SHIMMER Model

The numerical model SHIMMER (Bradley et al., 2015) is a novel microbial-biogeochemical model designed to simulate the initial stages of microbial community establishment during soil development in glacier forefields. SHIMMER has been previously developed and successfully used to quantify microbial and nutrient dynamics in a number of contrasting forefields in Switzerland, Canada, and Svalbard (Bradley et al., 2015, 2016b). It thus represents an ideal tool to quantitatively explore the potential response of these systems to projected climate change. The model is zero-dimensional and explicitly resolves the evolving dynamics of microbial biomass, labile and refractory organic substrate (S_1 and S_2 respectively), dissolved inorganic nitrogen (DIN) and dissolved inorganic phosphorus (DIP) along a chronosequence (Table S1). Microbes are categorized according to function: autotrophs (A_{1-3}) and heterotrophs (H_{1-3}) are further subdivided into glacial microbes (A_1 and H_1), soil microbes (A_2 and H_2) and nitrogen-fixing microbes (A_3 and H_3). Transformations in substrate and nutrients due to autotrophic and heterotrophic production and respiration, microbial growth, death and predation, exopolymetric-substance (EPS) production, and nitrogen fixation are also explicitly resolved (**Figure 1**). The following external forcings drive and regulate the model dynamics: (1) PAR (wavelength of approximately 400–700 nm) (W m^{-2}), (2) snow depth (m), (3) soil temperature (°C), and (4) allochthonous inputs of organic material, DIN and DIP (provided in µg g^{-1} d^{-1}). The Supplementary Information contains a detailed description of the model set-up used in this study, including a list of initial conditions (Table S1) and parameter values (Table S2).

Model Scenarios

Here, we used SHIMMER to quantitatively explore the response of the Midtre Lovénbreen glacier forefield to a number of different climate scenarios, described in detail below. A baseline scenario ("BASE") and five future climate change scenarios were designed to explore the response of microbial and biogeochemical dynamics to projected changes in the physical environment ("TEMP" and "TEMP&SNOW"), perturbations of external carbon and nitrogen inputs ("NITRO" and "SUBS"), as well as a combined scenario ("COMB"). This approach allows exploration into the individual effects of climate change driven variations in external forcings, as well as to assess the potential combined effects of these climate-driven variations on the forefield dynamics. Simulated future climate scenarios are described and evaluated as stand-alone simulations, or relative to the control model simulation run with baseline forcings.

"BASE" Scenario: Baseline Forcings

The baseline simulation was set up exactly as the optimized model simulation presented in Bradley et al. (2016b). Briefly, meteorological forcings were constrained by daily observations for the entire year 2013 and remained unchanged for the duration of the model run (**Figure 2**). Averaged daily soil temperature (at 1 cm depth) and PAR for 2013 were provided by the Alfred Wegener Institute for Polar and Marine Research (AWI) from the permafrost observatory near Ny-Ålesund, Svalbard, and the AWI meteorological station near Ny-Ålesund, Svalbard, respectively. Averaged daily snow depth for 2009–2013 is provided by the Norwegian Meteorological Institute (eKlima). The presence of snow on the ground attenuates sunlight and inhibits PAR from reaching the soil surface. This was accounted for in pre-processing of forcing data. Light attenuation was estimated according to the equation:

$$n = n_0 e^{-mx} \tag{1}$$

Whereby n is the irradiance (W m^{-2}), x is the snow depth (m) and m is the extinction coefficient for snow (Greenfell and Maykut, 1977; Bradley et al., 2015). Due to its high latitude, the study site experiences continual daylight for much of the summer and continual darkness for much of the winter. Forcing data was provided as daily averages, and linear interpolation was used between any (very infrequent) missing data points. Averaged daily allochthonous nutrients and carbon inputs were estimated based on the best available budget estimates of catchment hydrology and nutrients for Midtre Lovénbreen presented in Hodson et al. (2005), described and used in Bradley et al. (2016b) and also summarized in the Supplementary Information (Tables S3, S4).

"TEMP" Scenario: Increased Soil Temperature

Climate change is amplified in the Arctic region, and as such Svalbard is particularly susceptible to climate warming due to its northerly latitude and its geographical location at the northernmost reach of the North Atlantic Current (Overland et al., 2014). Air temperature warming at Kongsfjorden, the site of the Midtre Lovénbreen glacier, has occurred in the last two decades and is expected to continue with present emissions rates

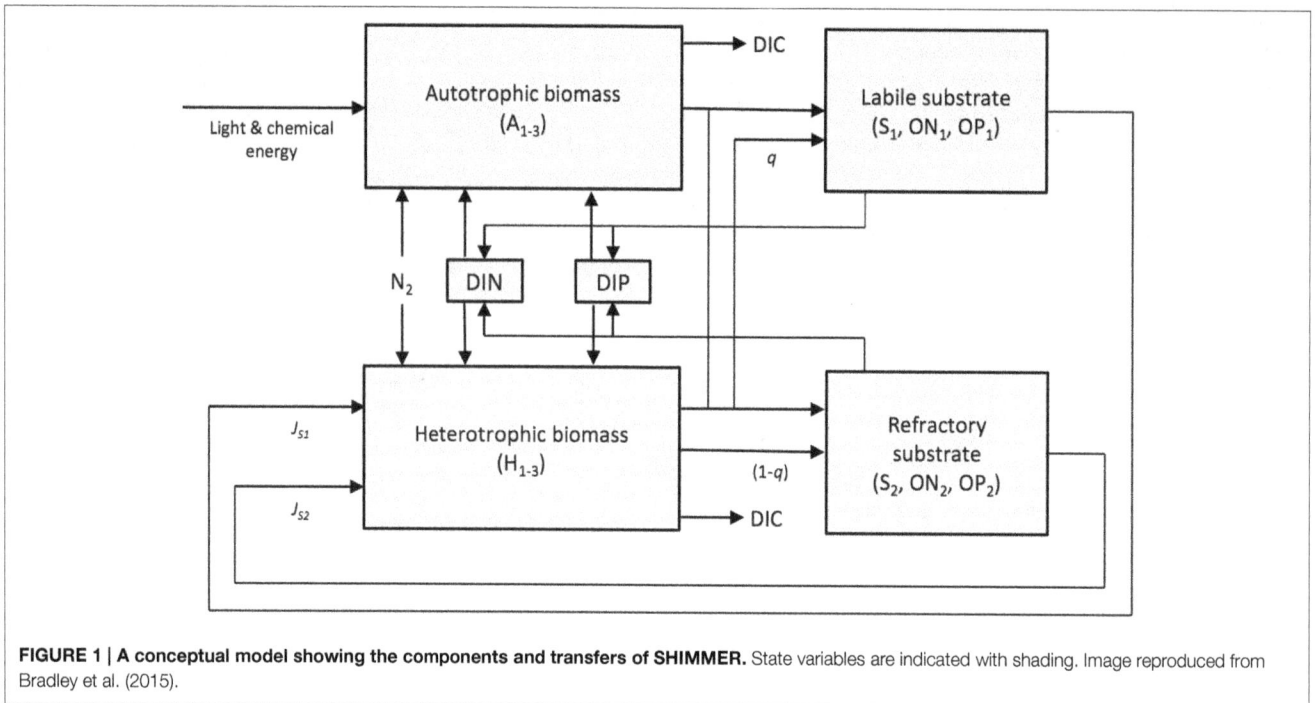

FIGURE 1 | A conceptual model showing the components and transfers of SHIMMER. State variables are indicated with shading. Image reproduced from Bradley et al. (2015).

(Maturilli et al., 2015). A very strong winter warming trend has been identified alongside positive linear temperature trends for spring, summer, and autumn (Forland et al., 2011). Therefore, TEMP explores the effect of increasing soil temperatures on the microbial and biogeochemical dynamics in the forefield. For TEMP, we used predicted monthly surface air temperature anomalies from the year 2000 to 2100 (relative to the 1981–2005 period mean) based on climate model predictions for the Arctic region (60–90°N) presented in Overland et al. (2014) (**Table 1**). Two end-member IPCC Representative Concentration Pathways (RCPs) that cover the entire range of potential warming are investigated: (1) the RCP 4.5 scenario ("mitigation" scenario) whereby CO_2 emissions increase only slightly before decline commences around 2040; and (2) RCP 8.5 scenario ("business-as-usual"/"extreme" scenario), the most extreme climate scenario whereby emissions are projected to increase and CO_2 concentration is projected to rise above 1,370 ppm by 2100 (Moss et al., 2010). Predicted surface air temperature warming was taken specifically from the Svalbard region in climate model predictions (Overland et al., 2014). We assumed that surface air temperature warming equated directly to an equal increase in soil temperature whenever there is an absence of snow cover (<1 cm). The changes (compared to the baseline scenario) to soil temperature (scenario RCP 8.5), snow depth and PAR after 100 years are illustrated in **Figure 2**.

"TEMP&SNOW" Scenario: Increased Soil Temperature and Earlier Spring Snow Melt

Observations indicate that there has been a persistent increase in the duration of snow-free conditions across Eurasia and North America for the past three decades, at a rate of 5 to 6 days per decade (Anisimov et al., 2007). TEMP&SNOW thus explores

the combined effect of increasing soil temperatures and earlier spring melt on the microbial and biogeochemical dynamics in the forefield. For TEMP&SNOW (RCP 4.5 and RCP 8.5), we prescribed an earlier spring snow melt rate of 1 day every 2 years to reflect observations (Anisimov et al., 2007), alongside soil temperature increases based on the (1) RCP 4.5 and (2) RCP 8.5 pathways explored in TEMP. For new snow-free days, soil temperature was set to the projected air temperature (based on RCP 4.5 and RCP 8.5). PAR attenuation was re-calculated for every simulated day to account for the new snow depth. The changes after 100 years (compared to the baseline scenario) are illustrated in **Figure 2**.

"NITRO" Scenario: Increased DIN Input

In recent decades, human activity has profoundly changed the biogeochemical cycling of reactive nitrogen in Arctic regions (Roberts et al., 2010). NO and NO_2 are released predominantly by burning fossil fuels (90%) and application of fertilizer (10%) (Geng et al., 2014). Enhanced winter transport of polluted air from both Europe and Russia has led to increased nitrogen deposition in the Arctic (Eneroth et al., 2003). Although a number of studies have attempted to assess the effect of increased nitrogen deposition on carbon cycling at both the regional as well as global scales (Holland et al., 1997; Nadelhoffer et al., 1999; Zaehle and Friend, 2010; Mahowald, 2011), no estimates of future deposition fluxes to the Artic, let alone their implications for Arctic carbon and nutrient cycling, currently exist. Local anthropogenic inputs of nitrogen to the biosphere can lead to changes in the productivity of nitrogen deficient ecosystems, that directly impact carbon uptake rates (Galloway et al., 2008). Evidence for anthropogenic nitrogen inputs to cryospheric ecosystems include northern hemisphere ice core

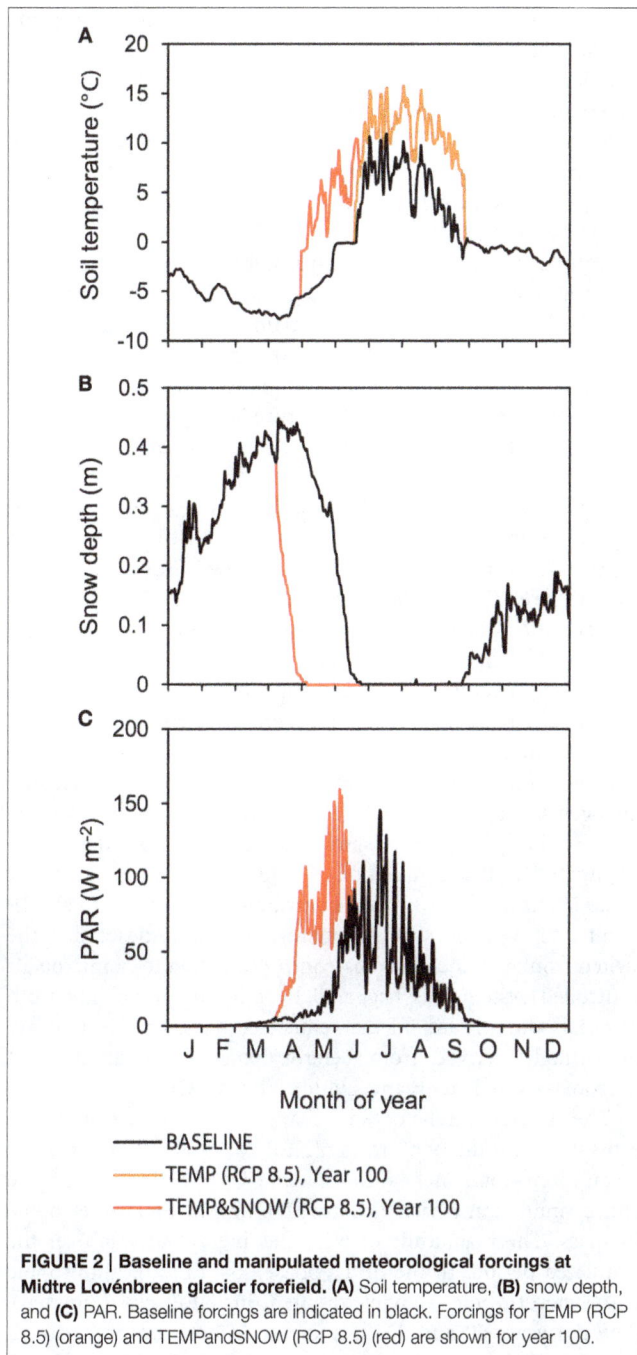

FIGURE 2 | Baseline and manipulated meteorological forcings at Midtre Lovénbreen glacier forefield. (A) Soil temperature, **(B)** snow depth, and **(C)** PAR. Baseline forcings are indicated in black. Forcings for TEMP (RCP 8.5) (orange) and TEMPandSNOW (RCP 8.5) (red) are shown for year 100.

TABLE 1 | Prescribed soil temperature increases for RCP 4.5 and 8.5 (from Overland et al., 2014).

Month	Temperature increase (°C year^{-1})	
	RCP 4.5	RCP 8.5
January	0.051	0.114
February	0.042	0.103
March	0.037	0.084
April	0.028	0.065
May	0.019	0.047
June	0.019	0.042
July	0.019	0.047
August	0.028	0.056
September	0.037	0.075
October	0.056	0.093
November	0.065	0.121
December	0.065	0.131

a doubling of NO_3^- concentrations in the last century, and a negative ^{15}N stable isotope excursion suggesting this increase is associated with anthropogenic sources (Geng et al., 2014). Furthermore, climatological data from Svalbard has shown that strong depositional events sometimes lead to reactive nitrogen deposition fluxes up to 3 times higher than the annual mean deposition flux (Kuhnel et al., 2013). However, in recent decades (since ~1970), nitrate deposition in the Arctic has stabilized due to North American air pollution mitigation strategies (Geng et al., 2014). However, despite recent mitigation policies, increased shipping in the Arctic due to sea-ice decline may further increase local reactive nitrogen inputs (Peters et al., 2011; Eckhardt et al., 2013). Therefore, overall trends of reactive deposition are uncertain, resulting from the interplay of emissions, atmospheric transport, chemistry, precipitation, and snowpack processes (Kuhnel et al., 2011). The NITRO simulation is thus designed to explore the effect of increased DIN input on the microbial and biogeochemical dynamics of the forefield. To account for the possible future range of DIN deposition in the Arctic due to anthropogenic factors, we carry out multiple simulations from 1.0 (BASE) to 4.0 times (extreme) nominal deposition flux. This range encapsulates the variability from observations from both ice core records (up to 2 times) (Geng et al., 2014) and climatological data (up to 3 times) (Kuhnel et al., 2011) and for the possible occurrence of "extreme" (up to 4 times) nitrogen deposition events in the future.

"SUBS" Scenario: Increased Input of Organic Substrate

Forefield soils accumulate organic carbon from autochthonous production, allochthonous deposition, and ancient sources that are mobilized during glacier retreat (Schulz et al., 2013). Similarly, glacier surfaces accumulate organic carbon from biological activity (e.g., *in situ* primary production) and the deposition of allochthonous organic material from terrestrial or anthropogenic sources (Hood and Berner, 2009; Singer et al., 2012; Stibal et al., 2012). It has been suggested that the

records (Goto-Azuma and Koerner, 2001; Isaksson et al., 2003; Hastings et al., 2009), Svalbard lake sediments (Birks et al., 2004), and field-based snowpack measurements and modeling (Bjorkman et al., 2013; Kuhnel et al., 2013). Differences in the magnitude and timing of periodic increases in nitrate deposition in the Arctic can be attributed to different source regions and pathways of pollutants (Goto-Azuma and Koerner, 2001; Geng et al., 2014). However, there is considerable uncertainty in predicting the magnitude of future reactive nitrogen deposition. Ice core analyses from the Greenland Summit station has shown

allochthonous flux of organic carbon to glacier forefields could increase over the next two centuries due to (1) increased glacier runoff and thus increased flux of organic carbon from glacier surfaces to the pro-glacial zone (Foreman et al., 2007; Hood et al., 2015), (2) a longer residence time of liquid water in the glacial snowpack supporting higher growth and carbon fixation rates of snow algae (Morgan-Kiss et al., 2006; Stibal et al., 2007; Takeuchi et al., 2009; Lutz et al., 2015), (3) the invasion of non-native birds and mammals (Jonsdottir, 2005; Kelly et al., 2010; Miller and Ruiz, 2014; Ware et al., 2014), (4) the invasion of new plant species as human activity and tourism increases (Ware et al., 2014), and (5) an enhanced input of feces-derived organic material to soils from increasing populations of Barnacle Geese, Reindeer herds and other biota in Svalbard in response to climate warming (Michelutti et al., 2009; Moe et al., 2009; Luoto et al., 2015). However, there is considerable uncertainty in predicting the magnitude of future substrate deposition and, to our knowledge, no studies have made direct predictions based on estimations or empirical data. Therefore, we have simulated a range of possible future scenarios from present day deposition rates increasing up to 4 times present day for the following reasons: (1) the chosen range encapsulates a substantial increase in organic matter deposition compared to the present day, and (2) it is consistent with the range investigated for DIN deposition, allowing for comparative sensitivity analysis between the two forcing scenarios.

"COMB" Scenario: Warming, Snow Melt, and Nitrogen and Substrate Deposition

COMB is based on a combination of all extreme scenarios to explore the maximum effect of climate change on microbial communities and biogeochemical dynamics in glacier forefield soils. Thus, COMB assumes:

- Soil temperature prescribed according to RCP 8.5-based predictions.
- Snow melting prescribed according to TEMP&SNOW.
- Allochthonous inputs of DIN and substrate prescribed to a maximum of 4 times nominal values.

RESULTS AND DISCUSSION

Baseline Scenario: Microbial and Biogeochemical Dynamics in the Midtre Lovénbreen Forefield

Figure 3 summarizes the simulated evolution of microbial and biogeochemical dynamics in the Midtre Lovénbreen forefield over a period of 200 years. Simulation results show that, while both autotrophic and heterotrophic biomass accumulated in the forefield, the microbial community was dominated by autotrophs (Figures 3A,B). The increase in biomass for all functional groups was first characterized by a lag phase (years 0–40), due to the slow accumulation of biomass from extremely low inocula concentrations. This was followed by a rapid growth phase (years 40–80), with exponential growth in bacterial populations due to increased biomass and nutrient availability. Finally, from year 80 onwards, microbial growth slowed down due to limiting

organic carbon availability in forefield soils, since soil organic carbon stocks were mostly refractory (Figure 3C). During initial soil development (20–50 years), nitrogen-fixers (A_3 and H_3) experienced more rapid growth than other functional groups because of their ability to overcome DIN-limitation by fixing atmospheric nitrogen. All other functional groups (A_{1-2} and H_{1-2}) were co-limited by DIN and DIP and therefore their biomass accumulated at a slower rate.

Carbon and nutrient dynamics were mainly controlled by allochthonous inputs during the lag phase (0–40 years), whereas microbial dynamics became the main control on carbon and nutrient stocks during and after the rapid growth phase (40–80 years). Refractory organic carbon (S_2) accumulated over the entire simulation period (Figure 3C), predominantly by contributions from allochthonous deposition (>50%) in years 0–114, and necromass (>50%) from year 114 onwards. In contrast, the concentration of labile substrate (S_1) increased slightly during the initial lag period (year 0–40), followed by a depletion due to rapid heterotrophic growth during years 50–100 (Figure 3C). Simulation results show that DIN and DIP concentrations increased from trace quantities (<4 μg g^{-1} and <2 μg g^{-1} respectively) to >157 μg g^{-1} and >201 μg g^{-1} respectively over two centuries. Figure 4A illustrates that the observed increase in nitrogen concentrations was mainly driven by allochthonous deposition (>50%) in young soils (years 0–38), while internal recycling by heterotrophic activity increased nitrogen stocks in older soils (contributing 90–95% in years 60–200). Similarly, Figure 5A shows that allochthonous deposition dominated DIP accumulation (up to 99%) from years 0–66, while heterotrophic recycling contributed 48–58% of total DIP input after year 66. Simulation results thus indicate that the switch from an allochthonous-controlled system to a microbial-controlled system was triggered by nutrient availability from external sources, and further enhanced by nutrients that are increasingly derived from heterotrophic remineralisation of necromass-supplied organic matter (Figure 3C).

The glacier forefield was characterized by negative net ecosystem production rates (NEP = total heterotrophic respiration—total net autotrophic CO_2 fixation) over the entire simulation period (Figure 3D), despite high autotrophic biomass. The magnitude of NEP was highly variable over the simulated period. In the initial phase, low rates of autotrophic and heterotrophic activity resulted in NEP close to zero. Between years 60 and 90, an increase in heterotrophic growth and degradation rates, supported mainly by initial stocks of organic carbon and the accumulation of allochthonous labile and refractory organic carbon, led to a decrease in NEP (< −8 μg C g^{-1} y^{-1}). Subsequently (years 110–200) NEP rates stabilized at −4 μg C g^{-1} y^{-1} (± 1 μg g^{-1} y^{-1}) as bacterial activity remained relatively constant. During this phase, nutrient limitation was alleviated (due to the accumulation of recycled DIN and DIP in the soil) and heterotrophic growth became limited by available organic matter (Figure 3C). In a previous study, measured heterotrophic growth efficiency for these soils was extremely low (Bradley et al., 2016b) (Table S2). Thus, the observed net heterotrophy and the associated net release of CO_2 to the atmosphere was mainly driven by substantial

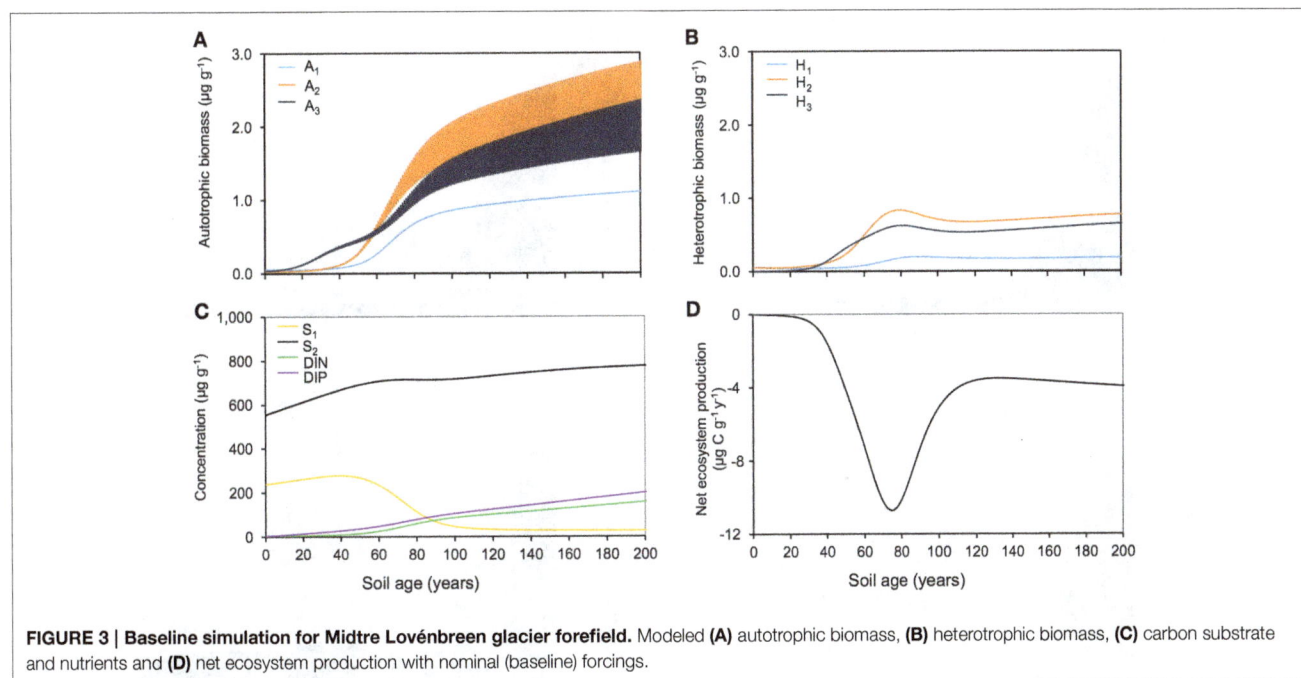

FIGURE 3 | Baseline simulation for Midtre Lovénbreen glacier forefield. Modeled **(A)** autotrophic biomass, **(B)** heterotrophic biomass, **(C)** carbon substrate and nutrients and **(D)** net ecosystem production with nominal (baseline) forcings.

degradation of initial carbon stocks and was sustained by allochthonous carbon inputs. Modeling therefore suggests that heterotrophic processing of initial and allochthonous inputs of organic substrate exert a major control on glacier forefield biogeochemistry. Predicted carbon fluxes in the Midtre Lovénbreen forefield are corroborated by data from the forefield of the Damma Glacier (Switzerland). Evidence from the Damma Glacier suggests high rates of CO_2 efflux from soils, particularly in the later stages of development (soil age >110 years) (Guelland et al., 2013b), and also suggest that organic matter becomes increasingly refractory in the later stages of development due to continual re-working and cycling by microbial communities (Goransson et al., 2011). However, empirical evidence from Midtre Lovénbreen is currently lacking, and this should be a particular focus of future experimental efforts. Heterotrophic activity in glacier forefield soils is sensitive to environmental conditions (Brankatschk et al., 2011; Goransson et al., 2011), indicating that projected future changes in temperature, snow cover and allochthonous inputs might have an important impact on regional carbon and nutrient cycling. The following sections explore the potential response of the Midtre Lovénbreen forefield to such changes.

Effect of Projected Climate Change on Microbial Dynamics in the Midtre Lovénbreen Forefield

General Response

Figure 6 summarizes the simulated response (relative to the baseline (BASE) scenario) of each individual functional group to each tested scenario of environmental change. Simulation results show that the entire forefield microbial community was responsive to the environmental changes imposed in all future

climate scenarios investigated. Model results show a peak in percentage biomass increase (relative to the baseline simulation) for all functional groups in response to all climatological forcings occurred between years 30 and 80, after which the percentage biomass increase stabilized at lower values. The observed peaks varied in height (abundance) and width (duration) depending on the specific forcing investigated and the functional group of interest, but in general they coincided with the exponential (rapid) growth phase of bacteria (see **Figure 3**). In younger soils, nutrient availability generally limited microbial growth (Bradley et al., 2014, 2016b). Thus, alleviating growth limitations by increasing allochthonous inputs or, to a lesser extent, improving physical conditions, triggered a pronounced growth response. Model results also show that generic soil autotrophs and heterotrophs (A_2 and H_2 respectively) reacted more strongly to an improvement in growth conditions than glacial bacteria (A_1 and H_1), which are better adapted to growing in oligotrophic low nutrient conditions, and nitrogen fixers (A_3 and H_3), which can source atmospheric nitrogen (N_2) in place of DIN. In older soils (year 80 onwards), bacterial abundance was high and the rate of new biomass accumulation was relatively low. These older soils generally contained nutrient concentrations well above limiting concentrations (K_N and K_P, Table S2) resulting in a stabilization of the system that approaches its ecological climax (although at greater abundance than in the baseline simulation). Therefore, in older forefield soils, the response of microbial biomass to environmental perturbations was weaker than in younger soils.

Response to Increased Temperature and Decreased Snow Cover (TEMP and TEMP&SNOW)

Warming promoted soil conditions that were more favorable to autotrophic and heterotrophic growth. Similarly, earlier snow melt allowed for a longer autotrophic growth season

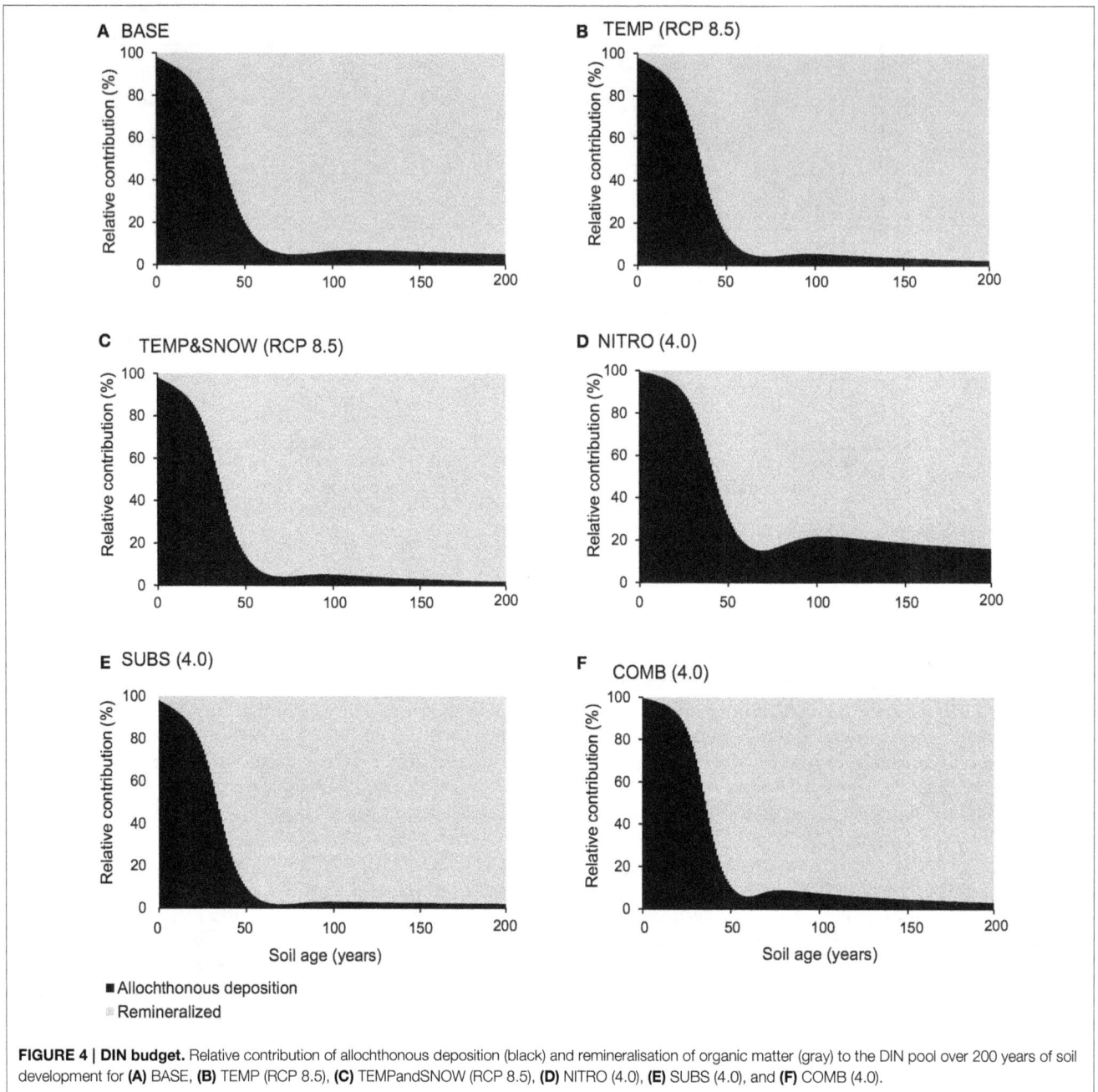

A BASE

B TEMP (RCP 8.5)

C TEMP&SNOW (RCP 8.5)

D NITRO (4.0)

E SUBS (4.0)

F COMB (4.0)

Soil age (years)

■ Allochthonous deposition
▨ Remineralized

FIGURE 4 | DIN budget. Relative contribution of allochthonous deposition (black) and remineralisation of organic matter (gray) to the DIN pool over 200 years of soil development for **(A)** BASE, **(B)** TEMP (RCP 8.5), **(C)** TEMPandSNOW (RCP 8.5), **(D)** NITRO (4.0), **(E)** SUBS (4.0), and **(F)** COMB (4.0).

due to increased exposure to PAR. Over century time-scales, temperature and snow cover play an important role in microbial community dynamics in glacier forefields. Model results show that the response of autotrophs (A_{1-3}) to modeled climatological changes was stronger than the response of heterotrophs (H_{1-3}) (**Figure 6**), suggesting that the positive effect of temperature increase and snow-free conditions on autotrophs did not linearly translate to heterotrophs. This observation can be explained by the reduced availability of labile organic carbon in these simulations, which limited heterotrophic growth. The improved thermal and light conditions induced a stronger long-term response in autotrophic populations (in particular A_2 and A_3)

than any other single driver (red line, **Figure 6**). Furthermore, the effects of temperature on biomass were enhanced with "extreme" temperature warming (RCP 8.5, light blue line, up to 33.9% increase in total biomass) compared to the "mitigation" climate scenario (RCP 4.5, yellow line, up to 25.0%). These trends are supported by empirical observations from soil manipulation experiments suggesting that a positive relationship exists between temperature and observed growth of Arctic and sub-Arctic microbial communities (Callaghan et al., 1999; Yergeau et al., 2012; Sistla et al., 2013; van der Wal and Stien, 2014; Lau et al., 2015; Bradley et al., 2016b; Newsham et al., 2016).

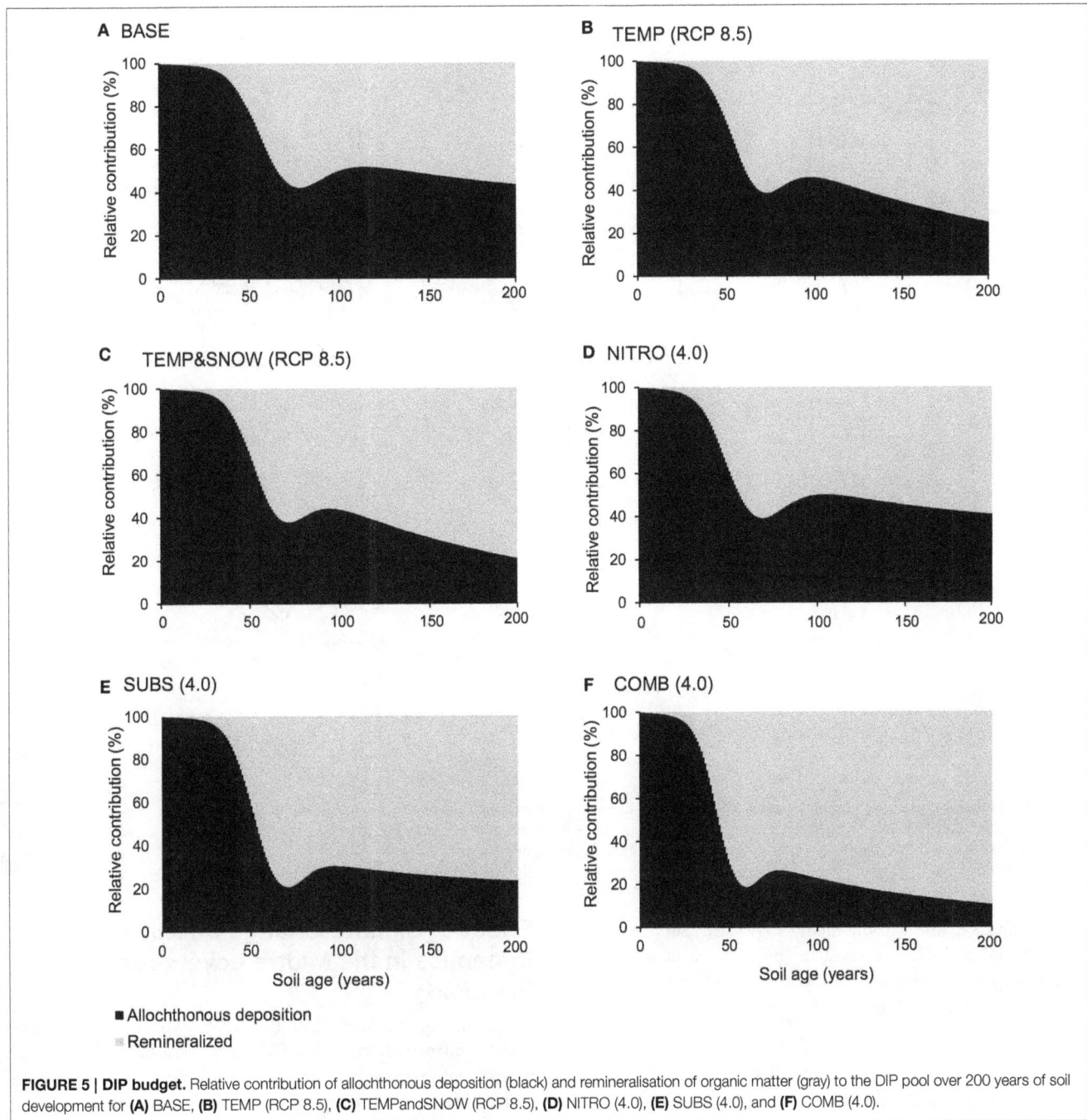

FIGURE 5 | DIP budget. Relative contribution of allochthonous deposition (black) and remineralisation of organic matter (gray) to the DIP pool over 200 years of soil development for **(A)** BASE, **(B)** TEMP (RCP 8.5), **(C)** TEMPandSNOW (RCP 8.5), **(D)** NITRO (4.0), **(E)** SUBS (4.0), and **(F)** COMB (4.0).

Response to Increased Nutrient and Organic Carbon Inputs (NITRO and SUBS)

Model results show that microbial growth responded more strongly to increases in allochthonous DIN (NITRO) and organic carbon input (SUBS) than climatological changes (TEMP and TEMP&SNOW), particularly in the very early stages of soil development (0–50 years) (**Figure 6**). DIN deposition exerted a greater effect on the non-nitrogen-fixing functional groups (A_1, A_2, H_1, H_2) reducing the pressure of DIN-limitation in young (0–50 year old) soils. Increased substrate input supported higher heterotrophic (H_{1-3}) (up to 111%) and to a lesser extent autotrophic (A_{1-3}) (up to 68%) biomass. Interestingly, the increased carbon inputs (SUBS) also caused long term increases in microbial biomass beyond the exponential growth phase, due to the enhanced supply of labile substrate to the older soils (100–200 years), which were typically depleted in labile organic matter. However, simulation results show that increased organic carbon and nutrient inputs did not trigger fundamental shifts in system behavior over the explored range. Total (autotrophic and heterotrophic) biomass responded linearly to increases

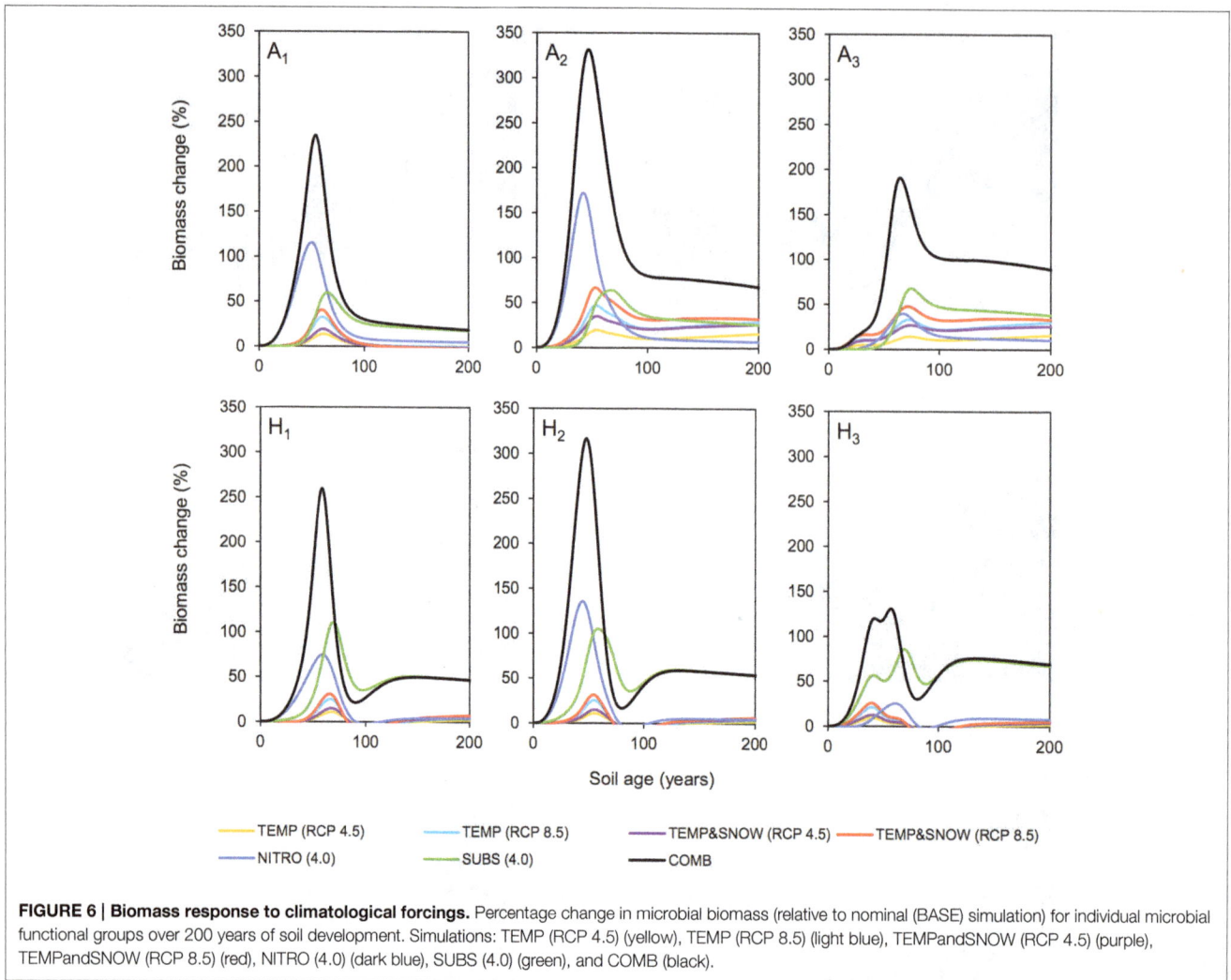

FIGURE 6 | Biomass response to climatological forcings. Percentage change in microbial biomass (relative to nominal (BASE) simulation) for individual microbial functional groups over 200 years of soil development. Simulations: TEMP (RCP 4.5) (yellow), TEMP (RCP 8.5) (light blue), TEMPandSNOW (RCP 4.5) (purple), TEMPandSNOW (RCP 8.5) (red), NITRO (4.0) (dark blue), SUBS (4.0) (green), and COMB (black).

in DIN and organic carbon input (**Figure S1**, Supplementary Information) and did not reveal tipping points, in contrast to studies on soils in temperate and desert regions (Zhou et al., 2012; Shcherbak et al., 2014; Scheer et al., 2016).

Response to Combined Changes (COMB)

In COMB we demonstrated that the combined effect of warming, snowmelt, and increased nitrogen and substrate input resulted in the largest increase in total microbial biomass (up to 190% relative to baseline forcings) throughout two centuries of soil development compared to the effects of individual environmental variations considered independently. Furthermore, the observed peak response in biomass (**Figure 6**) occurred earlier than other scenarios. Bacteria were rapidly alleviated of growth-limitations by the additional organic substrate and nutrients. Thus, the bacterial communities could respond quicker and more strongly to more favorable thermal and light conditions as a result of climate warming. Similarly, initial warming enabled soil microbial communities to take advantage of favorable nutrient and substrate concentrations.

Effect of Climate Change on Nutrient Dynamics in the Midtre Lovénbreen Forefield

Accumulating nutrients can be derived from allochthonous sources, nitrogen fixation (for DIN) or the release of nutrients by heterotrophic decomposition of organic matter. The deposition of reactive nitrogen in snow is a major source of nitrogen in Arctic regions (Hodson et al., 2005, 2010; Kuhnel et al., 2011; Bjorkman et al., 2013). In addition, many studies on microbial dynamics in glacier forefields also draw attention to microbially-mediated nitrogen fixation as an important source of nitrogen to glacier forefield soils (Deiglmayr et al., 2006; Duc et al., 2009a,b; Brankatschk et al., 2011; Strauss et al., 2012; Ansari et al., 2013). **Figure 7** illustrates the change in soil DIN (A) and DIP (B) stocks over 200 years of soil development for BASE, TEMP (RCP 8.5), TEMP&SNOW (RCP 8.5), NITRO (4.0), SUBS (4.0), and COMB. Additionally, **Figures 4, 5** show the relative contributions to soil DIN and DIP from allochthonous sources and organic carbon remineralisation, thus enabling separation of these two factors commonly observed only as a net outcome.

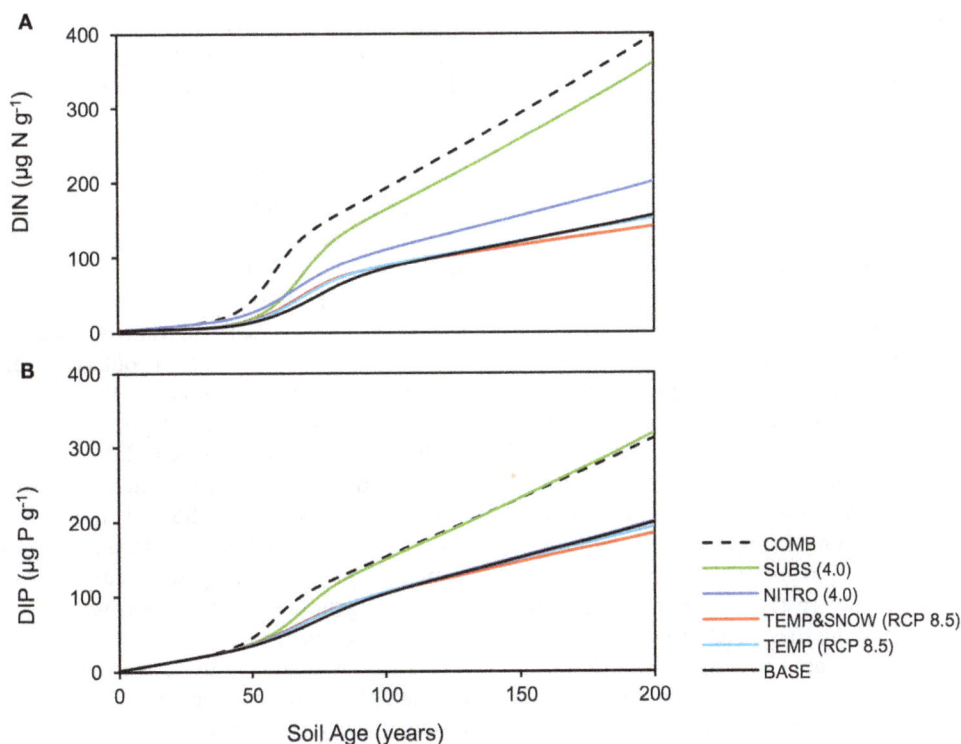

FIGURE 7 | Soil DIN and DIP response to climatological forcings. Response of soil **(A)** DIN and **(B)** DIP stocks to climatological forcings simulated in: BASE (solid black), TEMP (RCP 8.5) (light blue), TEMPandSNOW (RCP 8.5) (red), NITRO (4.0) (dark blue), SUBS (4.0) (green), and COMB (dashed black).

General Response

In all simulations, DIN and DIP accumulated from trace concentrations ($< 4 \, \mu g \, g^{-1}$) to $>150 \, \mu g \, g^{-1}$ over two centuries of soil development. In general, DIN and DIP fluxes were affected by all climate scenarios, but in particular, the greatest response was to organic carbon deposition, and the combination of all future changes (**Figure 7**). All climatological scenarios enhanced nutrient release by heterotrophic remineralisation of organic carbon. In TEMP and TEMP&SNOW, however, nutrient uptake was also enhanced, causing a minor depletion of nutrients relative to BASE, whereas in NITRO, SUBS, and COMB, enhanced organic matter remineralisation caused net accumulation of DIN and DIP. Based on the predicted growth of nitrogen-fixing autotrophs (A_3) and heterotrophs (H_3), modeling results suggest that nitrogen fixation contributed up to $0.03 \, \mu g \, N \, g^{-1} \, y^{-1}$ in intermediate and older soils (years 80–200). Nitrogen derived from nitrogen fixation is incorporated directly into biomass rather than soil DIN, however total nitrogen derived from nitrogen fixation would amount to only 2–4% of that from remineralisation. The effect of climatic and anthropogenic forcings explored in all future climate scenarios on the total nitrogen fixation was negligible (data not shown).

Response to Increased Temperature and Decreased Snow Cover (TEMP and TEMP&SNOW)

Model results show that the contribution of organic carbon remineralisation to soil DIN and DIP increased in response to increased thermal and light conditions (TEMP and

TEMP&SNOW) (**Figures 4, 5**). Heterotrophy is crucial to nutrient dynamics in glacier forefields, due to its role in the degradation of organic carbon into inorganic nutrients (Ingham et al., 1985). Under nominal conditions (BASE), DIN and DIP budgets were dominated by remineralisation rather than allochthonous deposition (>50% contribution) after 40 and 68 years of exposure respectively (see **Figures 4A, 5A**). In response to warming and earlier snowmelt (TEMP and TEMP&SNOW, RCP 8.5), the >50% threshold was reached 4 to 7 years earlier. Despite increased nutrient remineralisation, total nutrient stocks declined relative to BASE (**Figure 7**) due to enhanced bacterial growth. This is illustrated in **Figure 6** whereby climatological changes caused more favorable growth conditions which enhanced bacterial production (**Figure 6**). Thus, although organic carbon degradation and nutrient cycling was enhanced, the growth of biomass was, overall, the dominant control on nutrient budgets.

Response to Nutrient and Organic Carbon Inputs (NITRO and SUBS)

Model results show that increased allochthonous DIN deposition caused total soil DIN to increase (**Figure 7**). Predictably, the prescribed increased in external DIN in NITRO caused an increase in the proportion of DIN derived from allochthonous sources (black, **Figure 4D**) throughout the entire simulation, by ~1–15% compared to BASE. Further, the increased DIN deposition exerted a negligible effect on DIP budgets (**Figure 5D**), since the impact of DIN deposition on TOC decomposition, and thus DIP from remineralisation, was minimal (maximum relative change of −2%).

Model results show that the future delivery of allochthonous organic carbon to forefield soils may exert an important control on soil nutrient dynamics. Additional substrate [SUBS (4.0)] had the most pronounced effect on soil DIN and DIP stocks (**Figure 7**) for any single climatological forcing prescribed to the model, causing a maximum increase of 131% and 61% (respectively) relative to baseline forcings (BASE). Enhanced heterotrophic remineralisation of organic carbon substantially increased the relative contribution of nutrients derived from remineralisation (indicated in gray) to both the DIN pool (**Figure 4E**) and the DIP pool (**Figure 5E**). Consequently, this further stimulated the growth of all microbial groups (A_{1-3} and H_{1-3}) (green line, **Figure 6**). The abundance and bioavailability of phosphorus in glacier forefield soils has been causally linked to mineralogy, and particularly weathering of mineral surfaces (Anderson et al., 1997, 2000; Sattin et al., 2009; Prietzel et al., 2013). Here, we show through modeling, that the heterotrophic degradation of organic matter represents a major, potentially under-appreciated, process for sustaining sources of available nutrients to microbial communities in glacier forefield soils, particularly if organic carbon input to Arctic glacier forefields is enhanced in the future. The clear response of microbial growth to enhanced DIN and organic substrate deposition (NITRO and SUBS, **Figure 6**) is generally in line with the current perception of glacier forefields as nutrient-starved environments, and reinforces our result that nutrient availability is a major limiting factor on rates of bacterial growth and production (Jonsdottir et al., 1995; Duc et al., 2009b; Brankatschk et al., 2011; Schulz et al., 2013; Bradley et al., 2014).

Response to Combined Changes (COMB)

Model results suggested that when all climatological forcings were combined, total DIN and DIP stocks increased (**Figure 7**) and the contribution of remineralisation to DIN and DIP were elevated in comparison to the baseline forcings (**Figures 4F, 5F**). Despite the direct addition of DIN to soils from the COMB scenario, the effect of enhanced DIN delivery from remineralisation outweighed allochthonous DIN input. Warmer soils and increased exposure to PAR created more favorable conditions for bacterial growth. With strong allochthonous deposition of DIN and organic carbon, the limitations on microbial growth imposed by the availability of nutrients and labile substrate were alleviated in younger soils, and thus bacterial communities were able to respond more rapidly to the favorable thermal and light conditions created by climate change. Thus, heterotrophs grew rapidly (**Figure 6**) and nutrients were liberated from organic carbon at a much greater rate (**Figures 4F, 5F**) causing their net accumulation in soils (**Figure 7**).

Effect of Climatic Change on Carbon Dynamics and Net Ecosystem Production (NEP) in the Midtre Lovénbreen Forefield

Figure 8 summarizes the response of the total soil TOC pool ($S_1 + S_2$) to the BASE, TEMP (RCP 2.5 and 8.5), TEMP&SNOW (RCP 2.5 and 8.5), NITRO (1.2 to 4.0), SUBS (1.2 to 4.0) and

COMB scenarios. In addition, the seasonal evolution of daily-integrated TOC production and consumption processes over year 200 is also shown in **Figure 9** for the above-mentioned climate change scenarios to illustrate the variability of process rates. Net ecosystem production over 200 years of exposure is provided in **Table 2**.

General Response

Microbial dynamics exert an important control on carbon cycling and soil TOC content in glacier forefields (Guelland et al., 2013a,b; Schulz et al., 2013). Overall, the simulated response of glacier forefield soils to future climate change scenarios reinforce these observations. In contrast to the responses of soil nitrogen and phosphorous stocks, the responses of TOC stocks to the five climate and anthropogenic change scenarios were qualitatively and quantitatively different. Nitrogen deposition exerted a minimal effect on forefield TOC stocks, while climate change (TEMP and TEMP&SNOW) and, in particular, organic carbon deposition and the combined scenario (COMB) triggered a more pronounced response (**Figure 8**). As discussed earlier, forefield organic carbon dynamics are controlled by the balance between organic carbon inputs by microbial necromass, as well as allochthonous sources, and the consumption of organic carbon by heterotrophic degradation and respiration. Simulation results show that rates of organic carbon consumption and production were characterized by a strong seasonal variability, with higher process rates in summer and lower rates during the winter months (**Figure 9**) resulting in a net accumulation of TOC during summer (May-October) and a consumption of TOC stocks during the rest of the year in all scenarios. The relative significance of organic carbon inputs through necromass (light blue line, **Figure 9**) or allochthonous material (yellow line, **Figure 9**) was roughly equal in BASE and NITRO (4.0). However, in TEMP (RCP 8.5), TEMP&SNOW (RCP 8.5) and COMB, organic carbon inputs were dominated by necromass, while organic carbon inputs were dominated by external (allochthonous) deposition in the SUBS (4.0) scenario. Yet, the forefield soil was net heterotrophic (NEP < 0; heterotrophic respiration > autotrophic fixation) for all scenarios during all stages of soil development (**Table 2**), thus, emphasizing the importance of allochthonous organic carbon inputs on forefield ecosystem dynamics.

Response to Increased Temperature and Decreased Snow Cover (TEMP and TEMP&SNOW)

Heterotrophic respiration and growth rates, as well as the input of necromass, increased by a factor of eight in response to the temperature increase (TEMP RCP 8.5, **Figure 9B**). In addition, the earlier snowmelt allowed an earlier onset of microbial activity (mid-April compared to mid-June) (TEMP&SNOW RCP 8.5, **Figure 9C**). Simulation results indicate that soil warming and earlier onset of spring snow melt (TEMP and TEMP&SNOW) initially resulted in a slight depletion of soil organic carbon (relative to BASE) (**Figure 8A**), due to the increased consumption of labile substrate by enhanced heterotrophic degradation rates. However, on long time scales (>100 years), enhanced autotrophic growth and the associated increase in necromass (**Figures 9B,C**)

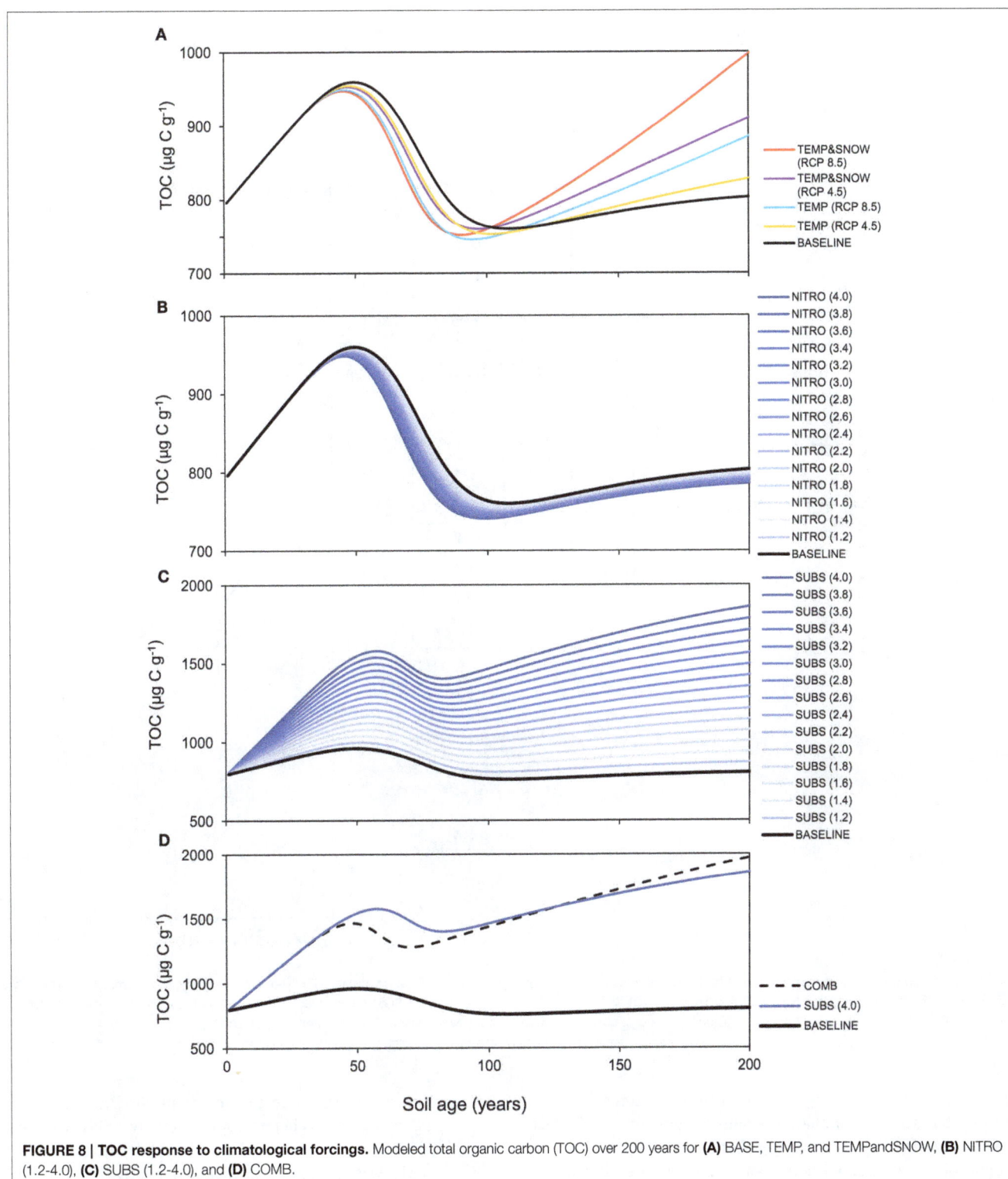

FIGURE 8 | TOC response to climatological forcings. Modeled total organic carbon (TOC) over 200 years for **(A)** BASE, TEMP, and TEMPandSNOW, **(B)** NITRO (1.2-4.0), **(C)** SUBS (1.2-4.0), and **(D)** COMB.

supported the accumulation of refractory organic carbon in the forefield (**Figure 9A**). In the most extreme warming and snowmelt scenarios, the long-term accumulation of refractory organic carbon resulted in a 24% increase in soil TOC compared to baseline values (red line, **Figure 8**). Enhanced autotrophic

growth (**Figure 6**) also increased NEP by 10.3% and 24.0% for TEMP (RCP 8.5) and TEMP&SNOW (RCP 8.5) respectively (**Table 2**). The positive response of autotrophic growth to warming thus theoretically renders a negative feedback possible, whereby CO_2-induced Arctic warming and snowmelt over large

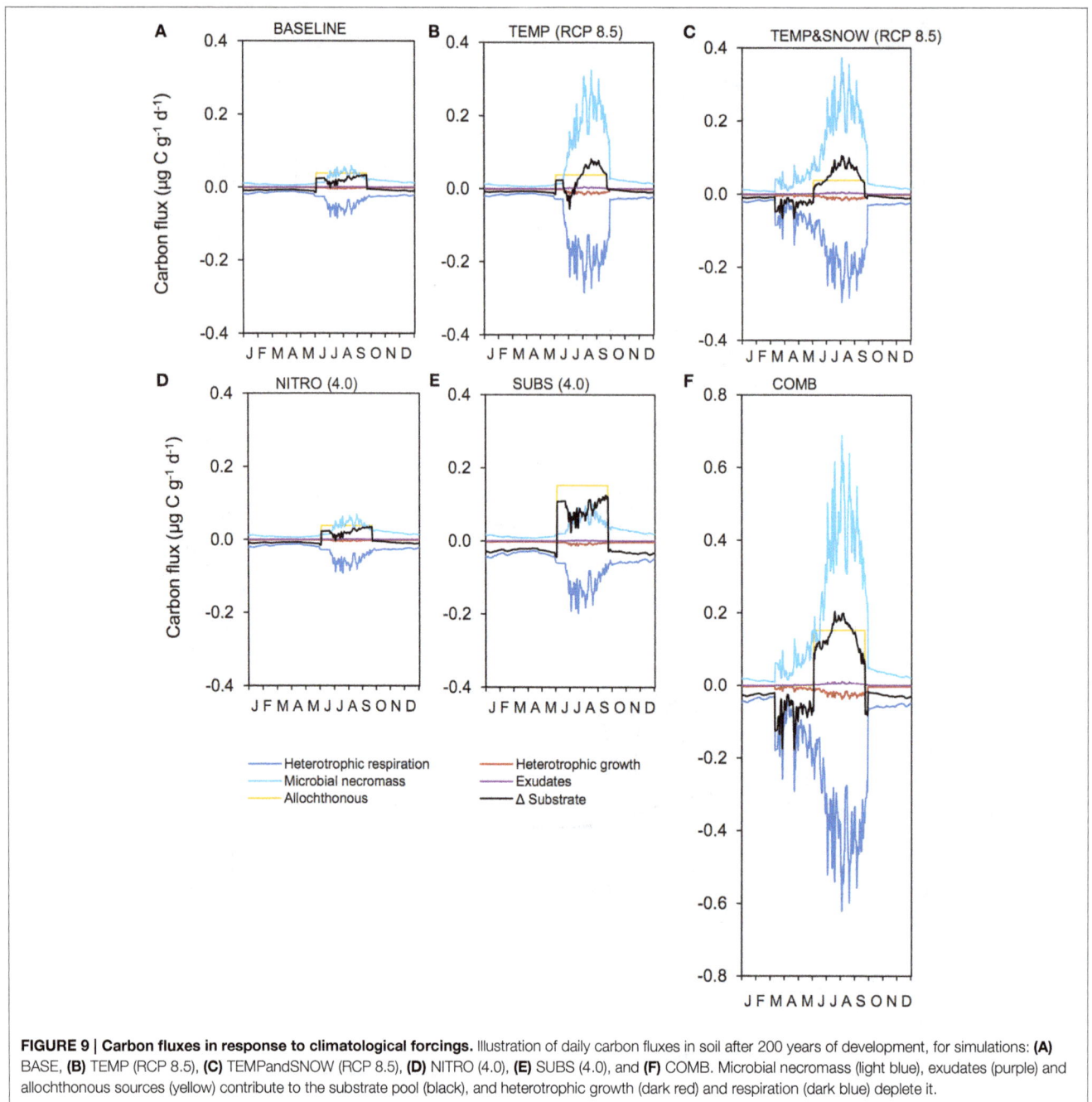

FIGURE 9 | Carbon fluxes in response to climatological forcings. Illustration of daily carbon fluxes in soil after 200 years of development, for simulations: **(A)** BASE, **(B)** TEMP (RCP 8.5), **(C)** TEMPandSNOW (RCP 8.5), **(D)** NITRO (4.0), **(E)** SUBS (4.0), and **(F)** COMB. Microbial necromass (light blue), exudates (purple) and allochthonous sources (yellow) contribute to the substrate pool (black), and heterotrophic growth (dark red) and respiration (dark blue) deplete it.

areas could enhance CO_2 fixation from the atmosphere by autotrophic activity. Simulation results suggest that this effect may be sustained in the later stages of soil development where biomass is greater, and where plants may become established. However, even under the extreme warming scenario simulated here the forefield remained net heterotrophic (NEP < 0).

Response to Nutrient and Organic Carbon Inputs (NITRO and SUBS)

Enhanced deposition of DIN (NITRO) alleviated nutrient limitation, resulting in a depletion of TOC relative to the baseline

scenario (years 50–200), due to the enhanced degradation of labile organic matter by heterotrophic activity (**Figure 6**). In general, however, the impact of enhanced DIN deposition on soil TOC dynamics was minimal, and resulted in relative reduction of −2% in TOC stocks (from 4 times DIN deposition) (**Figure 8B**). Moreover, the most extreme DIN deposition scenario [NITRO (4.0)] simultaneously enhanced both autotrophy (CO_2 fixation) and heterotrophy (CO_2 production), and thus exerted a negligible effect on NEP over 200 years (−2%) compared to other climate change scenarios (TEMP, TEMP&SNOW, and SUBS).

TABLE 2 | Net ecosystem production over 200 years of exposure.

Simulation	Net ecosystem production (μg C g^{-1})
BASE	−823
TEMP (RCP 4.5)	−797
TEMP (RCP 8.5)	−739
TEMP&SNOW (RCP 4.5)	−719
TEMP&SNOW (RCP 8.5)	−626
NITRO (4.0)	−842
SUBS (4.0)	−2310
COMB	−2195

Conversely, allochthonous inputs of organic carbon were an important factor in sustaining microbial growth in newly exposed glacier forefields by exerting a strong influence on forefield TOC stocks (**Figure 8C**). By integrating daily carbon fluxes over the 200-year simulation period, model predictions suggest that under extreme substrate input fluxes [SUBS (4.0)], allochthonous deposition contributed roughly 3 times more organic carbon (3.34 mg C g^{-1}) than microbial necromass (1.08 mg C g^{-1}). Yet, despite high substrate input (total = 4.42 mg C g^{-1}), soil TOC stocks increased by only 1.07 mg C g^{-1}, suggesting that most of the organic substrate delivered to the TOC pool (>75%) was rapidly utilized by heterotrophic activity. This caused a considerable decrease to NEP relative to the BASE scenario (−180.5%, **Table 2**). Overall, these findings reinforce previous experimental evidence suggesting that labile organic carbon is a limiting factor on bacterial activity in established glacier forefield soils (Goransson et al., 2011; Bradley et al., 2016b). Therefore, accurately predicting the likely magnitude of future organic carbon input to Arctic glacier forefields deserves attention.

Response to Combined Changes (COMB)

TOC accumulation in COMB mirrored high rates of TOC accumulation in SUBS (4.0) until the rapid growth phase (**Figure 8D**), upon which more favorable temperature and nutrient conditions enabled enhanced heterotrophic degradation, and TOC stocks declined moderately. The simulated carbon fluxes driven by COMB closely resembled TEMP&SNOW (RCP 8.5) (i.e., the period of increased microbial activity was extended earlier into the spring), but the overall magnitude of fluxes were substantially larger (**Figure 9F**). This resulted from increased biological activity due to the increased availability of organic carbon, aligning with the higher microbial abundance discussed earlier. Increased delivery of necromass contributed to the soil TOC pool (**Figure 9F**). Favorable conditions for phototrophic growth caused fixation of CO_2 such that total NEP for COMB (−2195 μg C g^{-1}) was higher than when considering deposition of organic carbon [SUBS (4.0)] alone (−2310 μg C g^{-1}) (**Table 2**). Overall, NEP was still substantially lower than the BASE simulation (−823 μg C g^{-1}), emphasizing the potential susceptibility of soil-atmosphere CO_2 exchange to future climate change (Billings, 1987; Oechel et al., 1993, 2000; Goulden et al., 1998).

CONCLUSIONS AND OUTLOOK

Arctic ecosystems are generally adapted to low availability of nutrients and relatively extreme changes in climate (Jonsdottir et al., 1995; Jonsdottir, 2005). We have shown that the anthropogenically driven emission of pollutants (including reactive nitrogen) and greenhouse gasses and its cascading effects on the Arctic environment (such as increases in soil temperature, the length of the melting season, and the availability of carbon and nutrients), has the potential to considerably impact the functioning of a basic soil microbial community in the initial stages of soil development in the present day and for centuries into the future. This is important in the context of increased glacier retreat and soil formation over large expanses of presently ice-covered regions. Our results thus demonstrate that future attempts to characterize soil development in glacier forefields must also consider changes to environmental, climatic and anthropogenic drivers. We have made important initial approximations of the sensitivity of microbiological and biogeochemical processes in forefield soils to future climate change by addressing the simulated responses to changes in temperature, snow cover and nutrient input. However, these initial approximations require further testing with in situ field-based experiments. The results of ecological models critically depend on the model used (Jackson et al., 2000; Meile and Jones, 2016). Although, SHIMMER has been carefully calibrated and tested based on field data from the Midtre Lovénbreen forefield (Bradley et al., 2016b) and thus is the most appropriate tool for the present study, further empirical measurements and experiments simulating soil warming and nutrient deposition may validate these predictions and strengthen the conclusions drawn here.

Critically, future studies must identify the role of nutrient availability on fungi and plant colonization (Insam and Haselwandter, 1989; Bernasconi et al., 2011; Knelman et al., 2012; Zumsteg et al., 2012; Schulz et al., 2013; Brown and Jumpponen, 2014). The SHIMMER model does not include a vegetation component and is thus not able to account for the effect of plants (Bradley et al., 2015). Whereas Alpine glacier forefields are usually abundant with vascular vegetation soon after ice retreat (Miniaci et al., 2007; Bernasconi et al., 2011), the initial young stages of the Midtre Lovénbreen forefield soils are characterized by almost a complete absence of plants. The Midtre Lovénbreen forefield is thus an ideal system to study the interactions between microbes and rock during soil formation. This forefield is also representative of a major Polar ecosystem, since most of the non-ice-covered surface land area of Antarctica and the high Arctic have very little plant coverage (Vanderpuye et al., 2002; Okuda et al., 2007, 2011; Birks, 2008). Plant colonization is likely to heavily re-structure the microbial community (Brown and Jumpponen, 2014), and the physical properties of the soil, including water retention, ultraviolet exposure, temperature fluctuations (Ensign et al., 2006; King et al., 2008) and nutrient status (Kastovska et al., 2005; Schutte et al., 2009). For example, Duc et al. (2009b) compared rhizosphere and bulk soils in the Damma Glacier, and found substantially higher total organic carbon concentrations in soils sampled in close proximity to plants. Thus, if plants were to become established within the simulated period, the results may be somewhat different. Although, the forcings investigated in the model caused increasing efflux of CO_2, the accelerated remineralisation and release of nutrients created favorable conditions for plants, which may encourage growth (Insam and Haselwandter, 1989; Bernasconi et al., 2011; Brown and Jumpponen, 2014). Plants may then cause a reduction in net soil CO_2 efflux by enhancing primary productivity (D'Amico et al., 2014), but may also generate more litter to fuel heterotrophy (Guelland et al., 2013a).

Our model analyses robustly indicate that microbial activity, nutrient cycling, and carbon cycling in the Midtre Lovénbreen forefield can respond strongly to future climate and anthropogenic impacts. Our simulations suggest that:

- Climate and anthropogenic changes generally enhance bacterial production and increase soil carbon and nutrient stocks. The strength of this response is highly variable between simulations.
- Nutrient availability is a major limiting factor for microbial activity in recently exposed soils and thus exerts a major control on the ability of forefield microbial communities to respond to climate change.
- Pioneer communities are most susceptible to climatological changes.
- Microbial and thus biogeochemical dynamics are mainly controlled by allochthonous nutrient inputs in younger soils. In developed soils, the heterotrophic recycling of allochthonous organic carbon and, to a lesser extent, necromass, causes sufficient nutrient accumulation over the course of a century of soil development to alleviate the dependence on allochthonous nutrient inputs.
- Climatological forcings encourage a more rapid switch between allochthonously-dominated processes in young soils to microbially-dominated processes in older soils.
- Factors that enhance heterotrophic activity, such as favorable temperatures and increased organic substrate, contribute additional nutrients via remineralization, which may facilitate earlier colonization by fungi and plants.
- The Midtre Lovénbreen forefield is net heterotrophic (NEP < 0) for the length of the simulated period and all climate change

scenarios. However, warming and reduced snow cover lower the net heterotrophy, while the increase in allochthonous inputs generally increase net heterotrophy.
- The Midtre Lovénbreen forefield is resilient to tipping points, in that, even with a combination of the most extreme climatological and anthropogenic changes, soil microbiology and biogeochemistry responded with an amplification of the baseline dynamics, rather than major fundamental shifts.

AUTHOR CONTRIBUTIONS

JB led the design of the study, assisted by AA and SA. The simulations were conducted by JB. All authors interpreted model results. JB wrote the manuscript with contributions from AA and SA.

FUNDING

This work was supported by Natural Environment Research Council (grant number NE/J02399X/1 to AA). JB was partially supported by a Postdoctoral Fellowship from the National Science Foundation Center for Dark Energy Biosphere Investigations (OCE-0939564). SA received funding from the European Union's Horizon 2020 research and innovation program under the Marie Sklodowska-Curie grant agreement number 643052 (C-CASCADES). This is the Center for Dark Energy Biosphere Investigations (C-DEBI) publication 363.

ACKNOWLEDGMENTS

We would like to thank the two reviewers who provided valuable comments on the manuscript.

REFERENCES

Anderson, S. P., Drever, J. I., Frost, C. D., and Holden, P. (2000). Chemical weathering in the foreland of a retreating glacier. *Geochim. Cosmochim. Acta* 64, 1173–1189. doi: 10.1016/S0016-7037(99)00358-0

Anderson, S. P., Drever, J. I., and Humphrey, N. F. (1997). Chemical weathering in glacial environments. *Geology* 25, 399–402. doi: 10.1130/0091-7613(1997)025<0399:CWIGE>2.3.CO;2

Anesio, A. M., and Laybourn-Parry, J. (2012). Glaciers and ice sheets as a biome. *Trends Ecol. Evol.* 27, 219–225. doi: 10.1016/j.tree.2011.09.012

Anisimov, O. A., Vaughan, D. G., Callaghan, T. V., Furgal, C., Marchant, H., Prowse, T. D., et al. (2007). "Polar regions (Arctic and Antarctic)," in *Climate Change 2007: Impacts, Adaptation and Vulnerability*, eds O. F. Canziani, J. P.

Palutikof, P. J. van der Linden, and C. E. Hanson (Cambridge: Cambridge University Press), 653–686.

Ansari, A. H., Hodson, A. J., Heaton, T. H. E., Kaiser, J., and Marca-Bell, A. (2013). Stable isotopic evidence for nitrification and denitrification in a High Arctic glacial ecosystem. *Biogeochemistry* 113, 341–357. doi: 10.1007/s10533-012-9761-9

Bernasconi, S. M., Bauder, A., Bourdon, B., Brunner, I., Bunemann, E., Christl, I., et al. (2011). Chemical and biological gradients along the damma glacier soil chronosequence, Switzerland. *Vadose Zo. J.* 10, 867–883. doi: 10.2136/vzj2010.0129

Billings, W. D. (1987). Carbon balance of alaskan tundra and taiga ecosystems - past, present and future. *Quat. Sci. Rev.* 6, 165–177. doi: 10.1016/0277-3791(87)90032-1

Birks, H. H. (2008). The late-quaternary history of arctic and alpine plants. *Plant Ecol. Divers.* 1, 135–146. doi: 10.1080/17550870802328652

Birks, H. J. B., Jones, V. J., and Rose, N. L. (2004). Recent environmental change and atmospheric contamination on Svalbard as recorded in lake sediments - synthesis and general conclusions. *J. Paleolimnol.* 31, 531–546. doi: 10.1023/B:JOPL.0000022550.81129.1a

Bjorkman, M. P., Kuhnel, R., Partridge, D. G., Roberts, T. J., Aas, W., Mazzola, M., et al. (2013). Nitrate dry deposition in Svalbard. *Tellus Ser. B. Chem. Phys. Meteorol.* 65:19071. doi: 10.3402/tellusb.v65i0.19071

Bradley, J. A., Anesio, A., and Arndt, S. (2016a). Bridging the divide: a model-data approach to Polar & Alpine microbiology. *FEMS Microbiol. Ecol.* 92:fiw015. doi: 10.1093/femsec/fiw015

Bradley, J. A., Anesio, A. M., Singarayer, J. S., Heath, M. R., and Arndt, S. (2015). SHIMMER (1.0): a novel mathematical model for microbial and biogeochemical dynamics in glacier forefield ecosystems. *Geosci. Model Dev.* 8, 3441–3470. doi: 10.5194/gmd-8-3441-2015

Bradley, J. A., Arndt, S., Šabacká, M., Benning, L. G., Barker, G. L., Blacker, J. J., et al. (2016b). Microbial dynamics in a High Arctic glacier forefield: a combined field, laboratory, and modelling approach. *Biogeosciences* 13, 5677–5696. doi: 10.5194/bg-13-5677-2016

Bradley, J. A., Singarayer, J. S., and Anesio, A. M. (2014). Microbial community dynamics in the forefield of glaciers. *Proc. Biol. Sci.* 281, 2793–2802. doi: 10.1098/rspb.2014.0882

Brankatschk, R., Towe, S., Kleineidam, K., Schloter, M., and Zeyer, J. (2011). Abundances and potential activities of nitrogen cycling microbial communities along a chronosequence of a glacier forefield. *ISME J.* 5, 1025–1037. doi: 10.1038/ismej.2010.184

Brown, S. P., and Jumpponen, A. (2014). Contrasting primary successional trajectories of fungi and bacteria in retreating glacier soils. *Mol. Ecol.* 23, 481–497. doi: 10.1111/mec.12487

Callaghan, T. V., Press, M. C., Lee, J. A., Robinson, D. L., and Anderson, C. W. (1999). Spatial and temporal variability in the responses of Arctic terrestrial ecosystems to environmental change. *Polar Res.* 18, 191–197. doi: 10.3402/polar.v18i2.6574

Ciais, P., Sabine, C., Bala, G., Bopp, L., Brovkin, V., Canadell, J., et al. (2013). "Carbon and other biogeochemical cycles," in *Climate Change 2013: The Physical Science Basis. Contribution of Working Group I to the Fifth Assessment Report of the Intergovernmental Panel on Climate Change*, eds T. F. Stocker, D. Qin, G. Plattner, M. Tignor, S. Allen, J. Boschung, A. Nauels, Y. Xia, V. Bex, and P. Midgley (Cambridge; New York, NY: Cambridge University Press), 465–570.

D'Amico, M. E., Freppaz, M., Filippa, G., and Zanini, E. (2014). Vegetation influence on soil formation rate in a proglacial chronosequence (Lys Glacier, NW Italian Alps). *Catena* 113, 122–137. doi: 10.1016/j.catena.2013.10.001

Deiglmayr, K., Philippot, L., Tscherko, D., and Kandeler, E. (2006). Microbial succession of nitrate-reducing bacteria in the rhizosphere of Poa alpina across a glacier foreland in the Central Alps. *Environ. Microbiol.* 8, 1600–1612. doi: 10.1111/j.1462-2920.2006.01051.x

Duc, L., Neuenschwander, S., Rehrauer, H., Wagner, U., Sobek, J., Schlapbach, R., et al. (2009a). Development and experimental validation of a nifH oligonucleotide microarray to study diazotrophic communities in a glacier forefield. *Environ. Microbiol.* 11, 2179–2189. doi: 10.1111/j.1462-2920.2009.01945.x

Duc, L., Noll, M., Meier, B. E., Burgmann, H., and Zeyer, J. (2009b). High diversity of diazotrophs in the forefield of a receding Alpine Glacier. *Microb. Ecol.* 57, 179–190. doi: 10.1007/s00248-008-9408-5

Eckhardt, S., Hermansen, O., Grythe, H., Fiebig, M., Stebel, K., Cassiani, M., et al. (2013). The influence of cruise ship emissions on air pollution in Svalbard - a harbinger of a more polluted Arctic? *Atmos. Chem. Phys.* 13, 8401–8409. doi: 10.5194/acp-13-8401-2013

Eneroth, K., Kjellstrom, E., and Holmen, K. (2003). A trajectory climatology for Svalbard; investigating how atmospheric flow patterns influence observed tracer concentrations. *Phys. Chem. Earth* 28, 1191–1203. doi: 10.1016/j.pce.2003.08.051

Ensign, K. L., Webb, E. A., and Longstaffe, F. J. (2006). Microenvironmental and seasonal variations in soil water content of the unsaturated zone of a sand dune system at Pinery Provincial Park, Ontario, Canada. *Geoderma* 136, 788–802. doi: 10.1016/j.geoderma.2006.06.009

Fleming, K. M., Dowdeswell, J. A., and Oerlemans, J. (1997). Modelling the mass balance of northwest Spitsbergen glaciers and responses to climate change. *Ann. Glaciol.* 24, 203–210. doi: 10.1017/s02603055000 12180

Foreman, C. M., Sattler, B., Mikucki, J. A., Porazinska, D. L., and Priscu, J. C. (2007). Metabolic activity and diversity of cryoconites in the Taylor Valley, Antarctica. *J. Geophys. Res.* 112:G04S32. doi: 10.1029/2006jg000358

Forland, E. J., Benestad, R., Hanssen-Bauer, I., Haugen, J. E., and Skaugen, T. E. (2011). Temperature and Precipitation Development at Svalbard 1900-2100. *Adv. Meteorol.* 2011:893790. doi: 10.1155/2011/893790

Fountain, A. G., Nylen, T. H., Tranter, M., and Bagshaw, E. (2008). Temporal variations in physical and chemical features of cryoconite holes on Canada Glacier, McMurdo Dry Valleys, Antarctica. *J. Geophys. Res.* 113:G01S92. doi: 10.1029/2007jg000430

Fountain, A. G., Tranter, M., Nylen, T. H., Lewis, K. J., and Mueller, D. R. (2004). Evolution of cryoconite holes and their contribution to meltwater runoff from glaciers in the McMurdo Dry Valleys, Antarctica. *J. Glaciol.* 50, 35–45. doi: 10.3189/172756504781830312

Galloway, J. N., Townsend, A. R., Erisman, J. W., Bekunda, M., Cai, Z., Freney, J. R., et al. (2008). Transformation of the nitrogen cycle: recent trends, questions, and potential solutions. *Science* 320, 889–892. doi: 10.1126/science.1136674

Geng, L., Alexander, B., Cole-Dai, J., Steig, E. J., Savarino, J., Sofen, E. D., et al. (2014). Nitrogen isotopes in ice core nitrate linked to anthropogenic atmospheric acidity change. *Proc. Natl. Acad. Sci. U.S.A.* 111, 5808–5812. doi: 10.1073/pnas.1319441111

Goransson, H., Venterink, H. O., and Baath, E. (2011). Soil bacterial growth and nutrient limitation along a chronosequence from a glacier forefield. *Soil Biol. Biochem.* 43, 1333–1340. doi: 10.1016/j.soilbio.2011.03.006

Goto-Azuma, K., and Koerner, R. M. (2001). Ice core studies of anthropogenic sulfate and nitrate trends in the Arctic. *J. Geophys. Res.* 106, 4959–4969. doi: 10.1029/2000JD900635

Goulden, M. L., Wofsy, S. C., Harden, J. W., Trumbore, S. E., Crill, P. M., Gower, S. T., et al. (1998). Sensitivity of boreal forest carbon balance to soil thaw. *Science* 279, 214–217. doi: 10.1126/science.279.5348.214

Greenfell, T. C., and Maykut, G. A. (1977). The optical properties of ice and snow in the Arctic basin. *J Glaciol.* 18, 445–463. doi: 10.1017/S0022143000021122

Guelland, K., Esperschutz, J., Bornhauser, D., Bernasconi, S. M., Kretzschmar, R., and Hagedorn, F. (2013a). Mineralisation and leaching of C from C-13 labelled plant litter along an initial soil chronosequence of a glacier forefield. *Soil Biol. Biochem.* 57, 237–247. doi: 10.1016/j.soilbio.2012.07.002

Guelland, K., Hagedorn, F., Smittenberg, R. H., Goransson, H., Bernasconi, S. M., Hajdas, I., et al. (2013b). Evolution of carbon fluxes during initial soil formation along the forefield of Damma glacier, Switzerland. *Biogeochemistry* 113, 545–561. doi: 10.1007/s10533-012-9785-1

Hastings, M. G., Jarvis, J. C., and Steig, E. J. (2009). Anthropogenic impacts on nitrogen isotopes of Ice-Core nitrate. *Science* 324, 1288. doi: 10.1126/science.1170510

Hodson, A. J., Mumford, P. N., Kohler, J., and Wynn, P. M. (2005). The High Arctic glacial ecosystem: new insights from nutrient budgets. *Biogeochemistry* 72, 233–256. doi: 10.1007/s10533-004-0362-0

Hodson, A., Roberts, T. J., Engvall, A. C., Holmen, K., and Mumford, P. (2010). Glacier ecosystem response to episodic nitrogen enrichment in Svalbard, European High Arctic. *Biogeochemistry* 98, 171–184. doi: 10.1007/s10533-009-9384-y

Holland, E. A., Braswell, B. H., Lamarque, J. F., Townsend, A., Sulzman, J., Muller, J. F., et al. (1997). Variations in the predicted spatial distribution of atmospheric nitrogen deposition and their impact on carbon uptake by terrestrial ecosystems. *J. Geophys. Res.* 102, 15849–15866. doi: 10.1029/96JD03164

Hood, E., Battin, T. J., Fellman, J., O'Neel, S., and Spencer, R. G. M. (2015). Storage and release of organic carbon from glaciers and ice sheets. *Nat. Geosci.* 8, 91–96. doi: 10.1038/ngeo2331

Hood, E., and Berner, L. (2009). Effects of changing glacial coverage on the physical and biogeochemical properties of coastal streams in southeastern Alaska. *J. Geophys. Res.* 114:G03001. doi: 10.1029/2009jg000971

Ingham, R. E., Trofymow, J. A., Ingham, E. R., and Coleman, D. C. (1985). Interactions of bacteria, fungi, and their Nematode Grazers : effects on nutrient cycling and plant growth. *Ecol. Monogr.* 55, 119–140. doi: 10.2307/1942528

Insam, H., and Haselwandter, K. (1989). Metabolic quotient of the soil microflora in relation to plant succession. *Oecologia* 79, 174–178. doi: 10.1007/BF00388474

Isaksson, E., Hermanson, M., Sheila, H. C., Igarashi, M., Kamiyama, K., Moore, J., et al. (2003). Ice cores from Svalbard - useful archives of past climate and pollution history. *Phys. Chem. Earth* 28, 1217–1228. doi: 10.1016/j.pce.2003.08.053

Jackson, L. J., Trebitz, A. S., and Cottingham, K. L. (2000). An introduction to the practice of ecological modeling. *Bioscience* 50, 694–706. doi: 10.1641/0006-3568(2000)050[0694:AITTPO]2.0.CO;2

Jonsdottir, I. S. (2005). Terrestrial ecosystems on Svalbard: heterogeneity, complexity and fragility from an Arctic island perspective. *Biol. Environ. Proc. R. Irish Acad.* 105B, 155–165. doi: 10.3318/BIOE.2005.105.3.155

Jonsdottir, I. S., Callaghan, T. V., and Lee, J. A. (1995). Fate of added nitrogen in a moss sedge arctic community and effects of increased nitrogen deposition. *Sci. Total Environ.* 160–61, 677–685. doi: 10.1016/0048-9697(95)04402-M

Kastovska, K., Elster, J., Stibal, M., and Santruckova, H. (2005). Microbial assemblages in soil microbial succession after glacial retreat in Svalbard (high Arctic). *Microb. Ecol.* 50, 396–407. doi: 10.1007/s00248-005-0246-4

Kelly, B., Whiteley, A., and Tallmon, D. (2010). The Arctic melting pot. *Nature* 468:891. doi: 10.1038/468891a

King, A. J., Meyer, A. F., and Schmidt, S. K. (2008). High levels of microbial biomass and activity in unvegetated tropical and temperate alpine soils. *Soil Biol. Biochem.* 40, 2605–2610. doi: 10.1016/j.soilbio.2008.06.026

Kirchman, D. L., Moran, X. A. G., and Ducklow, H. (2009). Microbial growth in the polar oceans - role of temperature and potential impact of climate change. *Nat. Rev. Microbiol.* 7, 451–459. doi: 10.1038/nrmicro2115

Knelman, J. E., Legg, T. M., O'Neill, S. P., Washenberger, C. L., Gonzalez, A., Cleveland, C. C., et al. (2012). Bacterial community structure and function change in association with colonizer plants during early primary succession in a glacier forefield. *Soil Biol. Biochem.* 46, 172–180. doi: 10.1016/j.soilbio.2011.12.001

Kuhnel, R., Bjorkman, M. P., Vega, C. P., Hodson, A., Isaksson, E., and Strom, J. (2013). Reactive nitrogen and sulphate wet deposition at Zeppelin Station, Ny-Alesund, Svalbard. *Polar Res.* 32:19136. doi: 10.3402/polar.v32i0.19136

Kuhnel, R., Roberts, T. J., Bjorkman, M. P., Isaksson, E., Aas, W., Holmen, K., et al. (2011). 20-year climatology of NO_3^- and NH_4^+ wet deposition at Ny-Ålesund, Svalbard. *Adv. Meteorol.* 2011:406508. doi: 10.1155/2011/406508

Larose, C., Dommergue, A., and Vogel, T. M. (2013a). Microbial nitrogen cycling in Arctic snowpacks. *Environ. Res. Lett.* 8:035004. doi: 10.1088/1748-9326/8/3/035004

Larose, C., Dommergue, A., and Vogel, T. M. (2013b). The dynamic arctic snow pack: an unexplored environment for microbial diversity and activity. *Biology* 2, 317–330. doi: 10.3390/biology2010317

Lau, M. C. Y., Stackhouse, B. T., Layton, A. C., Chauhan, A., Vishnivetskaya, T. A., Chourey, K., et al. (2015). An active atmospheric methane sink in high Arctic mineral cryosols. *ISME J.* 9, 1880–1891. doi: 10.1038/ismej.2015.13

Lazzaro, A., Hilfiker, D., and Zeyer, J. (2015). Structures of microbial communities in Alpine soils: seasonal and elevational effects. *Front. Microbiol.* 6:1330. doi: 10.3389/fmicb.2015.01330

Lefauconnier, B., Hagen, J. O., Orbaek, J. B., Melvold, K., and Isaksson, E. (1999). Glacier balance trends in the Kongsfjorden area, western Spitsbergen, Svalbard, in relation to the climate. *Polar Res.* 18, 307–313. doi: 10.3402/polar.v18i2.6589

Lenton, T. M. (2012). Arctic climate tipping points. *Ambio* 41, 10–22. doi: 10.1007/s13280-011-0221-x

Luoto, T. P., Oksman, M., and Ojala, A. E. K. (2015). Climate change and bird impact as drivers of High Arctic pond deterioration. *Polar Biol.* 38, 357–368. doi: 10.1007/s00300-014-1592-9

Lutz, S., Anesio, A. M., Edwards, A., and Benning, L. G. (2015). Microbial diversity on Icelandic glaciers and ice caps. *Front. Microbiol.* 6:307. doi: 10.3389/fmicb.2015.00307

Macdonald, R. W., Harner, T., and Fyfe, J. (2005). Recent climate change in the Arctic and its impact on contaminant pathways and interpretation of temporal trend data. *Sci. Total Environ.* 342, 5–86. doi: 10.1016/j.scitotenv.2004.12.059

Mahowald, N. (2011). Aerosol indirect effect on biogeochemical cycles and climate. *Science* 334, 794–796. doi: 10.1126/science.1207374

Maturilli, M., Herber, A., and Konig-Langlo, G. (2015). Surface radiation climatology for Ny-lesund, Svalbard (78.9A degrees N), basic observations for trend detection. *Theor. Appl. Climatol.* 120, 331–339. doi: 10.1007/s00704-014-1173-4

Meile, C., and Jones, C. (2016). "A mathematical perspective on microbial processes in Earth's biogeochemical cycles," in *Mathematical Paradigms of Climate Science*, eds F. Ancona, P. Cannarsa, C. Jones, and A. Portaluri (Cham: Springer International Publishing), 3–14.

Michelutti, N., Keatley, B. E., Brimble, S., Blais, J. M., Liu, H. J., Douglas, M. S., et al. (2009). Seabird-driven shifts in Arctic pond ecosystems. *Proc. R. Soc. B. Biol. Sci.* 276, 591–596. doi: 10.1098/rspb.2008.1103

Miller, A. W., and Ruiz, G. M. (2014). Arctic shipping and marine invaders. *Nat. Clim. Chang.* 4, 413–416. doi: 10.1038/nclimate2244

Mindl, B., Anesio, A. M., Meirer, K., Hodson, A. J., Laybourn-Parry, J., Sommaruga, R., et al. (2007). Factors influencing bacterial dynamics along a transect from supraglacial runoff to proglacial lakes of a high Arctic glacieri. *FEMS Microbiol. Ecol.* 59, 762. doi: 10.1111/j.1574-6941.2007.00295.x

Miniaci, C., Bunge, M., Duc, L., Edwards, I., Burgmann, H., and Zeyer, J. (2007). Effects of pioneering plants on microbial structures and functions in a glacier forefield. *Biol. Fertil. Soils* 44, 289–297. doi: 10.1007/s00374-007-0203-0

Moe, B., Stempniewicz, L., Jakubas, D., Angelier, F., Chastel, O., Dinessen, F., et al. (2009). Climate change and phenological responses of two seabird species breeding in the high-Arctic. *Mar. Ecol. Prog. Ser.* 393, 235–246. doi: 10.3354/meps08222

Moreau, M., Mercier, D., Laffly, D., and Roussel, E. (2008). Impacts of recent paraglacial dynamics on plant colonization: a case study on Midtre Lovenbreen foreland, Spitsbergen (79 degrees N). *Geomorphology* 95, 48–60. doi: 10.1016/j.geomorph.2006.07.031

Morgan-Kiss, R. M., Priscu, J. C., Pocock, T., Gudynaite-Savitch, L., and Huner, N. P. (2006). Adaptation and acclimation of photosynthetic microorganisms to permanently cold environments. *Microbiol. Mol. Biol. Rev.* 70, 222–252. doi: 10.1128/mmbr.70.1.222-252.2006

Moss, R. H., Edmonds, J. A., Hibbard, K. A., Manning, M. R., Rose, S. K., van Vuuren, D. P., et al. (2010). The next generation of scenarios for climate change research and assessment. *Nature* 463, 747–756. doi: 10.1038/nature08823

Nadelhoffer, K. J., Emmett, B. A., Gundersen, P., Kjonaas, O. J., Koopmans, C. J., Schleppi, P., et al. (1999). Nitrogen deposition makes a minor contribution to carbon sequestration in temperate forests. *Nature* 398, 145–148. doi: 10.1038/18205

Newsham, K. K., Hopkins, D. W., Carvalhais, L. C., Fretwell, P. T., Rushton, S. P., Odonnell, A. G., et al. (2016). Relationship between soil fungal diversity and temperature in the maritime Antarctic. *Nat. Clim. Chang.* 6, 182–186. doi: 10.1038/nclimate2806

Oechel, W. C., Hastings, S. J., Vourlitis, G., Jenkins, M., Riechers, G., and Grulke, N. (1993). Recent change of Arctic Tundra ecosystems from a net carbon-dioxide sink to a source. *Nature* 361, 520–523. doi: 10.1038/361520a0

Oechel, W. C., Vourlitis, G. L., Hastings, S. J., Zulueta, R. C., Hinzman, L., and Kane, D. (2000). Acclimation of ecosystem CO2 exchange in the Alaskan Arctic in response to decadal climate warming. *Nature* 406, 978–981. doi: 10.1038/35023137

Okuda, M., Imura, S., and Tanemura, M. (2007). Microtopographic analysis of plant distribution in polar desert. *Polar Sci.* 1, 113–120. doi: 10.1016/j.polar.2007.09.001

Okuda, M., Imura, S., and Tanemura, M. (2011). Microtopographic properties of sparse moss vegetation in the Antarctic polar desert. *Polar Sci.* 5, 432–439. doi: 10.1016/j.polar.2011.10.001

Overland, J. E., Wang, M. Y., Walsh, J. E., and Stroeve, J. C. (2014). Future Arctic climate changes: adaptation and mitigation time scales. *Earths Future* 2, 68–74. doi: 10.1002/2013EF000162

Paul, F., Frey, H., and Le Bris, R. (2011). A new glacier inventory for the European Alps from Landsat TM scenes of 2003: challenges and results. *Ann. Glaciol.* 52, 144–152. doi: 10.3189/172756411799096295

Peters, G. P., Nilssen, T. B., Lindholt, L., Eide, M. S., Glomsrod, S., Eide, L. I., et al. (2011). Future emissions from shipping and petroleum activities in the Arctic. *Atmos. Chem. Phys.* 11, 5305–5320. doi: 10.5194/acp-11-5305-2011

Prietzel, J., Dumig, A., Wu, Y. H., Zhou, J., and Klysubon, W. (2013). Synchrotron-based P K-edge XANES spectroscopy reveals rapid changes of phosphorus speciation in the topsoil of two glacier foreland chronosequences. *Geochim. Cosmochim. Acta* 108, 154–171. doi: 10.1016/j.gca.2013.01.029

Roberts, T. J., Hodson, A., Evans, C. D., and Holmen, K. (2010). Modelling the impacts of a nitrogen pollution event on the biogeochemistry of an Arctic glacier. *Ann. Glaciol.* 51, 163–170. doi: 10.3189/172756411795931949

Sattin, S. R., Cleveland, C. C., Hood, E., Reed, S. C., King, A. J., Schmidt, S. K., et al. (2009). Functional shifts in unvegetated, perhumid, recently-deglaciated soils do not correlate with shifts in soil bacterial community composition. *J. Microbiol.* 47, 673–681. doi: 10.1007/s12275-009-0194-7

Scheer, C., Grace, P., and Rowlings, D. (2016). Non-linear response of soil N2O emissions to nitrogen fertiliser in a cotton-fallow rotation in sub-tropical Australia. *Soil Res.* 54, 494–499. doi: 10.1071/SR14328

Schulz, S., Brankatschk, R., Dumig, A., Kogel-Knabner, I., Schloter, M., and Zeyer, J. (2013). The role of microorganisms at different stages of ecosystem development for soil formation. *Biogeosciences* 10, 3983–3996. doi: 10.5194/bg-10-3983-2013

Schutte, U. M., Abdo, Z., Bent, S. J., Williams, C. J., Schneider, G. M., Solheim, B., et al. (2009). Bacterial succession in a glacier foreland of the High Arctic. *ISME J.* 3, 1258–1268. doi: 10.1038/ismej.2009.71

Screen, J. A., and Simmonds, I. (2010). The central role of diminishing sea ice in recent Arctic temperature amplification. *Nature* 464, 1334–1337. doi: 10.1038/nature09051

Serreze, M. C., Walsh, J. E., Chapin, F. S., Osterkamp, T., Dyurgerov, M., Romanovsky, V., et al. (2000). Observational evidence of recent change in the northern high-latitude environment. *Clim. Change* 46, 159–207. doi: 10.1023/A:1005504031923

Shcherbak, I., Millar, N., and Robertson, G. P. (2014). Global metaanalysis of the nonlinear response of soil nitrous oxide (N2O) emissions to fertilizer nitrogen. *Proc. Natl. Acad. Sci. U.S.A.* 111, 9199–9204. doi: 10.1073/pnas.1322434111

Singer, G. A., Fasching, C., Wilhelm, L., Niggemann, J., Steier, P., Dittmar, T., et al. (2012). Biogeochemically diverse organic matter in Alpine glaciers and its downstream fate. *Nat. Geosci.* 5, 710–714. doi: 10.1038/ngeo1581

Sistla, S. A., Moore, J. C., Simpson, R. T., Gough, L., Shaver, G. R., and Schimel, J. P. (2013). Long-term warming restructures Arctic tundra without changing net soil carbon storage. *Nature* 497, 615–618. doi: 10.1038/nature12129

Stibal, M., Elster, J., Sabacka, M., and Kastovska, K. (2007). Seasonal and diel changes in photosynthetic activity of the snow alga *Chlamydomonas nivalis* (Chlorophyceae) from Svalbard determined by pulse amplitude modulation fluorometry. *FEMS Microbiol. Ecol.* 59, 265–273. doi: 10.1111/j.1574-6941.2006.00264.x

Stibal, M., Sabacka, M., and Zarsky, J. (2012). Biological processes on glacier and ice sheet surfaces. *Nat. Geosci.* 5, 771–774. doi: 10.1038/ngeo1611

Stocker, T. F., Qin, D., Plattner, G. K., Tignor, M., Allen, S. K., Boschung, J., et al. (2013). *IPCC, 2013: Climate Change 2013: The Physical Science Basis. Contribution of Working Group I to the Fifth Assessment Report of the Intergovernmental Panel on Climate Change.* Cambridge; New York, NY: Cambridge University Press.

Strauss, S. L., Garcia-Pichel, F., and Day, T. A. (2012). Soil microbial carbon and nitrogen transformations at a glacial foreland on Anvers Island, Antarctic Peninsula. *Polar Biol.* 35, 1459–1471. doi: 10.1007/s00300-012-1184-5

Takeuchi, N., Fujita, K., Nakazawa, F., Matoba, S., Nakawo, M., and Rana, B. (2009). A snow algal community on the surface and in an ice core of Rikha-Samba Glacier in Western Nepali Himalayas. *Bull. Glaciol. Res.* 27, 25–35. doi: 10.1657/1523-0430(2004)036

Vanderpuye, A., Arve, E., and Lennard, N. (2002). Plant communities along environmental gradients of high-arctic mires in Sassendalen, Svalbard. *J. Veg. Sci.* 13, 875–884. doi: 10.1111/j.1654-1103.2002.tb02117.x

van der Wal, R., and Stien, A. (2014). High-arctic plants like it hot: a long-term investigation of between-year variability in plant biomass. *Ecology* 95, 3414–3427. doi: 10.1890/14-0533.1

Vincent, W. F. (2010). Microbial ecosystem responses to rapid climate change in the Arctic. *ISME J.* 4, 1089–1090. doi: 10.1038/ismej.2010.108

Ware, C., Berge, J., Sundet, J. H., Kirkpatrick, J. B., Coutts, A. D. M., Jelmert, A., et al. (2014). Climate change, non-indigenous species and shipping: assessing the risk of species introduction to a high-Arctic archipelago. *Divers. Distrib.* 20, 10–19. doi: 10.1111/ddi.12117

Yergeau, E., Bokhorst, S., Kang, S., Zhou, J., Greer, C. W., Aerts, R., et al. (2012). Shifts in soil microorganisms in response to warming are consistent across a range of Antarctic environments. *ISME J.* 6, 692–702. doi: 10.1038/ismej.2011.124

Zaehle, S., and Friend, A. D. (2010). Carbon and nitrogen cycle dynamics in the O-CN land surface model: 1. Model description, site-scale evaluation, and sensitivity to parameter estimates. *Glob. Biogeochem. Cycles* 24:GB1005. doi: 10.1029/2009gb003521

Zhou, X., Zhang, Y., and Downing, A. (2012). Non-linear response of microbial activity across a gradient of nitrogen addition to a soil from the Gurbantunggut Desert, northwestern China. *Soil Biol. Biochem.* 47, 67–77. doi: 10.1016/j.soilbio.2011.05.012

Zumsteg, A., Luster, J., Goransson, H., Smittenberg, R. H., Brunner, I., Bernasconi, S. M., et al. (2012). Bacterial, Archaeal and fungal succession in the forefield of a Receding Glacier. *Microb. Ecol.* 63, 552–564. doi: 10.1007/s00248-011-9991-8

Conflict of Interest Statement: The authors declare that the research was conducted in the absence of any commercial or financial relationships that could be construed as a potential conflict of interest.

10

Stress Controls of Monogenetic Volcanism

Joan Martí[1]*, Carmen López[2], Stefania Bartolini[1], Laura Becerril[1] and Adelina Geyer[1]

[1] Group of Volcanology, Institute of Earth Sciences Jaume Almera, Agencia Estatal Consejo Superior de Investigaciones Científicasn CSIC, Barcelona, Spain, [2] Instituto Geográfico Nacional, Madrid, Spain

The factors controlling the preparation of volcanic eruptions in monogenetic fields are still poorly understood. The fact that in monogenetic volcanism each eruption has a different vent suggests that volcanic susceptibility has a high degree of randomness, so that accurate forecasting is subjected to a very high uncertainty. Recent studies on monogenetic volcanism reveal how sensitive magma migration is to the existence of changes in the stress field caused by regional and/or local tectonics or rheological contrasts (stratigraphic discontinuities). These stress variations may induce changes in the pattern of further movements of magma, thus conditioning the location of future eruptions. This implies that a precise knowledge of the stress configuration and distribution of rheological and structural discontinuities at crustal level of such volcanic systems would aid in forecasting monogenetic volcanism. This contribution reviews several basic concepts relative to the stress controls of magma transport into the brittle lithosphere, and uses this information to explain how magma migrates inside monogenetic volcanic systems and how it prepares to trigger a new eruption.

Keywords: monogenetic volcanism, magma ascent dynamics, stress field, hazard assessment, volcanic susceptibility

Edited by:
Geoffrey Wadge,
University of Reading, UK

Reviewed by:
Alessandro Tibaldi,
University of Milano-Bicocca, Italy
Greg A. Valentine,
University at Buffalo, USA

***Correspondence:**
Joan Martí
joan.marti@ictja.csic.es

Specialty section:
This article was submitted to
Volcanology,
a section of the journal
Frontiers in Earth Science

INTRODUCTION

Monogenetic volcanism is the most extended type of volcanic activity on Earth (Walker, 2000). It is commonly represented by volcanic fields containing tens to thousands of small volcanoes, each being the product of a single eruptive episode, in which different phases or pulses may occur (Walker, 2000; Valentine and Gregg, 2008; Németh, 2010; Németh and Kereszturi, 2015). Monogenetic fields may be active for several millions of years with eruption recurrences ranging from several tens to tens of thousands of years. They are usually mafic in composition and generate relatively small volume eruptions that produce cinder cones and lava flows, as well as occasional phreatomagmatic deposits when interaction between magma and surface water occurs (Lorenz, 1986; Valentine and Gregg, 2008). The distribution of volcanic cones in basaltic monogenetic fields is clearly controlled by regional and local tectonics (Wood, 1980; Pasquarè et al., 1988; Connor, 1990; Connor et al., 1992; Tibaldi, 1995; Walker, 2000; Valentine and Perry, 2007; Le Corvec et al., 2013b). The great variety of eruptive styles, edifice morphologies and deposits in monogenetic volcanoes is the result of a complex combination of internal (magma composition, gas content, rheology, volume) and external (regional and local stress fields, stratigraphic and rheological contrasts in substrate rock, hydrogeology) parameters that characterize each volcanic system (Tibaldi and Lagmay, 2006; Valentine and Gregg, 2008; Németh, 2010; Martí et al., 2011). Monogenetic volcanoes may also occur as flank eruptions in association with polygenetic volcanoes (e.g., El Teide, Martí et al., 2008; Etna, Cappello et al., 2012).

Central or composite volcanoes are characterized by the presence of a magma chamber located a few kilometers below the surface, which exerts a stress field on its surroundings that is superimposed on the regional stress field, thereby controlling potential pathways for magma to the surface (Pinel and Jaupart, 2004; Gudmundsson and Brenner, 2005; Martí and Geyer, 2009, **Figure 1**). On the contrary, in a monogenetic volcanic systems magma does not accumulate in such shallow reservoirs or chambers and tends to rise to the surface from greater depths, usually from intermediate reservoirs located deep in the crust, or even from the source region or shallower levels in the mantle. Thus, the stress field controlling the magma ascent will depend on the stress distribution inside the lithosphere and in particular, on the regional stress field and local stress barriers corresponding to rheological and/or structural discontinuities (Menand, 2008, 2011; Maccaferri et al., 2010, 2011; Gudmundsson, 2011a,b; Bolós et al., 2015). Knowing how these structural controls work and how they may change from one eruption to the next one is crucial to understand why in monogenetic eruptive vents produced under the same regional stress field (i.e., same age period) will tend to cluster in the same area. These volcanic clusters may have lifetimes of hundreds of thousands of years, so there is always a chance that a new volcano will come up in the same place than an old one, just out of random chance, not necessarily implying the initiation of a polygenetic behavior.

An important consequence of these different stress controls between central and monogenetic volcanoes is the accuracy in forecasting new eruptions. While in central volcanoes it is generally assumed that future eruptions will occur through the same vents that have been active in the past, in monogenetic systems forecasting the position of new vents is much more challenging due to this lack of a permanent shallow stress configuration. Spatial analysis addressed to infer the location of future vents (volcanic susceptibility, see Martí and Felpeto, 2010) in monogenetic volcanism generally assumes that the next eruption will occur close to the location of the previous ones (Connor, 1990; Connor et al., 1992, 2000; Ho, 1992, 1995; Martin et al., 1994; Ho and Smith, 1998; Connor and Conway, 2000; Alberico et al., 2002; Martí and Felpeto, 2010; Bebbington and Cronin, 2011; Cappello et al., 2012; Selva et al., 2012; Bartolini et al., 2013; Becerril et al., 2013a; Le Corvec et al., 2013a; Bevilacqua et al., 2015). The reason to make this assumption is based on the fact that in last eruptive episodes volcanoes had formed near previous ones (forming a cluster), so we assume that this behavior will continue. However, this does not necessarily mean they will not occur outside a cluster, just that the probability is weighted by the existence of the cluster.

The transport of magma occurs mostly through sheet intrusions and the conditions of flow in such magma-filled cracks will be governed by rock and fluid mechanics (Pollard, 1969, 1973; Pollard and Muller, 1976; Delaney and Pollard, 1981, 1982; Delaney et al., 1986; Pollard and Segall, 1987; Rubin, 1993a,b, 1995; Dahm, 2000; Gudmundsson, 2011a). Studies on monogenetic volcanism reveal how sensitive magma migration is to the existence of changes in the stress field produced by regional and/or local tectonics or rheological contrasts (stratigraphic discontinuities, sheet intrusions, tectonic fractures, Delaney et al., 1986; Dahm, 2000; Gudmundsson and Philipp, 2006; Gaffney et al., 2007; Menand, 2008, 2011; Taisne and Jaupart, 2009; Maccaferri et al., 2010, 2011; Taisne et al., 2011; Gudmundsson, 2011b; Le Corvec et al., 2013b,c; Rivalta et al., 2015). The presence of such stress barriers may induce stress rotation

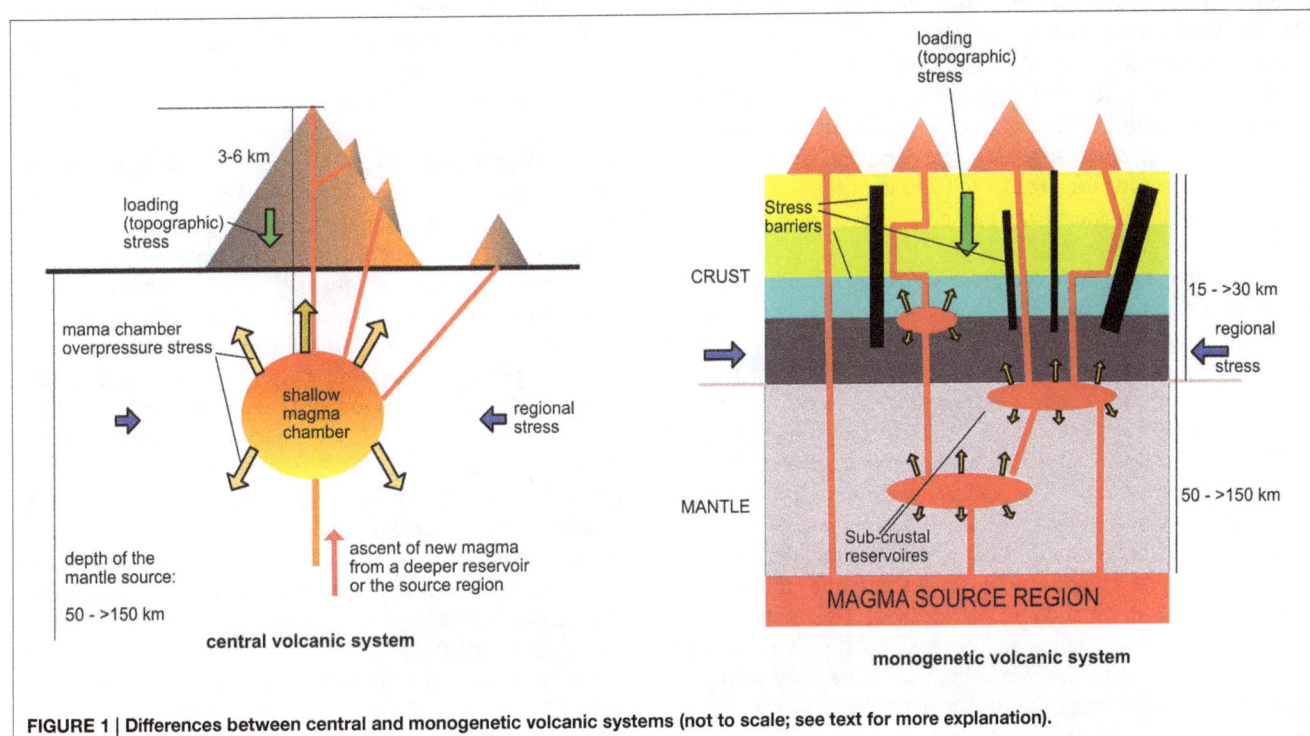

FIGURE 1 | Differences between central and monogenetic volcanic systems (not to scale; see text for more explanation).

and, consequently, changes in the direction of emplacement of magma. Therefore, knowledge of rock stress is crucial to understand how magma will move from its source regions up to the Earth's surface, and to forecast future eruptions.

In this review we will concentrate on the main concepts related to stress controls on magma transport in monogenetic fields, in order to offer a comprehensive picture on the paths that magma may follow inside the volcanic system and on why an eruption will occur from a particular point.

We do not pretend to discuss the different stress constraints that characterize monogenetic versus polygenetic volcanic systems, which would require a different approach. We basically concentrate on some important aspects of magma migration into the brittle lithosphere that certainly apply to monogenetic volcanism, and for which we include simple models specifically designed to understand magma migration in such volcanic systems. However, this does not exclude (for what concern the basic physics) that some of the considerations included here can also be aplicable to central volcanoes. Therefore, we will first review some basic concepts on rock stress, then we will focus our attention on the main physical controls on magma transport in monogenetic fields and, finally, we will discuss on the implications of the stress field on volcanic susceptibility and the forecast of monogenetic eruptions.

STRESS IN THE LITHOSPHERE

In the literature there are excellent experimental and theoretical approaches on magma transport and on the mechanics and fluid-dynamics of magma-filled cracks (e.g., Pollard, 1969, 1973; Pollard and Muller, 1976; Delaney and Pollard, 1981, 1982; Delaney et al., 1986; Pollard and Segall, 1987; Takada, 1989; Gudmundsson, 1990, 2011a,b; Lister and Kerr, 1991; Rubin, 1993a,b; Rubin, 1995; Dahm, 2000; Muller et al., 2001; Roman and Heron, 2007; Menand, 2008, 2011; Taisne and Jaupart, 2009; Maccaferri et al., 2010, 2011; Menand et al., 2010; Taisne et al., 2011; Gudmundsson, 2012; Le Corvec et al., 2013c; Rivalta et al., 2015), as well as on rock stress (e.g., Zang and Stephansson, 2010), and we address the reader to these contributions. In this section we only provide a basic background necessary to follow the rest of this review.

Knowledge of the state of stresses in the Earth's lithosphere is fundamental to understand how magma will migrate and accumulate inside it and, eventually, erupt at the Earth's surface. Stress in the lithosphere may have different origins: tectonic, gravitational, thermal, residual or fluid overpressure (Park, 1988; Zoback, 1992; Zoback and Zoback, 2002; Zang and Stephansson, 2010; Fossen, 2016). Tectonic stresses derive from the relative motion between mantle flow and plate motion, and may be subdivided into first order (plate scale), second order (regional scale), or third order (local scale), depending on the volume in which a stress component is supposed to be uniform in magnitude and orientation (Zoback, 1992; Heidbach et al., 2007; Zang and Stephansson, 2010). Gravitational (loading) stresses correspond to the overlaying rock mass; they increase with depth in the Earth's crust and consider also the stress resulting from Earth's topography near the surface (Zang and

Stephansson, 2010). Thermal stresses result from temperature changes in crustal rocks when they are buried, uplifted or exposed to local heat sources (e.g., magma, Turcotte and Schubert, 1982; Fossen, 2016). Residual stresses are those preserved in crustal rocks after the external force or stress field has been changed or removed (e.g., metamorphic transformations, cooling of magmatic intrusions, Fossen, 2016). Finally, we must also consider those stresses that can be imposed on crustal rocks by fluid overpressure, like when fluid is present in porous rocks trapped between non-permeable layers or due to magmatic intrusions (Gudmundsson, 2011a)

All these types of stresses will become components of the stress field that will characterize any point in the Earth's crust, the magnitude and orientation of which will depend on the spatial and temporal scales of observation and the corresponding relative value of each stress component. In a broad sense, it is important to know that first order tectonic stresses (plate scale stresses) are assumed to be constant in the whole thickness of the tectonic plate. Second order tectonic stresses (regional or intraplate scale) may vary over short distances depending on the relative position of the reference point inside a plate and the location of main geological structures, such as regional fault systems, mountain belts, or upwelling mantle plumes, or even the presence of collapse calderas, volcanic edifices or rift systems. Local tectonic stresses are also known as structural stresses (Jaeger and Cook, 1979) and correspond to stress field variations caused by active faults, local inclusions, magmatic intrusions, detachment horizons, and density and rheological contrasts. These forces act as major controls on the stress field orientations when the magnitudes of the horizontal stresses are close to isotropic (Heidbach et al., 2007).

Gravitational or loading stresses may also show significant variations depending on the spatial scale we are considering. At a regional scale it is considered that gravitational stresses increase progressively with depth, and that at a certain depth there are distributed more or less isotropically, thus giving rise to a stress state called lithostatic. However, at more local scales and much shallower depths gravitational stresses may differ significantly from one point to another depending on lateral stratigraphic changes and abrupt variations in topography (Muller et al., 2001; Gudmundsson, 2012). Thermal stresses at regional scales derive from the rheological changes that are produced on crustal rocks due to temperature variations caused by burial or uplift processes in sedimentary basins and orogens, as well as those caused by mantle upwelling in intraplate environments (Turcotte and Schubert, 1982; Schrank et al, 2012). At more local scale, these stresses may be significant around magma intrusions. Residual stresses appear in a rock if elastic strain remains after the external stress field is removed, as it may happen in cementation caused by overburden, metamorphic transformations, or tectonic deformation (Fossen, 2016). Finally, crustal stresses derived from fluid overpressure will appear in relation to fluid filled porous rocks, geothermal fields or magma intrusions (Gudmundsson, 2011a; Shapiro, 2015).

When evaluating the importance of each component on the stress field we should also consider the time scale at which they act or at which they may have a significant role. Stresses and the strain they may produce on the crustal rocks are time dependent, so when estimating the state of stress of a certain point we need to consider also the duration of the stresses acting on that point. For example, while first and second order tectonic stresses and gravitational stresses are more or less constant with regard to time, residual and thermal stresses will only be effective at long-term time scales (thousands to millions of years). In a similar way, fluid overpressure stresses may be effective in very short time scales (e.g., magma intrusions, pore-fluids in surface rock) or at longer time-scales when corresponding to buried fluid-saturated rock (e.g., oil reservoirs). Therefore, to understand how crustal stresses act on a specific point we will need to consider the spatial and temporal scales at which each stress component may operate.

Special attention is required for the local tectonic or structural stresses. These will depend on the mechanical characteristics of the rock mass we are considering. This may be classified as: (1) homogeneous, when the rock mass does not show rheological variations or structural discontinuities; (2) anisotropic, when rock properties vary with direction (i.e., there are rheological variations in the rock mass considered), and (3) heterogeneous, when inclusions of different rocks and/or structural discontinuities are present in the rock mass (Zang and Stephansson, 2010). The consequence of considering different rock mass characteristics is that the configuration of stresses in each of them may differ considerably from one to the other (**Figure 2**). Compared to homogenous rock masses in which the trajectories of the principal stresses will define a regular orthogonal pattern, anisotropic rocks formed by alternating stiff and soft materials will show a pattern oriented toward preferred anisotropy, as soft materials will accumulate higher strain while the stiff ones will attract higher stresses (Zang and Stephansson, 2010). In addition to anisotropies, crustal rock masses may show different scale heterogeneities caused by the presence of structural discontinuities.

MAGMA MOTION

Magma motion in the lithosphere will be basically controlled by the overpressure of magma over that of rock. How this magma overpressure is achieved will be discussed latter. Now it is sufficient to assume that magma is over-pressurized, so it will migrate through the lithosphere controlled by the regional and local stresses that act on it (Takada, 1994; Traversa et al., 2010).

In the source region, magma transport will be dominated by porous flow through a deformable and partially molten matrix, from which it will segregate by compaction of the mantle unmelted residuum (Spera, 1980; McKenzie, 1984, 1985; Rubin, 1993c, 1998). At such depths and when a sufficient volume of magma has been accumulated, thus becoming gravitationally unstable, it will tend to continue ascending in order to equilibrate its excess pressure. It has been assumed that basaltic magma in the mantle may ascent as buoyant diapirs when rocks surrounding magma may deform plastically due to their relative low viscosity (Spera, 1980; Rubin, 1998). This ascending movement of magma will continue until the rocks above behaves as a brittle solid, in moment magma will ascent through fractures in the host rock. The transition from porous flow to flow through magma driven cracks is not only a function of the temperature of the host rock, but also of the rate of strain and the stresses involved (Rubin, 1993c, 1998). The brittle-ductile transition in the crust may be located higher than initiation of dyke propagation, which may occur much deeper, even into the upper mantle. This is not because of the temperature differences in the host rock, but because the timescale of deformation around a dyke is small compared to the viscous relaxation time scale of the medium, so brittle behavior occurs (Rubin, 1993c, 1998). Magma ascent will be halted if the host rock does not fracture in response to the pushing action of magma (Turcotte and Schubert, 1982; Maaloe, 1985; Middleton and Wilcock, 1994). However, if magma overpressure exceeds the cohesive and confining stresses of host rock, this will be broken apart forming a fracture through which magma may intrude and continue its emplacement to shallower levels (Jaeger and Cook, 1979; Spera, 1980; Maaloe, 1985; Rubin, 1993c, 1998; Menand, 2011). This is the most effective mechanism of magma transport in the lithosphere, particularly in the crust, in comparison with the diapiric ascent. In fact, seismic evidence indicates that magma-filled cracks may start to form at depth of 40–50 km or even greater, as it is

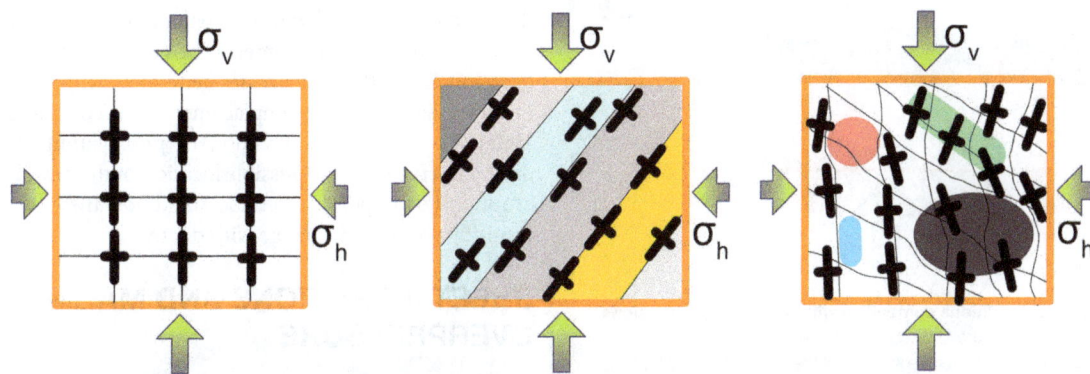

FIGURE 2 | Sketch of the stress field and stress trajectories (solid bars) in homogeneous (left), anisotropic (center), and heterogeneous (right) rocks (adapted from Zang and Stephansson, 2010).

suggested by the existence of earthquakes generated at these depths and by the presence of ophiolite peridotite bodies and other mantle inclusions erupted onto surface (Maaloe, 1985).

Magma-filled fractures or cracks are known as sheets intrusions, being named dykes when they cut across bedding or foliation/fabric in the host material, and sills when they are concordant with that. However, for the purposes of this paper, we will refer vertical or subvertical cases as dykes, and horizontal as sills, with the understanding that in detail it depends on the cross cutting relationships (**Figure 3**). The formation of sheet intrusions is regarded in the framework of hydraulic fracturing. In this context, a hydrofracture is a tension fracture in which the driving tensile stress is the fluid overpressure, so, in our case, the magma overpressure (see Gudmundsson, 2011a). Therefore, to start a sheet intrusion it will require to achieve the conditions for rupture of the rock and hydrofracture initiation (Jaeger and Cook, 1979; Gudmundsson et al., 1999; Gudmundsson, 2011a):

$$p_m \geq \sigma_3 + T \qquad (1)$$

where p_m is the total magma pressure, σ_3 is the minor principal compressive stress, and T is the local *in situ* tensile strength of the rock.

Magma migration will be controlled by the regional tectonics and the gravitational stresses but also by local stress barriers defined by crustal heterogeneities such as local tectonic structures and rheological changes in crustal rocks. This means that it will tend to follow a path normal to the minimum compressive stress. If in its ascent to the surface magma finds a rheological or structural contrast between rocks, magma may become arrested or intrude laterally forming a sill (Pasquarè and Tibaldi, 2007;

Tibaldi and Pasquarè, 2008; Tibaldi et al., 2008; Maccaferri et al., 2010; Menand, 2011; Gudmundsson, 2011b).

As we see in **Figure 3**, the propagation of a sheet intrusion as a dyke or a sill in an anisotropic medium will depend on the stress configuration at the front of the intrusion. If the total magma pressure (p_m) is greater than the principal horizontal stress (σ_h), which may coincide with σ_3, plus the tensile strength of the rock tested in extension parallel to the bedding ($T_{//}$), the intrusion will propagate as a dyke. However, if p_m is greater than the principal vertical stress (σ_v) plus the tensile strength of the rock tested in extension perpendicular to the bedding plane ($T_\perp$) the intrusion will propagate as a sill (Price et al., 1990). If the sill is fed by a dyke, as it is illustrated in **Figure 3**, both conditions must be satisfied simultaneously at the junction of the feeder dyke and the sill. Therefore, it follows that (Price et al., 1990):

$$(\sigma_v - \sigma_h) < (T_{//} - T_\perp) \qquad (2)$$

The tensile strength of a pile of stratified rocks will be determined by the strength of the individual bedding planes, so $T_\perp$ will be approximately zero (Price et al., 1990; Gudmundsson, 2011a,b). On the other hand, the tensile strength of most unjoined or unfractured rocks is of the order of 10 MPa or considerably smaller if the rock contains joints or fractures (Touloukian et al., 1989; Price et al., 1990; Gudmundsson, 2011a,b). So, we can deduce that a sill will only occur when the differences in magnitude between the horizontal and vertical stresses are very small (Pollard, 1973; Price et al., 1990). In many situations, if a sill is fed from a perfectly vertical dyke it will favor propagation laterally at both sides of the plane of weakness, while if the feeder dykes is inclined (i.e., oblique to the strata) the sill will tend to propagate only toward the side opposite to the dyke dip (see **Figure 4**). This generalized behavior, however, may change under particular stress configurations and rock rheology contrasts (e.g., Tibaldi and Pasquarè, 2008).

The orientation of the stress field may change (i.e., may rotate with respect to a reference position) depending on the exact contribution of each stress component at each point (Pollard and Muller, 1976; Pollard and Segall, 1987). At the time scale of magma propagation through a fracture (days to months) the near-tip stress field will be essentially controlled by the first and second order stresses, structural and gravitational stresses, the stresses associated with the magma overpressure, which will depend on the fracture geometry and total volume of magma, and occasionally by thermal stresses generated by thermal variations in the magma during emplacement (Turcotte and Schubert, 1982). To predict the direction of propagation of a magma-driven fracture and the possible location of the next monogenetic eruption it is important to understand how these stresses, or the resulting stress field, change with depth.

SHEET INTRUSIONS AND MAGMA OVERPRESSURE

The plumbing system of a monogenetic volcanic field (**Figure 1**) may include a source zone, where magma generates and migrates upwards by gravitationally-induced porous flow, and a series of

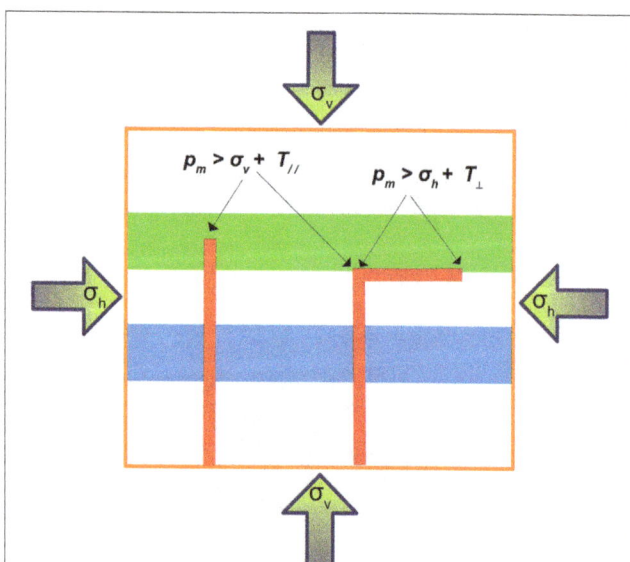

FIGURE 3 | Stress conditions compatible with dyke and sill intrusions. Dykes: Total magma pressure (pm) must be greater than the principal horizontal stress (σ_h) plus the tensile strength of the rock tested in extension parallel to the bedding ($T_{//}$). Sill: pm is greater than the principal vertical stress (σ_v) plus the tensile strength of the rock tested in extension perpendicular to the bedding plane ($T_\perp$) (adapted from Price et al., 1990).

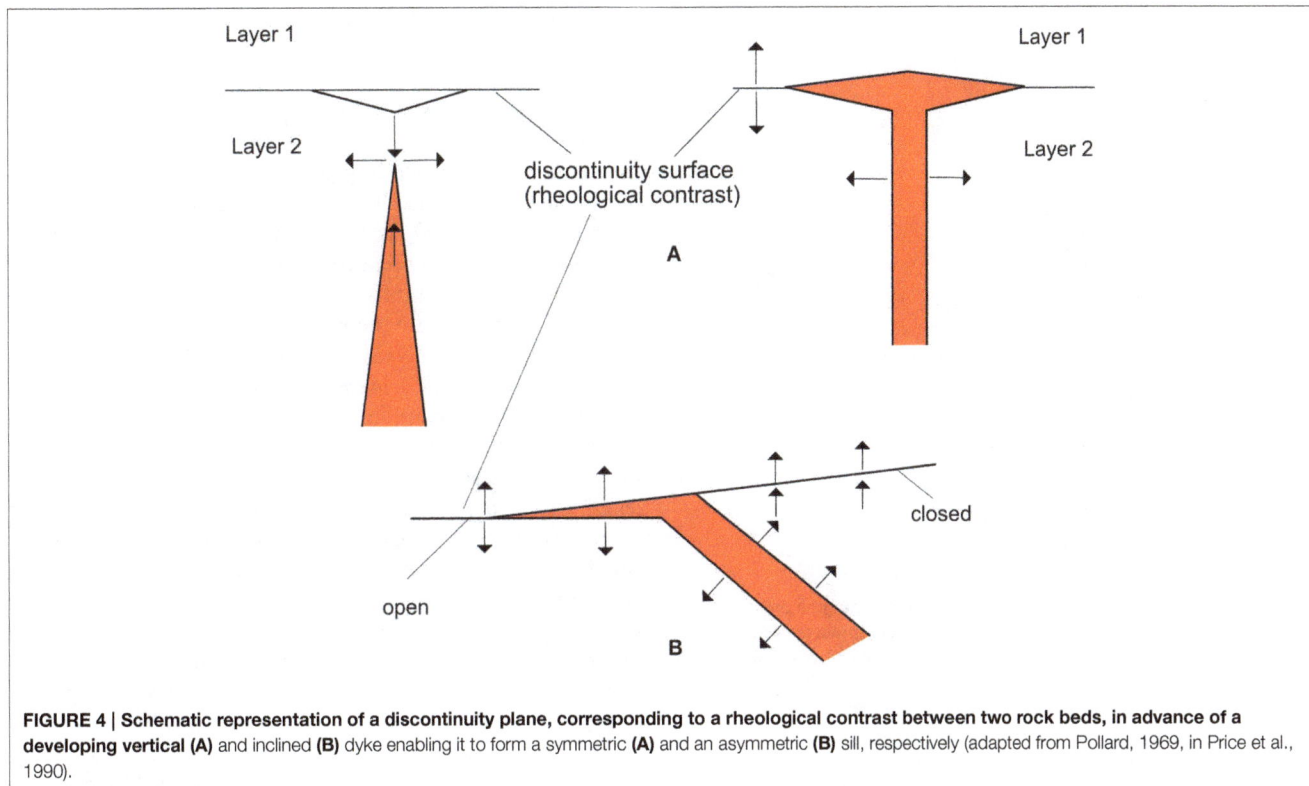

FIGURE 4 | Schematic representation of a discontinuity plane, corresponding to a rheological contrast between two rock beds, in advance of a developing vertical (A) and inclined **(B)** dyke enabling it to form a symmetric **(A)** and an asymmetric **(B)** sill, respectively (adapted from Pollard, 1969, in Price et al., 1990).

intermediate reservoirs where magma may stop and differentiate for a while before continuing its ascent to the surface. This is clearly indicated by the petrology and geochemistry of magmas from monogenetic eruptions, which evidence certain degree of differentiation in most cases and occasional assimilation of crustal rocks that occurred at different depths (Thirlwall et al., 2000; Klügel et al., 2005; Stroncik et al., 2009; Valentine and Hirano, 2010; Brenna et al., 2011; Rowe et al., 2011; Hernando et al., 2014; Albert et al., 2015, 2016; Klugel et al., 2015). These intermediate reservoir zones, which will normally be located at rheological or structural discontinuities inside the lithosphere, do not need to be stable or permanent along the whole history of the volcanic field. Also, there is evidence that in some cases magma erupts directly from the source region, without suffering any differentiation in its journey to the surface (Bacon et al., 1995; Garcia et al., 1995). The reasons magma will either stop at different depths before reaching the surface or will ascend straight from the source region or a deep reservoir, depend on the magma overpressure and the state of stresses inside the lithosphere. This balance between magma pressure and lithospheric stresses will decide whether magma will be able to follow a straight path to the surface or will stop at certain depth, arresting its ascent or continuing it until internal pressure conditions are favorable again. Also, it will control the exact path that magma will follow and, finally, the location of the new vent in case magma has been able to reach the surface. In fact, the proportion of magmatic intrusions that become feeder dykes is minimum compared to the total number of dykes that may be generated during the whole life of a volcanic system

(Gudmundsson et al., 1999). So, in order to know if a magma intrusion will reach the surface and cause an eruption we need to ask: (1) will magma have sufficient driving force (magma overpressure) to reach the surface)?, and (2) what is the path it will follow?

To answer the first question we need to understand which is the driving force of magma intrusions. It is obvious that for the same magma driving force, a different stress distribution may either reduce or enhance the possibilities for this magma to reach the surface. **Figure 5** illustrates the concepts of total magma pressure (p_m), magma excess pressure (p_e) and magma overpressure or driving pressure (p_o), which are fundamental to understand magma migration (see Dahm, 2000; Gudmundsson, 2012). We consider a lithostatic reference state, which is the simplest stress model for the interior of the lithosphere, so there is no differential stress at depth ($\sigma_1 = \sigma_3$). In such situation and in equilibrium the total pressure of a deep magma reservoir will be equal to the lithostatic pressure ($\rho_r g h$). If a new injection of magma from the source region (or from a deeper reservoir) enters into the magma reservoir, it will increase the total magma pressure ($p_m = p_l + p_e$) (see Blake, 1981). We assume that the resident and the new magmas are under-saturated in volatiles, so no free gas phase is present, which is a good assumption for basaltic magmas at depths of several tens of kilometers. The increase of magma pressure (excess pressure, p_e) inside the reservoir will force the volume of the reservoir to increase. Depending on the rigidity of the surrounding rocks they will be able to deform elastically to a certain limit, so the reservoir will expand a little bit. If the volume increase permitted by elastic

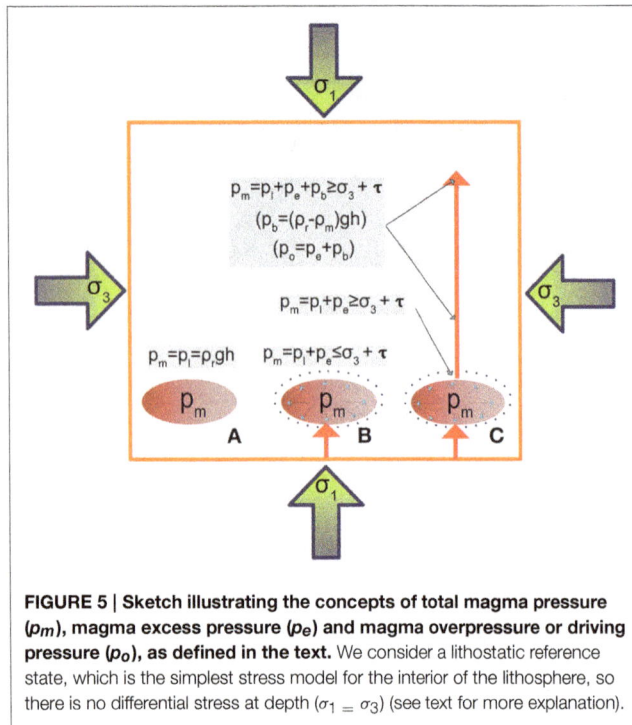

FIGURE 5 | Sketch illustrating the concepts of total magma pressure (p_m), magma excess pressure (p_e) and magma overpressure or driving pressure (p_o), as defined in the text. We consider a lithostatic reference state, which is the simplest stress model for the interior of the lithosphere, so there is no differential stress at depth ($\sigma_1 = \sigma_3$) (see text for more explanation).

expansion of the host rock is sufficient to accommodate the pressure increase, the situation will return to equilibrium until a new intrusion of magma arrives. However, if the elasticity the host rock is exceeded it will fracture due to the tensile stresses, which are generated around the reservoir walls, and magma will be injected into the host rock. Previous calculations have indicated that volume fractions of new injected magma of approximately 0.1% of the volume of the reservoir in the absence of any gas phase are sufficient to produce the excess pressure necessary to trigger a magma injection (Blake, 1981). This volume fraction may increase to approximately 1% of the volume of the reservoir when a gas phase is present in the resident magma (i.e., at much shallower depths) due to the higher compressibility of the resident magma (Bower and Woods, 1997; Folch and Martí, 1998).

The progression of a magma-driven fracture will depend on the total magma pressure, as indicated by Equation (1), which at the moment of the reservoir rupture can be rewritten as (Gudmundsson et al., 1999; Gudmundsson, 2012):

$$p_m = p_l + p_e \geq \sigma_3 + T \qquad (3)$$

where p_l is the lithostatic pressure and p_e is the excess pressure above the lithostatic necessary to initiate the rupture of the reservoir walls. At the moment in which magma starts to abandon the reservoir intruding the rock, and because at these depths magma tends to be less dense than the host rock, a buoyancy force (p_b) resulting from the difference between the average densities of the host rock and magma, will be added to the initial magma excess pressure to help driving the sheet intrusion, so the resulting

overpressure will be (Gudmundsson et al., 1999; Gudmundsson, 2012):

$$p_o = p_e + p_b = p_e + (\rho_r - \rho_m)gh + \sigma_d \qquad (4)$$

where ρ_r is the average host-rock density, ρ_m is the average magma density, g is acceleration due to gravity, h is the depth of the source, and σ_d is the differential stress ($\sigma_d = \sigma_1 - \sigma_3$), which in the case of an isotropic (lithostatic) situation will be 0. To continue the intrusion, magma overpressure must be large enough to fracture the rock and to overcome the viscous forces of resistance opposing to flow (Middleton and Wilcock, 1994). And this will be achieved if a sufficient volume of magma is available at the reservoir from which the intrusion is being pumped up. However, for simplification we have not considered in our calculations the viscous pressure dissipation along the length of the dyke due to magma flow. The exact volume of magma needed to ensure that an intrusion will reach the surface will depend on each case on the physical characteristic of magma and host rock (see Traversa et al., 2010, Equations 15 and 38). Also, magma intrusion must occur at sufficient high rate in order to avoid much cooling of magma that could increase its viscosity in excess, halting motion. If magma intrusion progresses enough to reach shallower levels, it may start degassing due to the decrease of lithostatic pressure. If this gas is retained at the tip of the sheet intrusion it will represent an additional increase of pressure at the interior of the intrusion, but if it escapes through the rock porosity, pressure will decrease as magma density will increase.

As indicated before, to know the exact path a magma intrusion will follow and where it will intersect the surface, we have to consider how the orientation of the stress field may change all along the magma pathway. This requires to know the internal structure of the system, including the location and size of stratigraphic, lithological, rheological and structural discontinuities, lateral and vertical extent of major tectonic features, horizontal distribution of deviatoric stresses, and distribution of loading stresses due to complex topographies. Obviously, this is not an easy task. In fact, in comparison with classical sedimentary basins, the internal geometry of volcanic systems is much more complex due to the irregular stratigraphic relationships shown by volcanic materials, their contrasting lithologies, their affectation by active tectonics, and the numerous magmatic intrusions that may be present. Imaging the interior of volcanic systems at lithospheric or crustal scales with geophysical methods does not provide models sufficiently detailed to detect dykes or sills. In recent years, the application of high resolution shallow geophysical methods has opened a new window to visualize in great detail the internal geology of volcanic systems (e.g., Mrlina et al., 2009; Cassidy and Locke, 2010; Bolós et al., 2012; Barde-Cabusson et al., 2013; Blaikie et al., 2014), but their resolution does not penetrate deeper than a few hundreds of meters. Therefore, obtaining a precise picture of the interior of a volcanic system and how a new injection of magma may cross it to reach the surface is, by now, still difficult. However, having a minimum knowledge at a lithospheric and crustal scale of the main stratigraphic units, the distribution of the main tectonic structures, the orientation of the current regional stress field, and the topography of the area, we can infer the main stress constraints that sheet intrusions may have.

DYNAMICS AND MECHANICS OF SHEETS INTRUSIONS IN THE LITHOSPHERE

We will start examining the conditions to drive a sheet intrusion (dyke) to the surface from four different reservoirs located at different depths, in an isotropic and homogeneous lithosphere. Each reservoir is recharged from below in order to cause the excess pressure necessary to initiate the sheet intrusion (**Figure 6**). The conditions for p_m, p_e, and p_o are as explained above and in **Figure 5**. For each case, we will study the critical influence of the magma reservoir size on eruption likelihood, quantifying the volume required for the creation and growth of a dyke from the reservoir to the Earth's surface, and the magma overpressure achieved in each case inside the dyke. In all cases, we will assume that the geometry of the reservoir allows tensile stresses to concentrate at the center of its upper part, so the sheet intrusion will propagate vertically toward the surface.

To compute the critical volume required for the creation and growth of a dyke we use the mathematical approximation performed by Traversa et al. (2010), which relates the magma reservoir volume, Vr, and the reservoir excess pressure variations due to dyke propagation:

$$V_r = \frac{\Delta V_r}{\exp\left(\Delta P_{rvar}(\frac{4G + 3K}{4GK})\right) - 1} \quad (5)$$

where ΔVr is the variation (decrease) in reservoir volume produced by the magma injected in the dyke, ΔP_{rvar} is the corresponding decrease of the reservoir excess pressure, and G and K are the shear and bulk moduli of the host rock.

Combining Equation (5) with Equation (4), which gives us the magma overpressure in the dyke, p_o, and assuming that the decrease of reservoir overpressure is (mainly) caused by the dyke propagation up to the surface, we may establish the approximation:

$$\Delta P_{rvar} \approx p_o \quad (6)$$

With this relation we can calculate the minimum reservoir volume, Vr, required for a dyke injection, ΔVr, for different reservoir depths and density contrast $(\rho_r - \rho_m)$ scenarios. **Figure 7** shows the results of model calculations for each scenario. We show different plots with the evolution of the magma overpressure at the dyke tip (in blue) propagating up from the reservoir roof (at 5, 15, 25, and 40 km depth) through the crust. Using Equations (4) and (5) we have calculated the corresponding minimum size of the reservoir at different heights of the dyke (in red) for a common priori magma injection $\Delta Vr = 1.0 \times 10^8$ m^3, which represents the maximum value of intrusions associated with the historical eruptions in the Canary Islands (Becerril et al., 2013b). As excess pressure, p_e, at the time of hydrofracturing formation is normally equal to the tensile strength of the rock (Gudmundsson, 2012), we used a constant = 3 MPa, that represents the most common value of the crustal rocks tensile strength (Gudmundsson, 2012). To estimate the magma overpressure at the dyke tip, p_o, we have also considered the contribution of magma buoyancy, p_b, as indicated in Equation (3). The results obtained applying this simple model (i.e., isotropic and homogeneous lithosphere) show that, depending on where it starts inside the lithosphere, any sheet intrusion will require a specific available volume of magma and a critical overpressure to ensure that it will arrive at the Earth's surface (**Figure 7**). A more realistic approach, even assuming an isotropic and homogeneous lithosphere, would have required considering the exact rheological behavior of the host rock (elastic or rigid) in front of the pressure changes in the reservoir, the rheological changes of magma due to pressure and temperature changes during dyke emplacement, and also the pressure drop due to viscous flow of the magma (see Turcotte and Schubert, 1982; Gudmundsson, 2011a). However, our model offers a first order approach that we consider valid for the purpose of this review. The implications derived from this model will not change in the case of a more realistic heterogeneous and anisotropic lithosphere. However, it will be necessary to know the exact distribution of potential stress barriers inside the lithosphere to be able to predict a magma path. In such cases, any stress barrier caused by a density or rheological contrast, presence of a tectonic structure, or existence of a differential stresses may induce stress rotation and make the sheet intrusion divert from the direction it was propagating, or even to arrest it in case the overpressure required to surpass that obstacle is not achieved. This may explain how dykes may divert into sills forming new magma reservoirs (Menand et al., 2010; Gudmundsson, 2011a), or how dykes or sills may propagate for tens of kilometers inside the crust before becoming vertical again and erupting at surface (e.g., Martí et al., 2013) or stopping before it reaches it (e.g., Wright et al., 2006; Ayele et al, 2007).

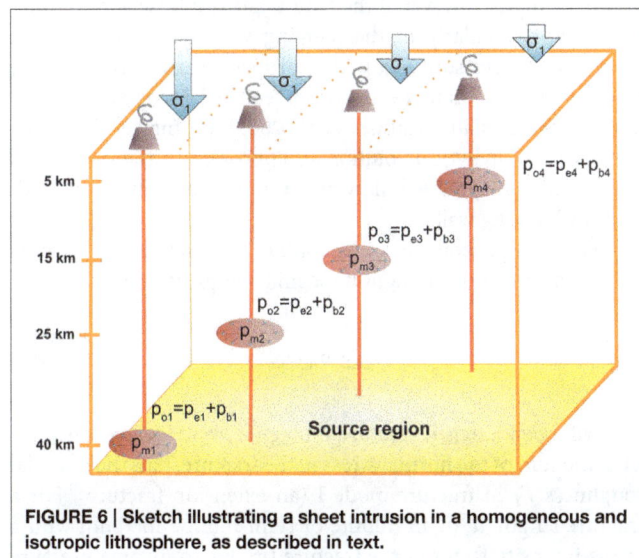

FIGURE 6 | Sketch illustrating a sheet intrusion in a homogeneous and isotropic lithosphere, as described in text.

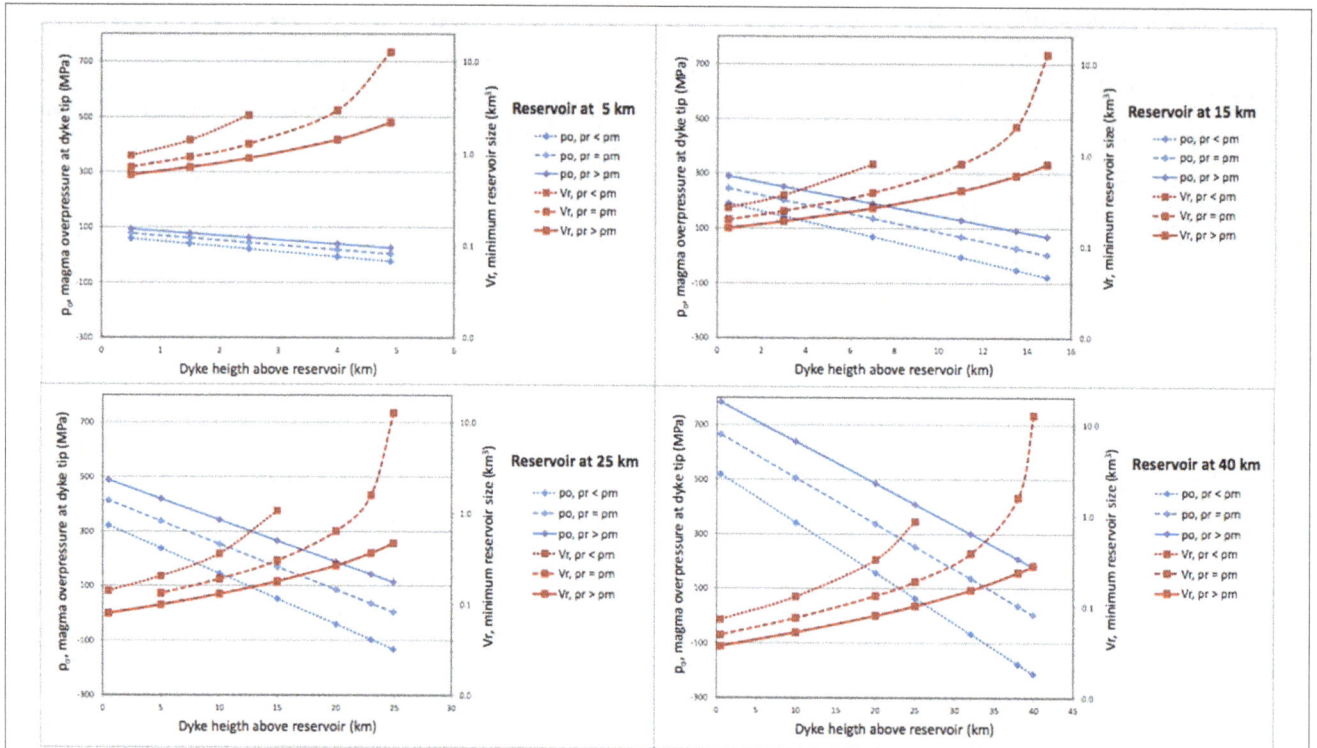

FIGURE 7 | Graph plots showing results concerning magma overpressures and magma volumes required in each scenario of Figure 5. We show different plots with the evolution of the magma overpressure at the dyke tip (in blue) propagating up from the reservoir roof (at 5, 15, 25 and 40 km depth) thought the crust, and the minimum size of the reservoir at different heights of the dyke (in red) for a common priori magma injection $\Delta V_r = 1.0 \times 10^8$ m^3. We have considered three possible cases of positive, null and negative density contrast between the magma and the host rock, with 2.55×10^3 kg/m^3 for a mean value of the magma density range (Murase and Mcbirney, 1973) and 3.0×10^3, 2.55×10^3 and 2.0×10^3 kg/m^3 for host rock, assuming that most crustal rock densities are in that range (Gudmundsson, 2012). We used $K = 1.0 \times 10^9$ Pa for the bulk modulus and $G = 1.13 \times 10^9$ Pa for the shear modulus.

Assuming similar conditions to the previous case for the magma reservoir, we consider now an heterogeneous and anisotropic lithosphere with rheological differences (layers with different color, **Figure 8**) or presence of faults or cracks (black bars, **Figure 8**), each one representing a different stress component in the total stress field. We also include an intermediate stop with the formation of a new intermediate reservoir for calculating what is needed in that case, first to stop magma migration for a while, and second to initiate and sustain a new sheet intrusion from that shallower position, assuming a continuous connection with the deeper reservoir. We assume that rock rheological contrasts hinder magma migration and that the presence of faults and fractures facilitate it (see Delaney et al., 1986; Gaffney et al., 2007; Le Corvec et al., 2013c).

With regard to structural discontinuities we have to differentiate between faults (discontinuity in an homogeneous rock mass that has undergone some relative shear displacement and that is assumed to have material toughness in fracture mode I (extension) equal or close to 0), and fissures, joints or cracks that would represent tensional pre-existing or newly formed rock rupture zones that present a considerable toughness. In other words, it is assumed that faults are already open while joints or cracks will offer a variable resistance to open and propagate under tensile stress depending on the material they form in.

Therefore, for a dyke to open a pre-existing fault magma pressure has to overcome the compressive normal stress on the fault (that keeps the fault closed). Once magma opens the fault, it will starts flowing inside and magma intrusion will continue through the fault plane while the mechanical conditions along it do not change. However, if the fault plane is intersected by other fractures, particularly in the hanging wall, these may capture magma if the pressure required to propagate any of the hanging fractures is less than the pressure required to open and flow along the fault plane (Gaffney et al., 2007). Gaffney et al. (2007), proposed an analytical solution to approximate the conditions under which magma will flow either along the fault or upwards into the hanging wall.

In a heterogeneous rock that contain faults and joints, magma will open a pre-existent fault if magma overpressure is:

$$p_o \geq \sigma_{n,fault} \tag{7}$$

and will open a tensile fracture if magma overpressure is greater than the sum of the normal stress in the fracture, plus the material toughness Γ_I in fracture mode I (an extension fracture). For a fracture length, a, in an infinite or semi-infinite medium with a stress-intensity factor at the fracture tip due to magma pressure

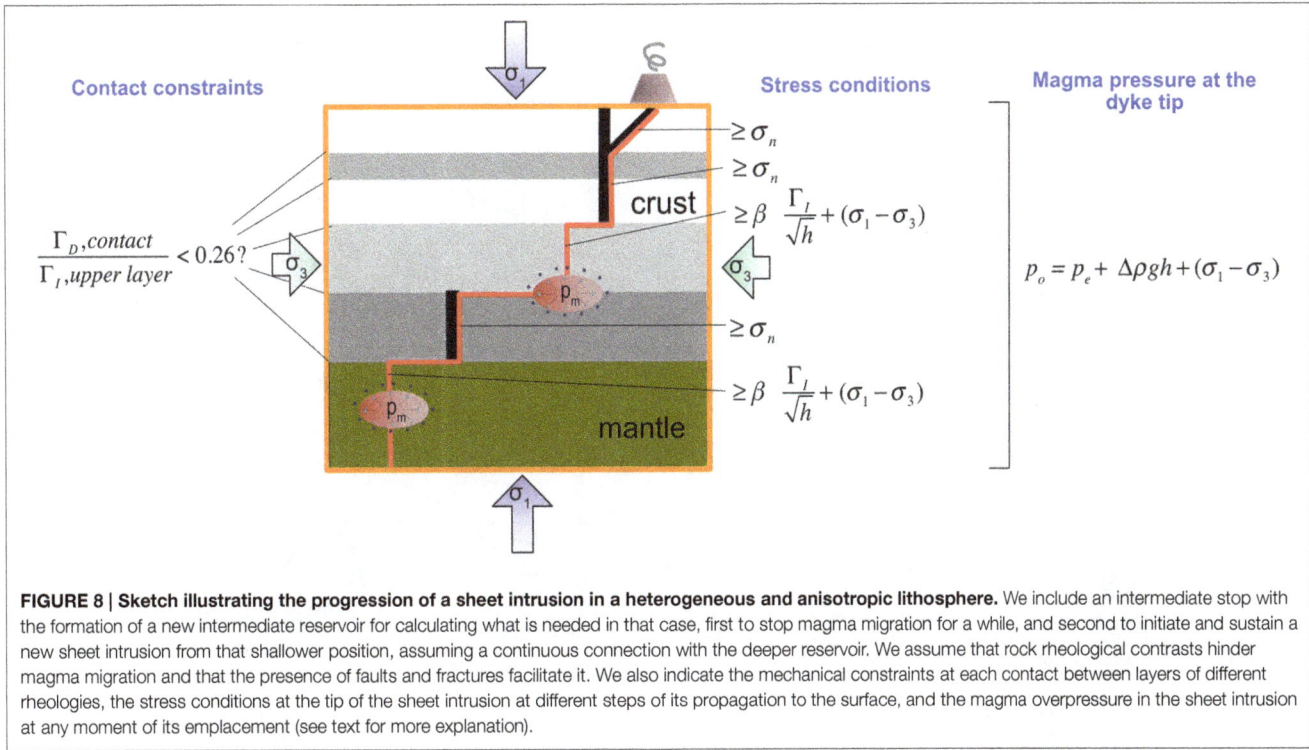

FIGURE 8 | Sketch illustrating the progression of a sheet intrusion in a heterogeneous and anisotropic lithosphere. We include an intermediate stop with the formation of a new intermediate reservoir for calculating what is needed in that case, first to stop magma migration for a while, and second to initiate and sustain a new sheet intrusion from that shallower position, assuming a continuous connection with the deeper reservoir. We assume that rock rheological contrasts hinder magma migration and that the presence of faults and fractures facilitate it. We also indicate the mechanical constraints at each contact between layers of different rheologies, the stress conditions at the tip of the sheet intrusion at different steps of its propagation to the surface, and the magma overpressure in the sheet intrusion at any moment of its emplacement (see text for more explanation).

inside the fracture, K_I, the condition can be written as follows:

$$P_m \geq \frac{K_I}{\beta\sqrt{\pi a}} + \sigma_{n,fracture} \qquad (8)$$

where the coefficient β accounts for the geometry of the fracture, being in the simplest form (uniform pressurized elliptical flat crack of radius, a, at the edge of a semi-infinite plane), $\beta = 1.12$ (Dundurs, 1969; Rice, 1980; Gaffney et al., 2007).

If the fault has a dip angle, α, the stress normal to the fault will be:

$$\sigma_n = \sigma_v \cos^2\alpha + \sigma_h \sin^2\alpha \qquad (9)$$

Assuming that the vertical stress is gravitational, $\sigma_v = \rho g d$ (ρ the density of the rock, g the gravity acceleration and d the depth) and that the horizontal stress, σ_h, (positive compressive, negative tensile) is proportional to the vertical stress ($\sigma_h = k.\sigma_v$), the minimum fracture length required for a dyke to propagate vertically in a medium where the normal stress on the fracture plane is approximated by the far-field stress, $\sigma = \sigma_h$, is:

$$a_c = \frac{\Gamma_I^2}{\pi(\beta\rho g d)^2 \left[k\left(\sin^2\alpha - 1\right) + \cos^2\alpha \right]^2} \qquad (10)$$

and the critical depth, d_c, for a given fracture length, a, greater the total depth, d, the dyke will propagate vertically:

$$d_c = \frac{\Gamma_I}{\beta\rho g \sqrt{\pi a}\left[k\left(\sin^2\alpha - 1\right) + \cos^2\alpha \right]} \qquad (11)$$

Figure 9 shows the results of these second calculations. These results are in good agreement with those obtained by Gaffney

et al. (2007). Both show that the minimum length of a vertically-oriented hanging-wall fracture needed to divert magma from a fault plane diminishes considerably with depth. Therefore, for a dyke to be captured by a pre-existing fault that it intersects, the fault will need to be either relatively high-angle, and/or the intersection will be at shallow depths (<3 km). In other words, it is not straightforward for magma to enter a fault in the first place. Once that has happened, then the analysis shown in **Figure 9** applies to the magma subsequently being diverted upward from the fault plane. Therefore, for magma to propagate horizontally at great depth, it will require specific structural and/or mechanical conditions that will force such type of movement instead of migrating vertically toward shallower levels. Another implication of the results shown in **Figure 9** is that, as earthquake magnitude is proportional to the fracture length, the smaller magnitude earthquakes (that would correspond to small fractures opening at the hanging wall) would be observed from very deep to close to the surface, while the greater magnitude earthquakes will only occur closer to the surface (**Figure 9D**).

In the case of a dyke intersecting a stratigraphic discontinuity that represents a rheological contrast between two rock layers, the dyke may become arrested, penetrate the contact, or be deflected along it (He and Hutchison, 1989; He et al., 1994; Gudmundsson, 2011a,b). The general stress conditions for the dyke to continue its propagation or to become a sill have been analyzed before (see **Figures 3, 4**, and Equation 2). We will examine here the influence of different mechanical properties at both sides of the rheological contact. Following the formulation presented by He and Hutchison (1989) and He et al. (1994) for the case of a dyke crossing a mechanical discontinuity, the condition for

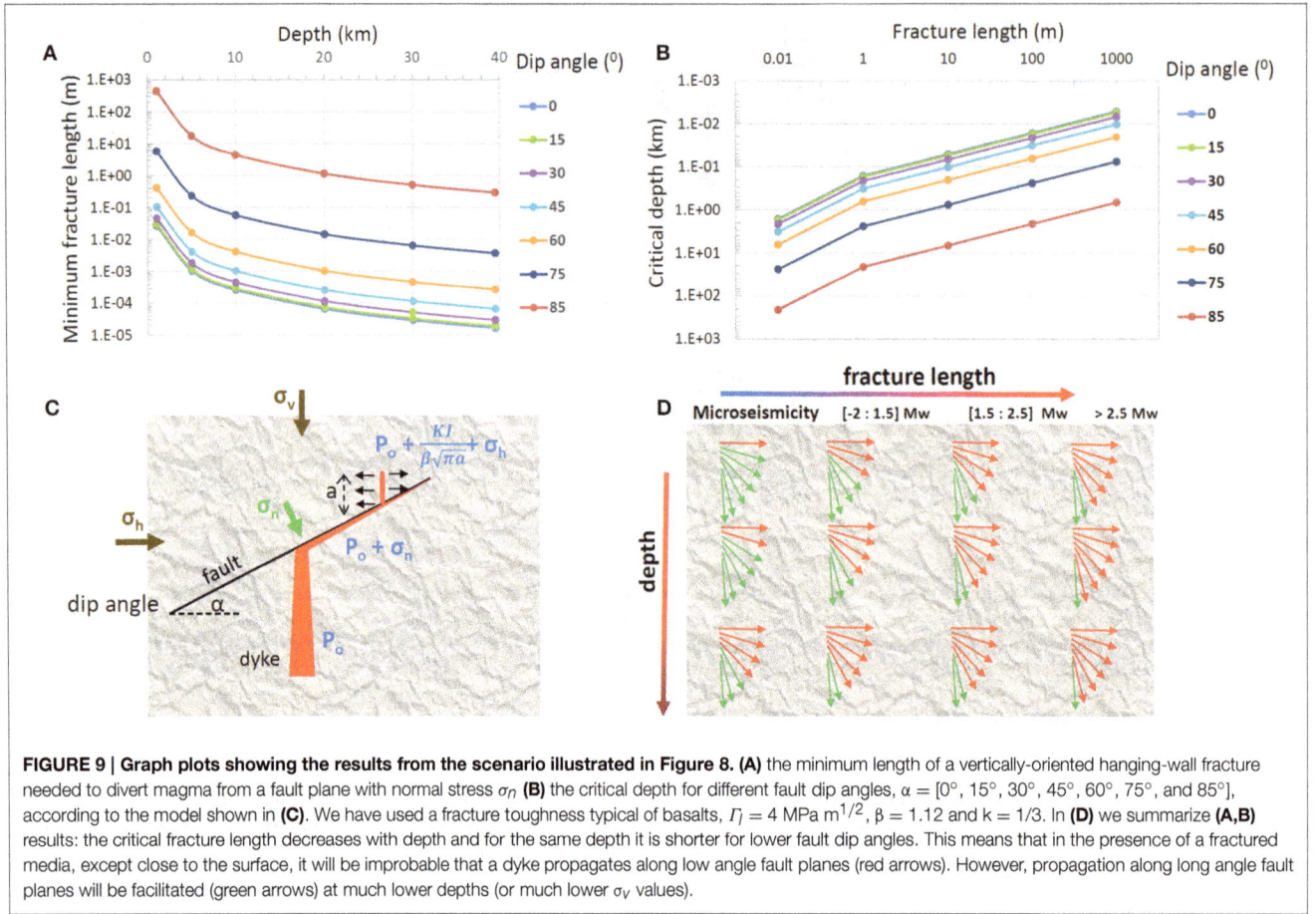

FIGURE 9 | Graph plots showing the results from the scenario illustrated in Figure 8. (A) the minimum length of a vertically-oriented hanging-wall fracture needed to divert magma from a fault plane with normal stress σ_n **(B)** the critical depth for different fault dip angles, α = [0°, 15°, 30°, 45°, 60°, 75°, and 85°], according to the model shown in **(C)**. We have used a fracture toughness typical of basalts, Γ_l = 4 MPa m$^{1/2}$, β = 1.12 and k = 1/3. In **(D)** we summarize **(A,B)** results: the critical fracture length decreases with depth and for the same depth it is shorter for lower fault dip angles. This means that in the presence of a fractured media, except close to the surface, it will be improbable that a dyke propagates along low angle fault planes (red arrows). However, propagation along long angle fault planes will be facilitated (green arrows) at much lower depths (or much lower σ_v values).

penetrating it depends on the rate of the strain energy release associated with dyke penetration:

$$G_p = \left(\frac{(1 - \upsilon_1)}{2\mu_1} \right) . K_I^2 \qquad (12)$$

and the strain energy rate associated with deflection (sill formation):

$$G_d = \left(\left(\frac{(1 - \upsilon_1)}{\mu_1} \right) + \left(\frac{(1 - \upsilon_2)}{\mu_2} \right) \right) .$$
$$\left(K_I^2 + K_{II}^2 \right) / \left(4.cosh^2 \ \pi\varepsilon \right) \qquad (13)$$

where, K_I , is the mode I (tensile) stress-intensity factor, K_{II} , the model II (shear) stress-intensity factor, which depends on the magma pressure inside the dyke and on the geometrical properties of the contact, and υ and μ area the Poisson's ratio and shear modulus, respectively, corresponding to layer 1 (upper) and layer 2 (lower), when the dyke in layer 2 tries to penetrate into layer 1.

The dyke is likely to continue its vertical path penetrating into layer 1 if:

$$\frac{G_d}{G_p} < \frac{\Gamma_D}{\Gamma_1} \qquad (14)$$

where Γ_1 is the layer 1 mode I toughness and Γ_D is the toughness on layer 2 for the combined models I and II.

For a given dyke-segment length a, reaching the contact, the rate G_d/G_p depends only on the relation of the mechanical properties of both layers and not on the magma pressure, so an increase in magma pressure will imply a longer distance reached by the sheet intrusion but will not influence the direction of emplacement. He and Hutchison (1989) showed that a dyke becomes deflected between two layers of contrasted mechanical properties only if the material toughness of layer 2 is less that 26% of the material toughness of layer 1. So, a dyke will deflect into a sill when Γ_D < 0.26 Γ_1 and will continue as a dyke crossing the rheological contact when Γ_D > 0.26 Γ_1. Moreover, when the stiffness (Young modulus, E) of the layer 1 (upper) is less that the stiffness of layer 2, there is generally much less tendency for deflection of a dyke into a sill (Gudmundsson, 2011b), and conversely, being improbable the dyke to penetrate in layer 1 when $E_1 >> E_2$.

Therefore, despite magma overpressure being the driving force for magma to move upwards, it will not control whether a dyke will be deflected into a sill at a rheological contrast or will continue straight. This will be determined by the mechanical contrast between the two rocks. This is an important concept, as higher magma overpressures will only represent longer emplacement distances but not necessarily higher capacities to cross heterogeneous lithologies and to reach the surface. This is particularly applicable to unfractured media. If an ascending dyke intersects a much stronger layer, it is still more likely to continue vertically if there are any vertical/subvertical fractures in that stronger layer, particularly if this occurs at depth as previously indicated by Gaffney et al. (2007), and now in this study (**Figure 9**). Faults and fractures may facilitate paths for magma ascent, as they represent zones of more favorable stress conditions for magma to penetrate into the host rock with lesser energy consumption, as it seems to be confirmed by the structural control in the location of vents observed in most of the monogenetic volcanic fields (e.g., Le Corvec et al., 2013b). However, the effectiveness of structural features to direct magma migration depends on the depth and dip angle of each structure, the pre-existing normal or transtensional faults being very effective in transporting magma at very shallow depths.

DISCUSSION

In this review we have considered the main aspects that govern magma migration in monogenetic volcanic systems. We have tried to offer a comprehensive review of what happens in the plumbing systems of such type of volcanoes and how they prepare for new eruptions. We have not intended to provide a complete review of monogenetic volcanism, for which there are already excellent studies (e.g., Wood, 1980; Tibaldi, 1995; Connor and Conway, 2000; Walker, 2000; Valentine and Gregg, 2008; Németh, 2010; Le Corvec et al., 2013b; Németh and Kereszturi, 2015), but we have tried to clarify the most relevant concepts that need to be understood when conducting hazard assessment in such volcanic fields and, in particular, during the probabilistic analysis of analysis of vent opening (volcanic susceptibility). In this sense, we have concentrated our attention on dyke ascent processes in response to the surrounding environment (stress fields, structures, etc.), but have not considered specifically the potential effect of magma flux and magmatism-induced stresses on the ambient ones. This effect was studied by Valentine and Perry (2007) and Le Corvec et al. (2013c), among others. We have considered it as a long term effect that will determine whether the magmatism responsible for the formation of the volcanic field will respond passively to or will actively overwhelm ambient tectonism, so its potential contribution should be added to the stress field. However, it will not significantly modify the stress configuration that governs sheet intrusion.

The reason why magma will erupt at one specific point and not at any other on the volcanic field depends on how the magma driving force will act against the stresses configuration inside the lithosphere, and not only at surface. This 3D stress configuration results from a combination of regional and local stress components that define for each point a resultant stress field that will determine whether magma will continue or halt its migration, and the path it will follow in the first case. Through the definition of a series of basic concepts and the application of two very simple models, we have tried to explain the first order physical requirements for magma to migrate inside the host rock and why that migration will follow a certain path or direction and not any other.

Monogenetic volcanic fields are well distributed all around the world and in most geodynamic environments. This implies that the geodynamic constraints (i.e., the regional tectonic stresses) do not determine whether or not such type of volcanism will be present in a specific tectonic setting, but contribute to the distribution and extent of monogenetic volcanic fields. A quantitative comparison of a large number of volcanic fields in different settings was provided by Le Corvec et al. (2013b). Monogenetic fields present different characteristics, including eruption frequency, total erupted volumes, or long term magma fluxes, and may correspond to different tectonic settings. However, magma migration processes are basically the same in all cases and differences in location of volcanic vents, extension of the volcanic fields, eruptive recurrence, or temporal and spatial evolution of vent clustering, may be easily interpreted as due to the different regional stresses (tectonics) governing each of them, as a consequence of the different geodynamic settings where they are located.

We have analyzed how magma acquires the necessary overpressure to cause a hydraulic fracture and migrate through it to shallower levels. We have assumed in all cases a scenario defined by the presence of a reservoir over-pressurized by intrusion of new magma from below, and it being this excess pressure that is the driving force to initiate and drive magma intrusion. Fracture propagation is then controlled by the magma overpressure that results from the excess pressure at the source reservoir and the buoyancy component derived from the density differences between magma and host rock. In this scenario there is always a connection between the source reservoir and the propagating sheet intrusion, so the overpressure required to ensure magma propagation is always depending on the source reservoir minimum volume. We have calculated this minimum volume and the resulting magma overpressures for different host rock configurations, and the results obtained show how shallower reservoirs require larger minimum volumes than deeper reservoirs to acquire the necessary magma overpressures to sustain sheet intrusion up to the surface. This is due to the fact that the buoyancy component of the magma overpressure is less effective at shallow depths, so the magma excess pressure (i.e., volume) in the reservoir needs to be higher. From these results it is also important to note the role of the density contrast between magma and host rock, as this also determines the magma excess pressure that will be necessary to ensure effective magma migration and, consequently, will determine the minimum volume the source reservoir will need to have. Magma migration through sheet intrusions in an anisotropic and heterogeneous host rock will be governed by the same principles but requiring higher overpressures to surpass stress barriers that may inhibit magma movement. This will normally imply changes in the direction of the intrusion, so making its paths toward shallower levels unpredictable, except the exact position of such stress barriers at the interior of the volcanic systems is well known. In a similar way, magma migration can change direction when finding structural discontinuities such as normal faults or tensional fractures that may trap magma facilitating its ascent toward shallower levels.

The scenario for magma intrusion that we have considered contrasts with other accepted models of magma propagation through fissures in which the sheet intrusion disconnects from its source as it migrates. In this case (see Rubin, 1995; Valentine and Gregg, 2008) it is assumed that the exsolved gas concentrates at the dyke tip and exerts the maximum overpressure, thus reducing the pressure at the base of the magma column causing the wall rocks to squeeze inward and push the magma upward. However, geophysical monitoring data recorded during recent eruptions (e.g., El Hierro, Martí et al., 2013; Bardarbunga, Gudmundsson et al., 2014; Sigmundsson et al, 2015) confirm that, at least in these cases, there was a continuous connection between the advancing sheet intrusion and an overpressure source during the whole event.

In this conceptualization of how magma migrates inside the host rock, we should not confuse the capacity for magma to create hydraulic fractures (i.e., magma overpressure) with the orientation that these fractures will have (determined by the external stress field components). From the models presented here we can deduce that only at very shallow depth (<3 km) magma overpressure will be the main component of the local stress field. This is an essential concept when dealing with volcano monitoring and eruption forecasting. Most volcanic models or methods that have been developed (Connor, 1990; Connor et al., 1992, 2000; Ho, 1992, 1995; Martin et al., 1994; Ho and Smith, 1998; Connor and Conway, 2000; Alberico et al., 2002; Martí and Felpeto, 2010; Cappello et al., 2012; Selva et al., 2012; Bartolini et al., 2013; Becerril et al., 2013a; Bevilacqua et al., 2015) consider the observable tectonic structures (eruptive fissures, joints, faults, dykes, sills, lineations, vent location) as indicators of paleostresses, so when they are combined with the age at which they formed we can obtain a picture of the stress evolution with time at surface. In this sense, the youngest structures will indicate the most recent stress configuration, and will define the areas with higher probabilities of hosting new vents. According to this assumption and regardless of the interpolation method used to estimate the spatial probabilities, the areas including the most recent structures will receive higher susceptibility values. And this would be a good approach and a reliable result if magma would ascent vertically from the source. However, none of these models consider the regional stress field or the variations of stress with depth. This implies that the result obtained may not be sufficiently precise according to the level of uncertainty that is acceptable in a hazard assessment. In fact, well monitored recent eruptions (e.g., Bardarbunga, El Hierro), even if they do not perfectly represent pure monogenetic eruptions, showed how magma may migrate horizontally for long distances inside the volcanic system, thus making the short term estimate of volcanic susceptibility (i.e., considering monitoring information, see Sobradelo and Martí, 2015) very challenging. Also, previous studies (e.g., Maccaferri et al., 2011; Menand, 2011; Taisne et al., 2011; Gudmundsson, 2011b) and the models presented here indicate how an ascending magma path may be diverted or arrested in a heterogeneous and anisotropic lithosphere, depending on how regional and local stress components will

be distributed. Moreover, it is worth mentioning that each new intrusion episode (ending or not with an eruption) may induce changes in the local stress distribution creating new stress barriers that did not exist before and that may affect magma movement in further intrusions.

Therefore, one of the aims of this review has been to add some basic physics to the geological record based hazard models that are being used in monogenetic volcanism, in order to help understanding how they work and the uncertainty that their results may have associated with them. In fact, if volcanoes are clustered or aligned, it is reasonable to use that information to weight spatial probabilities for future events, but it is also important to know that the conditions for such particular distribution of vents depend on how stresses distribute inside the volcanic system and not only at surface. Unfortunately, the uncertainty in the deep subsurface processes and material properties is quite large and largely irreducible. The very basic dyke models we have provided in this review may not help to reduce the uncertainty in forecasting monogenetic eruptions, but we hope they will help to better understand which is the source of that uncertainty.

Of course, if no changes occur in the distribution of regional and gravitation stresses in the area between two successive eruptions, we may consider that magma migration will follow a similar path in both if it starts from a source located in a similar position. So, consequently, the next eruption may occur close to the previous one. The difference in the final position of the new vent may be caused by the influence of very shallow stress barriers created by the intrusion(s) remaining from the last eruptive event. However, the occurrence of clusters of vents of different location and age suggests the existence of significant stress changes at a timescale longer than the eruptive recurrence of the system. Therefore, eruptions occurring under the same regional stress configuration will tend to vent one close to the other, but when tectonic changes have occurred the location of vents will probably change, clustering in another sector of the volcanic field, as it is observed in many monogenetic fields. So, if the time scale for a hazard forecast is short compared to the time scale for changes in the ambient stresses and material properties and magma generation at depth, then a hazard forecast based on a sufficient portion of the history of the volcanic field is reasonable. Anyway, it will be also necessary to assess whether the behavior of the volcanic field has changed over long time scales, and whether there is evidence for sufficient change in a very recent time scale that is not yet reflected in the pattern of volcanism (e.g., Connor et al., 2000, 2009, and references herein).

As we have explained before, when we conduct long term hazard assessment one of the first actions we have to undertake is the evaluation of spatial probability of vent opening. This task will essentially be undertaken based on structural indicators such as position of vents, eruptive fissures, fractures, faults, and dykes, which will be computationally weighted according to their relative age. It will be also important to consider the current configuration of regional horizontal stresses and gravitational stresses in case of abrupt topographies. However, in a volcanic crisis we will need to systematically update this information in real time as soon as monitoring information arrives (see

Sobradelo and Martí, 2015; Bartolini et al., 2016), as the evolution in the position of magma at depth and its potential arrival at surface may change with respect to what was predicted in the long-term susceptibility analysis. Anticipating possible changes of the unrest activity and, consequently, of the potential location of a future vent, will depend on the characteristics of our monitoring network, but also on the knowledge we may have of the internal structure of the volcanic system. The better this is, the more accurate (i.e., less uncertainty) will result our forecasting of the possible eruption.

FINAL REMARKS

Monogenetic volcanic fields are not easy to forecast due to the apparent random character of magma migration inside them. Even during unrest episodes, in which we have real time monitoring data, it is not an easy task to forecast well in advance where the new vent will form, as drastic changes in the direction of magma propagation may occur due the presence of unforeseen stress barriers at the interior of the volcanic system. This review has intended to clarify some basic aspects of magma transport in monogenetic fields, in order to help understanding the sources of uncertainty associated with eruption forecasting in such systems. It is obvious that the more information we will be able to provide on the internal structure of monogenetic fields, the better will be the interpretation of unrest episodes and the anticipation to future eruptions. Unfortunately, it is not easy to know how crustal stresses change with depth and where significant stress barriers may be located in such volcanic systems. Therefore, it is worth insisting on the need to combine geological studies, aimed at characterizing the nature and age of the main structural features observable at surface, together with geophysical studies (e.g., seismic, magnetotelluric, and electric tomographies, high resolution gravimetry, etc) imaging the interior of the volcanic systems, as well as with geodynamic models on regional stresses, to better characterize vents distribution in monogenetic fields. Moreover, the information provided by such multidisciplinary studies should be incorporated and computed into the long and short term susceptibility analysis of such volcanic systems in order to get more precise hazard assessments and, thus, to be able to forecast more accurately what may happen in case of new eruptions.

AUTHOR CONTRIBUTIONS

All authors have participated in the elaboration of this study. JM and CL have written the final version of the manuscript and have participated, together with SB in the elaboration of the models. SB, LB, and AG have elaborated the analysis of different natural examples and have contributed to the preparation of all preliminary drafts.

ACKNOWLEDGMENTS

This research was funded by the European Commission (FP7 Theme: ENV.2011.1.3.3-1; Grant 282759: VUELCO and EC ECHO Grant SI2.695524: VeTOOLS). AG thanks the support provided by the Ramón y Cajal research program (RYC-2012-11024). We thank Gregg Valentine and Alessandro Tibaldi for their useful and constructive reviews.

REFERENCES

Alberico, I., Lirer, L., Petrosino, P., and Scandone, R. (2002). A methodology for the evaluation of long-term volcanic risk from pyroclastic flows in Campi Flegrei (Italy). *J. Volcanol. Geotherm. Res.* 116, 63–78. doi: 10.1016/s0377-0273(02)00211-1

Albert, H., Costa, F., and Martí, J. (2015). Timing of magmatic processes and unrest associated with mafic historical monogenetic eruptions in Tenerife Island. *J. Petrol.* 56, 1945–1966. doi: 10.1093/petrology/egv058

Albert, H., Costa, F., and Martí, J. (2016). Years to weeks of seismic unrest and magmatic intrusions precede monogenetic eruptions. *Geology* 44, 211–214. doi: 10.1130/g37239.1

Ayele, A., Jacques, E., Kassim, M., Kidane, T., Omar, A., Tait, S., et al. (2007), The volcano-seismic crisis in Afar, Ethiopia, starting September 2005. *Earth Planet. Sci. Lett.* 255, 177–187.

Bacon, C. R., Bruggman, P. E., Christiansen, R. L., Clynne, M. A., Donnelly-Nolan, J. M., and Hildreth, W. (1995). Primitive magmas at fives Cascade volcanic fields: melts from hot, heterogeneous sub-arc mantle. *Can. Mineral.* 35, 397–423.

Barde-Cabusson, S., Bolós, X., Pedrazzi, D., Lovera, R., Serra, G., Martí, J., et al. (2013). Electrical resistivity tomography revealing the internal structure of monogenetic volcanoes. *Geophys. Res. Lett.* 40, 2544–2549. doi: 10.1002/grl.50538

Bartolini, S., Cappello, A., Martí, J., and Del Negro, C. (2013). QVAST: a new Quantum GIS plugin for estimating volcanic susceptibility. *Nat. Hazards Earth Syst. Sci.* 13, 3031–3042.

Bartolini, S., Sobradelo, R., and Martí, J. (2016). ST-HASSET for volcanic hazard assessment: a Python tool for evaluating the evolution of unrest indicators. *Comput. Geosci.* 93, 77–87. doi: 10.1016/j.cageo.2016.05.002

Bebbington, M. S., and Cronin, S. (2011). Spatio-temporal hazard estimation in the Auckland Volcanic Field, New Zealand, with a new event-order model. *Bull. Volcanol.* 73, 55–72. doi: 10.1007/s00445-010-0403-6

Becerril, L., Cappello, A., Galindo, I., Neri, M., and Del Negro, C. (2013a). Spatial probability distribution of future volcanic eruptions at El Hierro Island (Canary Islands, Spain). *J. Volcanol. Geotherm. Res.* 257, 21–30.

Becerril, L., Galindo, I., Gudmundsson, A., and Morales, J. M. (2013b). Depth of origin of magma in eruptions. *Sci. Rep.* 3:2762.

Bevilacqua, A., Isaia, R., Neri, A., Vitale, S., Aspinall, W. P., Bisson, M., et al. (2015). Quantifying volcanic hazard at Campi Flegrei caldera (Italy) with uncertainty assessment: 1. Vent opening maps. *J. Geophys. Res. Solid Earth* 120, 2309–2329. doi: 10.1002/2014JB011775

Blaikie, T. N., Ailleres, L., Betts, P. G., and Cas, R. A. F. (2014). A geophysical comparison of the diatremes of simple and complex maar volcanoes, Newer Volcanics Province, south-eastern Australia. *J. Volcanol. Geoth. Res.* 276, 64–81. doi: 10.1016/j.jvolgeores.2014.03.001

Blake, S. (1981). Volcanism and dynamics of open magma chambers. *Nature* 289, 783–785. doi: 10.1038/289783a0

Bolós, X., Martí, J., Becerril, L., Planagomà, L., Grosse, P., and Barde-Cabusson, S. (2015). Volcano-structural analysis of La Garrotxa Volcanic Field (NE Iberia): implications for the plumbing system. *Tectonophysics* 642, 58–70. doi: 10.1016/j.tecto.2014.12.013

Bolós, X., Barde-Cabusson, S., Pedrazzi, D., Martí, J., Casas, A., Himi, M., et al. (2012). Investigation of the inner structure of La Crosa de Sant Dalmai maar (Catalan Volcanic Zone, Spain). *J. Volcanol. Geotherm. Res.* 247–248, 37–48. doi: 10.1016/j.jvolgeores.2012.08.003

Bower, S., and Woods, A. (1997). Control of magma volatile content and chamber depth on the mass erupted during explosive volcanic eruptions. *J. Geophys. Res.* 102, 10273–10290. doi: 10.1029/96jb03176

Brenna, M., Cronin, S. J., Németh, K., Smith, I. E. M., and Sohn, Y. K. (2011). The influence of magma plumbing complexity on monogenetic eruptions, Jeju Island, Korea. *Terra Nova* 23, 70–75. doi: 10.1111/j.1365-3121.2010.00985.x

Cappello, A., Neri, M., Acocella, V., Gallo, G., Vicari, A., and Del Negro, C. (2012). Spatial vent opening probability map of Etna volcano (Sicily, Italy). *Bull. Volcanol.* 74, 2083–2094. doi: 10.1007/s00445-012-0647-4

Cassidy, J., and Locke, C. A. (2010). The Auckland volcanic field, New Zealand: geophysical evidence for structural and spatio-temporal relationships. *J. Volcanol. Geotherm. Res.* 195, 127–137. doi: 10.1016/j.jvolgeores.2010.06.016

Connor, C. B. (1990). Cinder cone clustering in the transMexican Volcanic Belt: implications for structural and petrologic models. *J. Geophys. Res.* 95, 19395–19405. doi: 10.1029/jb095ib12p19395

Connor, C. B., Chapman, N. A., and Connor, L. J. (2009). *Volcanic and Tectonic Hazard Assessment for Nuclear Facilities*. Cambridge, UK: Cambridge University Press. doi: 10.1017/cbo9780511635380

Connor, C. B., Condit, C. D., Crumpler, L. S., and Aubele, J. C. (1992). Evidence of regional structural controls on vent distribution: springerville Volcanic Field, Arizona. *J. Geophys. Res.* 97, 12349–12359. doi: 10.1029/92jb00929

Connor, C. B., and Conway, F. M. (2000). "Basaltic volcanic fields," in *Encyclopedia of Volcanoes*, ed H. Sigurdsson (New York, NY: Academic Press), 331–343.

Connor, C., Stamatakos, J. A., Ferrill, D. A., Hill, B. E., Ofoegbu, G. I., Conway, F. M., et al. (2000). Geologic factors controlling patterns of small-volume basaltic volcanism: application to a volcanic hazards assessment at Yucca Mountain, Nevada. *J. Geophys. Res.* 105, 417–432. doi: 10.1029/1999jb900353

Dahm, T. (2000). Numerical simulations of the propagation path and the arrest of fluid-filled fractures in the Earth. *Geophys. J. Int.* 141, 623–638. doi: 10.1046/j.1365-246x.2000.00102.x

Delaney, P. T., and Pollard, D. D. (1981). Deformation of host rocks and flow of magma during growth of minette dykes and breccia-bearing intrusions near Ship Rock, New Mexico, U.S. *Geol. Surv. Prof. Pap.* 1202, 1981.

Delaney, P. T., and Pollard, D. D. (1982). Solidification of basaltic magma during flow in a dyke. *Am. J. Sci.* 282, 856–885.

Delaney, P. T., Pollard, P. P., Ziony, J. I., and McKee, E. H. (1986). Field relations between dykes and joints: emplacement processes and paleostress analysis. *J. Geophys. Res.* 91, 4920–4938.

Dundurs, J. (1969). Edge-bonded dissimilar orthogonal wedges. *J. Appl. Mech.* 36, 650–652.

Folch, A., and Martí, J. (1998). The generation of overpressure in felsic magma chambers by replenishment. *Earth Planet. Sci. Lett.* 163, 301–314. doi: 10.1016/S0012-821X(98)00196-4

Fossen, H. (2016). *Structural Geology*. Cambridge: Cambridge University Press.

Gaffney, E. S., Damjanac, B., and Valantine, G. A. (2007). Localization of volcanic activity: 2. Effects of pre-existing structure. *Earth Planet. Sci. Lett.* 263, 323–338.

Garcia, M. O., Hulsebosch, T. P., and Rhodes, J. M. (1995). Olivine-rich submarine basalts from the southwest rift zone of Mauna Loa volcano: implications for magmatic processes and geo-chemical evolution. *Geophys. Monogr.* 92, 219–239. doi: 10.1029/gm092p0219

Gudmundsson, A. (1990). Emplacement of dykes, sills and crustal magma chambers at divergent plate boundaries. *Tectonophysics* 176, 257–275. doi: 10.1016/0040-1951(90)90073-H

Gudmundsson, A. (2011a). *Rock Fractures in Geological Processes*. Cambridge: Cambridge University Press. doi: 10.1017/CBO9780511975684

Gudmundsson, A. (2011b). Deflection of dykes into sills at discontinuities and magma-chamber formation *Tectonophysics* 500, 50–64.

Gudmundsson, A. (2012). Magma chambers: formation, local stresses, excess pressures, and compartments. *J. Volcanol. Geotherm. Res.* 237–238, 19–41. doi: 10.1016/j.jvolgeores.2012.05.015

Gudmundsson, A., and Brenner, S. L. (2005). On the conditions of sheet injections and eruptions in stratovolcanoes. *Bull. Volcanol.* 67, 768–782. doi: 10.1007/s00445-005-0433-7

Gudmundsson, A., Lecour, N., Mohajeri, N., and Thordarson, T. (2014). Dyke emplacement at Bardarbunga, Iceland, induces unusual stress changes, caldera deformation, and earthquakes. *Bull. Volcanol.* 76:869.

Gudmundsson, A., Marinoni, L. B., and Martí, J. (1999). Injection and arrest of dykes: implications for volcanic hazards. *J. Volcanol. Geotherm. Res.* 88, 1–13. doi: 10.1016/s0377-0273(98)00107-3

Gudmundsson, A., and Philipp, S. L. (2006). How local stress fields prevent volcanic eruptions. *J. Volcanol. Geotherm. Res.* 158, 257–268. doi: 10.1016/j.jvolgeores.2006.06.005

He, M. Y., Evans, A. G., and Hutchinson, J. W. (1994). Crack deflection at an interface between dissimilar elastic materials: role of residual stresses. *Int. J. Solids Struct.* 31, 3443–3455. doi: 10.1016/0020-7683(94)90025-6

He, M. Y., and Hutchison, J. W. (1989). Crack deflection at an interface between dissimilar elastic materials. *Int. J. Solids Struct.* 25, 1053–1067.

Heidbach, O., Reinecker, J., Tingay, M., Müller, B., Sperner, B., Fuchs, K. et al. (2007). Plate boundary forces are not enough: second- and third-order stress patterns highlighted in the World Stress Map database. *Tectonics* 26:TC6014. doi: 10.1029/2007TC002133

Hernando, I. R., Aragon, E., Frei, R., Gonzalez, P. D., and Spakman, W. (2014). Constraints on the origin and evolution of magmas in the Payún Matrú Volcanic Field, Quaternary Andean back-arc of Western Argentina. *J. Petrol.* 55, 209–239. doi: 10.1093/petrology/egt066

Ho, C. H. (1992). Risk assessment for the Yucca Mountain high-level nuclear waste repository site: estimation of volcanic disruption. *Math. Geol.* 24, 347–364. doi: 10.2172/196582

Ho, C. H. (1995). Sensitivity in volcanic hazard assessment for the Yucca Mountain high-level nuclear waste repository site: the model and the data. *Math. Geol.* 27, 239–258. doi: 10.1007/bf02083213

Ho, C. H., and Smith, E. I. (1998). A spatial-temporal/3-D model for volcanic hazard assessment: application to the Yucca Mountain region, Nevada. *Math. Geol.* 30, 497–510.

Jaeger, J. C., and Cook, N. G. W. (1979). *Fundamentals of Rock Mechanics, 3rd Edn.* London: Chapman ang Hall.

Klügel, A., Hansteen, T. H., and Galipp, K. (2005). Magma storage and underplating beneath Cumbre Vieja Volcano, La Palma (Canary Islands). *Earth Planet. Sci. Lett.* 236, 211–226. doi: 10.1016/j.epsl.2005.04.006

Klugel, A., Longpré, M. A., García-Cañada, L., and Stix, J. (2015). Deep intrusions, lateral magma transport and related uplift at ocean island volcanoes. *Earth Planet. Sci. Lett.* 431, 140–149. doi: 10.1016/j.epsl.2015.09.031

Le Corvec, N., Bebbington, M. S., Linsay, J. M., and McGee, L. E. (2013a). Age, distance, and geochemical evolution within a monogenetic volcanic field: analyzing patterns in the Auckland Volcanic Field eruption sequence. *Geochem. Geophys. Geosyst.* 14, 3648–3665.

Le Corvec, N., Spörli, K. B., Rowland, J., and Lindsay, J. (2013b). Spatial distribution and alignments of volcanic centers: clues to the formation of monogenetic volcanic fields. *Earth Sci. Rev.* 124, 96–114.

Le Corvec, N., Menand, T., and Lindsay, J. (2013c). Interaction of ascending magma with pre-existing crustal fractures in monogenetic basaltic volcanism: an experimental approach. *J. Geophys. Res. Solid Earth* 118, 968–984.

Lister, J., and Kerr, R. (1991). Fluid-mechanical models of crack propagation and their application to magma transport in dykes. *J. Geophys. Res.* 94, 10049–10077. doi: 10.1029/91jb00600

Lorenz, V. (1986). On the growth of maars and diatremes and its relevance to the formation of tuff rings. *Bull. Volcanol.* 48, 265–274. doi: 10.1007/bf01081755

Maaloe, S. (1985). Principles of Igneous Petrology. Berlin: Springer-Verlag. doi: 10.1007/978-3-642-49354-6

Maccaferri, F., Bonafede, M., and Rivalta, E. (2010). A numerical model of dyke propagation in layered elastic media. *Geophys. J. Int.* 180, 1107–1123. doi: 10.1111/j.1365-246X.2009.04495.x

Maccaferri, F., Bonafede, M., and Rivalta, E. (2011). A quantitative study of the mechanisms governing dyke propagation, dyke arrest and sill formation. *J. Volcanol. Geotherm. Res.* 208, 39–50.

Martí, J., and Felpeto, A. (2010). Methodology for the computation of volcanic susceptibility. An example for mafic and felsic eruptions on Tenerife (Canary Islands). *J. Volcanol. Geotherm. Res.* 195, 69–77.

Martí, J., and Geyer, A. (2009). Central vs flank eruptions at Teide–Pico Viejo twin stratovolcanoes (Tenerife, Canary Islands). *J. Volcanol. Geotherm. Res.* 181, 47–60.

Martí, J., Geyer, A., Andujar, J., Teixidó, F., and Costa, F. (2008). Assessing the potential for future explosive activity from Teide–Pico Viejo stratovolcanoes (Tenerife, Canary Islands). *J. Volcanol. Geotherm. Res.* 178, 529–542.

Martí, J., Pinel, V., López, C., Geyer, A., Abella, R., Tárraga, M., et al. (2013). Causes and mechanisms of the 2011–2012 El Hierro (Canary Islands) submarine eruption. *J. Geophys. Res. Solid Earth* 118, 823–839.

Martí, J., Planagumà, L., Geyer, A., Canal, E., and Pedrazzi, D. (2011). Complex interaction between Strombolian and phreatomagmatic eruptions in the Quaternary monogenetic volcanism of the Catalan Volcanic Zone (NE of Spain). *J. Volcanol. Geotherm. Res.* 201, 178–193.

Martin, A. J., Umeda, K., Connor, C. B., Weller, J. N., Zhao, D., and Takahashi, M. (1994). Modeling long-term volcanic hazards through Bayesian inference: an example from the Tohuku volcanic arc Japan. *J. Geophys. Res.* 109, B10208.

McKenzie, D. (1984). The generation and compaction of partially molten rock. *J. Petrol.* 25, 713–765. doi: 10.1093/petrology/25.3.713

McKenzie, D. (1985). The extraction of magma from the crust and mantle. *Earth Planet. Sci. Lett.* 74, 81–91.

Menand, T. (2008). The mechanics and dynamics of sills in layered elastic rocks and their implications for the growth of laccoliths and other igneous complexes. *Earth Planet. Sci. Lett.* 267, 93–99. doi: 10.1016/j.epsl.2007.11.043

Menand, T. (2011). Physical controls and depth of emplacement of igneous bodies: a review. *Tectonophysics* 500, 11–19. doi: 10.1016/j.tecto.2009.10.016

Menand, T., Daniels, K. A., and Benghiat, P. (2010). Dyke propagation and sill formation in a compressive tectonic environment. *J. Geophys. Res.* 115, B08201. doi: 10.1029/2009jb006791

Middleton, G. V., and Wilcock, P. R. (1994). *Mechanics in the Earth and Environmental Sciences*. Cambridge: Cambridge University Press.

Mrlina, J. H., Kämpf, C., Kroner, J., Mingram, M., Stebich, A., Seidl, M. et al. (2009). Discovery of the first Quaternary maar in the Bohemian Massif, Central Europe, based on combined geophysical and geological surveys. *J. Volcanol. Geoth. Res.* 182, 97–112. doi: 10.1016/j.jvolgeores.2009.01.027

Muller, J. R., Ito, G., and Martel, S. J. (2001). Effects of volcano loading on dyke propagation in an elastic half-space. *J. Geophys. Res.* 106, 11101–11113.

Murase, T., and Mcbirney, A. R. (1973). Properties of some common igneous rocks and their melts at high temperatures. *Geol. Soc. Amer. Bull.* 84, 3563–3592. doi: 10.1130/0016-7606(1973)84<3563:poscir>2.0.co;2

Németh, K. (2010). "Monogenetic volcanic fields: Origin, sedimentary record, and relationship with polygenetic volcanism," in *What Is a Volcano?* eds E. Cañón-Tapia and A. Szakács (Boulder, CO: Geological Society of America), 43–66. doi: 10.1130/2010.2470(04)

Németh, K., and Kereszturi, G. (2015). Monogenetic volcanism: personal views and discussion. *Int. J. Earth Sci.* 104, 2131–2146. doi: 10.1007/s00531-015-1243-6

Park, R. G. (1988). *Geological Structures and Moving Plates*. Glasgow: Blackie & Sons Ltd; Bishopbriggs. doi: 10.1007/978-94-017-1685-7

Pasquarè, G., Tibaldi, A., Attolini, C., and Cecconi, G. (1988). Morphometry, spatial distribution and tectonic control of Quaternary volcanoes in norhern Michoacan, Mexico. *Rend. Soc. It. Min. Petr.* 43, 1215–1225.

Pasquarè, F. A., and Tibaldi, A. (2007). Structure of a sheet-laccolith system revealing the interplay between tectonic and magma stresses at Stardalur Volcano, Iceland. *J. Volcanol. Geotherm. Res.* 161, 131–150.

Pinel, V., and Jaupart, C. (2004). Magma storage and horizontal dyke injection beneath a volcanic edifice. *Earth Planet. Sci. Lett.* 221, 245–262. doi: 10.1016/S0012-821X(04)00076-7

Pollard, D. D. (1969). *Aspects of the Mechanics of Sheets Intrusions*. Unpublished M. Sc. dissertation, University of London, UK

Pollard, D. D. (1973). Derivation and evaluation of a mechanical model for sheet intrusions. *Tectonophysics* 1, 233–269. doi: 10.1016/0040-1951(73)90021-8

Pollard, D. D., and Muller, O. H. (1976). The effects of gradients in regional stress and magma pressure on the form of sheet intrusions in cross section. *J. Geophys. Res.* 81, 975–984. doi: 10.1029/jb081i005p00975

Pollard, D. D., and Segall, P. (1987). "Theoretical displacements and stresses near fractures in rocks: With applications to faults, joints, veins, dykes, and solutions surfaces," in *Fracture Mechanics of Rock*, ed B. B. Atkinson (San Diego, CA; Academic), 277–349. doi: 10.1016/b978-0-12-066266-1.50013-2

Price, N. J., and Cosgrove, J., W (1990). *Analysis of Geological Structures*. Cambridge: Cambridge University Press.

Rice, J. R. (1980). "Mathematical analysis in the mechanics of fracture," in *Fracture, An Advanced Treatise*, Vol II, ed H. Liebowitz (New York, NY: Academic Press), 191–311.

Rivalta, E., Taisne, B., Bunger, A. P., and Katz, R. F. (2015). A review of mechanical models of dyke propagation: schools of thought, results and future directions. *Tectonophysics* 638, 1–42. doi: 10.1016/j.tecto.2014.10.003

Roman, D. C., and Heron, P. (2007). Effect of regional tectonic setting on local fault response to episodes of volcanic activity. *Geophys. Res. Lett.* 34, L13310. doi: 10.1029/2007gl030222

Rowe, M. C., Peate, D. W., and Ukstins-Peate, I. (2011). An investigation into the nature of the magmatic plumbing system at Paricutin volcano, Mexico. *J. Petrol.* 52, 2187–2220. doi: 10.1093/petrology/egr044

Rubin, A. M. (1993a). Tensile fracture of rock at high confining pressure: implications for dyke propagation. *J. Geophys. Res.* 98, 919–935.

Rubin, A. M. (1993b). On the thermal viability of dykes leaving magma chambers. *Geophys. Res. Lett.* 20, 257–260.

Rubin, A. M. (1993c). Dykes vs. diapirs in viscoelastic rock, Earth Planet. *Sci. Lett.* 119, 641–659.

Rubin, A. M. (1995). Propagation of magma-filled cracks. *Annu. Rev. Earth Planet. Sci.* 8, 287–336. doi: 10.1146/annurev.ea.23.050195.001443

Rubin, A. M. (1998). Dyke ascent in partially molten rock. J. Geophys. *Res.* 103, 20, 901–920.

Schrank, C. E., Fusseis, F., Karrech, A., and Regenauer-Lieb, K. (2012), Thermal-elastic stresses and the criticality of the continental crust. *Geochem. Geophys. Geosyst.* 13, Q09005.

Selva, J., Orsi, G., Di Vito, M., Marzocchi,W., and Sandri, L. (2012). Probability hazard map for future vent opening at the Campi Flegrei caldera, Italy. *Bull. Volcanol.* 74, 497–510. doi: 10.1007/s00445-011-0528-2

Shapiro, S. A. (2015). *Fluid Induced Seismicity*. Cambridge: Cambridge University Press. doi: 10.1017/cbo9781139051132

Sigmundsson, F. et al. (2015). Segmented lateral dyke growth in a rifting event at Bardarbunga volcanic system, Iceland. *Nature* 517, 191–195.

Sobradelo, R., and Martí, J. (2015). Short-term volcanic hazard assessment through Bayesian inference: retrospective application to the Pinatubo 1991 volcanic crisis. *J. Volcanol. Geotherm. Res.* 290:111. doi: 10.1016/j.jvolgeores.2014.11.011

Spera, F. J. (1980). "Aspects of magma transport." in *Physics of Magmatic Processes*, ed R. B Hargraves (Princeton, NJ: Princeton University Press), 265–323.

Stroncik, N. A., Klügel, A., and Hansteen, T. H. (2009). The magmatic plumbing system beneath El Hierro (Canary Islands): constraints from phenocrysts and naturally quenched basaltic glasses in submarine rocks. *Contrib. Mineral. Petrol.* 157, 593–607. doi: 10.1007/s00410-008-0354-5

Taisne, B., and Jaupart, C. (2009). Dyke propagation through layered rocks. *J. Geophys. Res.* 114, B09203.

Taisne, B., Tait, S., and Jaupart, C. (2011). Conditions for the arrest of a vertical propagating dyke. *Bull. Volcanol.* 73, 191–204. doi: 10.1007/s00445-010-0440-1

Takada, A. (1989). Magma transport and reservoir formation by a systems of propagating cracks. *Bull. Volcanol.* 52, 118–126. doi: 10.1007/BF00301551

Takada, A. (1994). The influence of regional stress and magmatic input on styles of monogenetic and polygenetic volcanism. *J. Geophys. Res.* 99, 563–513. doi: 10.1029/94jb00494

Thirlwall, M. F., Singer, B. S., and Marriner, G. F. (2000). 39Ar±40Ar ages and geochemistry of the basaltic shield stage of Tenerife, Canary Islands, Spain. *J. Volcanol. Geother. Res.* 103, 247–297.

Tibaldi, A. (1995). Morphology of pyroclastic cones and tectonics. *J. Geophys. Res.* 100, 24521–24535. doi: 10.1029/95jb02250

Tibaldi, A., and Lagmay, A. M. F. (2006). Interaction between volcanoes and their basement. *J. Volcanolo. Geother. Res.* 158, 1–5. doi: 10.1016/j.jvolgeores.2006.04.011

Tibaldi, A., and Pasquarè, F. (2008). A new mode of inner volcano growth: the "flower intrusive structure". *Earth Planet. Sci. Lett.* 271, 202–208. doi: 10.1016/j.epsl.2008.04.009

Tibaldi, A., Vezzoli, L., Pasquarè, F. A., and Rust, D. (2008). Strike-slip fault tectonics and the emplacement of sheet-laccolith systems: the thverfell case study (SW Iceland). *J. Struct. Geol.* 30, 274–290. doi: 10.1016/j.jsg.2007.11.008

Touloukian, Y. S., Judd,W. R., and Roy, R. F. (1989). "Physical properties of rocks and minerals," in *CINDAS Data Series on Material Properties, Group II Properties of special materials, Vol 1-2*. New York, NY: McGraw-Hill.

Traversa, P., Pinel, V., and Grasso, J. (2010). A constant influx model for dyke propagation: implications for magma reservoir dynamics. *J. Geophys. Res.* 115, B01201.

Turcotte, D. L., and Schubert, G. (1982). *Geodynamics: Applications of Continuum Physics to Geological Problems*. New York, NY: John Wiley.

Valentine, G. A., and Gregg, T. K. P. (2008). Continental basaltic volcanoes—Processes and problems. *J. Volcanol. Geother. Res.* 177, 857–873. doi: 10.1016/j.jvolgeores.2008.01.050

Valentine, G. A., and Hirano, N. (2010). Mechanisms of low-flux intraplate volcanic fields – basin and range (North America) and northwest Pacific Ocean. *Geology* 38, 55–58. doi: 10.1130/g30427.1

Valentine, G. A., and Perry, F. V. (2007). Tectonically controlled, time-predictable basaltic volcanism from a lithospheric mantle source (central Basin and Range Province, USA). *Earth Planet. Sci. Lett.* 261, 201–216. doi: 10.1016/j.epsl.2007.06.029

Walker, G. P. L. (2000). "Basaltic volcanoes and volcanic systems," in *Encyclopedia of Volcanoes*, ed H. Sigurdsson (San Francisco, CA: Academic Press), 283–289.

Wood, C. A. (1980). Morphometric evolution of cinder cones. *J. Volcanol. Geother. Res.* 7, 387–413. doi: 10.1016/0377-0273(80)90040-2

Wright, T. J., Ebinger, C., Biggs, J., Ayele, A., Yirgu, G., Keir, D., et al. (2006). Magma-maintained rift segmentation at continental rupture in the 2005 Afar dyking episode. *Nature* 442, 291294. doi: 10.1038/nature04978

Zang, A., and Stephansson, O. (2010). *Stress Field of the Earth's Crust*. Heidelberg: Springer. doi: 10.1007/978-1-4020-8444-7

Zoback, M. D., and Zoback, M. L. (2002). "Stress in the Earth's lithosphere," in *Encyclopedia of Physical Science and Technology, 3rd Edn.*, ed R. A. Meyers (San Diego, CA: Academic Press), 143–154.

Zoback, M. L. (1992). First- and second-order patterns of stress in the lithosphere: the world stress map project. *J. Geophys. Res.* 97, 11703–11728. doi: 10.1029/92jb00132

Conflict of Interest Statement: The authors declare that the research was conducted in the absence of any commercial or financial relationships that could be construed as a potential conflict of interest.

11

Initial Opening of the Eurasian Basin, Arctic Ocean

Kai Berglar*, Dieter Franke, Rüdiger Lutz, Bernd Schreckenberger and Volkmar Damm

Federal Institute for Geosciences and Natural Resources (BGR), Hannover, Germany

Analysis of the transition from the NE Yermak Plateau into the oceanic Eurasian Basin sheds light on the Paleocene formation of this Arctic basin. Newly acquired multichannel seismic data with a 3600 m long streamer shot during ice-free conditions enables the interpretation of crustal structures. Evidence is provided that no major compressional deformation affected the NE Yermak Plateau. The seismic data reveal that the margin is around 80 km wide and consists of rotated fault blocks, major listric normal faults, and half-grabens filled with syn-rift sediments. Taking into account published magnetic and gravimetric data, this setting is interpreted as a rifted continental margin, implying that the NE Yermak Plateau is of continental origin. The transition from the Yermak Plateau to the oceanic Eurasian Basin might be located at a prominent basement high, probably formed by exhumed mantle. In contrast to the Yermak Plateau margin, the North Barents Sea continental margin shows a steep continental slope with a relatively abrupt transition to the oceanic domain. Based on our limited data, we propose a working hypothesis speculating that the initial opening direction of the Eurasian Basin in the Arctic Ocean was highly oblique to the present day seafloor spreading direction.

Keywords: Yermak Plateau, Nansen Basin, Arctic Ocean, magma-poor rifting, continent-ocean transition, exhumed mantle, reflection seismic data

Edited by:
Øyvind Engen,
Statoil ASA, Norway

Reviewed by:
Valerio Acocella,
Roma Tre University, Italy
Deborah R. Hutchinson,
United States Geological Survey, USA

*Correspondence:
Kai Berglar
kai.berglar@bgr.de

Specialty section:
This article was submitted to
Structural Geology and Tectonics,
a section of the journal
Frontiers in Earth Science

INTRODUCTION

Two major oceanic basins are found in the Arctic Ocean (**Figure 1**). While the origin and evolution of the Amerasian Basin is still under discussion, the general evolution of the Eurasian Basin is much better known (e.g., Lawver et al., 2011; Shephard et al., 2013). Well-defined magnetic seafloor spreading anomalies in the Eurasian Basin of the Arctic Ocean from C24 and younger are merely undisputed (Srivastava and Tapscott, 1986; Lawver et al., 2002; Brozena et al., 2003). Accordingly, it is widely accepted that the continental Lomonosov Ridge (**Figure 1**), a major bathymetric elevation, separating the Eurasian from the Amerasian Basin (Jokat et al., 2013) was split off the Eurasia continent and migrated northward to its present position since ~53 Ma (e.g., Alvey et al., 2008). The Lomonosov Ridge previously might have formed a continuous structure with the Yermak Plateau (**Figure 1**) but the crustal nature of this major submarine plateau at the North Barents Sea continental margin is under debate (Jackson et al., 1984; Jokat et al., 2008; Geissler et al., 2011). This comes along with a considerable dispute, particularly about the earliest, Paleocene evolution of the Eurasian Basin. The split off of the elongated and about 1500 km long crustal splinter of the Lomonosov Ridge from the North Barents Sea continental margin and its eastern prolongation, the Kara Sea continental margin, is difficult to be explained with current rifting models. An episode of shear or oblique extension has been suggested before breakup to explain the observed narrow symmetric conjugate margins in the Eurasia Basin (Minakov et al., 2012, 2013).

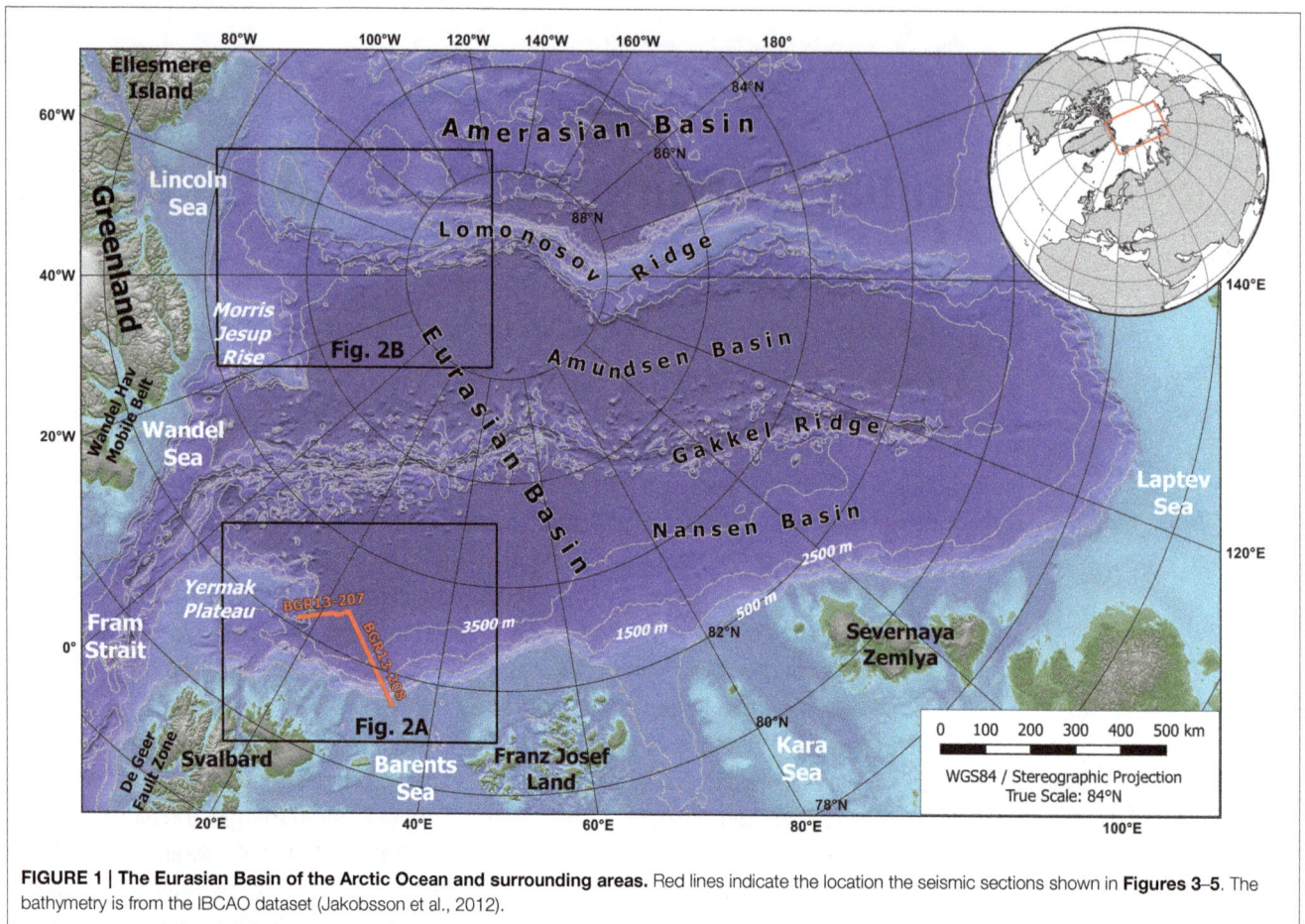

FIGURE 1 | The Eurasian Basin of the Arctic Ocean and surrounding areas. Red lines indicate the location the seismic sections shown in **Figures 3–5**. The bathymetry is from the IBCAO dataset (Jakobsson et al., 2012).

In addition, different interpretations are at hand about the presence of magnetic anomaly C25 (e.g., Brozena et al., 2003; Engen et al., 2008). A problem with this magnetic anomaly is that it may overlap with a major strike-slip fault east of the Yermak Plateau that is thought to have accommodated the northward migration of the Lomonosov Ridge. The presence of such a fault has been suggested on the basis of potential field data (e.g., Brozena et al., 2003). In addition, in settings with minor magma supply it is questionable if there is a sharp contact between continental and oceanic crust. As shown by Bronner et al. (2011) offshore Iberia there is a gradual transition from continental crust, to exhumed mantle, to oceanic crust and pulses of magmatism at breakup can generate anomalies that may be misinterpreted as magnetic seafloor spreading anomalies. The situation is further complicated by the Late Paleocene and early-middle Eocene compressional deformation (Eurekan) that affected wide portions of the Svalbard archipelago (e.g., Eldholm et al., 1987) and may have deformed the junction between the Yermak Plateau and the Eurasian Basin (Brozena et al., 2003; Døssing et al., 2013).

The NE Yermak Plateau and the North Barents Sea continental margin are poorly investigated areas because the nearly permanent ice cover hampers scientific investigations. In this study, newly acquired multichannel seismic reflection (MCS) data, supplemented by potential field data, are used to study the area of the eastern Yermak Plateau, the adjacent Eurasian Basin, and the North Barents Sea continental margin. Structural data reveal the margins architecture and allow speculations about the rifting and break-up of the Eurasian Basin at the junction of the Yermak Plateau with the North Barents Sea continental margin. This study shows that neither subduction nor substantial transpression affected the eastern Yermak Plateau. Rather it is suggested, that the initial extension of the Eurasian Basin took place between the Yermak Plateau and the Lomonosov Ridge at a high angle to the present day orientation of spreading.

GEOLOGICAL SETTING

Eurasian Basin of the Arctic Ocean

During Cretaceous times, North America, Greenland, and Eurasia, including the Lomonosov Ridge and Svalbard, were part of the common land mass of Laurasia. The oldest magnetic anomaly consistently interpreted along the margins of Eurasia and Greenland within the North Atlantic is C24 (e.g., Gaina et al., 2009) with an age of ∼53 Ma from the Paleocene-Eocene transition according to the time-scale after Gee and Kent (2007), which we use in this study. In the Eurasian Basin, Vogt et al. (1979) discussed the possibility of the presence of magnetic anomalies older than C24. However, they did not present a conclusive interpretation in between the North Barents Sea continental margin and magnetic anomaly C24. Several recent magnetic anomaly interpretations and plate models favor the presence of magnetic chron C25 (Brozena et al., 2003; Cochran et al., 2006; Døssing et al., 2014) with an age of ~56 Ma from the Late Paleocene. Brozena et al. (2003) identified anomalies C25 to C15 (56–34.8 Ma) terminating shortly before the NE Yermak Plateau and Morris Jesup Rise flanks, anomalies C12–C8 (30.8–25.9 Ma) spreading apart of the plateaus and extending for another 60–90 km to the SW, and anomaly C5 (9.8 Ma) continuing into the Fram Strait. This interpretation has been widely confirmed by Engen et al. (2008), although they do not interpret the existence of anomaly C25 in the Eurasian Basin. Here, we refer to the interpretation of Brozena et al. (2003). If indeed magnetic anomaly C25 is present in the Eurasian Basin (Brozena et al., 2003) then about 150 km of oceanic crust was created between the Lomonosov Ridge and the Eurasian margin by magnetic chron C24. This is in accordance with structural investigations, where, from the location of the transition to oceanic crust, and by assuming a spreading half-rate of 1 cm/yr, Geissler and Jokat (2004) calculate that seafloor spreading in the Eurasian Basin close to the Yermak Plateau started about 5 Ma before magnetic chron C24.

Close to the Yermak Plateau the magnetic isochrons C25 and C24 change direction from east-west to southwest-northeast (Brozena et al., 2003; **Figure 2A**). More prominently, this trend is observed in the magnetic data between the Morris Jesup Rise and the Lomonosov Ridge in the Lincoln Sea (**Figure 2B**), and is mirrored by gravity data (Døssing et al., 2014). Thus, the earliest magnetic chrons show an extension direction that, close to the Yermak Plateau and Morris Jesup Rise, considerably deviates from the younger chrons. On the Siberian side of the Eurasian Basin, the oldest anomaly is tentatively interpreted as chron C24 (Glebovsky et al., 2006). These observations imply that the opening of the Eurasian Basin commenced on the European side, and might have propagated toward the Siberian side (Cochran et al., 2006) where a major continental rift developed at the shelf, the Laptev Sea Rift (e.g., Franke and Hinz, 2005).

Seafloor spreading in the Eurasian Basin resulted in separation of a narrow continental microplate (Jokat et al., 2013), the Lomonosov Ridge, from the northeastern margin of Eurasia (Lawver et al., 2002; Gaina et al., 2009)—either as a part of the North American plate (Srivastava, 1985) or as an independent plate. Before the opening of the Eurasian Basin, the Lomonosov Ridge was located directly along the North Barents Sea continental margin, with the ridge's western end adjacent to the ancestral Yermak Plateau and Morris Jesup Rise (Brozena et al., 2003).

Prior to Oligocene times, the Yermak Plateau and the Morris Jesup Rise formed a contiguous plateau (Feden et al., 1979; Jackson et al., 1984; Engen et al., 2008), which has been placed northeast of Greenland, when reconstructed (e.g., Srivastava, 1985; Srivastava and Tapscott, 1986; Tessensohn and Piepjohn, 2000). Jointly with the opening of the North Atlantic, Late Paleocene plate motions led to a dextral transcurrent transfer of Svalbard, and probably also the Yermak Plateau and the Morris Jesup Rise, relative to north Greenland (e.g., Srivastava, 1985; Srivastava and Tapscott, 1986; von Gosen and Piepjohn,

2003; Tessensohn et al., 2008). However, not only strike-slip and extensional deformation occurred during the early Paleocene, but also compressional deformation. Late Paleocene and early-middle Eocene compression (Eurekan) north of the evolving North Atlantic Ocean basin created the West Svalbard fold belt (Eldholm et al., 1987). From a geometrical point of view, the Morris Jesup Rise and Yermak Plateau should have experienced the same compressional phase, and Døssing et al. (2013) suggest that significant Eurekan transpression, or possibly subduction (Brozena et al., 2003), took place along the boundary between the Morris Jesup Rise and the Yermak Plateau and the Eurasian Basin.

The youngest evolution of the area is merely undisputed. At 33–35 Ma, the conjugate Yermak Plateau and the Morris Jesup Rise started to break up by SW propagation of seafloor spreading—and magnetic anomaly C7 (24.8 Ma) is the oldest anomaly pair completely separating the two plateaus (Eldholm et al., 1987; Brozena et al., 2003; Engen et al., 2008; Geissler et al., 2011).

Yermak Plateau and Morris Jesup Rise

The crustal fabric of the Yermak Plateau and its conjugate, the Morris Jesup Rise, is not yet fully clear. However, the early interpretation that the plateaus were completely generated by massive extrusions of basalt during the lower and middle Tertiary, and that the origin of this basalt is related to hotspot activity (Feden et al., 1979), has been challenged in recent years. Seismic velocities of the acoustic basement underlying the southern portions of the Yermak Plateau were interpreted by Jackson et al. (1984) as thinned continental crust due to P-wave velocities between ~4.3 and 6.0 km/s. In the northeastern portion of the Yermak Plateau, Jackson et al. (1984) suggested the presence of over-thickened oceanic crust. In the western portion of the Yermak Plateau, up to 82° N, a crust with continental affinities has been inferred from seismic velocities (Ritzmann and Jokat, 2003) and also from the structural grain as derived from a dense grid of reflection seismic data (Jokat et al., 2008). Jokat et al. (2008) confirm that the southern and northwestern Yermak Plateau are made up of attenuated and locally intruded continental crust, based on preserved structures on the Yermak Plateau resulting from the strike-slip movements between North Greenland and Svalbard. Relatively weak magnetic anomalies over this area (Feden et al., 1979), and gravity modeling, support this interpretation and show that the southern Yermak Plateau consists of thinned, ~20 to 25 km thick continental crust (Geissler and Jokat, 2004). Based on sonobuoy-derived P-wave velocities, Geissler et al. (2011) speculate that the northern basement highs on the central Yermak Plateau are the continuation of Paleozoic Svalbard geology, or alternatively, were once part of the Cretaceous shelf in front of the northern tip of Greenland. From the sedimentary cover, these authors also propose that the acoustic basement of the plateau should be older than 33–35 Ma. This view was confirmed by Riefstahl et al. (2013) who suggest from analyzed dredge samples that the basement highs on the central Yermak Plateau consist of thinned continental crust of pre-Devonian age, forming a direct continuation of

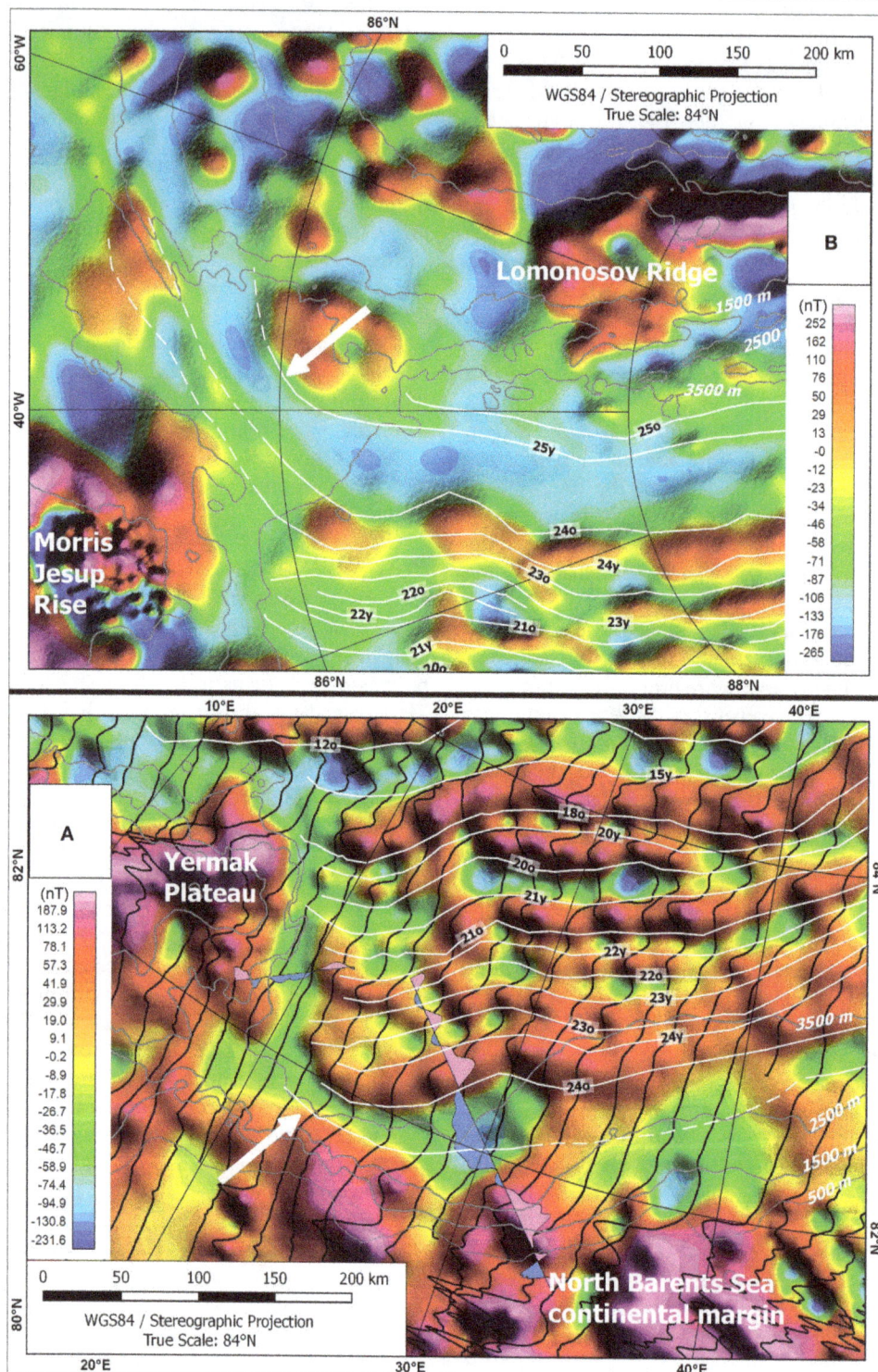

FIGURE 2 | Magnetic anomaly maps of the western Eurasian Basin with superimposed anomaly interpretation from Brozena et al. (2003, white lines).
White arrows indicate the northward bending of anomalies C25? and C24, gray lines the bathymetry from the IBCAO dataset (Jakobsson et al., 2012). **(A)** Newly gridded aeromagnetic line data (US Naval Research Laboratory Arctic surveys 1998/99, Brozena et al. (2003), downloaded from NCEI/NOAA) and raw data used for gridding (black wiggle traces). Also shown are the magnetic data measured along lines BGR13–207 and –208 as filled wiggle traces (see also **Figure 3B**). **(B)** Data from the CampGM compilation (Gaina et al., 2011).

the exposures on northern Svalbard. According to Riefstahl et al. (2013), the stretched continental crust has been strongly affected by alkaline magmatism that took place at ~51 Ma, and is probably also associated with the high-amplitude magnetic anomalies described in the northeastern Yermak Plateau. Thus, it appears plausible that the major portion of the Yermak Plateau up to 82°N is underlain by attenuated continental crust. Also, the basin south of the northeastern Yermak Plateau is characterized as rifted continental crust by Geissler and Jokat (2004). Engen et al. (2008) expanded the interpretation of continental crust and suggested that the outer Yermak Plateau and the conjugate Morris Jesup Rise also represent protrusions of stretched continental crust with high amplitude magnetic anomalies relating to magmatic intrusions. The Bouguer anomaly character of the inner Yermak Plateau continues, according to Engen et al. (2008), into the outer plateau. This is in agreement with a system of normal faults beneath the continental slope in a 15–20 km wide transitional area to the deep Nansen Basin in the north of the Yermak Plateau (Geissler and Jokat, 2004). One successful dredge at a basement high on the central Yermak Plateau (81° 38'N; 15° 32'E) revealed meta-sedimentary rocks, similar to outcropping lithologies along the northernmost Svalbard coastline (Hellebrand, 2000).

North Barents Sea Continental Margin

Geissler and Jokat (2004) identified two structural boundaries north of northeastern Svalbard: A hinge zone between the inner and outer shelf; and a system of normal faults beneath the continental slope of the North Barents Sea continental margin. The dislocation in the 15–20 km wide fault zone, with normal faults dipping at about 20°, reaches values up to 2000 m (Geissler and Jokat, 2004). Minakov et al. (2012) published a series of crustal-scale transects illustrating the architecture of the margin based on sparse seismic reflection lines and gravity modeling. Their gravity inversion supports a narrow and steep continent-ocean transition. A free-air gravity anomaly in the western part of the margin, close to the Yermak Plateau, is suggested to be caused by the exhumation of the lower crust and the continental upper mantle within the continent-ocean transition. They also suggested that an episode of shear or oblique extension before breakup is required to explain the observed narrow symmetric conjugate margins in the Eurasian Basin.

The conjugate continental margin of the Lomonosov Ridge consists of sets of rotated fault blocks stepping down to the basin over some tens of kilometers (Cochran et al., 2006; Langinen et al., 2009). Cochran et al. (2006) concluded from an extensive study of the Lomonosov Ridge that the continent-ocean transition zone at the Eurasian flank is much narrower than observed at other magma-poor margins worldwide. Sparse refraction seismic data across the Lomonosov Ridge show an abrupt crustal thinning from more than 25 km Moho depth beneath the Lomonosov Ridge to ~15 km in the Amundsen Basin over a few tens of kilometers (e.g., Artyushkov, 2010), a finding that is confirmed by regional gravity inversion (Alvey et al., 2008).

Reflection Seismic and Magnetic Data

The interpretation presented in this study are based on MCS data, acquired in 2013 during the PANORAMA-1 cruise of the Federal Institute for Geosciences and Natural Resources (BGR) with the research vessel OGS Explora. We used a 3600 m streamer with a total of 288 channels and an airgun array of 32.8 liter working on 140 bar. Shot distance was 25 m with a record length of 10 s. Data processing was carried out in the pre-stack domain including designature, multiple reduction (srme, TauP deconvolution, radon multiple prediction, and subtraction), common reflection surface, Kirchhoff migration, and time variant bandpass filtering. Magnetic data were measured along the reflection seismic lines with a towed gradient magnetometer. This method allows acquisition of variation-corrected data (Roeser et al., 2002).

RESULTS

The reflection seismic data presented here consist of a composite section running for about 125 km SW-NE from the northeastern Yermak Plateau into the Nansen Basin (BGR13-207), and for about 255 km NW-SE from the Nansen Basin to the North Barents Sea continental margin (BGR13-208) (**Figure 1**). The data depict the south-westernmost area of the Nansen Basin, which is a key region for understanding the initial opening of the Eurasian Basin, i.e., the detachment of the Lomonosov Ridge from the then adjunct Yermak Plateau and Morris Jesup Rise. They allows us a comparison of the continental margin types of the eastern Yermak Plateau and the North Barents Sea continental margin, as well as an interpretation of the sedimentary units.

Seismic Stratigraphy

For the sedimentary strata, we adopted the seismic stratigraphic units established by Engen et al. (2009), and tied our data to their interpretation of seismic line NPD-POLAR-16. Engen et al. (2009) divide the sedimentary strata of the Nansen Basin into four main units which we also identified in our seismic data (**Figure 3B**): Units NB-1A and NB-1B were deposited during the time interval from the opening of the Eurasian Basin until the separation of the Yermak Plateau and Morris Jesup Rise. The sediments of unit NB-1A were deposited during the initial opening, showing onlapping and draping of the underlying basement. This is followed by unit NB-2 deposited during the opening of the Fram Strait, which was the last major plate tectonic event before the establishment of the present regime in this area. Unit NB-3 represents sediments from this time until the onset of intensified glaciomarine deposition at about 2.6 Ma (Engen et al., 2009), named units NB-4A and NB-4B.

Next to the Yermak Plateau, we identified in addition the sedimentary unit NB-0 (**Figure 4**), underlying unit NB-1A. It is bounded at the top by a distinct unconformity, and the overlying reflectors of unit NB-1A onlap this unconformity. The maximum thickness is about 1.5 s (TWT), and this unit is characterized by a hummocky to contorted reflection pattern with upwards increasing continuity. Below unit NB-0, distinct southwest dipping reflectors indicate the top of

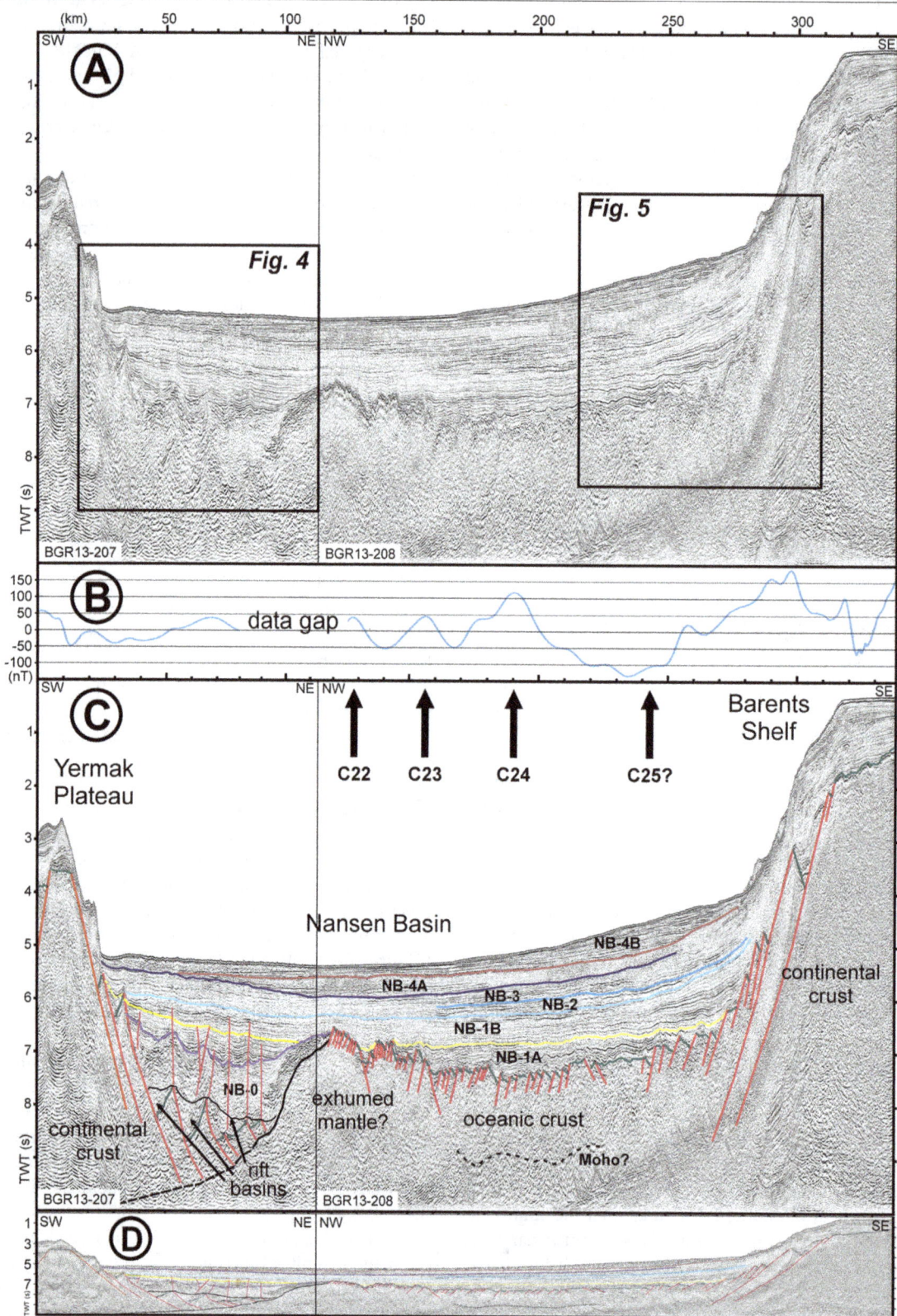

FIGURE 3 | Composite MCS line depicting the northeastern Yermak Plateau, the southwestern part of the Nansen Basin, and the North Barents Sea continental margin. The location is shown in **Figure 1**. **(A)** Reflection seismic sections BGR13–207 and –208. **(B)** Magnetic data measured along the composite MCS line. Data gap is present because of instrument recovery due to ice conditions. **(C)** Interpreted section with magnetic anomaly identifications from Brozena et al. (2003). At the line tie a basement high with distinct high amplitude reflections might be exhumed mantle material. The transition to the Barents Shelf is comparatively narrow and steep. **(D)** Interpreted section with lower vertical exaggeration (~3) to better depict the geometry of the basement structures.

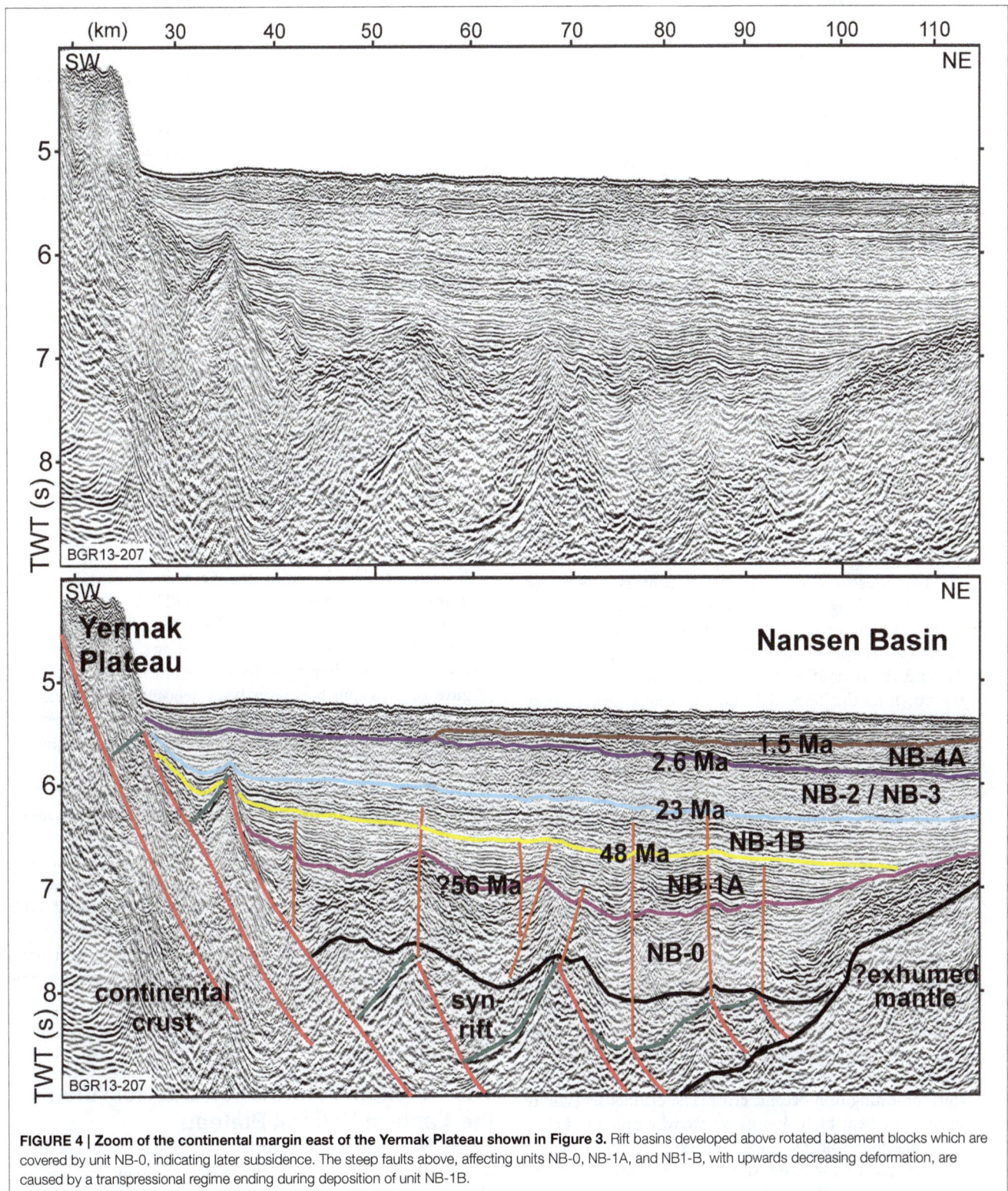

FIGURE 4 | Zoom of the continental margin east of the Yermak Plateau shown in Figure 3. Rift basins developed above rotated basement blocks which are covered by unit NB-0, indicating later subsidence. The steep faults above, affecting units NB-0, NB-1A, and NB1-B, with upwards decreasing deformation, are caused by a transpressional regime ending during deposition of unit NB-1B.

rotated fault blocks forming half grabens. These half grabens are filled by the wedge-shaped unit "syn-rift" (**Figure 4**), characterized by a contorted reflection pattern. This unit is bounded at the top by an unconformity indicated by toplaps visible in some places (e.g., **Figure 4**, km 45–55). Smaller-scale compressional faulting affects the sedimentary units from unit "syn-rift" to unit NB-1B, with upwards decreasing intensity.

Structural Interpretation

From the west to the east we divide the composite section into three characteristic basement domains (**Figure 3A**):

(I)   The most striking observation at the Yermak Plateau margin is an ~80 km wide domain of horizontal extension, made up of tilted basement blocks forming half graben structures (**Figure 4**). The domain is bounded to the west by the Yermak Plateau and to the east by a basement high with a distinct high impedance reflectivity. Using solely multi-channel seismic data, it is difficult to conclusively define the nature of the basement (Klimke et al., in press). However, the fabric of the oceanic crust in the abyssal plain is typically quite distinct from the block-faulted fabric of the continental domain (e.g., Taylor et al., 1999; Franke et al., 2011; Peron-Pinvidic et al., 2013). By analogy with the well-studied magma-poor margins of Iberia (e.g., Whitmarsh et al., 2001; Manatschal, 2004; Peron-Pinvidic et al., 2013), East India (Haupert et al., 2016), the South China Sea (e.g., Cullen et al., 2010; Franke et al., 2014; Ding et al., 2016), and southern Australia (Gillard et al., 2015), we suggest that the basement blocks are of continental origin. The basement blocks decrease in size toward the east, and we interpret these as continental crustal ribbons formed during the break-up of the Yermak Plateau and the Lomonosov Ridge. The detachment is interpreted as running along the top of the structure that forms a basement high further seaward. Thus, the Yermak Plateau margin is interpreted here as a rifted continental domain. An end-rift unconformity may be inferred, which seals the structures that formed during extensional deformation. After rifting, deposition of unit NB-0 indicates a phase of subsidence or sag. Further subsidence is indicated by the onlapping of unit NB-1A. A phase of smaller-scale compressive deformation affected the sedimentary units up to unit NB-1B, with upwards decreasing intensity, at about the Eocene-Oligocene transition, which is interpreted as transpression resulting from the northward movement of the Gakkel Ridge along the Yermak Plateau.

(II) Further to the east into the Nansen Basin, a basement high characterized by high impedance reflectivity is present (**Figure 3B**, km 110–150). From km 150–240 the basement is slightly deeper and has a lower reflectivity. Brittle deformation of the basement is indicated by small-scale normal faulting. Internal reflectivity is low, and a low frequency reflector about 1.8 s (TWT) below the basement reflector might represent the seismic Moho (km 170–220), thus indicating thin oceanic crust. The distinct reflectivity of the basement high, and its westward dipping to below tilted basement blocks, suggest an exhumation of mantle material during break-up, similar to that discovered and proved by scientific drilling along the Iberia magma-poor continental margin (e.g., Whitmarsh et al., 2001; Sibuet et al., 2007). Magnetic anomalies (**Figures 2A, 3**) are interpreted by Brozena et al. (2003) at the location of this proposed exhumed mantle, and might have developed during serpentinization. It is well known that mantle peridotite ridges can acquire significant amounts of magnetization during serpentinization, and can thus mimic linear magnetic anomalies (Sibuet et al., 2007).

(III) The North Barents Sea continental margin is characterized by narrow tilted basement blocks below a steep slope up to the shelf (**Figure 5**). Deeply penetrating faults are present along the margin at the boundary to the oceanic domain. The basement at the slope exhibits little internal reflectivity (partially because of the crossing of the seafloor multiple). In the seaward direction, disrupted basement blocks developed before the deposition of unit NB-1A (**Figures 3, 5**, km 240–280). The nature of the basement in this area is unclear. There are no indications for listric faulting or syn-tectonic sedimentation. In combination with the low frequency, high amplitude reflectivity this might indicate mafic or peridotite basement composition. Independent from the basement origin, the seismic data in this area (**Figures 3C, 5**) reveal that the transition from oceanic to continental crust is narrow: At maximum 40 km on our line, which runs at an oblique angle to the shelf.

Differences between the North Barents Sea and Yermak Plateau Margins

In terms of seismic stratigraphy (**Figure 6**), unit NB-0 is exclusively developed next to the Yermak Plateau (**Figure 4**). This unit, with a thickness of up to 1.5 s (TWT, ~3 km), covers the rift basins and is, from a confident tracing of the overlying sedimentary successions, interpreted as being significantly earlier deposited than the draping sedimentary units in the study area (**Figure 3**). This unit has never been deposited on the Barents side, because there are no indications for a large-scale erosion phase. The most distinct difference between the two margins are the large rift basins at the Yermak Plateau, which are missing at the North Barents Sea continental margin. These up to 15 km wide basins show indications for syn-tectonic sedimentary deposition and are bounded by deeply reaching listric normal faults (**Figure 4**). Top-lap truncations at the top of the syn-tectonic infill indicate an erosional phase, as it is widespread found in rift basins. This architecture is completely at odds with the North Barents Sea continental margin, where only a comparatively narrow area has been deformed by horizontal extension, the deformation is small-scale, and no syn-tectonic deposition took place at the slope (**Figure 5**).

DISCUSSION

The Eurekan Deformation Did Not Affect the Eastern Yermak Plateau

At the NE edge of the Yermak Plateau, our data reveal no significant compressional deformation. Rather we identify distinct rotated crustal blocks, interpreted to be bounded by major listric normal faults. Previously, gravity and magnetic anomaly lows immediately northeastward of the Yermak Plateau and Morris Jesup Rise, and trending about perpendicular to the seafloor spreading direction, have been interpreted as a major Eurekan fault zone, involving crustal shortening

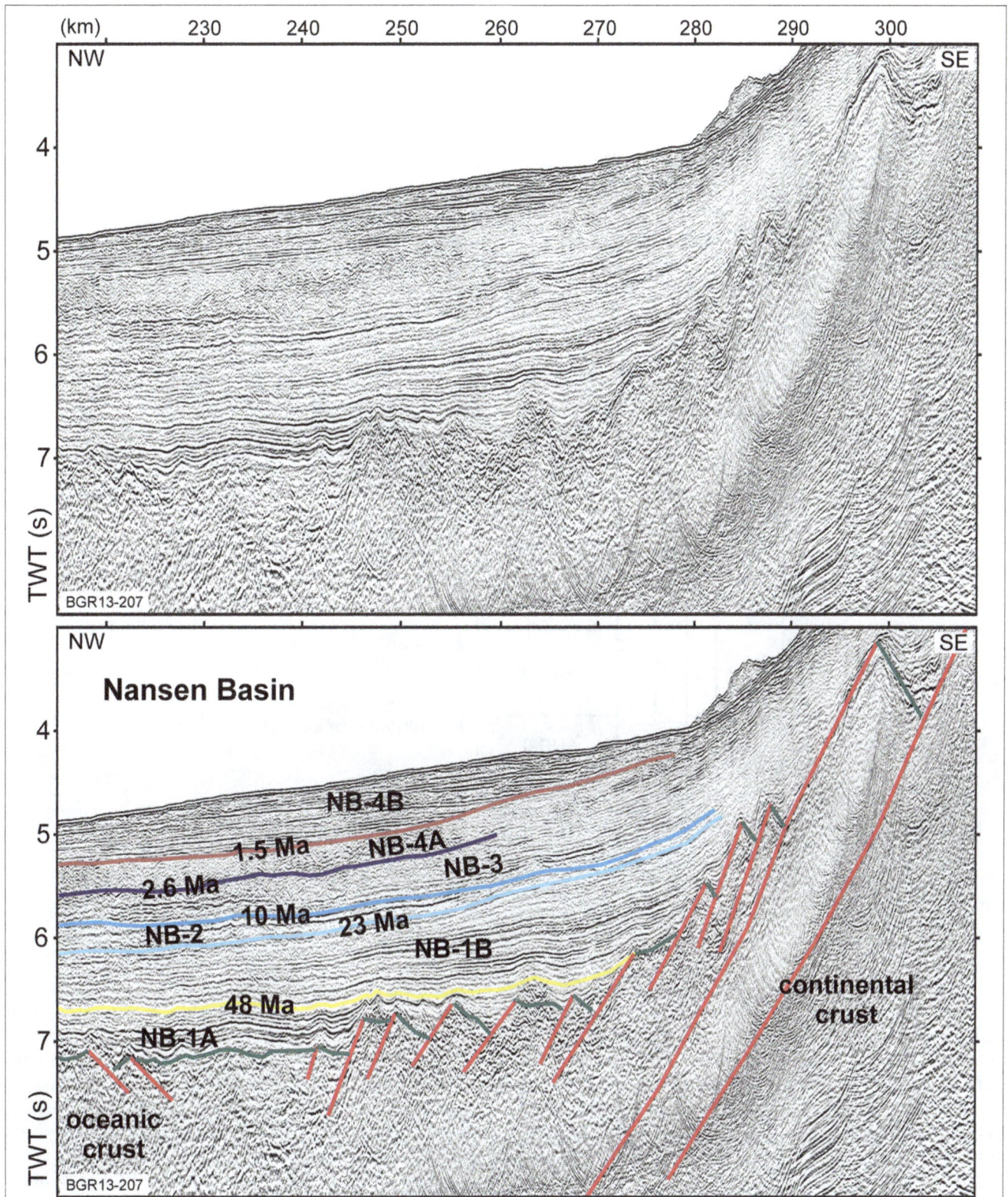

FIGURE 5 | Zoom of the continental margin north of the Barents Shelf shown in Figure 3, indicating a very narrow transition from oceanic to continental crust. Oldest sedimentary unit is NB-1A overlying the oceanic crust and infilling small basins formed by disrupted basement blocks (km 245–275).

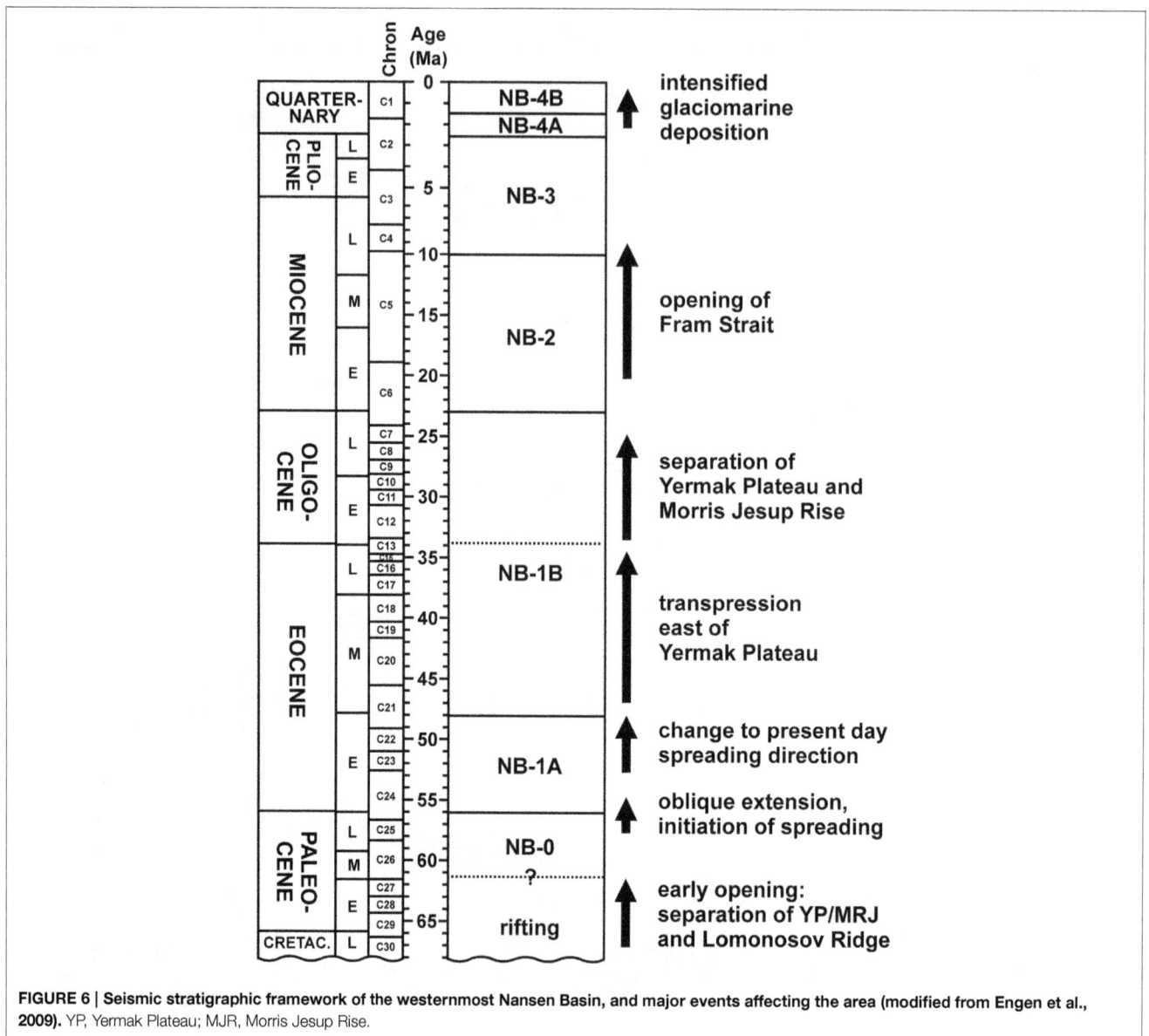

FIGURE 6 | Seismic stratigraphic framework of the westernmost Nansen Basin, and major events affecting the area (modified from Engen et al., 2009). YP, Yermak Plateau; MJR, Morris Jesup Rise.

(Døssing et al., 2013, 2014) and potentially subduction (Brozena et al., 2003), as consequence of the Paleogene northward convergence of Greenland. Plate tectonic models often compensate for the extension in the Labrador Sea and Baffin Bay by invoking a northeastward movement of Greenland. The oldest undisputed seafloor spreading magnetic anomaly in the Labrador Sea is magnetic chron C27 with an age of ~61 Ma (Chalmers and Pulvertaft, 2001) but seafloor spreading may have begun up to 10 My earlier. While the Eurekan N–S shortening is well-documented in Ellesmere Island, North Greenland and western Svalbard, the structural configuration north of Greenland remain unresolved. The North Greenland transform margin province includes sedimentary basins under the Wandel Sea and the Lincoln Sea (Sørensen et al., 2011). Structural analyses within different areas of the Wandel Hav Mobile Belt

of northeast Greenland suggest that the main compressive deformation was caused by a comparable dextral transpressive mechanism. According to von Gosen and Piepjohn (2003), it is probable that the Wandel Hav Mobile Belt represents the equivalent of the De Geer Fault Zone along which the Eurasian plate was dextrally displaced with respect to north Greenland. Thus, the relative motion between Greenland and Svalbard, including the Yermak Plateau, was mainly strike-slip with only a small component of compression (Srivastava, 1985). In addition, Tegner et al. (2011) found that the compression associated with the Eurekan deformation had affected the 85–60 Ma volcanic suite at the northern tip of Greenland—thus indicating that the late Cretaceous to Paleocene rifting preceded the predominantly Eocene Eurekan transpressional-compressional deformation. These interpretations correspond well with our findings that

Paleocene and Eocene compressional tectonics did not extend across the Yermak Plateau into the Eurasian Basin. We link the smaller scale Eocene compressional features, observed in our seismic section (**Figure 4**), to the northward displacement of the Lomonosov Ridge (**Figures 7C,D**).

The North Barents Sea Continental Margin

Continental rifting and the detachment of a narrow and elongated crustal splinter, like the Lomonosov Ridge, from the North Barents Sea continental margin, is not easily explained given the strength of cold continental lithosphere that typically exceeds the available forces. Müller et al. (2001) suggest that prolonged periods of asymmetries in oceanic crustal accretion, as well as the mechanical and thermal effects of excess magmatism, are preconditions for detaching fragments of continental crust. However, as pointed out by Minakov et al. (2013), the magma-starved evolution of the Eurasian Basin implies the occurrence of another mechanism to enable the detachment of the Lomonosov Ridge microcontinent. Their preferred explanation, underpinned by numerical modeling, is a combination of strike-slip deformation and shear heating. These results correspond well with other modeling studies which show that a narrow zone of deformation, of the kind observed along the conjugate Lomonosov Ridge and the North Barents Sea continental margin, indicates highly oblique rifting (e.g., Autin et al., 2010; Brune et al., 2012).

The architecture of the North Barents Sea continental margin with a narrow and steep continent-ocean transition

(Geissler and Jokat, 2004; Minakov et al., 2012, this study), and the mirrored image at the conjugate Eurasian Basin margin of the Lomonosov Ridge (Cochran et al., 2006; Langinen et al., 2009) are also indicative of highly oblique movements during the initial formation of these continental margins, rather than typical magma-poor extension. Our interpretation of an initial rift-axis, highly oblique to the later spreading axis (**Figure 7A**), is well in accordance with the previous interpretation of initial strike-slip faulting along the North Barents Sea continental margin (Minakov et al., 2013). A transfer fault, segmenting this early rift might have enabled the split off of the Lomonosov Ridge from the present-day North Barents Sea continental margin (**Figure 7B**). In addition such a setting offers an explanation for the observed narrow continent-ocean transition zone.

The Early Eurasian Basin—a Rift at a High Angle to the Present Day Gakkel Ridge Seafloor Spreading System

In contrast to earlier interpretations, we suggest that spreading in the Eurasian Basin did not initiate at a triple-junction north of Greenland between North America, Greenland and Eurasia. Our single composite reflection seismic line indicate the presence of a (?) Late Cretaceous to Paleocene rifted continental margin at the NE edge of the Yermak Plateau (and consequently also at the Morris Jesup Rise), and highly oblique deformation at the boundary between the North Barents Sea continental margin to the north of Svalbard and the Eurasian Basin. By acknowledging

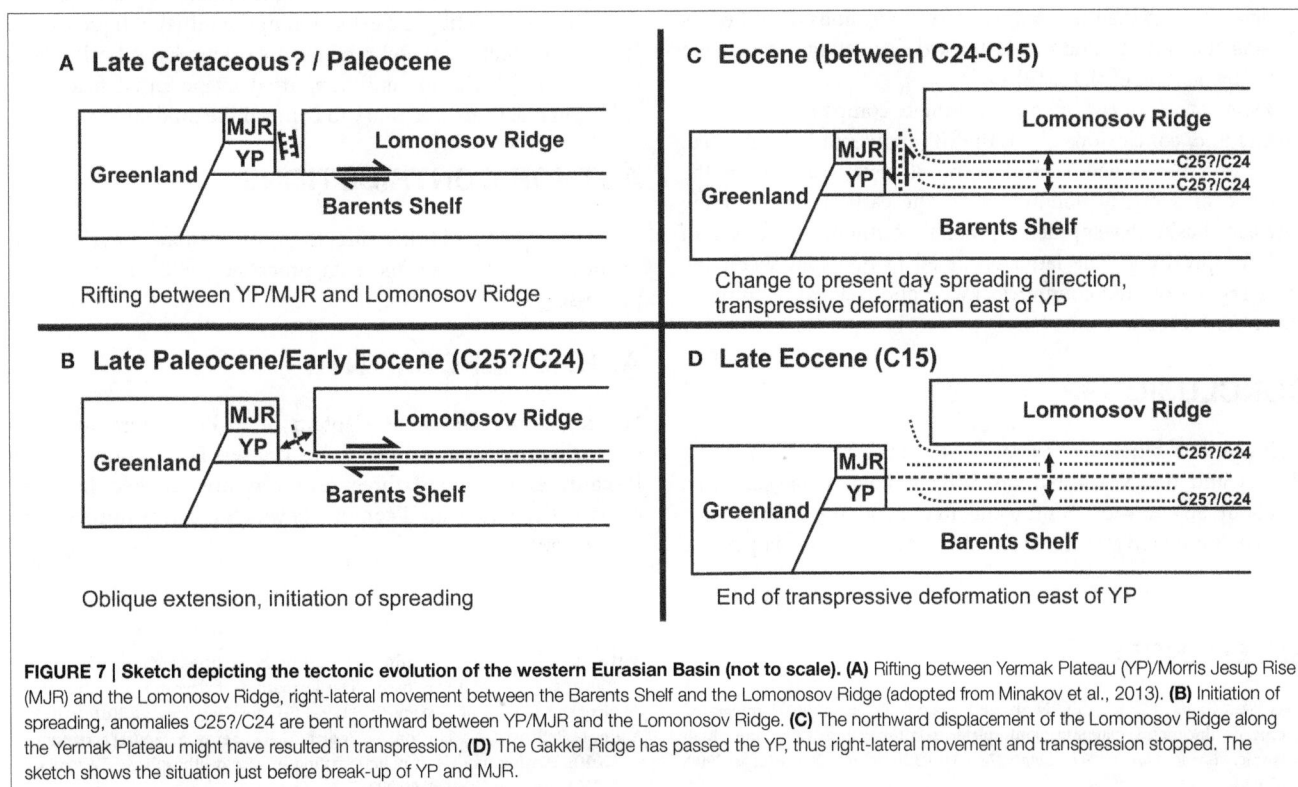

FIGURE 7 | Sketch depicting the tectonic evolution of the western Eurasian Basin (not to scale). (A) Rifting between Yermak Plateau (YP)/Morris Jesup Rise (MJR) and the Lomonosov Ridge, right-lateral movement between the Barents Shelf and the Lomonosov Ridge (adopted from Minakov et al., 2013). **(B)** Initiation of spreading, anomalies C25?/C24 are bent northward between YP/MJR and the Lomonosov Ridge. **(C)** The northward displacement of the Lomonosov Ridge along the Yermak Plateau might have resulted in transpression. **(D)** The Gakkel Ridge has passed the YP, thus right-lateral movement and transpression stopped. The sketch shows the situation just before break-up of YP and MJR.

the gravity and magnetic signals running roughly perpendicular to the seafloor spreading fabric along the NE Yermak Plateau and Morris Jesup Rise, we speculate that this is in fact the trend of the rift axis, with the North Barents Sea continental margin being deformed by a rift transform which is highly oblique to the present day rift axis. The bend toward the north in the earliest magnetic isochrons close to the Yermak Plateau, as well as a prominent bend in the magnetic data between the Morris Jesup Rise and the Lomonosov Ridge in the Lincoln Sea (**Figure 2B**, Brozena et al., 2003; Engen et al., 2008) may be the manifestation of this early extensional direction. The basement below the small basin located between the northeastern Yermak Plateau and the North Barents Sea continental margin (**Figure 1**) is described as rifted continental crust (Geissler and Jokat, 2004). The timing of crustal thinning in this small basin area is unknown, it could also be related to a later stage of opening of the Eurasian Basin.

Further north, at the transition from the Eurasia to the Amerasian Basin, a crustal-scale wide-angle refraction seismic line confirms a northward shallowing of the Moho from the Lincoln Sea toward the edge of the Lomonosov Ridge (Jackson et al., 2010). Jackson et al. (2010) propose that the thinning was inherited from a rifting event prior to seafloor spreading in the Eurasian Basin. This interpretation of rifting is appealing because it may be correlated with the Late Cretaceous to Paleocene alkaline volcanic suites of northern Canada and north Greenland that probably formed in a failed rift zone preceding seafloor spreading in the Eurasian Basin (Estrada and Henjes-Kunst, 2004; Tegner et al., 2011). Considering an early extensional phase with a trend roughly perpendicular to the present day seafloor spreading fabric in the Eurasian Basin, this speculative rift might be linked to extension in the Lincoln Sea resulting in the deposition of the thick sedimentary successions described by Døssing et al. (2014), and the northward shallowing of the Moho shown by Jackson et al. (2010).

Being constrained by only a single composite reflection seismic line, conclusions about the rift architecture are limited. However, from a geometrical point of view, we suggest the presence of a highly oblique rift in the earliest stages of the Eurasian Basin development, probably without the onset of seafloor spreading. The latter occurred in our view when the stress regime rotated counterclockwise and enabled opening in the present direction.

CONCLUSIONS

Multichannel reflection seismic data from the NE Yermak Plateau and the North Barents Sea continental margin show an about 80 km wide continental margin at the NE Yermak Plateau dominated by rotated fault blocks and bounded by major

listric normal faults. The corresponding half-grabens are filled with syn-rift sedimentary strata. We interpret this structural setting as the expression of a magma-poor rifted continental margin. This indicates that this portion of the Yermak Plateau is underlain by continental crust, similar to the other portions of the plateau. There are negligible indications for compressional deformation, and we exclude the possibility of subduction in this area. We thus conclude that the Late Paleocene and early-middle Eocene compressional deformation well known from North Greenland and Svalbard (Eurekan) did not extend across the Yermak Plateau.

The eastward transition from the rift system into the oceanic Nansen Basin is formed by a basement high, characterized in the seismic data by high impedance reflectors, which is possibly formed by exhumed mantle. At the steep slope of the North Barents Sea continental margin no rift basins are observed and only a comparatively narrow area has been deformed by horizontal extension.

By acknowledging the structural configuration as derived from the MCS data and the potential field data we speculate about the presence of a ?Late Cretaceous to Paleocene rift between the Yermak Plateau and the Lomonosov Ridge, the latter having been still attached to the Eurasian continent. This rift is suggested as striking at a high angle to the present-day seafloor spreading fabric in the Eurasian Basin. We suggest that the initial formation of the Eurasian Basin took place along this rift, between the plateaus of the Morris Jesup Rise-Yermak Plateau and the Lomonosov Ridge, and possibly connected to shallowing of the Moho in the northern Lincoln Sea. Within this hypothesis, the later breakup of the Lomonosov Ridge from the North Barents Sea continental margin initiated by a transform fault, accommodating the extension of the initial oblique rift. So far, this conceptual model remains a working hypothesis, based only on one MCS line and supported by potential field data. Additional data are necessary to confirm the model.

AUTHOR CONTRIBUTIONS

KB, DF, RL, BS, VD: Interpretation, discussions, development of ideas, text. KB: Seismic data processing. BS: Magnetic data processing.

ACKNOWLEDGMENTS

We thank the crew and Captain of OGS Explora for their professional support during the cruise. The editors Alexey Piskarev and ØE and three reviewers are thanked for their constructive and helpful comments which greatly improved this manuscript.

REFERENCES

Alvey, A., Gaina, C., Kusznir, N. J., and Torsvik, T. H. (2008). Integrated crustal thickness mapping and plate reconstructions for the high Arctic. *Earth Planet. Sci. Lett.* 274, 310–321. doi: 10.1016/j.epsl.2008.07.036

Artyushkov, E. V. (2010). Continental crust in the Lomonosov Ridge, Mendeleev Ridge, and the Makarov basin. The formation of deep-water basins in the Neogene. *Russ. Geol. Geophy.* 51, 1179–1191. doi: 10.1016/j.rgg.2010.10.003

Autin, J., Bellahsen, N., Husson, L., Beslier, M.-O., Leroy, S., and D'acremont, E. (2010). Analog models of oblique rifting in a cold lithosphere. *Tectonics* 29, TC6016. doi: 10.1029/2010tc002671

Bronner, A., Sauter, D., Manatschal, G., Peron-Pinvidic, G., and Munschy, M. (2011). Magmatic breakup as an explanation for magnetic anomalies at magma-poor rifted margins. *Nat. Geosci.* 4, 549–553. doi: 10.1038/ngeo1201

Brozena, J. M., Childers, V. A., Lawver, L. A., Gahagan, L. M., Forsberg, R., Faleide, J. I., et al. (2003). New aerogeophysical study of the Eurasia Basin and Lomonosov Ridge: implications for basin development. *Geology* 31, 825–828. doi: 10.1130/g19528.1

Brune, S., Popov, A. A., and Sobolev, S. V. (2012). Modeling suggests that oblique extension facilitates rifting and continental break-up. *J. Geophys. Res.* 117, B08402. doi: 10.1029/2011jb008860

Chalmers, J. A., and Pulvertaft, T. C. R. (2001). Development of the continental margins of the Labrador Sea: a review. *Geol. Soc. Lond. Special Publ.* 187, 77–105. doi: 10.1144/gsl.sp.2001.187.01.05

Cochran, J. R., Edwards, M. H., and Coakley, B. J. (2006). Morphology and structure of the Lomonosov Ridge, Arctic Ocean. *Geochem. Geophys. Geosyst.* 7, Q05019. doi: 10.1029/2005gc001114

Cullen, A., Reemst, P., Henstra, G., Gozzard, S., and Ray, A. (2010). Rifting of the South China Sea: new perspectives. *Petroleum Geosci.* 16, 273–282. doi: 10.1144/1354-079309-908

Ding, W., Li, J., and Clift, P. D. (2016). Spreading dynamics and sedimentary process of the Southwest Sub-basin, South China Sea: constraints from multi-channel seismic data and IODP Expedition 349. *J. Asian Earth Sci.* 115, 97–113. doi: 10.1016/j.jseaes.2015.09.013

Døssing, A., Hansen, T. M., Olesen, A. V., Hopper, J. R., and Funck, T. (2014). Gravity inversion predicts the nature of the amundsen basin and its continental borderlands near greenland. *Earth Planet. Sci. Lett.* 408, 132–145. doi: 10.1016/j.epsl.2014.10.011

Døssing, A., Hopper, J. R., Olesen, A. V., Rasmussen, T. M., and Halpenny, J. (2013). New aero-gravity results from the Arctic: linking the latest Cretaceous-early Cenozoic plate kinematics of the North Atlantic and Arctic Ocean. *Geochem. Geophys. Geosyst.* 14, 4044–4065. doi: 10.1002/ggge.20253

Eldholm, O., Faleide, J. I., and Myhre, A. M. (1987). Continent-ocean transition at the western Barents Sea/Svalbard continental margin. *Geology* 15, 1118–1122. doi: 10.1130/0091-7613(1987)15<1118:ctatwb>2.0.co;2

Engen, Ø., Faleide, J. I., and Dyreng, T. K. (2008). Opening of the fram strait gateway: a review of plate tectonic constraints. *Tectonophysics* 450, 51–69. doi: 10.1016/j.tecto.2008.01.002

Engen, Ø., Gjengedal, J. A., Faleide, J. I., Kristoffersen, Y., and Eldholm, O. (2009). Seismic stratigraphy and sediment thickness of the Nansen Basin, Arctic Ocean. *Geophys. J. Int.* 176, 805–821. doi: 10.1111/j.1365-246X.2008.04028.x

Estrada, S., and Henjes-Kunst, F. (2004). Volcanism in the Canadian High Arctic related to the opening of the Arctic Ocean. *Z. Dtsch. Geol. Ges.* 154, 579–603. Available online at: https://www.schweizerbart.de/papers/zdgg_alt/detail/154/74983/Volcanism_in_the_Canadian_High_Arctic_related_to_the_opening_of_the_Arctic_Ocean

Feden, R. H., Vogt, P. R., and Fleming, H. S. (1979). Magnetic and bathymetric evidence for the "Yermak hot spot" northwest of Svalbard in the Arctic Basin. *Earth Planet. Sci. Lett.* 44, 18–38. doi: 10.1016/0012-821x(79)90004-9

Franke, D., Barckhausen, U., Baristeas, N., Engels, M., Ladage, S., Lutz, R., et al. (2011). The continent-ocean transition at the southeastern margin of the South China Sea. *Mar. Petroleum Geol.* 28, 1187–1204. doi: 10.1016/j.marpetgeo.2011.01.004

Franke, D., and Hinz, K. (2005). The structural style of sedimentary basins on the shelves of the Laptev Sea and the western East Siberian Sea, Siberian Arctic. *J. Petroleum Geol.* 28, 269–286. doi: 10.1111/j.1747-5457.2005.tb00083.x

Franke, D., Savva, D., Pubellier, M., Steuer, S., Mouly, B., Auxietre, J.-L., et al. (2014). The final rifting evolution in the South China Sea. *Mar. Petroleum Geol.* 58, 704–720. doi: 10.1016/j.marpetgeo.2013.11.020

Gaina, C., Gernigon, L., and Ball, P. (2009). Palaeocene–Recent plate boundaries in the NE Atlantic and the formation of the Jan Mayen microcontinent. *J. Geol. Soc. Lond.* 166, 601–616. doi: 10.1144/0016-76492008-112

Gaina, C., Werner, S. C., Saltus, R., Maus, S., and the, CAMP-GM GROUP (2011). Circum-Arctic mapping project: new magnetic and gravity anomaly maps of the Arctic. *Geol. Soc. Lond. Mem.* 35, 39–48. doi: 10.1144/m35.3

Gee, J. S., and Kent, D. V. (2007). "5.12–Source of oceanic magnetic anomalies and the geomagnetic polarity timescale," in *Treatise on Geophysics*, ed G. Schubert (Amsterdam: Elsevier), 455–507.

Geissler, W. H., and Jokat, W. (2004). A geophysical study of the northern Svalbard continental margin. *Geophys. J. Int.* 158, 50–66. doi: 10.1111/j.1365-246X.2004.02315.x

Geissler, W. H., Jokat, W., and Brekke, H. (2011). The Yermak Plateau in the Arctic Ocean in the light of reflection seismic data—implication for its tectonic and sedimentary evolution. *Geophys. J. Int.* 187, 1334–1362. doi: 10.1111/j.1365-246X.2011.05197.x

Gillard, M., Autin, J., Manatschal, G., Sauter, D., Munschy, M., and Schaming, M. (2015). Tectonomagmatic evolution of the final stages of rifting along the deep conjugate Australian-Antarctic magma-poor rifted margins: constraints from seismic observations. *Tectonics* 34, 753–783. doi: 10.1002/2015tc003850

Glebovsky, V. Y., Kaminsky, V. D., Minakov, A. N., Merkur'ev, S. A., Childers, V. A., and Brozena, J. M. (2006). Formation of the Eurasia Basin in the Arctic Ocean as inferred from geohistorical analysis of the anomalous magnetic field. *Geotectonics* 40, 263–281. doi: 10.1134/s0016852106040029

Haupert, I., Manatschal, G., Decarlis, A., and Unternehr, P. (2016). Upper-plate magma-poor rifted margins: stratigraphic architecture and structural evolution. *Mar. Petroleum Geol.* 69, 241–261. doi: 10.1016/j.marpetgeo.2015.10.020

Hellebrand, E. (2000). "Petrology," in *Berichte zur Polarforschung (Reports on Polar Research), The Expedition ARKTIS-XV/2 of 'Polarstern'in 1999*, ed W. Jokat (Bremerhaven: AWI), 59–70.

Jackson, H. R., Dahl-Jensen, T., and The, L. W. G. (2010). Sedimentary and crustal structure from the Ellesmere Island and Greenland continental shelves onto the Lomonosov Ridge, Arctic Ocean. *Geophys. J. Int.* 182, 11–35. doi: 10.1111/j.1365-246X.2010.04604.x.

Jackson, H. R., Johnson, G. L., Sundvor, E., and Myhre, A. M. (1984). The Yermak Plateau: formed at a triple junction. *J. Geophys. Res.* 89, 3223–3232. doi: 10.1029/JB089iB05p03223.

Jakobsson, M., Mayer, L., Coakley, B., Dowdeswell, J. A., Forbes, S., Fridman, B., et al. (2012). The International Bathymetric Chart of the Arctic Ocean (IBCAO) Version 3.0. *Geophys. Res. Lett.* 39:L12609. doi: 10.1029/2012GL052219

Jokat, W., Geissler, W., and Voss, M. (2008). Basement structure of the north-western Yermak Plateau. *Geophys. Res. Lett.* 35, 1–6. doi: 10.1029/2007GL032892

Jokat, W., Ickrath, M., and O'Connor, J. (2013). Seismic transect across the Lomonosov and Mendeleev Ridges: constraints on the geological evolution of the Amerasia Basin, Arctic Ocean. *Geophys. Res. Lett.* 2013, GL057275. doi: 10.1002/grl.50975

Klimke, J., Franke, D., Gaedicke, C., Schreckenberger, B., Schnabel, M., Stollhofen, H., et al. (in press). How to identify oceanic crust–evidence for a complex break-up in the Mozambique Channel, off East Africa. *Tectonophysics* 17. doi: 10.1016/j.tecto.2015.10.012

Langinen, A. E., Lebedeva-Ivanova, N. N., Gee, D. G., and Zamansky, Y. Y. (2009). Correlations between the Lomonosov Ridge, Marvin Spur and adjacent basins of the Arctic Ocean based on seismic data. *Tectonophysics* 472, 309–322. doi: 10.1016/j.tecto.2008.05.029

Lawver, L. A., Gahagan, L. M., and Norton, I. (2011). Chapter 5 Palaeogeographic and tectonic evolution of the Arctic region during the Palaeozoic. *Geol. Soc. Lond. Mem.* 35, 61–77. doi: 10.1144/m35.5

Lawver, L. A., Grantz, A., and Gahagan, L. M. (2002). Plate kinematic evolution of the present Arctic region since the Ordovician. *Geol. Soc. Am. Special Pap.* 360, 333–358. doi: 10.1130/0-8137-2360-4.333

Manatschal, G. (2004). New models for evolution of magma-poor rifted margins based on a review of data and concepts from West Iberia and the Alps. *Int. J. Earth Sci.* 93, 432–466. doi: 10.1007/s00531-004-0394-7

Minakov, A., Faleide, J. I., Glebovsky, V. Y., and Mjelde, R. (2012). Structure and evolution of the northern Barents–Kara Sea continental margin from integrated analysis of potential fields, bathymetry and sparse seismic data. *Geophys. J. Int.* 188, 79–102. doi: 10.1111/j.1365-246X.2011.05258.x

Minakov, A. N., Podladchikov, Y. Y., Faleide, J. I., and Huismans, R. S. (2013). Rifting assisted by shear heating and formation of the Lomonosov Ridge. *Earth Planet. Sci. Lett.* 373, 31–40. doi: 10.1016/j.epsl.2013.04.042

Müller, R. D., Gaina, C., Roest, W. R., and Hansen, D. L. (2001). A recipe for microcontinent formation. *Geology* 29, 203–206. doi: 10.1130/0091-7613(2001)029<0203:arfmf>2.0.co;2

Peron-Pinvidic, G., Manatschal, G., and Osmundsen, P. T. (2013). Structural comparison of archetypal Atlantic rifted margins: a review of observations and concepts. *Mar. Petroleum Geol.* 43, 21–47. doi: 10.1016/j.marpetgeo.2013.02.002

Riefstahl, F., Estrada, S., Geissler, W. H., Jokat, W., Stein, R., Kämpf, H., et al. (2013). Provenance and characteristics of rocks from the Yermak Plateau, Arctic Ocean: petrographic, geochemical and geochronological constraints. *Mar. Geol.* 343, 125–145. doi: 10.1016/j.margeo.2013.06.009

Ritzmann, O., and Jokat, W. (2003). Crustal structure of northwestern Svalbard and the adjacent Yermak Plateau: evidence for Oligocene detachment tectonics and non-volcanic breakup. *Geophys. J. Int.* 152, 139–159. doi: 10.1046/j.1365-246X.2003.01836.x

Roeser, H. A., Steiner, C., Schreckenberger, B., and Block, M., (2002). Structural development of the Jurassic Magnetic Quiet Zone off Morocco and identification of Middle Jurassic magnetic lineations. *J. Geophys. Res.* 107, 2207. doi: 10.1029/2000JB000094

Shephard, G. E., Müller, R. D., and Seton, M. (2013). The tectonic evolution of the Arctic since Pangea breakup: integrating constraints from surface geology and geophysics with mantle structure. *Earth Sci. Rev.* 124, 148–183. doi: 10.1016/j.earscirev.2013.05.012

Sibuet, J.-C., Srivastava, S., and Manatschal, G. (2007). Exhumed mantle-forming transitional crust in the Newfoundland-Iberia rift and associated magnetic anomalies. *J. Geophys. Res.* 112, B06105. doi: 10.1029/2005jb003856

Sørensen, K., Gautier, D., Pitman, J., Jackson, H. R., and Dahl-Jensen, T. (2011). Geology and petroleum potential of the Lincoln Sea Basin, offshore North Greenland. *Geol. Soc. Lond. Mem.* 35, 673–684. doi: 10.1144/m35.44

Srivastava, S. P. (1985). Evolution of the Eurasian Basin and its implications to the motion of Greenland along Nares Strait. *Tectonophysics* 114, 29–53. doi: 10.1016/0040-1951(85)90006-x

Srivastava, S. P., and Tapscott, C. (1986). "Plate kinematics of the North Atlantic," in *The Geology of North America, Vol. M, The Western Atlantic Region. (A Decade of North American Geology)*, eds B. E. Tucholke and P. R. Vogt (Boulder, CO: Geological Society of America), 379–404.

Taylor, B., Goodliffe, A. M., and Martinez, F. (1999). How continents break up: insights from Papua New Guinea. *J. Geophys. Res.* 104, 7497–7512.

Tegner, C., Storey, M., Holm, P. M., Thorarinsson, S. B., Zhao, X., Lo, C. H., et al. (2011). Magmatism and Eurekan deformation in the High Arctic Large Igneous Province: 40Ar-39Ar age of Kap Washington Group volcanics, North Greenland. *Earth Planet. Sci. Lett.* 303, 203–214. doi: 10.1016/j.epsl.2010.12.047

Tessensohn, F., and Piepjohn, K. (2000). Eocene compressive deformation in Arctic Canada, North Greenland and Svalbard and its plate tectonic causes. *Polarforschung* 68, 121–124. doi: 10.2312/polarforschung.68.121

Tessensohn, F., Von Gosen, W., Piepjohn, K., Saalmann, K., and Mayr, U. (2008). "Narers transform motion and Eurekan compression along the northeast coast of Ellesmere island," in *Geology of Northeast Ellesmere Island Adjacent to Kane Basin and Kennedy Channel*, ed U. Mayr (Nunavut: Geological Survey of Canada), 227–243. doi:10.4095/226146

Vogt, P. R., Taylor, P. T., Kovacs, L. C., and Johnson, G. L. (1979). Detailed aeromagnetic investigation of the Arctic Basin. *J. Geophys. Res.* 84, 1071–1089.

von Gosen, W., and Piepjohn, K. (2003). Eurekan transpressive deformation in the Wandel Hav Mobile Belt (northeast Greenland). *Tectonics* 22, 1039. doi: 10.1029/2001tc901040

Whitmarsh, R. B., Manatschal, G., and Minshull, T. A. (2001). Evolution of magma-poor continental margins from rifting to seafloor spreading. *Nature* 413, 150–154. doi: 10.1038/35093085

Conflict of Interest Statement: The authors declare that the research was conducted in the absence of any commercial or financial relationships that could be construed as a potential conflict of interest.

12

Modeling the Controls on the Front Position of a Tidewater Glacier in Svalbard

Jaime Otero[1]*, Francisco J. Navarro[1], Javier J. Lapazaran[1], Ethan Welty[2], Darek Puczko[3] and Roman Finkelnburg[1]

[1] Department of Applied Mathematics, Universidad Politécnica de Madrid, Madrid, Spain, [2] Institute of Arctic and Alpine Research, University of Colorado Boulder, Boulder, CO, USA, [3] Institute of Biochemistry and Biophysics, Polish Academy of Sciences, Warsaw, Poland

Calving is an important mass-loss process at ice sheet and marine-terminating glacier margins, but identifying and quantifying its principal driving mechanisms remains challenging. Hansbreen is a grounded tidewater glacier in southern Spitsbergen, Svalbard, with a rich history of field and remote sensing observations. The available data make this glacier suitable for evaluating mechanisms and controls on calving, some of which are considered in this paper. We use a full-Stokes thermomechanical 2D flow model (Elmer/Ice), paired with a crevasse-depth calving criterion, to estimate Hansbreen's front position at a weekly time resolution. The basal sliding coefficient is re-calibrated every 4 weeks by solving an inverse model. We investigate the possible role of backpressure at the front (a function of ice mélange concentration) and the depth of water filling crevasses by examining the model's ability to reproduce the observed seasonal cycles of terminus advance and retreat. Our results suggest that the ice-mélange pressure plays an important role in the seasonal advance and retreat of the ice front, and that the crevasse-depth calving criterion, when driven by modeled surface meltwater, closely replicates observed variations in terminus position. These results suggest that tidewater glacier behavior is influenced by both oceanic and atmospheric processes, and that neither of them should be ignored.

Keywords: tidewater glacier, Hansbreen, Svalbard, calving, terminus position, modeling

Edited by:
Timothy C. Bartholomaus,
University of Idaho, USA

Reviewed by:
Andrew John Sole,
University of Sheffield, UK
Chris Borstad,
University Centre in Svalbard, Norway

*Correspondence:
Jaime Otero
jaime.otero@upm.es

Specialty section:
This article was submitted to
Cryospheric Sciences,
a section of the journal
Frontiers in Earth Science

INTRODUCTION

Iceberg calving is one of the most important and least understood mechanisms of ice loss at ice sheet and marine-terminating glacier margins accounting for about half of the mass loss from the Greenland and Antarctic ice sheets (Cuffey and Paterson, 2010; Rignot et al., 2013). Although a third of the world's glaciated area (excluding the ice sheets) presently drains into the ocean (Gardner et al., 2013), very few estimates of frontal ablation (the overall mass loss due to iceberg calving and submarine melt) have been made for glaciers (Huss and Hock, 2015).

Benn et al. (2007) introduced a calving criterion based on the modeled penetration depth of surface crevasses, in turn a function of longitudinal stresses near the glacier terminus. In their model, calving occurs when crevasses reach the waterline (CDw model), a criterion supported by observations at many glaciers (e.g., Benn and Evans, 2010). The subaerial part of the calving face will typically calve first, followed by calving of the submerged buoyant ice toe (Motyka, 1997).

The crevasse-depth criterion was incorporated into a three-dimensional, full-Stokes glacier model of a tidewater glacier on Livingston Island, Antarctica (Otero et al., 2010). Their model could successfully predict the observed terminus position for a given glacier surface, bed geometry, and boundary conditions, but the glacier's evolution through time was not investigated. The CDw model was also applied to flowline modeling of Columbia Glacier, Alaska, by Cook et al. (2012). These studies have suggested that the calving rate is highly dependent on the depth of water filling the surface crevasses near the calving front.

Nick et al. (2010) implemented a modified crevasse-depth model in which the new calving front is defined as the point where water-filled surface crevasses and basal crevasses penetrate the full thickness of the glacier (CD model). Their conclusion was that both models, CDw and CD, produce qualitatively similar behavior.

Rather than introducing new calving criteria, other contributions to the calving problem—such as those by Amundson and Truffer (2010) and Bassis (2011)—have established frameworks that embrace existing calving models and serve as a guide to develop new ones.

Water filling crevasses is known to play an important role in calving processes, favoring calving through hydrofracturing (e.g., Scambos et al., 2003; Cook et al., 2012, 2014). Some modeling experiments have explored the influence of crevasse water depth, as well as other environmental variables (and associated processes) like basal water pressure, undercutting of the terminus by submarine melt, and backstress from ice mélange. Of these four variables, only crevasse water depth and basal water pressure were found by Cook et al. (2014) to have a significant effect on the terminus position of Helmhein Glacier, Greenland, when applied at realistic magnitudes. In contrast, Todd and Christoffersen (2014), whose study focused on the effect of ice mélange and submarine melting of Store Gletscher, Greenland, found that ice mélange was the primary driver of the observed seasonal advance of the glacier front. Luckman et al. (2015), in turn, studied two tidewater glaciers in Svalbard and found a statistical correlation between frontal ablation (the mass loss from both calving and submarine melting) and ocean temperatures between 20 and 60 m depth, suggesting that submarine melting may dominate frontal ablation. In the case of Hansbreen, thermo-erosional undercutting at the sea waterline, has been shown to play a role in calving (Petlicki et al., 2015).

Ice mélange, a heterogeneous mixture of sea ice and calved ice, can freeze solid and provide a stress opposing the flow of the glacier. This stress maintains the integrity of the calving margin, preventing calving (Amundson et al., 2010) and potentially slowing glacier flow (Walter et al., 2012). To our knowledge, Walter et al. (2012) is the only study that has measured the stress exerted on the front of a glacier by ice mélange, estimating a backstress of 30–60 kPa over the full calving face of Store Gletscher, Greenland.

Recently, Bondzio et al. (2016) presented a theoretical and technical framework for a level-set method (an implicit boundary tracking scheme) which they applied to Jakobshavn Isbræ, Greenland, using prescribed calving rates instead of a calving law. Morlighem et al. (2016) used this level-set method to model Store Gletscher, Greenland with a new calving law adapted from a von Mises yield criterion; their results suggested that calving is triggered by ocean-induced submarine melting.

In this paper, we present results from a numerical model developed using the open-source finite-element software Elmer/Ice (Gagliardini et al., 2013) and coupled to a crevasse-depth calving criterion. We use this model to investigate the seasonal dynamics of Hansbreen Glacier, a grounded tidewater glacier located near the Hornsund Polish Polar Station in southern Spitsbergen, Svalbard, with the aim of reproducing the terminus positions observed over a period of 132 weeks beginning September 2008 (assuming a week as a 1/48 of a year). We explore the sensitivity of the model to crevasse water depth (in turn a function of surface melt) and ice mélange backpressure. Although these two processes alone allow us to explain the observed seasonal variations of the glacier front, some other mechanism, such as ocean-induced submarine melting, remain to be investigated. As shown by Luckman et al. (2015) for nearby glaciers, the latter factor could be important also for Hansbreen at the end of the summer, when warmer water flows from the open ocean into Hornsund fjord.

Some previous modeling work has been applied to Hansbreen: Vieli et al. (2002) developed a flowline model with a prescribed seasonal calving rate and a modified flotation criterion, while Oerlemans et al. (2011) applied a "minimal model" to qualitatively understand Hansbreen's dynamics in broad terms. Our work represents a step forward, as it uses improved dynamical and calving models and avoids prescribing an a-priori calving rate.

GEOGRAPHICAL SETTING AND AVAILABLE DATA

Hansbreen is a polythermal tidewater glacier which flows into Hornsund fjord in southern Spitsbergen (**Figure 1**). It is about 16 km long and covers an area of 57 km^2 from 0 to 500 m above sea level (a.s.l.). The glacier terminus is about 2.5 km wide, the central 1.5 km of which sits in water. The ice thickness of the central flowline at the terminus is about 100 m, of which 55 m are submerged. The glacier lies on a reverse-sloping bed for the first 4 km up-glacier from the terminus and the center of the fjord lies below sea level as far as 10 km up-glacier. The maximum ice thickness is about 400 m. Further, details on the glacier surface and bed morphology can be found in Grabiec et al. (2012).

Glacier Geometry

To account for gentle surface slopes, glacier surface elevations were taken from the SPIRIT Digital Elevation Model (DEM) V1 (whose correlation parameters are set for gentle slopes), based on SPOT5 Stereoscopic Survey of Polar Ice imagery acquired on 1 September 2008. The DEM has a 40 m resolution and a 30 m root-mean-square (RMS) absolute horizontal precision (http://polardali.spotimage.fr:8092/wstools/IPY/).

Bed topography was derived from ground-penetrating radar (GPR) data (Grabiec et al., 2012; Navarro et al., 2014) and depth

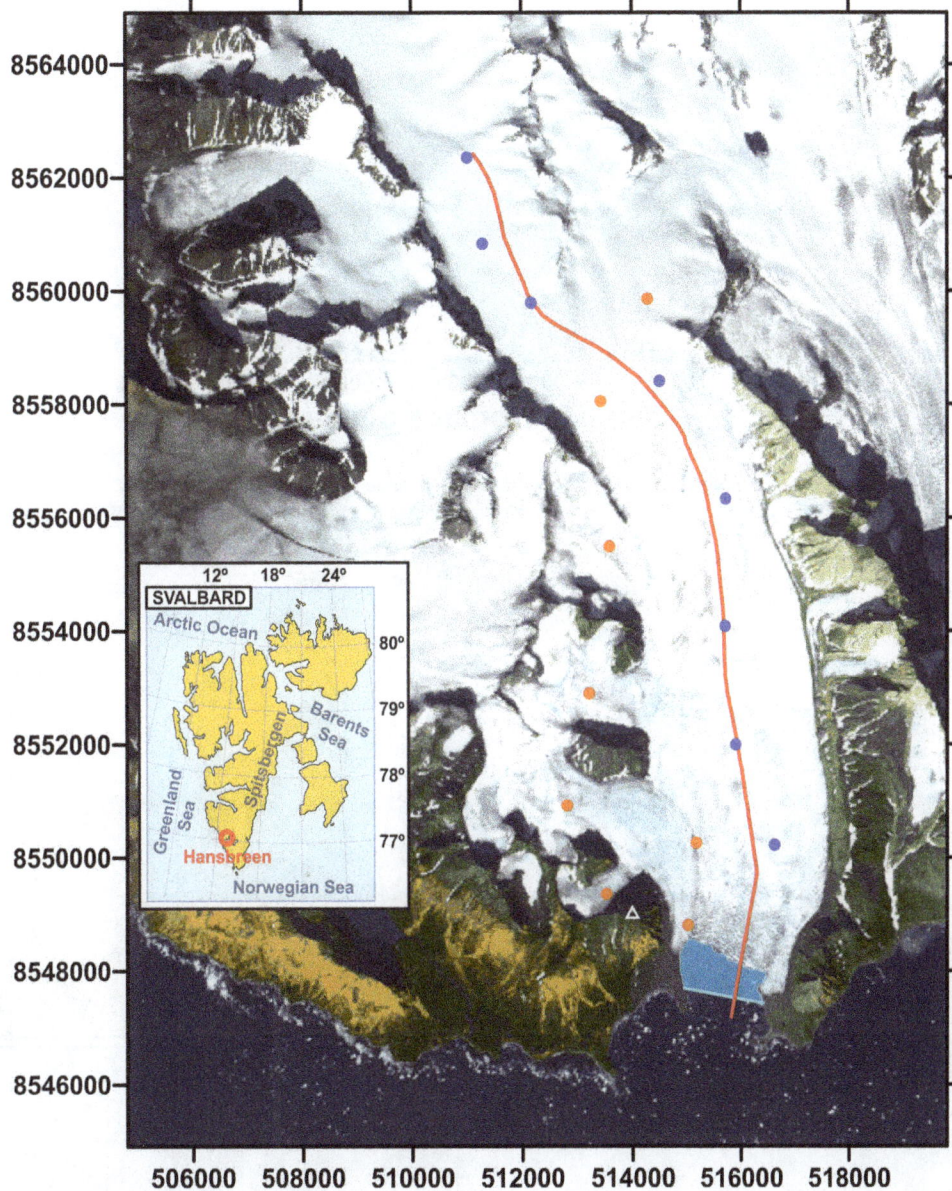

FIGURE 1 | Location of Hansbreen in Spitsbergen, Svalbard (inset). ASTER image of Hansbreen taken in 2011 showing the location of the modeled flowline (red line) and the locations of the stakes for velocity measurements (colored circles; the blue ones were used in this paper). The white triangle indicates the position of Fugleberget Peak. The blue polygon indicates the portion of open water over which the relative coverage of mélange/sea ice was quantified. UTM coordinates for zone 33X are included.

soundings in the glacier forebay (Vieli et al., 2002). The resulting initial geometry of the modeled glacier is shown in **Figure 2**.

Surface Velocity

Surface velocities were measured daily at 16 stakes (**Figure 1**) between May 2005 and April 2011, using differential GPS (Puczko, 2012). In this study, we focus on a subset of eight stakes close to the modeled flowline (**Figure 3**). We additionally use velocities of the calving front measured in 2009 by terrestrial laser scanning (TLS; data generously provided by Jacek Jania

[University of Silesia] from surveying and data processing by Jacek Krawiec [Laser 3D], Artur Adamek [Warsaw University of Technology], and Jacek Jania). Since a value for the velocity at the calving front is needed at each time step during the entire simulation period, but we only have TLS measurements for 2009, we assume a constant ratio between the frontal velocities and those at the closest stake.

The near-terminus surface velocities exhibit a seasonal pattern overlaid by strong interannual variability. Each year has a spring speed up, followed by a rapid slow down, followed sometimes

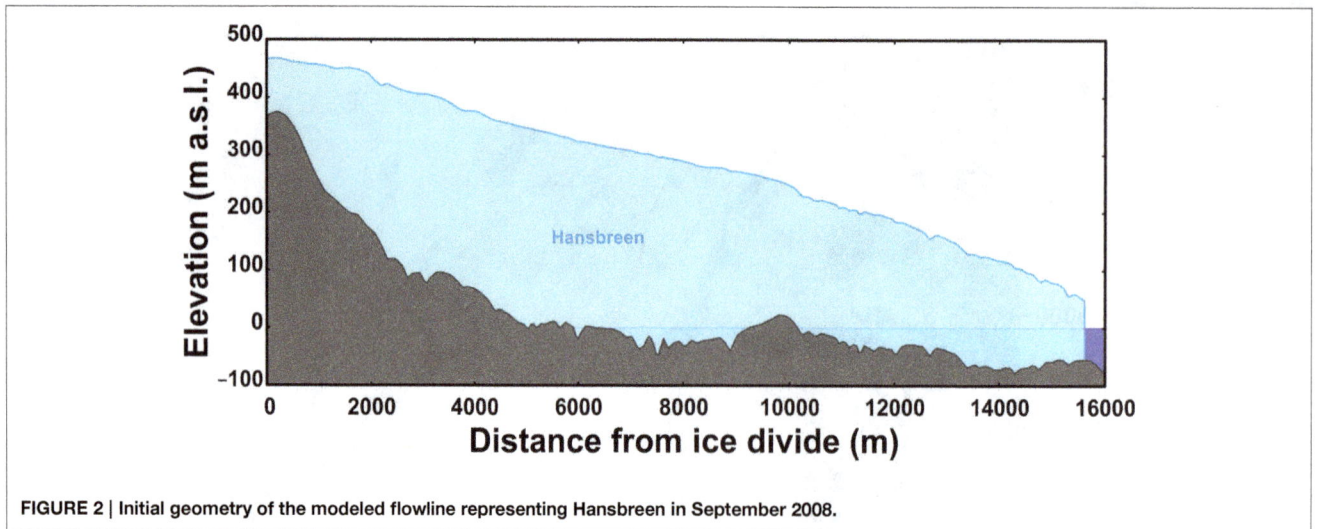

FIGURE 2 | Initial geometry of the modeled flowline representing Hansbreen in September 2008.

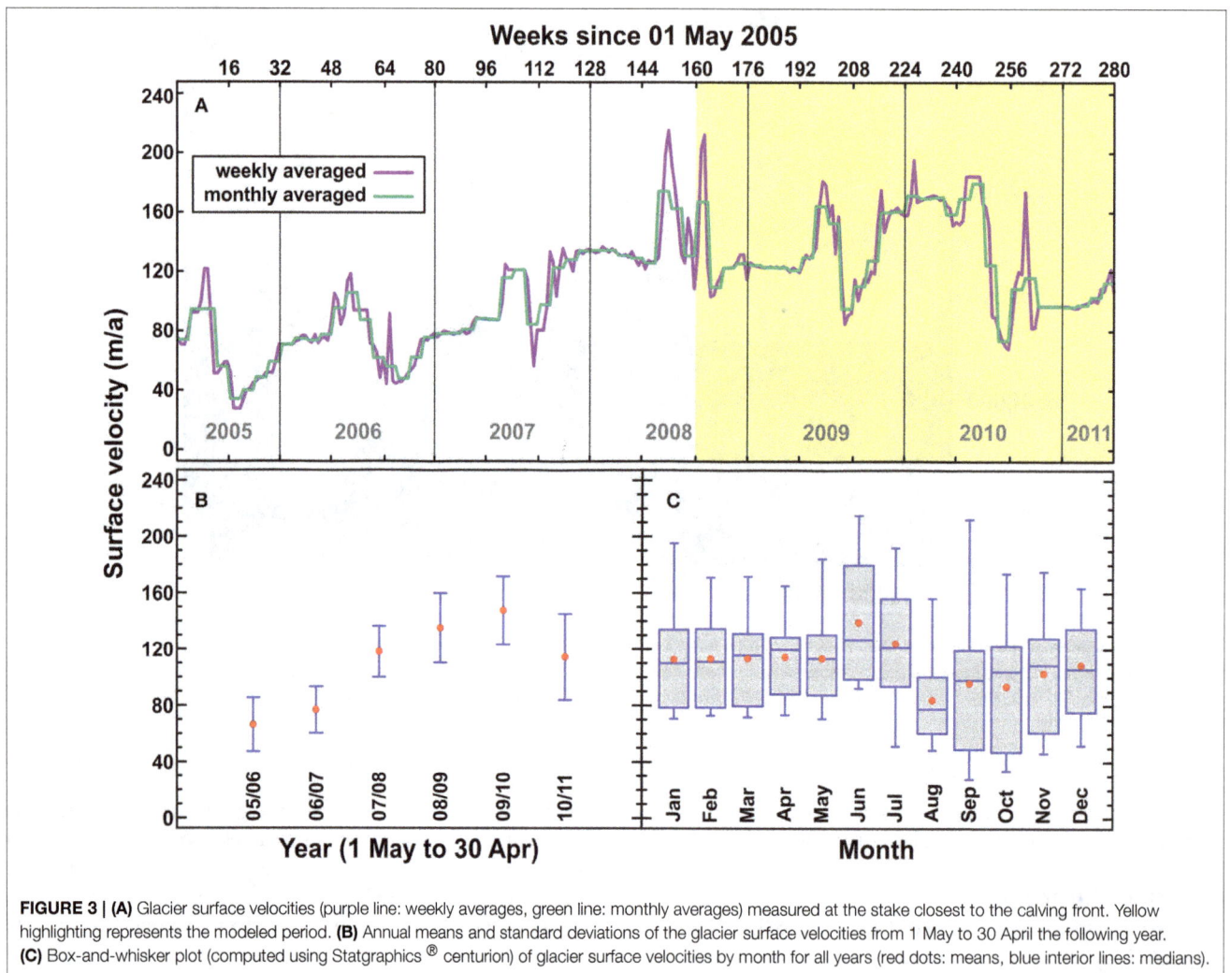

FIGURE 3 | (A) Glacier surface velocities (purple line: weekly averages, green line: monthly averages) measured at the stake closest to the calving front. Yellow highlighting represents the modeled period. (B) Annual means and standard deviations of the glacier surface velocities from 1 May to 30 April the following year. (C) Box-and-whisker plot (computed using Statgraphics ® centurion) of glacier surface velocities by month for all years (red dots: means, blue interior lines: medians).

by a gradual speed up through the winter (**Figure 3A**). In 2005–2006, the highest velocities occurred in summer, while in 2007 the highest velocities occurred in November–December. In

2010, velocities remained very high through winter and spring before dropping dramatically during the summer. In 2008–2011, velocities were much higher than in 2005–2008. The mean varied considerably between years (**Figure 3B**), discouraging the use of a single mean velocity in the model. Similarly, the very large interannual variability discourages the use of summer and winter means for summer and winter modeled periods (**Figure 3C**).

Front Positions

Interannual observations of Hansbreen's terminus position over the last decades reveal a generally smooth retreat with occasional abrupt changes (e.g., Vieli et al., 2002). Seasonal variations of the terminus position have also been observed (Blaszczyk et al., 2013). The "weekly" (assuming a week as a 1/48 of a year) terminus positions used in this paper were derived from time-lapse photographs of the Hansbreen terminus taken ca. every 3 h from December 2009 to September 2011 by three different cameras installed at surveyed positions on the eastern slope of Fugleberget (**Figure 1**). The two Canon EOS 1000D cameras (each equipped with a Canon EF-S 18–55 mm lens) were calibrated from images of a grid pattern using the Camera Calibration Toolbox for Matlab (www.vision.caltech.edu/bouguetj/calib_doc/), while the Canon Powershot A530 camera (which no longer existed at the time of the analysis) had to be calibrated and oriented simultaneously from multiple images of ground control points surveyed by L. Kolondra (unpublished report, 2011). Terminus positions were mapped by tracing the waterline along the terminus in each image, then projecting those pixel coordinates onto a horizontal plane at an altitude of 0 m (tidal heights were not available) to convert them to world coordinates. Standard errors of 0.47 m in water level due to tides (Zagórski et al., 2015; Michał Ciepły, personal communication, 2012), 0.68 pixels due to uncertainties in terminus tracing, and 3.23, 5.70, and 10.53 pixels for each of the three cameras due to uncertainties in camera calibration and orientation result in standard errors for width-averaged glacier length of 3.79, 3.37, and 14.54 m, respectively.

Ice Mélange

The ice mélange in the glacier forebay was qualitatively evaluated as either "complete," "partial," or "free" from the same time-lapse photographs used to measure terminus position. For the remaining period, we used the values of the nearest cell

in a 25-km resolution time series of sea ice concentrations derived from Nimbus-7 SMMR and DMSP SSM/I-SSMIS passive microwave data (Cavalieri, 1996, updated yearly). These values were used taking into account the partial overlap of the grid cell with Hornsund fjord mouth, and their comparable sizes (**Figure 4**).

Surface Mass Balance and Surface Meltwater

We apply a dynamical downscaling method (which uses a modified version of the Polar WRF 3.4.1 model) to produce—from a regional climate dataset consisting of meteorological, sea-surface temperature and sea-ice concentration data—input data for glacier thermomechanical modeling of Hansbreen (Finkelnburg et al., in preparation).

Surface mass balance (SMB) was obtained from European Arctic Reanalysis (EAR) data, with 2 km horizontal resolution and hourly temporal resolution, constrained by automatic weather stations (one in Hornsund and two in Hansbreen) and stake observations (Finkelnburg, 2013). First, ablation was calculated from the surface energy balance (SEB), which is resolved in the EAR by an optimized version of the unified NOAA Land Surface Model (Chen and Dudhia, 2001) of the Polar WRF 3.4.1 model. The algorithm solving for the SEB takes into account net radiation, sensible heat flux, latent heat flux, and ground heat flux, and encompasses all heat fluxes involved in melt and refreezing processes within the snowpack. Second, accumulation was obtained as the solid (frozen) precipitation of the Morrison bulk microphysics scheme for cloud physics used by the EAR. Finally, monthly mean SMB and surface meltwater (SMW) at each flowline point was calculated by applying bilinear interpolation to the available 2-km resolution hourly accumulation and ablation data (**Figure 4**).

MODEL DESCRIPTION

Dynamical Model Equations and Flow Law

Ice is treated as an incompressible viscous fluid. The Stokes system of equations describing the dynamical model is composed

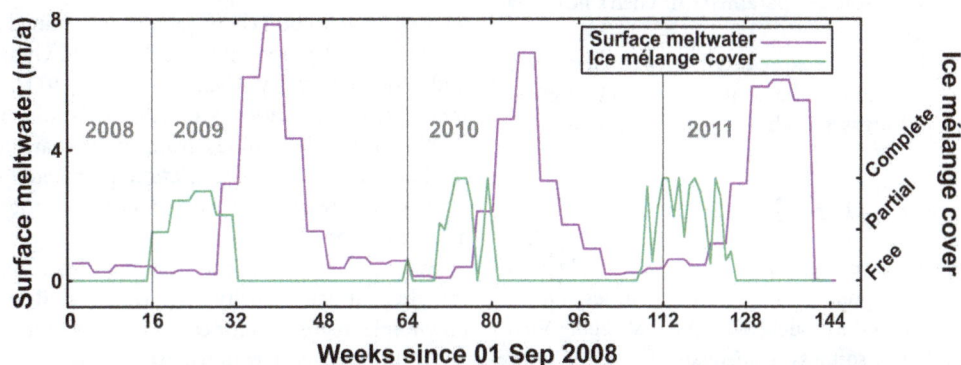

FIGURE 4 | Modeled surface meltwater near the calving front (purple line) and observed ice mélange coverage (weekly average of daily observations) in the glacier forebay (green). Week means here 1/48 of a year.

of equations describing the steady conservation of linear momentum and the conservation of mass of an incompressible continuous medium:

$$div\,(\sigma) + \rho g + f = 0, \tag{1}$$
$$div\,(u) = 0,$$

where σ is the Cauchy stress tensor, u is the velocity vector, g is the gravity force vector and ρ is the density. The body force f is added to account, in our 2-dimensional (2-D) model, for the friction on the lateral side of the glacier. To this end, the concept of shape factor (Nye, 1965) is here extended to the full-Stokes formulation by defining the body force f as (Jay-Allemand et al., 2011):

$$f = -\rho g \cdot t \left(1 - f\right) t, \tag{2}$$

where the shape factor $f = f(x)$ is a scalar function of the transversal shape of the glacier and t is the unit vector tangent to the upper surface. Jay-Allemand et al. (2011) evaluated $f(x)$ by assuming that the transverse shape of the bedrock is a parabola, and they found an empirical estimate of the shape factor:

$$f = \frac{2}{\pi}\tan^{-1}\left(\frac{0.186w}{h}\right), \tag{3}$$

where $h\,(x) = z_s\,(x) - b(x)$ is the ice thickness (expressed as the difference between the surface and bed elevations) and $w(x)$ is the half-width at the glacier surface.

As the constitutive relation, we adopt Nye's generalization of Glen's flow law (Glen, 1955; Nye, 1957):

$$\tau = 2\eta\dot{\varepsilon} \tag{4}$$

This equation links the deviatoric stress τ to the strain rate $\dot{\varepsilon}$. The effective viscosity η is written as

$$\eta = \frac{1}{2}(EA)^{-1/n}I_{\dot{\varepsilon}_2}^{(1-n)/n}, \tag{5}$$

where $I_{\dot{\varepsilon}_2}^2$ represents the square of the second invariant of the strain rate tensor, A is the softness parameter in Glen's flow law and E is an enhancement factor.

The constitutive relation (4) is expressed in terms of deviatoric stresses, while the conservation of linear momentum (1) is given in terms of full (Cauchy) stresses. Both stresses are linked through the equation

$$\sigma = \tau - pI\,,\ p = -tr(\sigma)/3, \tag{6}$$

where p is the pressure (compressive mean stress). Typical values for the flow-law exponent ($n = 3$) and softness ($A = 0.1\ \text{bar}^{-3}\text{a}^{-1}$) were used in the model (Albrecht et al., 2000; Vieli et al., 2002). This value of softness is adequate (see e.g., Cuffey and Paterson, 2010) for a polythermal glacier like Hansbreen, composed mostly of temperate ice except for a thin upper layer of cold ice in the ablation zone (Jania et al., 1996).

Continuum Damage Mechanics Model

We introduce a scalar damage variable D that quantifies the loss of load-bearing surface area due to fractures, known as fracture-induced softening (Borstad et al., 2012). This softening is taken into account through the introduction of the damage within Glen's law. Following Borstad et al. (2012) and Krug et al. (2014), the enhancement factor E can be linked to the damage D as

$$E = \frac{1}{(1-D)^n} \tag{7}$$

For undamaged ice ($D = 0$), $E = 1$ and the flow regime is unchanged. As damage increases ($D > 0$), $E > 1$, ice viscosity decreases, and flow velocity increases.

In this study, we use a very simple function for D. Damage is nonzero only in the lowermost 2 km of the glacier (near-terminus heavily crevassed area) and increases linearly toward the terminus, where it reaches a maximum value of 0.4 (Krug et al., 2015). This value provided a good fit to observed velocities in preliminary experiments (not shown here) of the sensitivity of the modeled velocities to changes in damage.

Free Surface Evolution

The time evolution of the glacier surface is calculated by solving the free-surface evolution equation

$$\frac{\partial z_S}{\partial t} = b - u_S\frac{\partial z_S}{\partial x} + v_S, \tag{8}$$

where z_S is the surface elevation, t is time, u_S and v_S are the horizontal and vertical components of the flow velocity at the surface, respectively, and b is the surface mass balance.

Boundary Conditions

The upper surface of the glacier is a traction-free zone with unconstrained velocities. At the ice divide at the head of the glacier, horizontal velocity and shear stresses are set to zero.

For boundary conditions at the bed, we use a friction law that relates the sliding velocity to the basal shear stress in such a way that the latter is not set as an external condition but part of the solution:

$$Cu_t = \sigma_{nt} \tag{9}$$

where C, the friction coefficient, is determined using the inverse Robin method proposed by Arthern and Gudmundsson (2010) and modified by Jay-Allemand et al. (2011). The latter study includes a regularization parameter, λ, for which we have adopted a value of 0.4 determined from preliminary tests. The inverse method infers the basal friction parameter by reducing the mismatch between observed and modeled surface velocities using a cost function.

Since the inversion procedure requires a continuous function for the surface velocity, we calculate it as a sixth-degree polynomial regression for each 4-week period.

At the glacier terminus, we set backstress to zero above sea level and equal to the water-depth-dependent hydrostatic pressure below sea level. In model runs with ice mélange, the additional backstress is applied to the calving face, in the opposite direction of ice flow (negative X). In the absence of further data, we assumed the range of 30–60 kPa estimated by Walter et al. (2012) for Store Gletscher as indicative of the order of magnitude that we could expect for Hansbreen, despite their different settings. While Store Gletscher is buttressed by a rigid proglacial mélange, with a thickness reaching 75 m (Todd and Christoffersen, 2014), Hansbreen presents a thinner layer of ice mélange made up of a mixture of growlers, bergy bits and small icebergs bonded by sea ice. As no measurements of ice mélange thickness in Hansbreen forebay are available, but the sea-ice maximum thickness in Hornsund fjord is known to be around 1 m (Kruszewski, 2012), we assume a mean ice mélange freeboard height of 0.5 m and a mean thickness of 4.5 m.

Calving Model

The CDw calving criterion (Benn et al., 2007) assumes that calving is triggered by the downward propagation of transverse surface crevasses near the calving front as a result of the extensional stress regime. Following Nye (1957), crevasse depth is calculated as the depth where the longitudinal tensile strain rate tending to open the crevasse equals the creep closure resulting from the ice overburden pressure. This procedure incorporates the full stress solution into the crevasse depth criterion. In Benn's model, calving is assumed to occur when surface crevasses reach the waterline.

Following Todd and Christoffersen (2014), crevasse depths are calculated from the balance of forces:

$$\sigma_n = 2\tau_e sgn(\tau_{xx}) - \rho_i gd + P_w \qquad (10)$$

where σ_n, the "net stress," is positive for extension and negative for compression. The first term on the right-hand side of Equation (10) represents the opening force of longitudinal stretching, adapted by Todd and Christoffersen (2014) from Otero et al. (2010); τ_e represents the effective stress, $\tau_e^2 = \tau_{xx}^2 + \tau_{zx}^2$. τ_e is multiplied by the sign function of the longitudinal deviatoric stress, τ_{xx}, to ensure that crevasse opening is only produced under longitudinal extension ($\tau_{xx} > 0$). The second term on the right-hand side is the ice overburden pressure, which leads to creep closure, where ρ_i is the density of glacier ice, g is the acceleration of gravity and d is the crevasse depth. The last term represents the water pressure which contributes to open the crevasse, which is a function of the depth of water filling the crevasse.

Numerical Solution

At each time step, the glacier is divided into a rectangular mesh with 10 vertical layers and a horizontal grid size of ca. 50 m in the upper glacier and ca. 25 m near the terminus. The Stokes system of equations (1) is solved by a finite element method using Elmer/Ice and the 2-D stress and velocity fields are computed along the central flowline (**Figure 1**). The new surface elevations are computed from the surface mass-balance input and the solved surface velocities using the free-surface evolution equation and the grid nodes are shifted vertically to fit the new geometry.

At the terminus, the grid nodes are shifted down-glacier according to the velocity vector and the length of the time step and the terminus position is updated according to the calving criterion.

Prognostic model runs were carried out with a 1-week (1/48 of a year) time step, starting from the 2008 glacier geometry. Every 4 weeks (four time steps), we ran an initialization process which consisted of solving the Robin problem (Jay-Allemand et al., 2011) to force a best-fit friction coefficient to be used for those four model runs. This forcing was done to minimize the misfit between the observed and modeled velocities. The choice of the initialization time step was made as a compromise between the time resolution needed for capturing the sudden changes in velocity and an acceptable computational cost.

NUMERICAL EXPERIMENTS AND THEIR RESULTS

Our aim was to investigate the influence of ice mélange backstress and crevasse water depth on terminus position. Given the absence of field measurements, we parameterized crevasse water depth in terms of surface meltwater.

First, we analyzed the effect of crevasse water depth held fixed throughout the entire modeled period. Under this scenario, it was not possible to replicate the observed terminus position variations; instead, we constrained the magnitude of the crevasse water depth to that which best approximates the observed terminus positions. Using this best-fit crevasse water depth, we ran a similar sensitivity analysis for ice mélange backstress and determined the backstress that best fits the observed terminus positions. Finally, using this best-fit ice mélange backstress, we ran the model with a time-varying crevasse water depth d_w expressed as a linear function of the surface meltwater M_w (units meters per week) predicted by the SEB model, i.e., $d_w = k M_w$, where k is a tuning coefficient.

This experiment was repeated for a range of values for the linear coefficient, and the results corresponding to the best-fitting value very closely matched the observed terminus position variations.

Crevasse Water Depth

Given the difficulties of measuring the depth of water in crevasses, we ran the model for a range of crevasse water depths (from 6 to 12 m) to evaluate the sensitivity of the model to this parameter. We found that calving rate is highly dependent on the depth of water in crevasses, with an increase of just a few meters causing the glacier to switch from advance to retreat (**Figure 5**). If a constant water depth is prescribed, the model is unable to reproduce the observed terminus position fluctuations (**Figure 5**), although the results allow us to select the water depth which, on average, best fits the observations (10 m, as illustrated by **Figure 5**).

Ice Mélange

To test the effect of ice mélange on calving rate and terminus position, we varied ice mélange backstress from 0 to 70 kPa (based on Walter et al., 2012, as discussed in the Introduction) in steps of 10 kPa and multiplied by the fraction of ice mélange coverage (weekly average) in the glacier forebay, varying from 0 when no ice mélange is present to 1 when ice mélange completely fills the glacier forebay. When there is only partial coverage of ice mélange, ice mélange remains concentrated near the margins of the glacier and the fjord and therefore continues to exert some backstress on the glacier front. This supports the prescription of backstress even at low ice mélange concentrations. In this experiment, we prescribed a constant crevasse water depth of 10 m, the best-fit value from the previous experiment.

We found that the effect of ice mélange backstress on glacier front position was significant, even under low stresses (**Figure 6**), and that 50 kPa (for full mélange coverage), applied to the front of the glacier, yielded the best fit to observations during the winter when ice mélange was present. The effect of a backstress of 50 kPa on the modeled longitudinal deviatoric stress profile is shown in **Figure 7**.

Surface Meltwater

As discussed by Todd and Christoffersen (2014), the relationship between surface melt rate and crevasse water depth depends on many factors, including the distribution, shape, and depth of crevasses, the melting and refreezing on crevasse walls, and the potential drainage of water from crevasses into englacial, subglacial, and proglacial water bodies. Since observations of these processes are very scarce and the water in crevasses starts as surface meltwater, we chose to parameterize crevasse water depth

in terms of the available surface meltwater predicted by the SEB model (**Figure 8**).

Use of such a parameterized time-varying crevasse water depth, in combination with the best-fit ice mélange backstress from the previous experiment, yielded terminus positions in very good agreement with the observations (**Figure 8**), reproducing the winter advance and the subsequent summer retreat.

DISCUSSION

In tidewater glacier modeling, it is common practice to use mean annual surface velocities (e.g., Cook et al., 2012, 2014) or seasonal means (e.g., Vieli et al., 2002; Todd and Christoffersen, 2014) to tune the free parameters of the model. In this study, we use a 4-week mean of daily velocity observations to infer a sliding parameter for each 4-week period, with the aim of obtaining more realistic modeled velocities and, consequently, a more realistic calving rate. An accurate representation of velocities is key to glacier models which use a crevasse-depth calving criterion, since such a calving criterion relies on the stress field derived from the velocity field and its associated strains. In particular, higher temporal resolutions for velocities and other model parameters are necessary to capture speed-up events such as those observed at Hansbreen (**Figure 3A**). Because of the short duration of these events, they have a negligible effect on mean

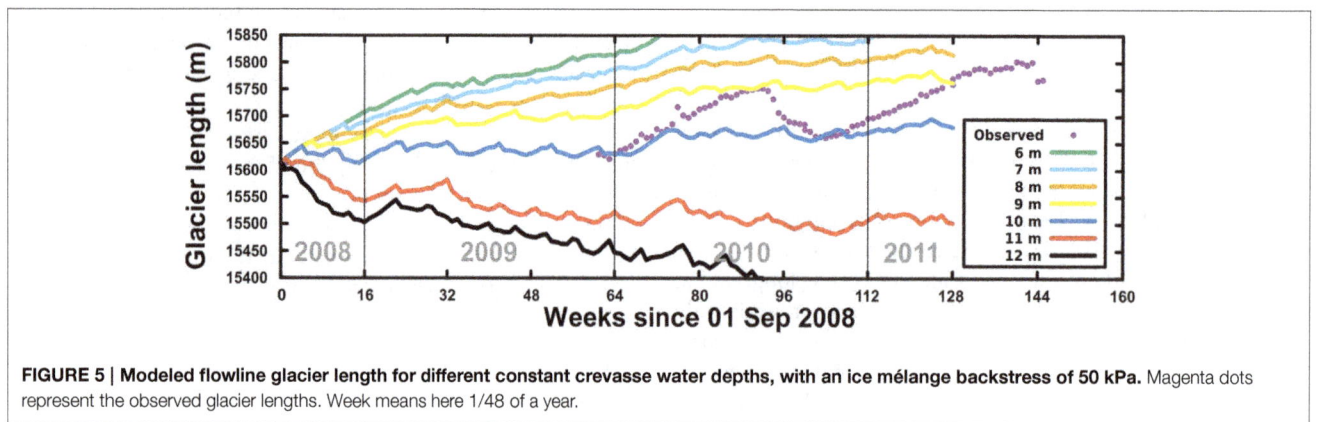

FIGURE 5 | Modeled flowline glacier length for different constant crevasse water depths, with an ice mélange backstress of 50 kPa. Magenta dots represent the observed glacier lengths. Week means here 1/48 of a year.

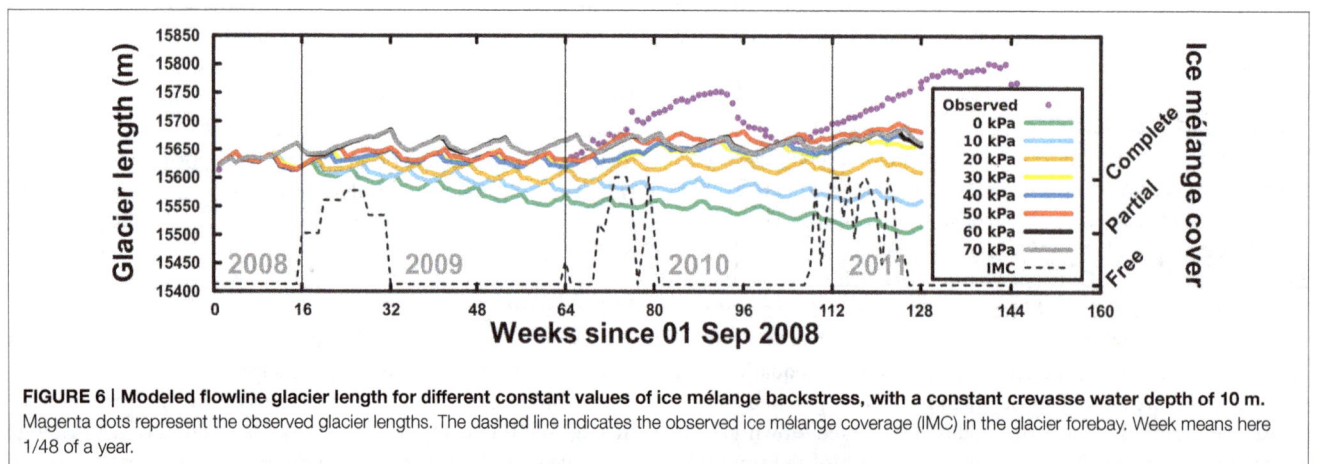

FIGURE 6 | Modeled flowline glacier length for different constant values of ice mélange backstress, with a constant crevasse water depth of 10 m. Magenta dots represent the observed glacier lengths. The dashed line indicates the observed ice mélange coverage (IMC) in the glacier forebay. Week means here 1/48 of a year.

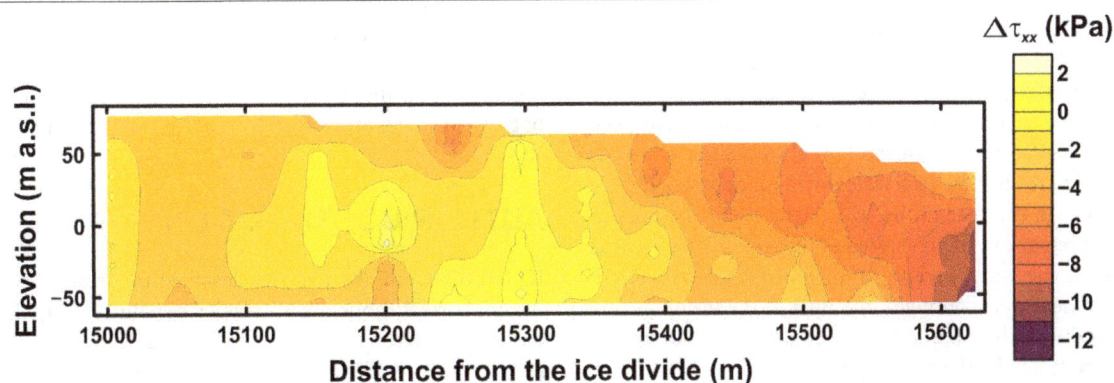

FIGURE 7 | Change in longitudinal deviatoric stress, in the terminal part of the glacier, resulting from an increase in ice mélange backstress from 0 to 50 kPa at the calving front.

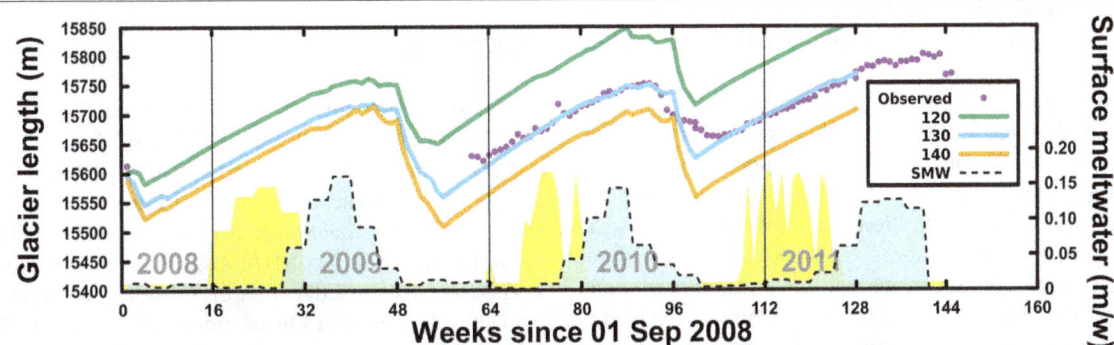

FIGURE 8 | Modeled glacier length for crevasse water depths parameterized as a linear function of the available surface meltwater for a range of values of the linear coefficient *k* (shown in box), with an ice mélange backstress of 50 kPa. Magenta dots represent the observed glacier lengths. The dashed line indicates the modeled surface meltwater near the calving front. Yellow shaded areas represent the observed ice mélange coverage in the glacier forebay. Week means here 1/48 of a year.

annual and seasonal velocities, but have a significant impact on the stress regime and therefore the calving rate.

Our results demonstrate that our model is capable of reproducing the seasonal fluctuations of the terminus position of Hansbreen, provided that the key model variables are adequately tuned and parameterized.

The modeled terminus positions are shown to be highly sensitive to changes in crevasse water depth, in agreement with previous studies (e.g., Cook et al., 2012). In particular, we found that a small change in depth, from 10 to 11 m, resulted in a switch from advance to retreat (**Figure 5**). As small changes to parameter values can lead to large changes in the model results (i.e., a mathematical instability), extreme care should be taken when implementing crevasse-depth criteria in prognostic models. Besides, when a constant crevasse water depth is applied the results show a several-month periodicity (**Figures 5, 6**). This is a consequence of the CDw criterion, as at each time step the glacier terminal zone thins by ablation until the threshold for calving is reached, and then process restarts.

The water in crevasses is mostly produced by melting at the glacier surface, so the mean water depth in crevasses can be parameterized in terms of surface melting, which can be

modeled either using air temperature or temperature-radiation index models (e.g., Jonsell et al., 2012) or energy-balance models (e.g., Hock and Holmgren, 2005) trained by observed and/or modeled data (e.g., Finkelnburg, 2013). This allows us to replace a parameter inherently difficult to measure in the field by another that can be based on easier field observations, models, or both.

In contrast with one previous modeling study (Cook et al., 2014), but in agreement with another (Todd and Christoffersen, 2014), our results suggest that ice mélange backstress is an important control on calving and should not be ignored. A change from 0 to 50 kPa in backstress resulted in a change of ca. 170 m in glacier length over 2.5 years (**Figure 6**). In addition, our simulations show how the presence of ice mélange may prevent calving, as others have established from observations (Amundson et al., 2010; Howat et al., 2010). When a backstress of 50 kPa of ice mélange is present, the glacier advances but does not calve. Therefore, any higher backstress (up to 70 kPa) has no further effect on calving and has only a very minor effect on front position. Calving, accompanied by ice front retreat, is only occasionally produced during the warmer periods, in absence of ice mélange.

Both our model and the observations show frontal retreat beginning soon after the peak in surface meltwater, although the maximum calving occurs a few weeks later, suggesting a delayed response by the glacier system to meltwater input. The source of this lag could be two-fold. On one hand the cumulative effect of thinning by ablation in both the real glacier and the model, which helps the crevasses to penetrate down to the waterline. On the other hand, in the case of the real glacier there is a buildup of the water pressure in the crevasses as meltwater accumulates; alternatively, if the water escapes from the crevasses there will also be a cumulative weakening of the bulk of the terminal zone of the glacier due to the enlargement of the conduits and fissures by melting promoted by the escaping water.

Even though our model does a good job reproducing the observed front positions, it does not consider other possible controls on the calving process, specifically ocean-induced melting. Adequately incorporating this mechanism would require developing a fjord circulation model to estimate the subaqueous melt rates and couple the fjord and glacier systems.

In regards to possible shortcomings of the model, we note that we added a body force term to the Stokes system (Equation 1) to take into account, in our 2-D flowline model, of the lateral drag on the glacier sidewalls. However, this body force term does not take into consideration the effect of ice flow from tributaries on the central flowline dynamics. This effect is expected to be significant at Hansbreen, which has three tributary glaciers flowing into the main branch near the terminus (**Figure 1**). To properly model the lateral drag, and take into account the tributary glaciers, a 3-D model would be necessary.

CONCLUSIONS

In this study, we investigate the relative importance of some proposed controls on calving—namely, crevasse water depth and ice mélange backstress—and evaluate their influence on the terminus position changes of a tidewater glacier.

Our results suggest that ice mélange backstress plays an important role in regulating the seasonal advance and retreat of the terminus, mostly by preventing calving when the mélange chokes the fjord. The model results also indicate that calving and the associated terminus position changes are highly sensitive to the amount of water filling near-terminus crevasses, itself a function of surface meltwater availability. The sensitivity of calving rate to crevasse water depth suggests that calving is strongly affected by atmospheric forcing. These results, taken together, show that tidewater glacier dynamics are influenced by both oceanic and atmospheric processes, and that neither of them should be ignored.

AUTHOR CONTRIBUTIONS

JO led the study. JO and FN designed the experiments. JO did the numerical modeling of glacier dynamics and RF provided regional climate modeling and SEB data. JL provided GPR data, DP ice velocity and AWS data, and EW terminus positions and sea ice coverage data. JO, FN, and JL contributed to the discussion of the results. JO and FN wrote the initial draft of the paper primarily, JL made the figures. All authors contributed to and approved the final manuscript.

FUNDING

This research was funded by Spanish State Plan for Research and Development (R&D) project CTM2014-56473-R. Field data collection was funded by grants EUI2009-04096 and CTM2011-28980 of the Spanish Programmes of Euro-Research and R&D, respectively, and the Polish National Science Centre within statutory activities No3841/E-41/S/2014 of the Ministry of Science and Higher Education of Poland. The regional climate modeling data was produced under grant no. SCHE 750/3-1 of the German Research Foundation (DFG) and grant no. 03F0623A of the German Federal Ministry of Education and Research (BMBF).

ACKNOWLEDGMENTS

This research was carried out under the frames of the International Arctic Science Committee Network on Arctic Glaciology (IASC-NAG) and the European Science Foundation PolarCLIMATE programme's SvalGlac project. The satellite images used in this paper were provided by the SPIRIT Program © CNES (2008), Spot Image, and ASTER © METI and NASA (2011), all rights reserved. The surface velocity data used in the paper were collected based on Stanislaw Siedlecki Polish Polar Station in Hornsund.

REFERENCES

Albrecht, O., Jansson, P., and Blatter, H. (2000). Modelling glacier response to measured mass-balance forcing. *Ann. Glaciol.* 31, 91–96. doi: 10.3189/172756400781819996

Amundson, J. M., Fahnestock, M., Truffer, M., Brown, J., Lüthi, M. P., and Motyka, R. J. (2010). Ice mélange dynamics and implications for terminus stability, Jakobshavn Isbræ, Greenland. *J. Geophys. Res.* 115:F01005. doi: 10.1029/2009JF001405

Amundson, J. M., and Truffer, M. (2010). A unifying framework for iceberg-calving models. *J. Glaciol.* 56, 822–830. doi: 10.3189/002214310794457173

Arthern, R., and Gudmundsson, G. (2010). Initialization of ice-sheet forecasts viewed as an inverse Robin problem, *J. Glaciol.* 56, 527–533. doi: 10.3189/002214310792447699

Bassis, J. N. (2011). The statistical physics of iceberg calving and the emergence of universal calving laws. *J. Glaciol.* 57, 3–16. doi: 10.3189/002214311795306745

Benn, D. I., and Evans, D. J. A. (2010). *Glaciers and Glaciation, 2nd Edn.* London; New York, NY: Hodder Education.

Benn, D. I., Hulton, N. R. J., and Mottram, R. H. (2007). ≪Calving laws≫, ≪sliding laws≫ and the stability of tidewater glaciers. *Ann. Glaciol.* 46, 123–130. doi: 10.3189/172756407782871161

Blaszczyk, M., Jania, J. A., and Kolondra, L. (2013). Fluctuations of tidewater glaciers in Hornsund Fjord (Southern Svalbard) since the beginning of the 20 th century. *Polish Polar Res.* 34, 327–352. doi: 10.2478/popore-2013-0024

Bondzio, J. H., Seroussi, H., Morlighem, M., Kleiner, T., Rückamp, M., Humbert, A., et al. (2016). Modelling calving front dynamics using a level-set method: application to Jakobshavn Isbræ, West Greenland. *Cryosphere* 10, 497–510. doi: 10.5194/tc-10-497-2016

Borstad, C. P., Khazendar, A., Larour, E., Morlighem, M., Rignot, E., Schodlok, M. P., et al. (2012). A damage mechanics assessment of the Larsen B ice shelf prior to collapse: toward a physically-based calving law. *Geophys. Res. Lett.* 39, L18502. doi: 10.1029/2012gl053317

Cavalieri, D. J., Parkinson, C. L., Gloersen, P., and Zwally, H. J. (1996). updated yearly. *Sea Ice Concentrations from Nimbus-7 SMMR and DMSP SSM/I-SSMIS Passive Microwave Data, Version 1.* Boulder, CO: USA. NASA National Snow and Ice Data Center Distributed Active Archive Center.

Chen, F., and Dudhia, J. (2001). Coupling an advanced land surface-hydrology model with the Penn State-NCAR MM5 Modeling System. Part I: Model Implementation and Sensitivity. *Monthly Weather Rev.* 129, 569–585. doi: 10.1175/1520-0493(2001)129<0569:CAALSH>2.0.CO;2

Cook, S., Rutt, I. C., Murray, T., Luckman, A., Zwinger, T., Selmes, N., et al. (2014). Modelling environmental influences on calving at Helheim Glacier in eastern Greenland. *Cryosphere* 8, 827–841. doi: 10.5194/tc-8-827-2014

Cook, S., Zwinger, T., Rutt, I. C., O'Neel, S., and Murray, T. (2012). Testing the effect of water in crevasses on a physically based calving model. *Ann. Glaciol.* 53, 90–96. doi: 10.3189/2012AoG60A107

Cuffey, K., and Paterson, W. (2010). *The Physics of Glaciers. Elsevier, 4th Edn.* Oxford: Elsevier.

Finkelnburg, R. (2013). *Climate Variability of Svalbard in the First Decade of the 21st Century and its Impact on Vestfonna ice cap, Nordaustlandet - An Analysis based on Field Observations, Remote Sensing and Numerical Modeling.* Doctoral thesis, Technische Universität Berlin. Berlin.

Gagliardini, O., Zwinger, T., Gillet-Chaulet, F., Durand, G., Favier, L., de Fleurian, B., et al. (2013). Capabilities and performance of Elmer/Ice, a new-generation ice sheet model. *Geosci. Model Dev.* 6, 1299–1318. doi: 10.5194/gmd-6-1299-2013

Gardner, A. S., Moholdt, G., Cogley, J. G., Wouters, B., Arendt, A. A., Wahr, J., et al. (2013). A reconciled estimate of glacier contributions to sea level rise: 2003 to 2009. *Science* 340, 852–857. doi: 10.1126/science.1234532

Glen, J. W. (1955). The creep of polycrystalline ice. *Proc. R. Soc. Lond. A Math. Phys. Sci.* 228, 519–538. doi: 10.1098/rspa.1955.0066

Grabiec, M., Jania, J. A., Puczko, D., Kolondra, L., and Budzik, T. (2012). Surface and bed morphology of Hansbreen, a tidewater glacier in Spitsbergen. *Polish Polar Res.* 33, 111–138. doi: 10.2478/v10183-012-0010-7

Hock, R., and Holmgren, B. (2005). A distributed surface energy-balance model for complex topography and its application to Storglaciären, Sweden. *J. Glaciol.* 51, 25–36. doi: 10.3189/172756505781829566

Howat, I. M., Box, J. E., Ahn, Y., Herrington, A., and McFadden, E. M. (2010). Seasonal variability in the dynamics of marine-terminating outlet glaciers in Greenland. *J. Glaciol.* 56, 601–613. doi: 10.3189/002214310793146232

Huss, M., and Hock, R. (2015). A new model for global glacier change and sea-level rise. *Front. Earth Sci.* 3:54. doi: 10.3389/feart.2015.00054

Jania, J., Mochnacki, D., and Gadek, B. (1996). The thermal structure of Hansbreen, a tidewater glacier in southern Spitsbergen, Svalbard. *Polar Res.* 15, 53–66. doi: 10.3402/polar.v15i1.6636

Jay-Allemand, M., Gillet-Chaulet, F., Gagliardini, O., and Nodet, M. (2011). Investigating changes in basal conditions of Variegated Glacier prior to and during its 1982-1983 surge. *Cryosphere* 5, 659–672. doi: 10.5194/tc-5-659-2011

Jonsell, U. Y., Navarro, F. J., Bañón, M., Lapazaran, J. J., and Otero, J. (2012). Sensitivity of a distributed temperature-radiation index melt model based on AWS observations and surface energy balance fluxes, Hurd Peninsula glaciers, Livingston Island, Antarctica. *Cryosphere* 6, 539–552. doi: 10.5194/tc-6-539-201210.5194/tc-6-539-2012

Krug, J., Durand, G., Gagliardini, O., and Weiss, J. (2015). Modelling the impact of submarine frontal melting and ice mélange on glacier dynamics. *Cryosphere* 9, 989–1003. doi: 10.5194/tc-9-989-2015

Krug, J., Weiss, J., Gagliardini, O., and Durand, G. (2014). Combining damage and fracture mechanics to model calving. *Cryosphere* 8, 2101–2117. doi: 10.5194/tc-8-2101-2014

Kruszewski, G. (2012). Zlodzenie Hornsundu i wód przyległych (Spitsbergen) w sezonie zimowym 2010-2011 (Ice conditions in Hornsund and adjacent waters (Spitsbergen) during winter season 2010-2011), *Problemy Klimatologii Polarnej* 22, 69–82.

Luckman, A., Benn, D. I., Cottier, F., Bevan, S., Nilsen, F., and Inall, M. (2015). Calving rates at tidewater glaciers vary strongly with ocean temperature. *Nat. Commun.* 6:8566. doi: 10.1038/ncomms9566

Morlighem, M., Bondzio, J., Seroussi, H., Rignot, E., Larour, E., Humbert, A., et al. (2016). Modeling of Store Gletscher's calving dynamics, West Greenland, in response to ocean thermal forcing. *Geophys. Res. Lett.* 43, 2659–2666. doi: 10.1002/2016GL067695

Motyka, R. J. (1997). Deep-water calving at Le Conte Glacier, southeast Alaska. *Byrd Polar Res. Cent. Rep.* 15, 115–118.

Navarro, F. J., Martín-Español, A., Lapazaran, J. J., Grabiec, M., Otero, J., Vasilenko, E. V., et al. (2014). Ice volume estimates from ground-penetrating radar surveys, Wedel Jarlsberg land glaciers, Svalbard. *Arct. Antarct. Alp. Res.* 46, 394–406. doi: 10.1657/1938-4246-46.2.394

Nick, F. M., van der Veen, C. J., Vieli, A., and Benn, D. I. (2010). A physically based calving model applied to marine outlet glaciers and implications for the glacier dynamics. *J. Glaciol.* 56, 781–794. doi: 10.3189/002214310794457344

Nye, J. F. (1957). The distribution of stress and velocity in glaciers and ice-sheets. *Proc. R. Soc. A* 239, 113–133. doi: 10.1098/rspa.1957.0026

Nye, J. F. (1965). The flow of a glacier in a channel of rectangular, elliptic or parabolic cross-section. *J. Glaciol.* 5, 661–690. doi: 10.1017/S002214300001 8670

Oerlemans, J., Jania, J., and Kolondra, L. (2011). Application of a minimal glacier model to Hansbreen, Svalbard. *Cryosphere* 5, 1–11. doi: 10.5194/tc-5-1-2011

Otero, J., Navarro, F. J., Martin, C., Cuadrado, M. L., and Corcuera, M. I. (2010). A three-dimensional calving model: numerical experiments on Johnsons Glacier, Livingston Island, Antarctica. *J. Glaciol.* 56, 200–214. doi: 10.3189/002214310791968539

Petlicki, M., Cieply, M., Jania, J. A., Prominska, A., and Kinnard, C. (2015). Calving of a tidewater glacier driven by melting at the waterline. *J. Glaciol.* 61, 851–863. doi: 10.3189/2015JoG15J062

Puczko, D. (2012). *Czasowa I Przestrzenna Zmienność Ruchu Spitsbergeńskich Lodowców Uchodzących Do Morza Na Przykładzie Lodowca Hansa. Temporal And Spatial Variability Of Tidewater Glaciers Movement Based On Example Of Hansbreen.* Doctoral thesis, Institute of Geophysics, Polish Academy of Sciences, Warsaw.

Rignot, E., Jacobs, S., Mouginot, J., and Scheuchl, B. (2013). Ice-Shelf Melting Around Antarctica. *Science* 341, 266–270. doi: 10.1126/science.1235798

Scambos, T., Hulbe, C., and Fahnestock, M. (2003), "Climate-induced ice shelf disintegration in the Antarctic Peninsula," in *Antarctic Peninsula Climate Variability: Historical and Paleoenvironmental Perspectives,* eds E. Domack, A. Levente, A. Burnet, R. Bindschadler, P. Convey, and M. Kirby (Washington, DC: American Geophysical Union), 79–92. doi: 10.1029/AR079p0079

Todd, J., and Christoffersen, P. (2014). Are seasonal calving dynamics forced by buttressing from ice mélange or undercutting by melting? Outcomes from full-Stokes simulations of Store Glacier, West Greenland. *Cryosphere* 8, 2353–2365. doi: 10.5194/tc-8-2353-2014

Vieli, A., Jania, J., and Kolondra, L. (2002). The retreat of a tidewater glacier: observations and model calculations on Hansbreen, Spitsbergen. *J. Glaciol.* 48, 592–600. doi: 10.3189/172756502781831089

Walter, J. I., Box, J. E., Tulaczyk, S., Brodsky, E. E., Howat, I. M., Ahn, Y., et al. (2012). Oceanic mechanical forcing of a marine-terminating Greenland glacier. *Ann. Glaciol.* 53, 181–192. doi: 10.3189/2012AoG60A083

Zagórski, P., Rodzik, J., Moskalik, M., Strzelecki, M. C., Lim, M., Błaszczyk, M., et al. (2015). Multidecadal (1960–2011) shoreline changes in Isbjørnhamna (Hornsund, Svalbard). *Polish Polar Res.* 36, 369–390. doi: 10.1515/popore-2015-0019

Conflict of Interest Statement: The authors declare that the research was conducted in the absence of any commercial or financial relationships that could be construed as a potential conflict of interest.

Transformations and Decomposition of MnCO$_3$ at Earth's Lower Mantle Conditions

Eglantine Boulard[1][*][†], Yijin Liu[2], Ai L. Koh[3], Mary M. Reagan[1], Julien Stodolna[4], Guillaume Morard[5], Mohamed Mezouar[6] and Wendy L. Mao[1]

[1] Geological Sciences, Stanford University, Stanford, CA, USA, [2] Stanford Synchrotron Radiation Lightsource, SLAC National Accelerator Laboratory, Menlo Park, CA, USA, [3] Stanford Nano Shared Facilities, Stanford University, Stanford, CA, USA, [4] EDF Lab Les Renardieres, Dpt MMC, Moret sur Loing, France, [5] Centre National de la Recherche Scientifique, UMR Centre National de la Recherche Scientifique 7590, Institut de Minéralogie, de Physique des Matériaux et de Cosmochimie, IRD, Sorbonne Universités—Université Pierre et Marie Curie, Muséum National d'Histoire Naturelle, Paris, France, [6] European Synchrotron Radiation Facility (ESRF), Grenoble, France

Edited by:
Benjamin Alexander Black,
City College of New York, USA

Reviewed by:
Matteo Alvaro,
University of Pavia, Italy
Zachary M. Geballe,
Geophysical Laboratory (CIS), USA

*Correspondence:
Eglantine Boulard
eglantine.boulard@synchrotron-soleil.fr

†Present Address:
Eglantine Boulard,
Synchrotron SOLEIL, L'Orme les
Merisiers, St. Aubin, France

Specialty section:
This article was submitted to
Geochemistry,
a section of the journal
Frontiers in Earth Science

Carbonates have been proposed as the principal oxidized carbon-bearing phases in the Earth's interior. Their phase diagram for the high pressure and temperature conditions of the mantle can provide crucial constraints on the deep carbon cycle. We investigated the behavior of MnCO$_3$ at pressures up to 75 GPa and temperatures up to 2200 K. The phase assemblage in the resulting run products was determined in situ by X-ray diffraction (XRD), and the recovered samples were studied by analytical transmission electron microscopy (TEM) and X-ray absorption near edge structure (XANES) imaging. At moderate temperatures below 1400 K and pressures above 50 GPa, MnCO$_3$ transformed into the MnCO$_3$-II phase, with XANES data indicating no change in the manganese oxidation state in MnCO$_3$-II. However, upon heating above 1400 K at the same pressure conditions, both MnCO$_3$ and MnCO$_3$-II undergo decomposition and redox reactions which lead to the formation of manganese oxides and reduced carbon.

Keywords: carbonate, phase transition, redox reaction, deep carbon cycle, high pressure

INTRODUCTION

Carbonates represent the main oxidized carbon-bearing phases which are transported into the mantle during subduction. The high-pressure behavior of carbonates can provide insight on crystal chemistry of carbon-bearing phases relevant to Earth's deep interior. In particular, the stability of carbonates vs. their decomposition and melting provides critical constraints for understanding the global carbon cycle. For all these reasons, the thermodynamic properties and phase diagrams for relevant carbonate compositions are needed down to core-mantle boundary conditions, i.e., megabar pressures and temperatures up to 3000 K.

The high-pressure behavior of various divalent cation-bearing carbonates has been the subject of a large number of studies. Systematic differences in compressibility and high pressure and high temperature polymorphs that depend on cation type have been observed in previous studies demonstrating that the polymorphism of carbonates is likely more complex than previously thought (see Shatskiy et al., 2015 and references therein). No single structural parameter or electronic property of the cation can account for the behavior of carbonates, suggesting that a combination of factors must be considered in explaining compressibility trends among members of the calcite-structure type (Zhang et al., 1998; Zhang and Reeder, 1999). With a Mn^{2+} cation size that lies between those of Mg^{2+} and Ca^{2+}, rhodochrosite ($MnCO_3$) represents a potential model compound for understanding the differences in the high-pressure behavior of the two main carbonate compositions (Mg and Ca carbonates). The interplay between these two species has indeed been the subject of many studies (e.g., de Capitani and Peters, 1981; Wang et al., 2011). At ambient conditions, $MnCO_3$ crystallizes with the calcite-type structure, R-3c. Santillán and Williams (2004) reported that $MnCO_3$ is stable in its rhombohedral calcite-type structure up to 50 GPa. However, more recently, Farfan et al. (2013) reported a possible electronic transition in rhodochrosite in the pressure interval 25–40 GPa which may be related to fine structural changes in the MnO_6 octahedra in the rhombohedral structure (Merlini et al., 2015). Evidence of a first order phase transition above ~40 GPa at room temperature in which $MnCO_3$ transforms into a $CaCO_3$-VI structure was reported by Boulard et al. (2015a) and confirmed by Merlini et al. (2015). Finally, Ono (2007) reported a phase transition at 50 GPa, after heating at 1500–2000 K and proposed an orthorhombic symmetry, the structure of which, however, could not be refined. To clarify the high pressure and high temperature behavior of $MnCO_3$, we combined *in situ* XRD using laser-heated diamond anvil cells and *ex situ* analyses using analytical TEM and XANES tomography for conditions up to ~75 GPa and 2200 K, the results of which are reported in this work.

MATERIALS AND METHODS

Sample Preparation

Powdered samples of rhodochrosite were loaded between two thermal insulating layers into symmetric diamond anvil cells (DAC) with 300 μm diamonds culets. Both NaCl and KCl were used as the thermal insulators and also served as pressure calibrants using the thermal equations of state from Dorogokupets and Dewaele (2007) and Dewaele et al. (2012), respectively. Hot spots with a diameter larger than 20 μm (FWHM) were obtained by two YAG lasers with excellent power stability aligned on both sides of the sample. The X-ray spot, spectrometer entrance and the heating laser spot were carefully aligned before the experiments. Temperatures were obtained by fitting the sample thermal emission spectrum from the central $2 \times 2 \ \mu m^2$ of the hotspot to the Planck's function using the wavelength range 600–900 nm. Reflective lenses were used for measurement in order to prevent any chromatic aberration (Benedetti and Loubeyre, 2004). Temperature stability was checked by measuring continuously during heating and X-ray acquisition. Uncertainties are of about 13% (Dewaele et al., 2012) on the pressure and 150 K on the temperature (Morard et al., 2008). For each run, the sample was first compressed at ambient temperature to the target pressure and then heated between 1000 and 2200 K using double-sided infrared laser heating. Because of their high stiffness, the high-pressure phase of NaCl and KCl do not guarantee hydrostatic conditions. Therefore, high temperature annealing of the sample have been performed in

order to partly reduce the contribution of deviatoric stresses which may have built up at high pressure. Samples were heated between 5 and 10 min at each temperature step. Some runs were performed with the addition of platinum black to the $MnCO_3$ sample in order to heat at a lower temperature.

In situ X-Ray Diffraction

Angle-dispersive XRD spectra were collected *in situ* at high pressure and high-temperature at beamline 12.2.2 of the Advanced Light Source (ALS), Lawrence Berkeley National Laboratory (LBNL), and beamline ID27 of the European Synchrotron Radiation Facility (ESRF) using a monochromatic incident X-ray beam with a wavelength of 0.4959 or 0.6199 Å at ALS and 0.3738 Å at ESRF. The monochromatic X-ray beam was focused to a smaller size than the laser heating spot in order to reduce both the radial and axial temperature gradients, typically: $3 \times 3 \ \mu m$ at ID27 beamline and $10 \times 10 \ \mu m$ at 12.2.2 beamline. The diffraction images were integrated with the Fit2d software (Hammersley et al., 1996), and the one-dimensional diffraction patterns were treated with the General Structure Analysis System (GSAS) software package (Larson and Von Dreele, 2004) using the LeBail method to identify the different phases and refine their lattice parameters.

X-Ray Absorption Near Edge Structure (XANES)

In order to constrain possible changes in the redox state of Mn, XANES spectra were collected on the recovered sample at the Mn-K edge using the nanoscale X-ray transmission microscope (nanoTXM) at the beamline 6-2c of the Stanford Synchrotron Radiation Lightsource (SSRL), SLAC National Accelerator Laboratory. This microscope is equipped with optics optimized for photon energies ranging from ~5 to 14 keV, it provides a spatial resolution as high as 30 nm and a single flat field of view of $30 \times 30 \ \mu m$. Depth of focus is ~50 μm. More details about the instrument can be found in Andrews et al. (2009) and Liu et al. (2011). NanoTXM is capable of nondestructive spectroscopic imaging, and the use of hard x-rays allows it to image the entire sample thickness avoiding possible contamination which could occur if special sample preparation were required. Transmission images were collected at small energy steps from the pre-edge region through the electronic edge (from 6470 to 6693 eV), enabling us to map out the oxidation state for Mn from its XANES signal.

Focused Ion Beam and Transmission Electron Microscopy

In order to be analyzed by TEM, the recovered samples were thinned to electron transparency (~150 nm) with a Ga^+ focused ion beam (FIB) operating at 30 kV and currents from 20 nA to 1 pA for final thinning. FIB milling was performed with a FEI Helios NanoLab 600i DualBeam FIB/SEM at the Stanford Nano Shared Facilities. FIB thin sections were extracted from the center of laser heated spots. Analytical TEM was carried out on the FIB thin sections in order to help with phase identification and to obtain chemical analyses on individual phases. TEM was performed with FEI Titan 80–300 operated at 300 keV,

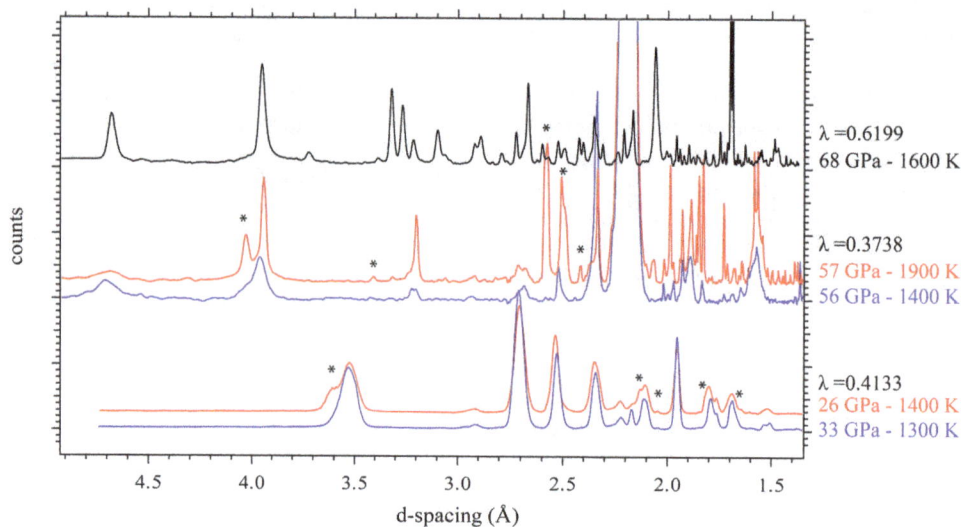

FIGURE 1 | *In situ* high *P-T* XRD patterns collected in the three pressure ranges: *P* < 33 GPa, 49 < *P* < 65 GPa, *P* > 65 GPa at lower and higher temperature. Stars indicate the main new diffraction peaks observed upon increase in temperature.

equipped with an extra high brightness field emission gun. Semi-quantitative information on the sample chemical composition was obtained by energy dispersive X-ray spectrometry (EDXS).

RESULTS

Lower Pressure Experiments (*P* < 33 GPa)

Figure 1 shows XRD patterns collected *in situ* at the different pressure ranges and at low or high temperature. These XRD patterns show that the rhombohedral $MnCO_3$ structure—rhodochrosite—is stable up to 33 GPa. For simplicity we will refer to this phase as $MnCO_3$-I. When heating at relatively low temperature (~1300 K) only $MnCO_3$-I is observed; however upon heating above 1400 K, the XRD patterns reveal the presence of another phase which does not correspond to the high pressure phase of $MnCO_3$ (marked by stars in **Figure 1**), MnO, cubic α-Mn_2O_3 (bixbyite), or Mn_3O_4 (hausmanite) manganese oxides. In a recent study, Ovsyannikov et al. (2013) reported the transformation of α-Mn_2O_3 into a perovskite-like structure (ζ-Mn_2O_3) at these pressure and temperature (*P-T*) conditions. Although they used a supercell in order to fit their diffraction patterns, here we found that a single cell of an orthorhombic perovskite structure can be used to index the new diffraction peaks (**Figure 2** and **Table 1**).

Medium Pressure Experiments (49 < *P* < 65 GPa)

Upon compressing $MnCO_3$ at ambient temperature and pressures above ~45 GPa, the XRD patterns change drastically, as the rhombohedral $MnCO_3$-I transforms into the high-pressure phase $MnCO_3$-II which is triclinic. This is in good agreement with previous high pressure studies performed at ambient

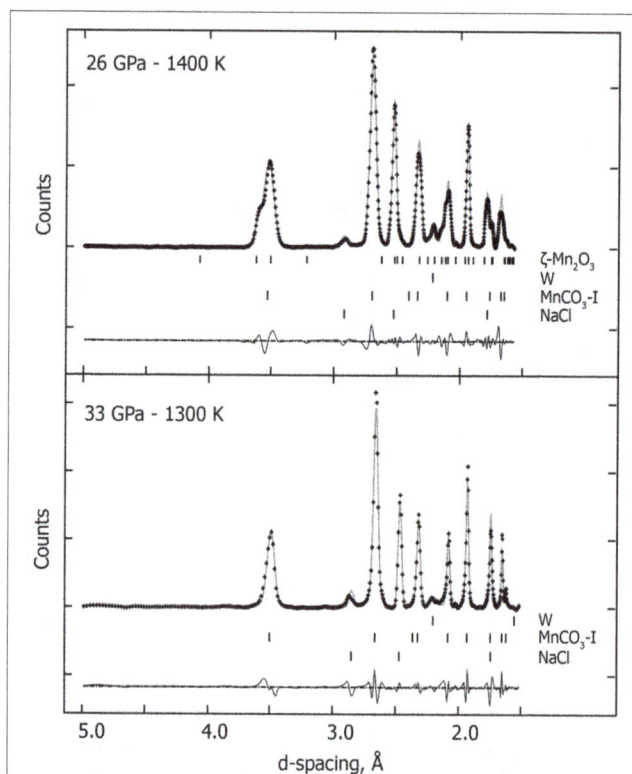

FIGURE 2 | *In situ* high *P-T* XRD patterns. Crosses represent observed data (background subtracted), and the gray solid lines show the profile refinements. Residuals between experiment and fit are shown below the diffraction pattern. At 33 GPa and 1300 K, we used an assemblage of NaCl and W (gasket) and rhodochrosite ($MnCO_3$-I). At higher temperature (26 GPa and 1400 K), we used the same assemblage with an additional high pressure manganese oxide polymorph (ζ-Mn_2O_3).

TABLE 1 | Crystallographic data.

P-T conditions	Lattice parameters				
	MnCO$_3$-I a, c (Å)		V(Å^3)	**ζ-Mn$_2$O$_3$** a, b, c (Å)	V(Å^3)
33 GPa−1300 K	4.6368(5)		264.17(5)		
	14.1876(33)				
26 GPa−1400 K	4.6667(4)		273.10(25)	4.9935(23)	184.49(9)
	14.4798(14)			5.2481(16)	
				7.0401(30)	
	MnCO$_3$-II a, b, c (Å)	α, β, γ	V(Å^3)	**δ-Mn$_2$O$_3$** a, b, c (Å)	V(Å^3)
56 GPa−1400 K	2.8116(3)	101.79(1)	72.11(1)		
	4.8083(6)	91.95(1)			
	5.4533(7)	88.45(1)			
57 GPa−1900 K	2.7759(6)	102.42(2)	70.16(2)	2.7137(6)	139.53(6)
	4.7620(8)	92.04(2)		8.0501(2)	
	5.4408(1)	87.92(2)		6.3973(3)	

temperature (Boulard et al., 2015a; Merlini et al., 2015). Similar to the lower pressure results, when heated at moderately low temperature, i.e., 1300 K at 57 GPa, a single carbonate phase, MnCO$_3$-II is observed (**Figure 1**). However upon heating above 1400 K, new XRD peaks appear which can be indexed to the high pressure manganese oxide phase, δ-Mn$_2$O$_3$, with a post-perovskite like structure described by Santillán et al. (2006). We identified this assemblage (MnCO$_3$-II + δ-Mn$_2$O$_3$) up to 57 GPa and 2100 K (**Figure 3** and **Table 1**). When recovered to ambient pressure and temperature, XRD patterns show the back transformation of MnCO$_3$-II into MnCO$_3$-I as well as δ-Mn$_2$O$_3$ into the cubic α-Mn$_2$O$_3$ phase.

Highest Pressure Experiments (*P* > 65 GPa)

Above 65 GPa, new changes in the diffraction patterns are observed as a new phase appears together with the previous assemblage (**Figure 1**). Above 70 GPa, the diffraction peaks from δ-Mn$_2$O$_3$ disappear and we observed an assemblage of MnCO$_3$-II plus this new phase (**Figure 4**). This new phase may correspond to a new high-pressure polymorph of Mn$_2$O$_3$. To our knowledge, no experimental studies exist on Mn$_2$O$_3$ at such high pressure and temperature conditions and theoretical studies support the stability of δ-Mn$_2$O$_3$ up to at least 120 GPa (Xu et al., 2015). Therefore, the composition and structure of this new phase remain unresolved. After transformation 68 GPa and 1600 K, diffraction patterns were collected on the sample once recovered to ambient pressure and temperature, and MnCO$_3$-I, α-Mn$_2$O$_3$ as well as another unknown phase were observed.

Further, *ex situ* analyses were performed on these samples to understand the highest pressure phase assemblage. XANES imaging at the manganese K-edge (6545 eV) was conducted in order to get insight on the manganese oxidation state. These measurements were collected in the heating spot, the area that has

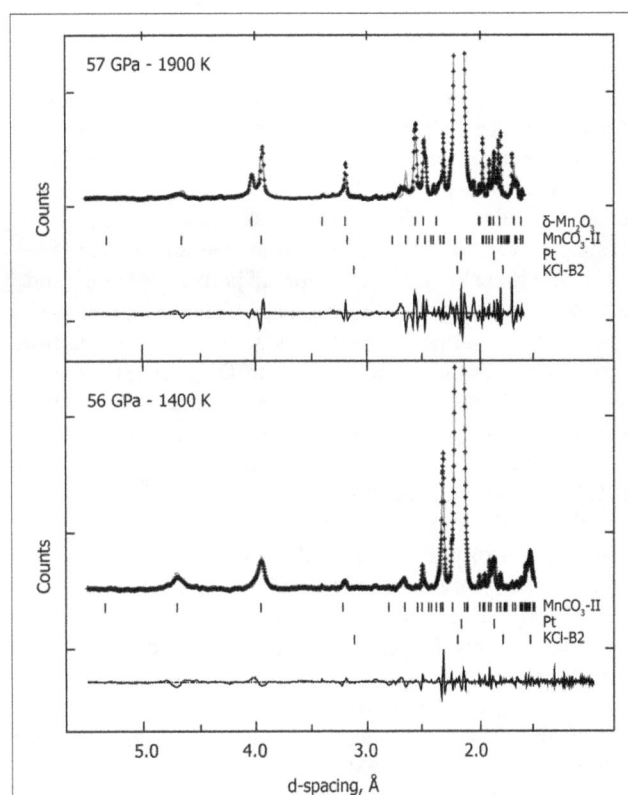

FIGURE 3 | *In situ* high *P-T* XRD patterns collected at 56 GPa−1400 K and at 57 GPa−1900 K. We used the following assemblage: the high-pressure polymorph MnCO$_3$-II, KCl-B2 high pressure polymorph and W (gasket), and the post-perovskite like high pressure polymorph of the manganese oxide (δ-Mn$_2$O$_3$). Crosses represent observed data (background subtracted), and the solid lines show the profile refinements. Residuals between experiment and fit are shown below the diffraction pattern.

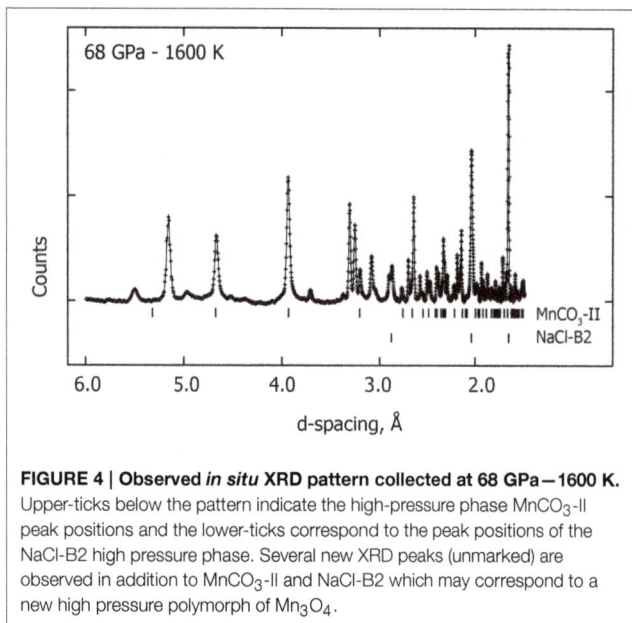

FIGURE 4 | Observed *in situ* XRD pattern collected at 68 GPa—1600 K.
Upper-ticks below the pattern indicate the high-pressure phase $MnCO_3$-II
peak positions and the lower-ticks correspond to the peak positions of the
NaCl-B2 high pressure phase. Several new XRD peaks (unmarked) are
observed in addition to $MnCO_3$-II and NaCl-B2 which may correspond to a
new high pressure polymorph of Mn_3O_4.

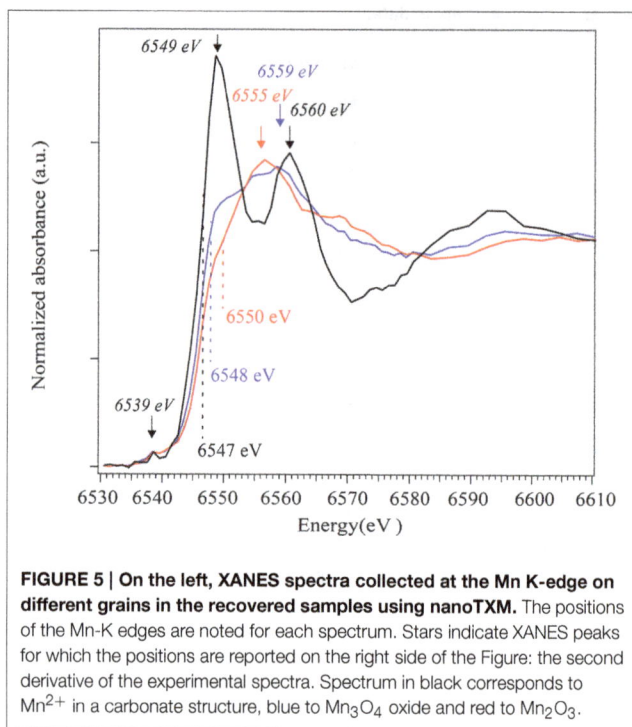

FIGURE 5 | On the left, XANES spectra collected at the Mn K-edge on
different grains in the recovered samples using nanoTXM. The positions
of the Mn-K edges are noted for each spectrum. Stars indicate XANES peaks
for which the positions are reported on the right side of the Figure: the second
derivative of the experimental spectra. Spectrum in black corresponds to
Mn^{2+} in a carbonate structure, blue to Mn_3O_4 oxide and red to Mn_2O_3.

been laser heated during the *in situ* experiments, directly on the
sample that had been recovered from 68 GPa and 1600 K without
special sample preparation. The results indicate three areas with
different absorption contrast characterized by distinct Mn-K edge
spectra. The background corrected and normalized spectra are
presented in **Figure 5**. A spectrum from the first area (shown in
black in **Figure 5**) reveals a peak at 6549 eV, another at 6560 eV
as well as a weak pre-edge peak at 6539 eV. This spectrum is
consistent with Mn^{2+} in octahedral sites as in a rhodochrosite
$MnCO_3$-I structure. In the spectrum for the second area (show
in blue in **Figure 5**), one can see a broad peak at ∼6559 eV and a
shift to higher energy of the Mn K-edge (from 6547 to 6548 eV)
which indicate an increase in Mn oxidation state. This spectrum
is consistent with the Mn-K edge for Mn_3O_4 (Ressler et al., 1999;
Jiao and Frei, 2010). Finally the spectrum collected on the third
area (in red in **Figure 5**) show an Mn-K edge at 6550 eV and
broad peak at 6555 eV which is consistent with Mn_2O_3 (Nam
et al., 2007; Jiao and Frei, 2010).

A thin section of the same sample was then prepared by FIB
in order to perform analytical TEM. A global high-angle annular
dark-field imaging scanning transmission electron microscopy
(HAADF-STEM) image of the thin section is presented in
Figure 6A. In this image, the grayscale is related to the mean
Z of each phase. It highlights the presence of four different
phases: a manganese and carbon rich phase that corresponds to
the carbonate (**Figure 6B**); a carbon rich phase (**Figure 6C**) that
appears in dark in the STEM image indicating the presence of
a reduced form of carbon that could not be detected by *in situ*
XRD; finally TEM-XEDS analyses confirm the presence of two
manganese oxides (**Figures 6D,E**). Semi-quantitative analyses
show that the experimental spectrum in **Figure 6D** has an O/Mn
ratio of 1.68 while that from the spectrum in **Figure 6E** is 1.22.
These numbers are consistent with the elemental ratio expected
in Mn_2O_3 (theoretical O/Mn = 1.5) and Mn_3O_4 (theoretical
O/Mn=1.33), respectively.

DISCUSSION

New High Pressure and Temperature Phase Diagram

The run products for all high-temperature experiments are
summarized in **Figure 7**. The Mn_2O_3 phase diagram from
Ovsyannikov et al. (2013) is also represented. To our knowledge,
no study exists on manganese oxides above 25 GPa at high
temperature. The *P-T* conditions of the different polymorphs α-
Mn_2O_3, ζ-Mn_2O_3, and δ-Mn_2O_3 (respectively red, yellow, and
purple *P-T* field in **Figure 6**) reported in the present study are in
very good agreement with Ovsyannikov et al. (2013). In addition,
our study shows that δ-Mn_2O_3 is stable up to at least 75 GPa—
1600 K. This is in good agreement with theoretical prediction of
the stability of the post-perovskite structure δ-Mn_2O_3 up to 120
GPa from Xu et al. (2015). Above 65 GPa—1700 K, we report
evidences of an additional manganese oxide: a high pressure
polymorph of Mn_3O_4. The presence of a fourth phase Mn_3O_4 as
observed by TEM and XANES imaging is in agreement with the
in situ diffraction which show an unknown structure in addition
to $MnCO_3$-II and δ-Mn_2O_3: a new high-pressure polymorph
of Mn_3O_4 which does not back transform into the tetragonal
hausmanite (Mn_3O_4). The structure of this polymorph is beyond
the scope of this study and was not resolved. Further, studies on
Mn_3O_4 at high pressure would be necessary in order to elucidate
this new phase.

The phase transition boundary $MnCO_3$ → $MnCO_3$-II is
indicated in gray on **Figure 7**. $MnCO_3$-I is observed up to 33
GPa—1300 K (circles in **Figure 7**). The high pressure polymorph
$MnCO_3$-II appears for *P-T* condition above 49 GPa—1500 K
(squares in **Figure 7**). This is in good agreement with previous

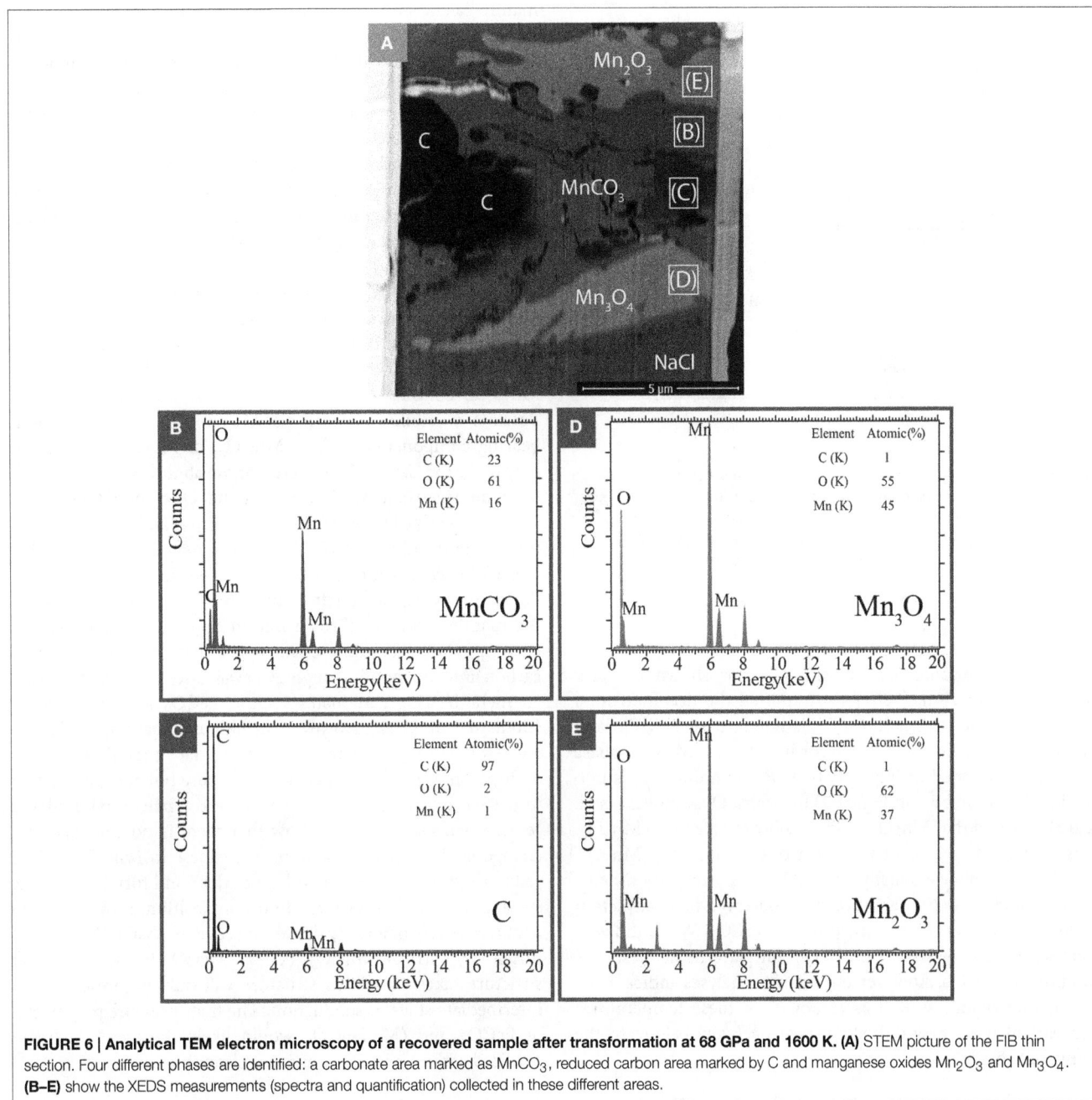

FIGURE 6 | Analytical TEM electron microscopy of a recovered sample after transformation at 68 GPa and 1600 K. (A) STEM picture of the FIB thin section. Four different phases are identified: a carbonate area marked as MnCO₃, reduced carbon area marked by C and manganese oxides Mn₂O₃ and Mn₃O₄. **(B–E)** show the XEDS measurements (spectra and quantification) collected in these different areas.

studies performed at room temperature that reported the phase transition of rhodochrosite into the triclinic structure $MnCO_3$-II at pressure above ~40 GPa (Boulard et al., 2015a and Merlini et al., 2015). Here, we show that $MnCO_3$-II is also observed at high temperature and up to 75 GPa—1700 K. This phase is isostructural to $CaCO_3$-VI which is metastable for the calcium composition ($CaCO_3$) and only observed at room temperature. We did not observe any evidence of the aragonite or post-aragonite structures upon heating. In fact, Oganov et al. (2006) showed that the several metastable structures of $CaCO_3$ are similar in energy with aragonite and are almost as stable as aragonite and post-aragonite at these conditions. Due to the large cation site in aragonite it is not

surprising that $MnCO_3$ adopts the $CaCO_3$-VI structure instead of aragonite.

Our study show that the carbonate as well as its high pressure polymorph $MnCO_3$-II are stable for temperature bellow 1400 K. Above 1400 K, we observed the decomposition of $MnCO_3$ and the reduction of carbon corresponding to the two reactions:

$$MnCO_3 \rightarrow MnO + CO_2 \qquad (1)$$

$$4MnO + CO_2 \rightarrow 2Mn_2O_3 + C \qquad (2)$$

The persistence of a carbonate phase in our experiments may indicate that the decomposition reaction was not complete by the time we stopped laser-heating.

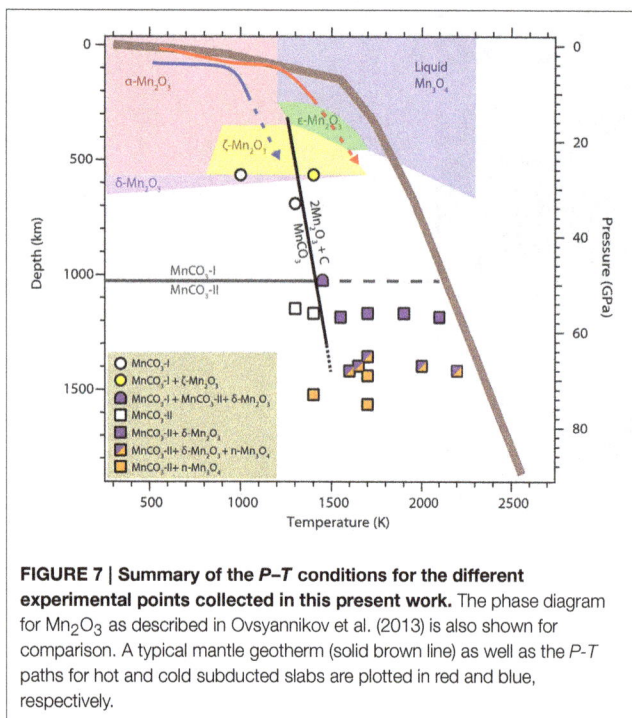

FIGURE 7 | Summary of the *P–T* conditions for the different experimental points collected in this present work. The phase diagram for Mn$_2$O$_3$ as described in Ovsyannikov et al. (2013) is also shown for comparison. A typical mantle geotherm (solid brown line) as well as the *P-T* paths for hot and cold subducted slabs are plotted in red and blue, respectively.

The corresponding decomposition line is shown in black in **Figure 7**. Liu et al. (2001) also reported the decomposition of MnCO$_3$ and formation of diamond at 6–8 GPa and temperatures above 2300 K. According to the Mn$_2$O$_3$ phase diagram represented in **Figure 6**, their *P-T* conditions falls into the liquid area (in blue in **Figure 7**) for which Ovsyannikov et al. (2013) reported that Mn$_2$O$_3$ is irreversibly reduced into Mn$_3$O$_4$. This is in good agreement with their observations of a Mn$_3$O$_4$, C and O$_2$ assemblage. Finally, Ono (2007) reported experiments at 54 GP and 1500/2000 K in which he observed new diffraction peaks and proposed a new structure for MnCO$_3$. We found very good agreement between our diffraction patterns at these *P-T* conditions and his, however our *ex situ* analyses indicate the presence of oxides as well as carbonate at these temperatures and part of these new diffraction peaks actually belong to the manganese oxide.

Implications for the Deep Carbon Cycle

Carbon in the Earth may exist in various forms including carbides, diamond, graphite, lonsdaleite, hydrocarbons, CO$_2$, and carbonates depending on the oxygen fugacity and *P-T* conditions (Dasgupta and Hirschmann, 2010; Hazen et al., 2013; Jones et al., 2016). Although it is generally thought that the conditions of the Earth's deep interior and of the subducting slab materials may be not compatible with the stability of carbonates or carbonate-rich liquid (Anzolini et al., 2016; Thomson et al., 2016), the observation of carbonate inclusions in diamonds potentially brought up to the Earth's surface from the deep mantle indicates that carbonates can exist at least locally in the mantle (e.g., Stachel et al., 1998, 2000; Leost et al., 2003; Bulanova et al., 2010). Carbonate compositions at the surface of the planet are

mainly calcite and dolomite. When transported into the deep Earth via subducting slabs, dolomite breaks down into aragonite and magnesite at a depth of around 250 km (e.g., Hammouda et al., 2011). These two end members display very different behavior at high pressure: magnesite is now known to be stable down to about 2000 km depth, below which it transforms into a tetrahedrally coordinated CO$_4$ structure (Boulard et al., 2015b) while aragonite undergoes a phase transition into a CO$_3$-bearing post-aragonite structure at about 1000 km depth that remains stable down to core mantle boundary conditions (Ono et al., 2007).

Together with previous studies on MnCO$_3$-II, we show that metastable structures of CaCO$_3$ such as CaCO$_3$-VI may play an important role in the transport of carbon into our planet's deep interior as it can host intermediate sized cations such as Mn^{2+} in contrast to aragonite and post-aragonite structures. Possible miscibility between CaCO$_3$ and other smaller cation-bearing components, such as MnCO$_3$, or MgCO$_3$ and FeCO$_3$ may make the CaCO$_3$-VI structure more abundant in the planet's deep interior than previously thought. One should note that MnCO$_3$ and CaCO$_3$ display miscibility gap at ambient conditions (de Capitani and Peters, 1981) and the possibility of a miscibility in CaCO$_3$-VI structure merits further investigation. However, as the slabs undergo subduction and reach temperatures closer to the mantle geotherm (brown line in **Figure 6**), this phase will decompose into manganese oxides and diamonds. Reduction of carbon into diamond at high *P-T* was also reported for other 3d metal-bearing carbonates such as FeCO$_3$ and (Mg,Fe)CO$_3$ (Boulard et al., 2012). However, formation of diamond in FeCO$_3$ and (Mg,Fe)CO$_3$ system is due to the fact that Fe is mainly incorporated as Fe^{3+} in their high-pressure polymorphs leading to partial reduction of carbon and a coexistence of oxidized and reduced carbon. Here, we show that there is no evidence of a change in the oxidation state of manganese in MnCO$_3$-II. The redox reaction (2) that would lead to diamond formation in the subducted slabs is in fact due to decomposition of MnCO$_3$ into MnO and CO$_2$ oxides. It therefore seems probable that even if miscibility between MnCO$_3$, FeCO$_3$, MgCO$_3$ in the CaCO$_3$-VI structure takes place, this structure will only be present as an intermediate stage of subduction. The high-pressure polymorph of FeCO$_3$ and (Mg,Fe)CO$_3$ would likely represent the main oxidized carbon host at lower mantle conditions.

CONCLUSIONS

Our study brings new insight into the phase diagram of MnCO$_3$ at high pressure and temperature. Rhombohedral MnCO$_3$-I as well as its high pressure polymorph MnCO$_3$-II are stable at temperatures up to 1400 K. At higher temperatures however, both MnCO$_3$-I and MnCO$_3$-II break down into oxides and redox reactions take place which lead to the formation of manganese oxides such as Mn$_2$O$_3$ and Mn$_3$O$_4$ and diamond. These reactions occur at *P-T* conditions close to the mantle geotherm. The CaCO$_3$-VI structure that can host small cations at relatively high pressures can only be encountered at an intermediate stage of subduction, thus, the high pressure tetrahedral carbonate phase of $FeCO_3$ and $(Mg,Fe)CO_3$ would represent the main mineral host for oxidized carbon in the deep Earth.

AUTHOR CONTRIBUTIONS

EB and WM designed the research project; EB, YL, AK, MR, JS, GM, and MM performed experiments and analysis; EB and YL analyzed data; and EB and WM wrote the paper with input from all co-authors.

ACKNOWLEDGMENTS

EB and WM acknowledge support from the Deep Carbon Observatory. WM is supported by NSF-EAR-1055454. Portion of the XRD work was performed at the high-pressure beamline 12.2.2, ALS which is supported by the DOE-BES under contract DE-AC02-05CH11231. We thank J. Yan, J. Knight, and A. MacDowell for their assistance with XRD experiments at ALS. We also thank M. Scott for providing the rhodochrosite sample.

REFERENCES

Andrews, J. C., Brennan, S., Pianetta, P., Ishii, H., Gelb, J., Feser, M., et al. (2009). Full-field transmission x-ray microscopy at SSRL. *J. Phys.* 186:12002. doi: 10.1088/1742-6596/186/1/012002

Anzolini, C., Angel, R. J., Merlini, M., Derzsi, M., Tokár, K., Milani, S., et al. (2016). Depth of formation of $CaSiO_3$-walstromite included in super-deep diamonds. *Lithos.* 265, 138–147. doi: 10.1016/j.lithos.2016.09.025

Benedetti, L. R., and Loubeyre, P. (2004). Temperature gradients, wavelength-dependent emissivity, and accuracy of high and very-high temperatures measured in the laser-heated diamond cell. *High Press. Res.* 24, 423–445. doi: 10.1080/08957950412331331718

Boulard, E., Goncharov, A. F., Blanchard, M., and Mao, W. L. (2015a). Pressure-induced phase transition in $MnCO_3$ and its implications on the deep carbon cycle. *J. Geophys. Res. Solid Earth* 120, 4069–4079. doi: 10.1002/2015JB011901

Boulard, E., Menguy, N., Auzende, A. L., Benzerara, K., Bureau, H., Antonangeli, D., et al. (2012). Experimental investigation of the stability of Fe-rich carbonates in the lower mantle. *J. Geophys. Res.* 117, B02208. doi: 10.1029/2011jb008733

Boulard, E., Pan, D., Galli, G., Liu, Z., and Mao, W. L. (2015b). Tetrahedrally coordinated carbonates in Earth's lower mantle. *Nat. Commun.* 6, 6311. doi: 10.1038/ncomms7311

Bulanova, G. P., Walter, M. J., Smith, C. B., Kohn, S. C., Armstrong, L. S., Blundy, J., et al. (2010). Mineral inclusions in sublithospheric diamonds from Collier 4 kimberlite pipe, Juina, Brazil: subducted protoliths, carbonated melts and primary kimberlite magmatism. *Contrib. Minerl. Petrol.* 160, 489–510. doi: 10.1007/s00410-010-0490-6

Dasgupta, R., and Hirschmann, M. M. (2010). The deep carbon cycle and melting in Earth's interior. *Earth Planet. Sci. Lett.* 298, 1–13. doi: 10.1016/j.epsl.2010.06.039

de Capitani, C., and Peters, T. (1981). The solvus in the System MnCO3-CaCO. *Contrib. Mineral. Petrol.* 76, 394–400, doi: 10.1007/bf00371481

Dorogokupets, P. I., and Dewaele, A. (2007). Equations of state of MgO, Au, Pt, NaCl-B1, and NaCl-B2: internally consistent high-temperature pressure scales. *High Pressure Res.* 27, 431–446. doi: 10.1080/08957950701659700

Farfan, G. A., Boulard, E., Wang, S., and Mao, W. L. (2013). Bonding and electronic changes in rhodochrosite at high pressure. *Am. Mineral.* 98, 1817–1823. doi: 10.2138/am.2013.4497

Dewaele, A., Belonoshko, A., Garbarino, G., Occelli, F., Bouvier, P., Hanfland, M., et al. (2012). High-pressure–high-temperature equation of state of KCl and KBr. *Phys. Rev.* 85, 1–7. doi: 10.1103/physrevb.85.214105

Hammersley, A. P., Svensson, S. O., Hanfland, M., Fitch, A. N., and Hausermann, D. (1996). Two-dimensional detector software: from real detector to idealised image or two-theta scan. *High Press. Res.* 14, 235–248.

Hammouda, T., Andrault, D., Koga, K., Katsura, T., and Martin, A. M. (2011). Ordering in double carbonates and implications for processes at subduction zones. *Contrib. Mineral. Petrol.* 161, 439–450, doi: 10.1007/s00410-010-0541-z

Hazen, R. M., Downs, R. T., Jones, A. P., and Kah, L. (2013). Carbon Mineralogy and Crystal Chemistry. *Rev. Mineral. Geochem.* 75, 7–46. doi: 10.2138/rmg.2013.75.2

Jiao, F., and Frei, H. (2010). Nanostructured manganese oxide clusters supported on mesoporous silica as efficient oxygen-evolving catalysts. *Chem. Commun.* 46, 2920–2922. doi: 10.1039/b921820c

Jones, A. P., McMillan, P. F., Salzmann, C. G., Alvaro, M., Nestola, F., Prencipe, M., et al. (2016). Structural characterization of natural diamond shocked To 60GPa; implications for earth and planetary systems. *Lithos.* 265, 214–221. doi: 10.1016/j.lithos.2016.09.023

Larson, A. C., and Von Dreele, R. B. (2004). *General Structure Analysis System (GSAS).* Los Alamos National Laboratory Report LAUR, 86–748.

Leost, I., Stachel, T., Brey, G. P., Harris, J. W., and Ryabchikov, I. D. (2003). Diamond formation and source carbonation: mineral associations in diamonds from Namibia. *Contrib. Mineral. Petrol.* 145, 15–24. doi: 10.1007/s00410-003-0442-5

Liu, L., Lin, C., and Yang, Y. (2001). Formation of diamond by decarbonation of MnCO3. *Solid State Commun.* 118, 195–198. doi: 10.1016/s0038-1098(01)00068-0

Liu, Y., Andrews, J. C., Wang, J., Meirer, F., Zhu, P., Wu, Z., et al. (2011). Phase retrieval using polychromatic illumination for transmission X-ray microscopy. *Opt. Express* 19, 540–545. doi: 10.1364/OE.19.000540

Merlini, M., Hanfland, M., and Gemmi, M. (2015). The $MnCO_3$ -II high-pressure polymorph of rhodochrosite. *Am. Mineral.* 100, 2625–2629. doi: 10.2138/am-2015-5320

Morard, G., Andrault, D., Guignot, N., Sanloup, C., Mezouar, M., Petitgirard, S., et al. (2008). *In situ* determination of Fe–Fe3S phase diagram and liquid structural properties up to 65 GPa. *Earth Planet. Sci. Lett.* 272, 620–626. doi: 10.1016/j.epsl.2008.05.028

Nam, K., Kim, M. G., and Kim, K. (2007). *In Situ* Mn K-edge X-ray absorption spectroscopy studies of electrodeposited manganese oxide films for electrochemical capacitors. *J. Phys. Chem. C* 111, 749–758. doi: 10.1021/jp063130o

Oganov, A. R., Glass, C. W., and Ono, S. (2006). High-pressure phases of CaCO3. Crystal structure prediction and experiment. *Earth Planet. Sci. Lett.* 241, 95–103. doi: 10.1016/j.epsl.2005.10.014

Ono, S. (2007). High-pressure phase transformation in $MnCO_3$. a synchrotron XRD study. *Mineral. Mag.* 71, 105–111. doi: 10.1180/minmag.2007.071.1.105

Ono, S., Kikegawa, T., and Ohishi, Y. (2007). High-pressure transition of CaCO3. *Am. Mineral.* 92, 1246–1249. doi: 10.2138/am.2007.2649

Ovsyannikov, S. V., Abakumov, A. M., Tsirlin, A. A., Schnelle, W., Egoavil, R., Verbeeck, J., et al. (2013). Perovskite-like Mn2O3. A path to new manganites. *Angew. Chem. Int. Ed. Eng.* 52, 1494–1498. doi: 10.1002/anie.201208553

Ressler, T., Brock, S. L., Wong, J., and Suib, S. L. (1999). Multiple-Scattering EXAFS analysis of tetraalkylammonium manganese oxide colloids. *J. Phys. Chem. B* 103, 6407–6420. doi: 10.1021/jp9835972

Santillán, J., Shim, S. H., Shen, G., and Prakapenka, B. (2006). High-pressure phase transition in Mn2O3: application for the crystal structure and preferred orientation of the CaIrO3 type. *Geophys. Res. Lett.* 33, 1–5. doi: 10.1029/2006GL026423

Santillán, J., and Williams, Q. (2004). A high-pressure infrared and X-ray study of FeCO3 and MnCO3: comparison with CaMg(CO3)2-dolomite. *Phys. Earth Planet. Interiors* 143–144, 291–304. doi: 10.1016/j.pepi.2003.06.007

Shatskiy, A. F., Litasov, K. D., and Palyanov, Y. N. (2015). Phase relations in carbonate systems at pressures and temperatures of lithospheric mantle: review of experimental data. *Russ. Geol. Geophys.* 56, 113–142. doi: 10.1016/j.rgg.2015.01.007

Stachel, T., Brey, G. P., and Harris, J. W. (2000). Kankan diamonds (Guinea) I: from the lithosphere down to the transition zone. *Contrib. Mineral. Petrol.* 140, 1–15. doi: 10.1007/s004100000173

Stachel, T., Harris, J. W., and Brey, G. P. (1998). Rare and unusual mineral inclusions in diamonds from Mwadui, Tanzani. *Contrib. Mineral. Petrol.* 132, 34–47, doi: 10.1007/s004100050403

Thomson, A. R., Walter, M. J., Kohn, S. C., and Brooker, R. A. (2016). Slab melting as a barrier to deep carbon subduction. *Nature* 529, 76–79. doi: 10.1038/nature16174

Wang, Q., Grau-Crespo, R., and de Leeuw, N. H. (2011). Mixing thermodynamics of the calcite-structured (Mn,Ca)CO3 solid solution: a computer simulation study. *J. Phys. Chem. B* 115, 13854–13861. doi: 10.1021/jp200378q

Xu, C., Xu, B., Yang, Y., Dong, H., Oganov, A. R., Wang, S., et al. (2015). Prediction of a stable post-post-perovskite structure from first principles. *Phys. Rev. B* 91, 1–5. doi: 10.1103/physrevb.91.020101

Zhang, J., Martinez, I., Guyot, F., and Reeder, R. J. (1998). Effects of Mg-Fe 2+ substitution in calcite-structure carbonates: thermoelastic properties. *Am. Mineral.* 83, 280–287.

Zhang, J., and Reeder, R. J. (1999). Comparative compressibilities of calcite-structure carbonates: deviations from empirical relations. *Am. Mineral.* 84, 861–870.

Conflict of Interest Statement: The authors declare that the research was conducted in the absence of any commercial or financial relationships that could be construed as a potential conflict of interest.

14

Orientation of the Eruption Fissures Controlled by a Shallow Magma Chamber in Miyakejima

Nobuo Geshi * and Teruki Oikawa

Institute of Earthquake and Volcano Geology, Geological Survey of Japan, National Institute of Advanced Industrial Science and Technology (AIST), Tsukuba, Japan

Edited by:
Roberto Sulpizio,
University of Bari, Italy

Reviewed by:
Alessandro Tibaldi,
University of Milano-Bicocca, Italy
Thomas R. Walter,
GFZ Potsdam, Germany

*Correspondence:
Nobuo Geshi
geshi-nob@aist.go.jp

Specialty section:
This article was submitted to
Volcanology,
a section of the journal
Frontiers in Earth Science

Orientation of the eruption fissures and composition of the lavas of the Miyakejima volcano is indicative of the competitive processes of the regional tectonic stress and the local stress generated by the activity of a magma plumbing system beneath the volcano. We examined the distributions and magmatic compositions of 23 fissures that formed within the last 2800 years, based on a field survey and a new dataset of ^{14}C ages. The dominant orientation of the eruption fissures in the central portion of the volcano was found to be NE-SW, which is perpendicular to the direction of regional maximum horizontal compressive stress (σ_{Hmax}). Magmas that show evidence of mixing between basaltic and andesitic compositions erupted mainly from the eruption fissures with a higher offset angle from the regional σ_{Hmax} direction. The presence of a dike pattern perpendicular to the direction of maximum compression σ_{Hmax} is an unusual and uncommon feature in volcanoes. Here we investigate the conditions possibly controlling this unexpected dike pattern. The distribution and magmatic compositions of the eruption fissures in Miyakejima volcano highlight the tectonic influence of shallow magma chamber on the development of feeder dikes in a composite volcano. The presence of a shallow dike-shaped magma chamber controls the eccentric distribution of the eruption fissures perpendicular to the present direction of σ_{Hmax}. The injection of basaltic magma into the shallow andesitic magma chamber caused the temporal rise of internal magmatic pressure in the shallow magma chamber which elongates in NE-SW direction. Dikes extending from the andesitic magma chamber intrude along the local stress field which is generated by the internal excess pressure of the andesitic magma chamber. As the result, the eruption fissures trend parallel to the elongation direction of the shallow magma chamber. Some basaltic dikes from the deep-seated magma chamber reach the ground surface without intersection with the andesitic magma chamber. These basaltic dikes develop parallel to the regional compressive stress in NW-SE direction. The patterns of the eruption fissures can be modified in the future as was observed in the case of the destruction of the shallow magma chamber during the 2000 AD eruption.

Keywords: fissure eruption, feeder dike, magma chamber, local stress, Miyakejima

INTRODUCTION

Propagation of a dike is controlled by the stress field in its host rock; the orientation of a dike is basically parallel to the maximum compressive axis ($\sigma 1$) and perpendicular to the minimum compressive axis ($\sigma 3$). Therefore, the orientation of dikes has been used as an indicator of the stress field in volcanic fields (Nakamura, 1977; Yamaji and Sato, 2011). Dikes will be oriented parallel to the maximum stress axis in a homogeneous stress field. However, the distribution of dikes and eruption fissures in volcanoes display radial, circumferential, and curvature patterns reflecting the local stress in the volcanic edifice (e.g., Chadwick and Howard, 1991). Local disturbances in the stress field of a volcano are caused by the prominent topography of a volcanic edifice (e.g., Tibaldi et al., 2014), flank instability (e.g., Walter et al., 2005), unloading by collapse (Corbi et al., 2015), active faults near the volcanic system (e.g., Seebeck and Nicol, 2009), mechanical heterogeneity in the volcanic edifice and basement (e.g., Letourneur et al., 2008), and the magmatic activity within the volcano (e.g., Chadwick and Dieterich, 1995; Takada, 1997). Recent observations of the dike intrusion events reveal that the stress field in the host rock is affected by the emplacement of intrusions, and consequently feedbacks to the growth of the next dikes (e.g., Bagnardi et al., 2013; Falsaperla and Neri, 2015).

The local pressure source within a volcano is one of the major causes of a local disturbance in the stress field within the volcano (Gudmundsson, 2006). Magma injection into a local magma reservoir results in a rise of magma pressure, and it consequently has a strong influence on the distribution pattern of an eruptive fissure (e.g., Tibaldi, 2015 and the references therein). Therefore, utilizing the knowledge of local disturbances in a stress field within a volcano, we can predict the distribution pattern of fissure eruptions.

Predicting the distribution pattern of flank fissure eruptions is not only pertinent to volcanology but is also of critical importance to hazard management. As the site and orientation of fissure eruptions are controlled by the orientation and propagation direction of the feeder dikes, understanding of the stress field within a volcano and the propagation direction of feeder dikes are crucial to evaluating the risk of flank fissure eruptions (e.g., Gudmundsson, 2006). Flank fissure eruptions are one of the major causes of volcanic disasters (Cappello et al., 2015).

The distribution of the flank fissure eruptions in Miyakejima Island is indicative of the competitive processes of the regional tectonic stress and the local stress generated by the activity of a magma plumbing system beneath the volcano (Nakamura, 1977). The development of a rift zone perpendicular to the regional horizontal compressive axis (σ_{Hmax}) provides a good example of the role of a local stress field within the volcanic edifice on the distribution of the eruption fissures. A combination of the inferences made from geological and petrological investigations and the observations from a recent caldera collapse event reveal the influence of shallow magma chambers on the development of eruption fissures. Here, we examined the distribution patterns of recent eruption fissures, with a new dataset of ^{14}C ages and petrological analysis to reveal the influence of a shallow magma chamber.

BACKGROUND OF MIYAKEJIMA

Tectonic Setting

The northernmost part of the Izu-Bonin volcanic arc, including Miyakejima, is compressed in the NW-SE direction owing to the subduction of the Philippine Sea Plate to the Eurasian Plate (**Figure 1A**). The orientations of several volcanic chains in the northernmost part of the Izu-Bonin volcanic arc are consistent with the NW-SE compression in this area (Nakamura, 1977). Focal mechanisms of the earthquakes in the vicinity of Miyakejima also indicate that the σ_{Hmax} is aligned in the NW-SE direction and σ_{Hmin} in the NE-SW direction (Nishimura, 2011). Orientations of volcanic fissures in the Izu-Tobu monogenic volcanoes, ~100 km north of Miyakejima, are predominantly NW-SE (Koyama and Umino, 1991). The Izu-Oshima volcano, ~60 km north of Miyakejima, also has volcanic chains on its flank trending in the WNW-ESE direction in the northwestern flank and in the NW-SE direction in the southeastern flank (Ishizuka et al., 2014). The Hachijojima volcano, ~100 km south of Miyakejima, has volcanic chains trending in the NW-SE direction (Ishizuka et al., 2008). A regional dike trending in the N45°W direction intruded the submarine area between Miyakejima Island and Kozushima Island during the 2000 AD eruption (**Figure 1A**: Ito and Yoshida, 2002; Yamaoka et al., 2005).

The dominant orientation of the eruption fissures in the Miyakejima volcano is, however, different from its neighboring volcanoes (Nakamura, 1977). The eruption fissures in the Miyakejima volcano are concentrated in the NE-SW direction near the summit, giving rise to a NE-SW rift zone (**Figure 1B**). The eruption fissures tend to curve to trend in the NW-SE direction in the coastal, peripheral part of the island (Nakamura, 1977; Aramaki et al., 1986).

Geology

The Miykejima volcano is a basaltic–andesitic composite volcano sitting on the Izu-Bonin volcanic arc on the Philippine Sea Plate (**Figure 1A**). Miyakejima has a conical volcanic edifice, which extends ~700 m high above sea level and ~500 m below sea level (**Figure 1B**). Four eruptions were recorded in the last 100 years, two in the northeastern flank (the 1940 and 1962 eruptions) and two in the southwestern flank (the 1983 eruption and the 2000 submarine eruption).

The development of chains of scoria cones on the flanks of the island indicates the frequent lateral fissure eruptions during the development of the volcano (Tsukui and Suzuki, 1999). Though fissure eruptions occurred throughout the history of the volcano, we focused on the fissure eruptions after the formation of Hachodaira caldera because the formation of the Hachodaira caldera is one of the key events in the development of the volcano. Phreatomagmatic eruptions during the caldera formation produced the Hachodaira ash layer, which provides good key bed in all part of the island. Our new ^{14}C data indicate that the Hachodaira caldera was formed at 2700–2800 years ago.

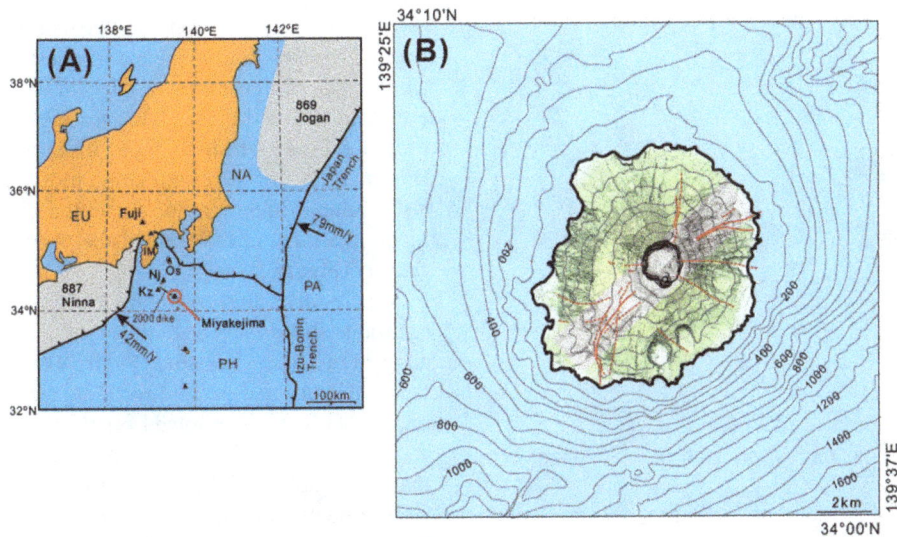

FIGURE 1 | (A) Tectonic setting of the Miyakejima Island. PH, Philippine Sea Plate; PA, Pacific Plate; NA, North America Plate; EU, Eurasia Plate Arrows indicate the direction and speed of the subduction at the trench. Solid triangles show the location of active volcanoes. IM, Izu-Tobu Monogenetic Volcanoes; Os, Izu-Oshima; Nj, Niijima; Kz, Kozushima. Gray areas show the approximate source area of the 869AD Jogan earthquake and the 887AD Ninna-Nankai earthquake. **(B)** Distribution of the eruption fissures. Solid red lines show the eruption fissure in the nineteenth and twentieth centuries. Historical records confirm their ages. Broken red lines show the eruption fissures younger than the seventh century. Their ages are suggested by the ^{14}C age. The shadowed area shows the approximate area of the NE-SW rift zone. Counters are 100 m interval.

A new collapse caldera, which is 1.7 km across, was formed at the summit during the 2000 AD eruption due to the draining of the shallow andesitic magma chamber through a large lateral intrusion (Geshi et al., 2002). Temporal changes of the juvenile material during the eruption show that the dike intrusion to the northwest of the island withdrew andesitic magma from a shallow chamber, which was then replaced by basaltic magma in the later stage of eruption (Geshi et al., 2002).

Magma Plumbing System

The magmas that erupted from Miyakejima volcano within the last 2800 years have been basaltic to andesitic in composition, with whole-rock SiO_2 contents ranging from 50 to 62 wt% (**Figure 2**). The disequilibrium mineral assemblage and the wide and linear variations of the whole-rock composition of the basaltic-andesite of Miyakejima indicate mixing between the basaltic and andesitic end components (Amma-Miyasaka and Nakagawa, 2003; Kuritani et al., 2003). Judging by the narrow diffusion texture at the rim of the olivine phenocrysts, magma mixing occurred just before the eruption (Niihori et al., 2003).

The basaltic magma, which is fed from the deeper chamber, has the whole-rock SiO_2 content of ~50–51 wt%. Equilibrium relationships between minerals and the groundmass suggest that the less-fractionated magmas with basaltic compositions are stored in a magma chamber at around ~200 MPa. Whereas the fractionating magmas with andesitic compositions are stored at a shallower depth at around ~100 MPa (Kuritani et al., 2003; Amma-Miyasaka et al., 2005). These pressures correspond to depth of ~10 km for the basaltic magma and ~5 km for the andesitic magma, respectively.

RECONSTRUCTION OF ERUPTION FISSURES

Distribution and Age of the Eruption Fissures

The distribution and ages of the eruption fissures in Miyakejima were re-examined to reveal the temporal distribution patterns of the flank fissure eruptions in the volcano. The locations and orientations of the eruption fissures were confirmed, based on aerial photographs, a digital elevation model (5 m grid provided by the Geospatial Information Authority of Japan), and our field survey. The ages of each of the eruption fissures were determined by both the stratigraphic relationships of the tephra layers and the ^{14}C dating of the carbonized plant fragments at the base of the tephra layer.

In this study, we focus on the eruption fissures younger than the 2.7–2.8 ka caldera-forming event. We confirmed 23 fissure eruptions in this period (**Table 1**). Among them, sixteen fissure eruptions were identified as being younger than the seventh century Suoana-Kazahaya eruption (**Figure 3** and **Table 1**). The eruption fissures younger than the seventh century are mainly distributed in the NE-SW rift zone. Since the seventh century, four fissure eruptions occurred in the northeastern sector of the volcano, and ten occurred in the southwestern sector of the volcano. Two fissure eruptions (Ninth century and sixteenth century) occurred in the eastern flank. Though the exposure of the eruption fissures before the seventh century is limited owing to the coverage by the younger deposits, we also confirmed the locations and ages of six fissure eruptions between 2.5 ka and the seventh century (**Table 1**). Among the six eruptions,

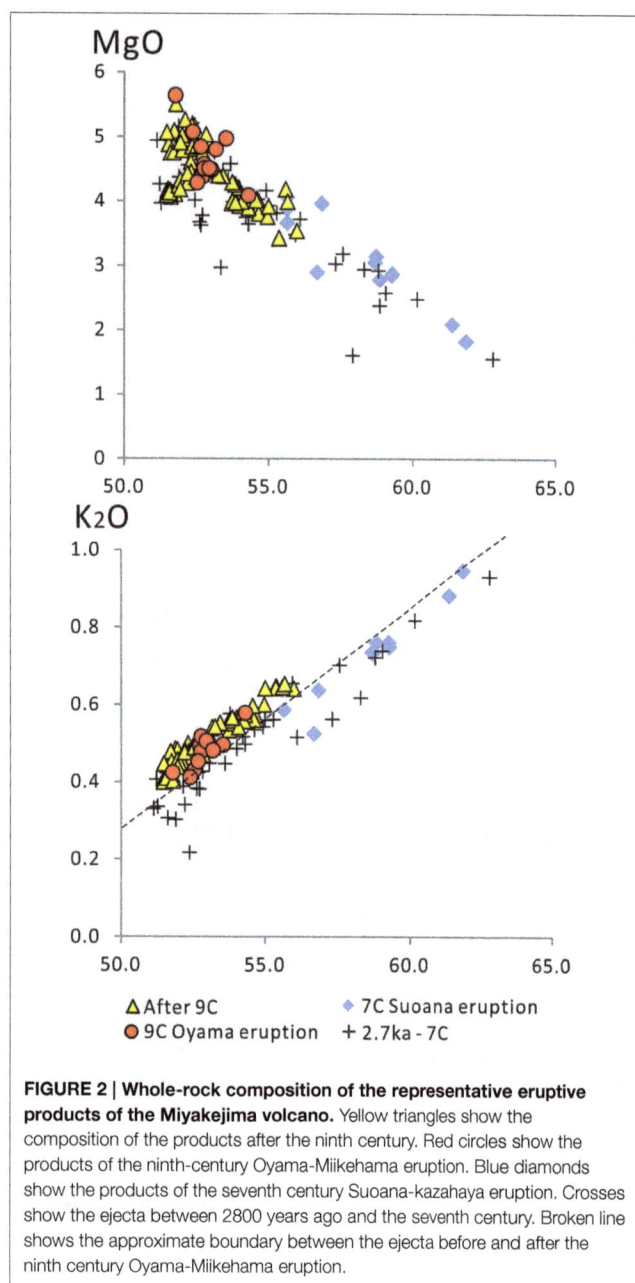

FIGURE 2 | Whole-rock composition of the representative eruptive products of the Miyakejima volcano. Yellow triangles show the composition of the products after the ninth century. Red circles show the products of the ninth-century Oyama-Miikehama eruption. Blue diamonds show the products of the seventh century Suoana-kazahaya eruption. Crosses show the ejecta between 2800 years ago and the seventh century. Broken line shows the approximate boundary between the ejecta before and after the ninth century Oyama-Miikehama eruption.

Magma Composition

The relationship between the orientation of eruption fissures and the composition of magma is investigated. We analyzed the whole-rock compositions of lavas and/or scoria of the fissure eruptions within the last ~2800 years. Whole-rock compositions of the erupted products form a linear compositional trend (**Figure 2**). A detailed examination indicated that the whole-rock K_2O contents of the products after the ninth century are slightly higher than that of the products between 2.8 ka and the seventh century (**Figure 2**).

Whole-rock SiO_2 contents of the lavas from the flank eruptions show wider variations with the increase of the offset angle of the eruption fissure (**Figure 4B**). Basaltic magmas (SiO_2 ~51 wt%) erupted in all directions, whereas the magmas with higher SiO_2 content erupted from the fissures with a higher offset angle (**Figure 4B**). The eruption fissures with offset angles less than 30° erupted basaltic andesite with a <53 wt% of whole-rock SiO_2 content. Andesite with SiO_2 >56 wt% was erupted from two fissures; the seventh century fissure in the northern flank with an offset angle of ~60° and the 2.7–2.8 ka eruption from the southern flank with an offset angle of ~85°.

DISCUSSION

A combined analysis of the location, age, and magmatic composition of the lavas of the eruption fissures in Miyakejima revealed the following facts: (i) The eruption fissures with higher offset angles from the regional compressive direction occurred within the last ~2800 years and formed the NE-SW rift, and (ii) Magmas with significant evidence of magma mixing erupted mainly from the eruption fissures with larger offset angles.

The limited distribution of hybrid magma in the NE-SW rift zone confirms the presence of an andesitic magma chamber beneath the NE-SW rift zone in the summit area. Several previous works on the petrological analysis of Miyakejima (Amma-Miyasaka and Nakagawa, 2003; Kuritani et al., 2003; Niihori et al., 2003; Amma-Miyasaka et al., 2005) show that the magmas that erupted within the last ~2800 years are hybrid with basaltic and andesitic end-components. The mixed end components were stored in a duplicated magma chamber system consisting of a deep-seated basaltic chamber and a shallow andesitic chamber. An analysis of the volatile component concentration of the melt inclusion in the ejecta of the 2000 AD eruption indicates that the equilibrium depth of basaltic magma is ~8 km for basaltic magma and ~3–5 km for andesitic magma (Saito et al., 2005, 2010).

Geophysical and geochemical observations also support the theory of a duplex magma chamber system beneath the Miyakejima before and during the 2000 AD eruption. The existence of a deep-seated basaltic magma chamber is suggested by the presence of the inflation source at around 9.5 km below sea level prior to the eruption (Nishimura et al., 2001). The existence of a shallow magma chamber at ~3–5 km below sea level is suggested by the position of the deflation source during the 2000 eruption (Nishimura et al., 2001), the source depth of the very-long-period seismic signal (Kumagai et al., 2001; Kobayashi et al.,

only two (Hatchodaira and Daihannyayama) occurred in the NE-SW rift.

Here, we introduce the "offset angle" of eruption fissures to evaluate the direction of the eruption fissures (**Figure 4A**). Offset angle is defined as the angle between the strike of the eruption fissure at the near side and the direction of the maximum horizontal compressive stress (σ_{Hmax}). The offset angle varies from 0° (parallel to the σ_{Hmax}) to 90° (perpendicular to the σ_{Hmax}). The direction of σ_{Hmax} is assumed to be N135°E, based on the strike of the regional dike intruded during the 2000 AD eruption (Ito and Yoshida, 2002; Yamaoka et al., 2005). The eruption fissures in the NE-SW rift zone have high offset angles ranging from 40° to 90°.

TABLE 1 | Age, offset angle, and whole-rock composition of fissure eruptions.

Name of eruption	Age	14C age	Location	Offset angle	Whole-rock SiO$_2$ wt.%
Hatchodaira	2.7–2.8 ka		SW flank	45	56.1–58.8
Kanaso-minami	2.1 ka	2110 ± 30 BP	E flank	50	52.2
Tairayama	2.1 ka		NW flank	19	52.7–53.4
Tairoike	2.0 ka	1950 ± 20 BP	S flank	20	54.0
Daihannyayama	1.8 ka	>1880 ± 30 BP	NE flank	60	54.2
Anegakata	~2 ka		NW flank	10	50.0
Suoana-Kazahaya	Seventh century	1410 ± 30 yrBP, 1360 ± 20 yrBP	N flank	60	54.0–61.4
Oyama-Miikehama	Ninth century	1240 ± 20 yrBP	E-W summit-flank	50	51.8–54.3
Kamane	0.9 ka	860 ± 20 yrBP	SW flank	80	52.0
Nanto	1085 AD?	750 ± 20 yrBP	SW flank	65	55.0
Hinoyama	1154 AD?	700 ± 20 yrBP, 690 ± 20 yrBP	NE flank	85	53.7
Enokizawa	1469 AD?		W flank	40	52.6–53.2
Sonei-Bokujo			Summit	–	52.2–54.1
Benkene	1535 AD?	340 ± 20 yrBP, 380 ± 20 yrBP, 390 ± 20 yrBP	E flank	15	51.8–52.8
Jinanyama	1595 AD?	315 ± 20 yrBP, 320 ± 20 yrBP	SW flank	85	52.1–52.8
Koshiki-Imasaki	1643 AD	250 ± 20 yrBP	SW flank	70	52.2–52.9
Shinmio	1712 AD?		SW flank	85	55.0
Tatsune	1763 AD?		SW flank	90	52.2
Bunka	1811AD		Summit?	–	
Kasaji-Kannon	1835 AD		W flank	55	56.0
1874 AD (Meiji)	1874 AD		N flank	50	51.9–54.3
1940 AD	1940 AD		NE flank and summit	80	50.8–55.7*
1962 AD	1962 AD		NE flank	80	52.2–54.7*
1983 AD	1983 AD		SW flank	85	52.5–54.7**
2000 AD	2000 AD		SW flank (submarine)	45	51.7–54.0

*Amma-Miyasaka and Nakagawa (2003), **Kuritani et al. (2003).
Ages with "?" are not supported by the historic records.

2009), and the location of the dike injection at the initial stage of eruption (Ueda et al., 2005; Irwan et al., 2006). Source analysis of the very-long-period seismic signal and tilt change reveals the existence of a vertically aligned and NE-SW elongated ellipsoidal magma chamber at around 2.6 km beneath the south flank of the edifice (Munekane et al., 2016).

These petrological and geophysical observations indicate the existence of a duplicated magma plumbing system with a deep (~10 km) basaltic chamber and shallow (~2–5 km) andesitic magma chamber. Intermittent injections of basaltic magma into the shallow andesitic magma chamber formed hybrid magmas with higher SiO$_2$ concentrations. The injection of basaltic magma into a shallow magma chamber also caused rapid rise of internal magmatic pressure in the shallow magma chamber. This rapid increase of internal magmatic pressure may cause the dike propagation from the shallow magma chamber to feed a fissure eruption. Narrow diffusion profile observed in the titanomagnetite crystals of the andesitic ejecta suggests that the rupturing of the magma chamber and fissure eruption occurred within a few hours after the injection of basaltic magma into the andesitic system (Kuritani et al., 2003). A shallow magma chamber with a NE-SW trend and an elongated shape will form a stress field to accommodate the NE-SW trending vertical dikes above the magma chamber (**Figure 5A**). A feeder dike, which

propagated from the shallow magma chamber, reached the NE-SW rift zone and erupted magmas with a signature of magma mixing between andesite and basalt.

Development of feeder dike system controlled by a duplex magma chamber proposed for the Miyakejima's system is similar to the feeding system of the 2005 eruption of Fernandina volcano, Galapagos (Chadwick et al., 2011). Petrological evidences of the lavas support the tapping of magma from the shallow magma chamber (Chadwick et al., 2011). The path of the feeder dike followed the maximum compressive stress generated by the overpressurized sill-like chamber (Chadwick and Dieterich, 1995) and also the existence of caldera (Corbi et al., 2015) to form a circumferential fissure. Development of the circumferential inclined sheets in Fernandina is controlled by the shape of the shallow magma chamber (horizontal sill), whereas the vertical dike-like shape of the shallow chamber in Miyakejima forms preferred oriented vertical feeder dikes. Difference of the regional tectonic setting [almost isotropic horizontal stress in Fernandina (Chadwick and Dieterich, 1995) and strong horizontal differential stress in Miyakejima (Nishimura, 2011)] may control the difference of the orientation of the shallow chambers.

Development of the compressive stress field around the shallow andesitic magma chamber is limited at the time of

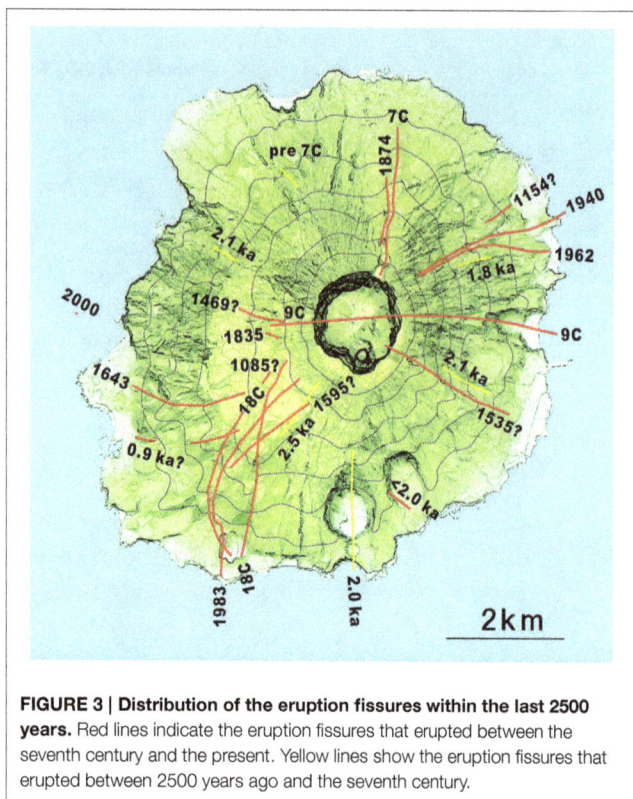

FIGURE 3 | Distribution of the eruption fissures within the last 2500 years. Red lines indicate the eruption fissures that erupted between the seventh century and the present. Yellow lines show the eruption fissures that erupted between 2500 years ago and the seventh century.

basaltic injection because pressure in the shallow andesitic magma chamber drops with the withdrawal of magma to the eruption fissure and/or the relaxation of the host rock surrounding the magma chamber. Some dikes from a deep-seated magma chamber could propagate directly to the ground surface and caused fissure eruptions of basaltic lavas (e.g., the 1535 eruption). In this case, the shallow magma chamber is not affected by the injection and, therefore, the magma chamber did not have any remarkable stress effect on the propagation of the basaltic dike (**Figure 5B**). Propagation of these basaltic dikes is mainly controlled by the regional stress field, and/or local stress field generated by the deep-seated basaltic magma chamber.

Our dataset of the ^{14}C dating shows that the development of the NE-SW rift zone became significant after the ninth century. This suggests that the shallow andesitic magma chamber was formed after the ninth century. The Oyama-Miikehama in the middle of the ninth century eruption erupted a volume of ∼0.08 km^3 DRE of basaltic magma (Tsukui and Suzuki, 1999) from the summit crater and an eruption fissure ∼5 km in length. The Oyama-Miikehama eruption is the largest eruption within the last ∼2800 years after the formation of the Hatchodaira caldera. A part of the basaltic magma, which was intruded during this eruption, formed a secondary magma chamber at a shallow depth and produced andesitic magma with fractional crystallization. A shift from the compositional trend of the eruption products since the ninth century (**Figure 2**) suggests the formation of a new magma system at the Oyama-Miikehama eruption. The existence of a similar secondary magma chamber is also recognized in the neighboring basaltic volcanoes (e.g., source of the 1986 fissure eruption in Izu-Oshima; Nakano

and Yamamoto, 1991). NE-SW elongated shape of the shallow magma chamber may control the emplacement direction of the dikes, though the reason why the shallow magma chamber was formed perpendicular to the regional σ_{Hmax} direction is still open to debate. One of the possibilities is the disturbance of stress field by coseismic extensional deformation at the plate boundary earthquake (e.g., The 2011 Tohoku earthquake; Ozawa and Fujita, 2013). Two large trench-type earthquakes occurred in the ninth century (the 869AD Jogan Earthquake M>8.3 at Japan Trench and the 887AD Ninna-Nankai Earthquake Mw>8.6 at Nankai Trough; **Figure 1A**). These earthquakes might change the stress field from the NW-SE compressive state to the NW-SE extensive one in which a NE-SW elongated magma chamber can be formed. Tectonic disturbance is also suggested by frequent and intensive eruptions in the ninth century in the northern part of the Izu area (e.g., eruptions of Kozushima 838AD, Niijima 886AD, Mt. Fuji 864AD; Tsukui et al., 2006).

In the future, distribution pattern of the fissure eruptions in the future can change to a radial or NW-SE trend depending on the properties of the regional tectonic stress. During the caldera formation of the 2000 eruption, the shallow magma chamber of andesitic magma was emptied and collapsed (Geshi et al., 2002). Basaltic ejecta in the later stage of the eruption indicate that the shallow andesitic magma chamber was completely replaced by the basaltic magma. Therefore, the feeder dikes can intrude directly from the deep-seated basaltic magma chamber without any tectonic influence from the shallow magma chamber.

The example of Miyakejima suggests that the location, shape, and orientation of a shallow magma chamber will control the orientation and propagation direction of the feeder dikes. Similar cases where the stress of a magma plumbing system influence dike propagation are recognized in several basaltic volcanoes (e.g., Stromboli: Corazzato et al., 2008). The effect of local stress generated by an additional intrusion can be temporal because the stress relaxation weakens the local stress field in the host rock. The timescale of the stress relaxation is, in general, difficult to evaluate because it depends on the macroscopic viscoelastic behavior of the volcanic edifice which is controlled by the mechanical and thermal structure and the magnitude of the local stress field. Some field observations can suggest the time scale; in the case of Fernandina volcano, the local stress field generated by the emplacement of a sill in 1995 can remain and control the development of the eruption fissure of 2005 (Bagnardi et al., 2013). The local distribution of the eruption fissures may suggest frequent intrusions with a relatively short interval as the case of Miyakejima (2–3 events per 100 years).

The propagation direction of dikes in a volcanic edifice can be also affected by the topographic contrast, especially in an asymmetric edifice (e.g., Tibaldi et al., 2014; Corbi et al., 2015; Rivalta et al., 2015 and references therein). The volcanic edifice of Miyakejima, including its submarine part, has a conical shape without any remarkable bulge along the rift zone (**Figure 1B**). The development of the NE-SW rift zone became significant within the last 2800 years. However, no remarkable change in the shape of the edifice was noted in this period. These facts suggest that the topographic control on the rift development is limited in the case of Miyakejima.

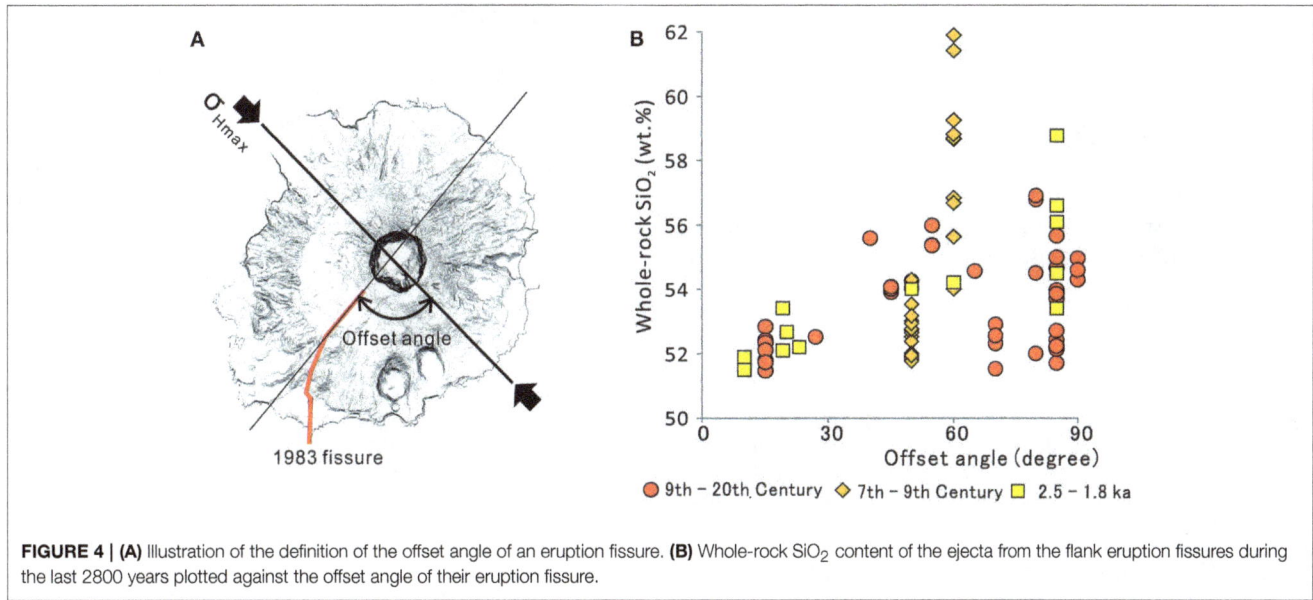

FIGURE 4 | (A) Illustration of the definition of the offset angle of an eruption fissure. **(B)** Whole-rock SiO_2 content of the ejecta from the flank eruption fissures during the last 2800 years plotted against the offset angle of their eruption fissure.

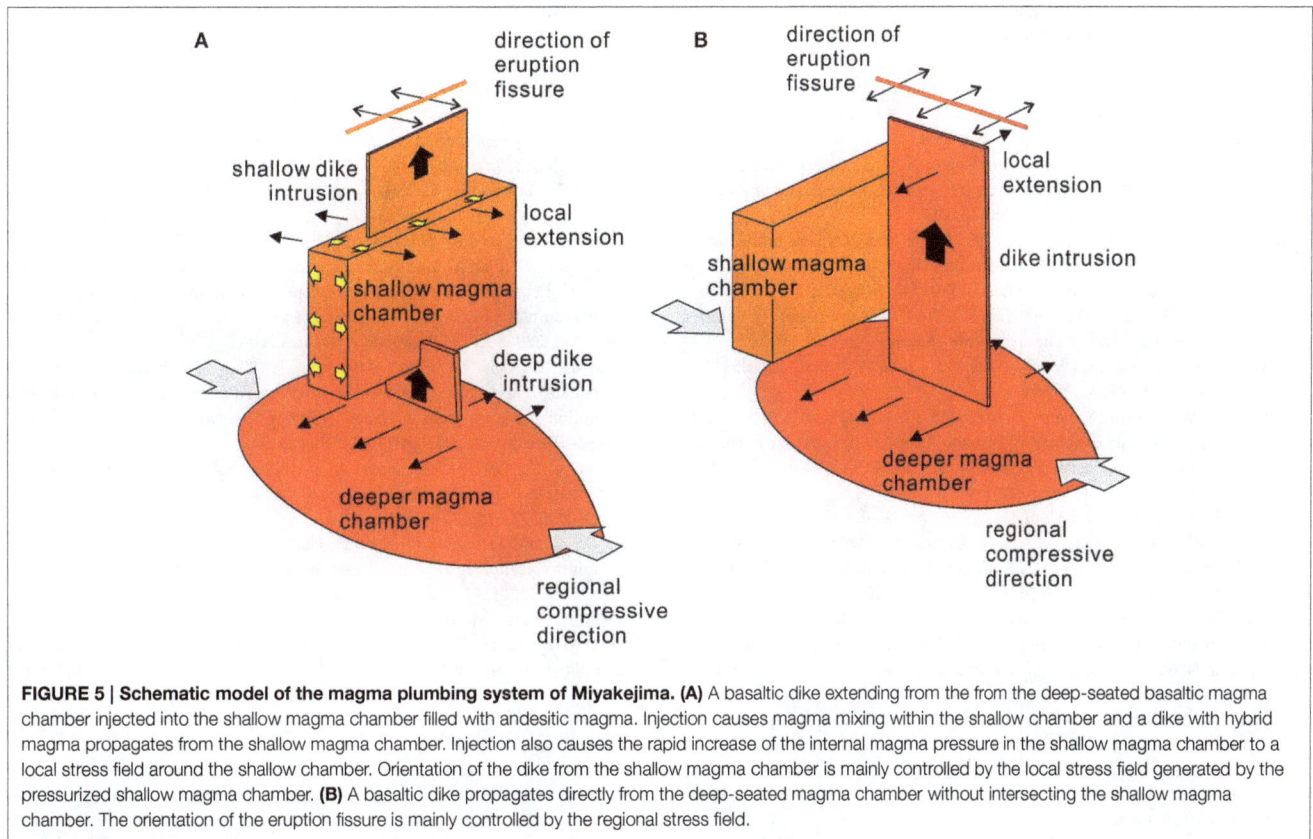

FIGURE 5 | Schematic model of the magma plumbing system of Miyakejima. (A) A basaltic dike extending from the from the deep-seated basaltic magma chamber injected into the shallow magma chamber filled with andesitic magma. Injection causes magma mixing within the shallow chamber and a dike with hybrid magma propagates from the shallow magma chamber. Injection also causes the rapid increase of the internal magma pressure in the shallow magma chamber to a local stress field around the shallow magma chamber. Orientation of the dike from the shallow magma chamber is mainly controlled by the local stress field generated by the pressurized shallow magma chamber. **(B)** A basaltic dike propagates directly from the deep-seated magma chamber without intersecting the shallow magma chamber. The orientation of the eruption fissure is mainly controlled by the regional stress field.

CONCLUSIONS

Combined analyses of the distribution patterns of the eruption fissures and the magmatic composition of the lavas within the last ~2800 years reveal that a shallow andesitic magma chamber beneath the summit of Miyakejima volcano controls the orientation of eruption fissures. The injection of basaltic magma into the shallow andesitic magma chamber caused a rapid increase in the internal magmatic pressure in this magma chamber. The local compressive stress field around the andesitic magma chamber controls the orientation of feeder dikes in the NE-SW rift zone. Development of the NE-SW rift zone after a major eruption in the ninth century indicates that the shallow magma chamber is a remnant magma chamber formed during the ninth century eruption.

AUTHOR CONTRIBUTIONS

NG and TO prepares all the field data. NG did the petrological analysis. Reconstruction of the eruption history is mainly by TO. NG and TO build the manuscript.

ACKNOWLEDGMENTS

The field survey in Miyakejima was supported by the village of Miyake and the Japan Metrological Agency. The authors thank Hiroshi Shinohara, Karoly Nemeth, and Rina Noguchi for their assistance during the field survey. The authors also appreciate the reviewers for their critical review comments. The authors also appreciate Roberto Sulpizio and Valerio Acocella for their suggestions.

REFERENCES

Amma-Miyasaka, M., and Nakagawa, M. (2003). Evolution of deeper basaltic and shallower andesitic magmas during the ad 1469-1983 eruptions of Miyake-Jima Volcano, Izu– Mariana Arc: inferences from temporal variations of mineral compositions in crystal-clots. *J. Petrol.* 44, 2113-2138. doi: 10.1093/petrology/egg072

Amma-Miyasaka, M., Nakagawa, M., and Nakada, S. (2005). Magma plumbing system of the 2000 eruption of Miyakejima Volcano, Japan. *Bull. Volcanol.* 67, 254-267. doi: 10.1007/s00445-004-0408-0

Aramaki, S., Hayakawa, Y., Fujii, T., Nakamura, K., and Fukuoka, T. (1986). The October 1983 eruption of Miyakejima volcano. *J. Volcanol. Geotherm. Res.* 29, 203-229. doi: 10.1016/0377-0273(86)90045-4

Bagnardi, M., Amelung, F., and Poland, M. P. (2013). A new model for the growth of basaltic shields based on deformation of Fernandina volcano, Galápagos Islands. *Earth Planet. Sci. Lett.*, 377-378, 358-366. doi: 10.1016/j.epsl.2013.07.016

Cappello, A., Geshi, N., Neri, M., and Del Negro, N. (2015). Lava flow hazards – An impending threat at Miyakejima volcano, Japan. *J. Volcanol. Geotherm. Res.* 308, 1-9. doi: 10.1016/j.jvolgeores.2015.10.005

Chadwick, W. W., and Dieterich, J. H. (1995). Mechanical modeling of circumferential and radial dike intrusion on Galapagos volcanoes. *J. Volcanol. Geotherm. Res.* 66, 37-52. doi: 10.1016/0377-0273(94)00060-T

Chadwick, W. W., and Howard, K. A. (1991). The pattern of pattern of circumferential and radial eruptive fissures on the volcanoes of Fernandina and Isabela islands, Galapagos. *Bull. Volcanol.* 53, 259-275. doi: 10.1007/BF00414523

Chadwick, W. W., Jónsson, S., Geist, D. J., Poland, M., Johnson, D. J., Batt, S., et al. (2011). The May 2005 eruption of Fernandina volcano, Galápagos: The first circumferential dike intrusion observed by GPS and InSAR. *Bull. Volcanol.* 73, 679-697. doi: 10.1007/s00445-010-0433-0

Corazzato, C., Francalanci, L., Menna, M., Petrone, C., Renzulli, A., Tibaldi, A., et al. (2008). What does it guide sheet intrusion in volcanoes? Petrological and structural characters of the Stromboli sheet complex, Italy. *J. Volcanol. Geotherm. Res.* 173, 26-54. doi: 10.1016/j.jvolgeores.2008.01.006

Corbi, F., Rivalta, E., Pinel, V., Maccaferri, F., Bagnardi, M., and Acocella, V. (2015). How caldera collapse shapes the shallow emplacement and transfer of magma in active volcanoes. *Earth Planet Sci. Lett.* 431, 287-293. doi: 10.1016/j.epsl.2015.09.028

Falsaperla, S., and Neri, M. (2015). Seismic footprints of shallow dyke propagation at Etna, Italy. *Sci. Rep.* 5:11908. doi: 10.1038/srep11908

Geshi, N., Shimano, T., Nagai, M., and Nakada, S. (2002). Caldera collapse during the 2000 eruption of Miyakejima volcano, Japan. *Bull. Volcanol.* 64, 55-68. doi: 10.1007/s00445-001-0184-z

Gudmundsson, A. (2006). How local stresses control magma-chamber ruptures, dyke injections, and eruptions in composite volcanoes. *Earth Sci Rev.* 79, 1-31. doi: 10.1016/j.earscirev.2006.06.006

Irwan, M., Kimata, F., and Fujii, N. (2006). Time dependent modeling of magma intrusion during the early stage of the 2000 Miyakejima activity. *J. Volcanol. Geotherm. Res.* 150, 102-112. doi: 10.1016/j.jvolgeores.2005.07.014

Ishizuka, O., Geshi, N., Itoh, J., Kawanabe, Y., and TuZino, T. (2008). The magmatic plumbing of the submarine Hachijo NW volcanic chain, Hachijojima, Japan: long-distance magma transport? *J. Geophys. Res.* 113:B08S08. doi: 10.1029/2007jb005325

Ishizuka, O., Geshi, N., Kawanabe, Y., Ogitsu, I., Taylor, R. N., Tuzino, T., et al. (2014). Long-distance magma transport from arc volcanoes inferred from the submarine eruptive fissures offshore Izu-Oshima volcano, Izu–Bonin arc. *J. Volcanol. Geotherm. Res.* 285, 1-17. doi: 10.1016/j.jvolgeores.2014.08.006

Ito, T., and Yoshida, S. (2002). A dike intrusion model in and around Miyakejima, Niijima and Kozushima in 2000. *Tectonophy* 359, 171-187. doi: 10.1016/S0040-1951(02)00510-3

Kobayashi, T., Ohminato, T., Ida, Y., and Fujita, E. (2009). Very long period seismic signals observed before the caldera formation with the 2000 Miyake-jima volcanic activity, Japan. *J. Geophys. Res.* 114:B02211. doi: 10.1029/2007jb005557

Koyama, M., and Umino, S. (1991). Why does the Higashi-Izu monogenetic volcano group exist in the Izu Peninsula? Relationships between late Quaternary volcanism and tectonics in the northern tip of the Izu-Bonin arc. *J. Phys. Earth.* 39, 391-420. doi: 10.4294/jpe1952.39.391

Kumagai, H., Ohminato, T., Nakano, M., Ooi, M., Kubo, A., Inoue, H., et al. (2001). Very-long-period seismic signals and caldera formation at Miyake Island, Japan, *Science* 293, 687-690. doi: 10.1126/science.1062136

Kuritani, T., Yokoyama, T., Kobayashi, K., and Nakamura, E. (2003). Shift and rotation of composition trends by magma mixing: 1983 Eruption at Miyake-jima Volcano, Japan. *J. Petrol.* 44, 1895-1916. doi: 10.1093/petrology/egg063

Letourneur, L., Peltier, A., Staudacher, T., and Gudmundsson, A. (2008). The effects of rock heterogeneities on dyke paths and asymmetric ground deformation: The example of Piton de la Fournaise (Réunion Island). *J. Volcanol. Geotherm. Res.* 173, 289-302. doi: 10.1016/j.jvolgeores.2008.01.018

Munekane, H., Oikawa, J., and Kobayashi, T. (2016). Mechanisms of step-like tilt changes and very long period seismic signals during the 2000 Miyakejima eruption: insights from kinematic GPS. *J. Geophy. Res. Solid Earth* 121, 2932-2946. doi: 10.1002/2016JB012795

Nakamura, K. (1977). Volcanoes as possible indicators of tectonic stress orientation — principle and proposal. *J. Volcanol. Geotherm. Res.* 2, 1-16. doi: 10.1016/0377-0273(77)90012-9

Nakano, S., and Yamamoto, T. (1991). Chemical variations of magmas at Izu-Oshima volcano, Japan: plagioclase-controlled and differentiated magmas. *Bull. Volcanol.* 53, 112-120. doi: 10.1007/BF00265416

Niihori, K., Tsukui, M., and Kawanabe, Y. (2003). Evolution of magma and magma plumbing system of Miyakejima Volcano in the last 10,000 years. *Bull. Volcanol. Soc. Japan* 48, 387-405.

Nishimura, T. (2011). Back-arc spreading of the northern Izu–Ogasawara (Bonin) Islands arc clarified by GPS data. *Tectonophy* 512, 60-67. doi: 10.1016/j.tecto.2011.09.022

Nishimura, T., Ozawa, S., Murakami, M., Sagiya, T., Tada, T., Kaizu, M., et al. (2001). Crustal deformation caused by magma migration in the northern Izu Islands, Japan. *Geophys. Res. Lett.* 28, 3745-3748. doi: 10.1029/2001GL013051

Ozawa, T., and Fujita, E. (2013). Local deformations around volcanoes associated with the 2011 off the Pacific coast of Tohoku earthquake. *J. Geophy. Res. Solid Earth* 118, 390-405. doi: 10.1029/2011JB009129

Rivalta, E., Taisne, B., Bunger, A., and Katz, R. (2015). A review of mechanical models of dike propagation: schools of thought, results and future directions. *Tectonophysics* 638, 1-42. doi: 10.1016/j.tecto.2014.10.003

Saito, G., Morishita, Y., and Shinohara, H. (2010). Magma plumbing system of the 2000 eruption of Miyakejima volcano, Japan, deduced from volatile and major component contents of olivine-hosted melt inclusions. *J. Geophys. Res.* 115:B11202. doi: 10.1029/2010jb007433

Saito, G., Uto, K., Kazahaya, K., Shinohara, H., Kawanabe, Y., and Satoh, H. (2005). Petrological characteristics and volatile content of magma from the 2000 eruption of Miyakejima Volcano, Japan. *Bull. Volcanol.* 67, 268–280. doi: 10.1007/s00445-004-0409-z

Seebeck, H., and Nicol, H. (2009). Dike intrusion and displacement accumulation at the intersection of the Okataina Volcanic Centre and Paeroa Fault zone, Taupo Rift, New Zealand. *Tectonophy* 475, 575–585. doi: 10.1016/j.tecto.2009.07.009

Takada, A. (1997). Cyclic flank-vent and central-vent eruption patterns. *Bull. Volcanol.* 58, 539–556. doi: 10.1007/s004450050161

Tibaldi, A. (2015). Structure of volcano plumbing systems: a review of multi-parametric effects. *J. Volcanol. Geotherm. Res.* 298, 85–135. doi: 10.1016/j.jvolgeores.2015.03.023

Tibaldi, A., Bonali, F. L., and Corazzato, C. (2014). The diverging volcanic rift system. *Tectonophy* 611, 94–113. doi: 10.1016/j.tecto.2013.11.023

Tsukui, M., Saito, K., and Hayashi, K. (2006). Frequent and intensive eruptions in the 9th century, Izu Islands, Japan: revisions of volcano-stratigraphy based on tephras and historical document. *Bul. Volcanol. Soc. Japan* 51, 327–338.

Tsukui, M., and Suzuki, Y. (1999). Eruptive history of Miyakejima Volcano during the last 7000 years. *Bull. Volcanol. Soc. Japan* 43, 149–166.

Ueda, H., Fujita, E., Ukawa, M., Yamamoto, E., Irwan, M., and Kimata, F. (2005). Magma intrusion and discharge process at the initial stage of the 2000 activity of Miyakejima, central Japan, inferred from tilt and GPS data. *Geophys. J. Int.* 161, 891–906. doi: 10.1111/j.1365-246X.2005.02602.x

Walter, T. R., Troll, V. R., Cailleau, B., Belousov, A., Schmincke, H.-U., Amelung, F., et al. (2005). Rift zone reorganization through flank instability in ocean island volcanoes: an example from Tenerife, Canary Islands. *Bull. Volcanol.* 67, 281–291. doi: 10.1007/s00445-004-0352-z

Yamaji, A., and Sato, K. (2011). Clustering of fracture orientations using a mixed Bingham distribution and its application to paleostress analysis from dike or vein orientations. *J. Struct. Geol.* 33, 1148–1157. doi: 10.1016/j.jsg.2011.05.006

Yamaoka, K., Kawamura, M., Kimata, F., Fujii, N., and Kudo, T. (2005). Dike intrusion associated with the 2000 eruption of Miyakejima Volcano, Japan. *Bull. Volcanol.* 67, 231–242. doi: 10.1007/s00445-004-0406-2

Conflict of Interest Statement: The authors declare that the research was conducted in the absence of any commercial or financial relationships that could be construed as a potential conflict of interest.

15

Molecular and Optical Properties of Tree-Derived Dissolved Organic Matter in Throughfall and Stemflow from Live Oaks and Eastern Red Cedar

Aron Stubbins[1]*, Leticia M. Silva[1], Thorsten Dittmar[2] and John T. Van Stan[3]

[1] Department of Marine Sciences, Skidaway Institute of Oceanography, University of Georgia, Savannah, GA, USA,
[2] Research Group for Marine Geochemistry, Institute for Chemistry and Biology of the Marine Environment, University of Oldenburg, Oldenburg, Germany, [3] Department of Geology and Geography, Georgia Southern University, Statesboro, GA, USA

Edited by:
Richard G. Keil,
University of Washington, USA

Reviewed by:
David Christopher Podgorski,
Florida State University, USA
Christian Schlosser,
GEOMAR Kiel, Germany

*Correspondence:
Aron Stubbins
aron.stubbins@skio.uga.edu

Specialty section:
This article was submitted to
Marine Biogeochemistry,
a section of the journal
Frontiers in Earth Science

Studies of dissolved organic matter (DOM) transport through terrestrial aquatic systems usually start at the stream. However, the interception of rainwater by vegetation marks the beginning of the terrestrial hydrological cycle making trees the headwaters of aquatic carbon cycling. Rainwater interacts with trees picking up tree-DOM, which is then exported from the tree in stemflow and throughfall. Stemflow denotes water flowing down the tree trunk, while throughfall is the water that drips through the leaves of the canopy. We report the concentrations, optical properties (light absorbance) and molecular signatures (ultrahigh resolution mass spectrometry) of tree-DOM in throughfall and stemflow from two tree species (live oak and eastern red cedar) with varying epiphyte cover on Skidaway Island, Savannah, Georgia, USA. Both stemflow and throughfall were enriched in DOM compared to rainwater, indicating trees were a significant source of DOM. The optical and molecular properties of tree-DOM were broadly consistent with those of DOM in other aquatic ecosystems. Stemflow was enriched in highly colored DOM compared to throughfall. Elemental formulas identified clustered the samples into three groups: oak stemflow, oak throughfall and cedar. The molecular properties of each cluster are consistent with an autochthonous aromatic-rich source associated with the trees, their epiphytes and the microhabitats they support. Elemental formulas enriched in oak stemflow were more diverse, enriched in aromatic formulas, and of higher molecular mass than for other tree-DOM classes, suggesting greater contributions from fresh and partially modified plant-derived organics. Oak throughfall was enriched in lower molecular weight, aliphatic and sugar formulas, suggesting greater contributions from foliar surfaces. While the optical properties and the majority of the elemental formulas within tree-DOM were consistent with vascular plant-derived organics, condensed aromatic formulas were also identified. As condensed aromatics are generally interpreted as deriving from partially combusted organics, some of the tree-DOM may have derived from the atmospheric deposition of thermogenic and other windblown organics. These initial findings should prove useful as future studies seek to track tree-DOM across the aquatic gradient from canopy roof, through soils and into fluvial networks.

Keywords: Tree-DOM, dissolved organic matter (DOM), carbon, CDOM, deposition, dissolved organic carbon (DOC), stemflow, throughfall

INTRODUCTION

In forested catchments, trees represent the first interceptors of precipitation and the first potential source of dissolved organic matter (DOM) to the aquatic carbon cycle. The earliest trees appear in the fossil record approximately 385 million years ago (Stein et al., 2007), since when they have fundamentally altered terrestrial (Algeo et al., 2001; Gensel, 2001) and wetland ecosystems (Greb et al., 2006). Forests are estimated to have covered close to 50 million km^2 of the planet 5,000 years ago (FAO, 2016) equivalent to approximately 1/3rd of the earth's land surface. Just as forests transformed the global ecosystem, humans now have a similarly profound influence upon global ecology and biogeochemistry. Deforestation during the Anthropocene (Crutzen, 2002) has seen forest land cover reduced by approximately 50% to 31.7 million km^2 as of 2005 (Hansen et al., 2010) and was continuing at a rate of approximately 1.5 million km^2 $year^{-1}$ between 2000 and 2012 (Hansen et al., 2013).

Despite the vast and rapidly changing expanse of land covered and volume of precipitation intercepted by trees, only modest attention has been focused upon the DOM delivered by trees to downstream ecosystems (Kolka et al., 1999; Michalzik et al., 2001; Neff and Asner, 2001; Levia et al., 2011; Inamdar et al., 2012). Once intercepted, rainwater takes one of two hydrological flow paths to the forest floor: throughfall (water that drips from the canopy or falls directly through canopy gaps) and stemflow (water funneled by the canopy to the stem). Both stemflow (5–200 mg-C L^{-1}) (Moore, 2003; Tobón et al., 2004; Levia et al., 2011) and throughfall (1–100 mg-C L^{-1}) (Michalzik et al., 2001; Neff and Asner, 2001; Le Mellec et al., 2010; Inamdar et al., 2012) are enriched in DOM relative to rainwater (0.3–2 mg-C L^{-1}) (Willey et al., 2000).

The fate of exported tree-derived DOM (tree-DOM) will depend upon the chemistry of the tree-DOM, the nature of the receiving ecosystem and hydrological considerations. Significant losses and alteration of tree-DOM occurs as stemflow and throughfall enter soils due to sorption to mineral soils, which preferentially retain hydrophobic DOM fractions (Jardine et al., 1989; Kaiser and Zech, 1998, 2000). The spatially and temporally uneven delivery of biolabile tree-DOM to soil ecosystems during storms may fuel biogeochemical hot spots and hot moments (McClain et al., 2003; Vidon et al., 2010) and microbial utilization of biolabile organics in soils further modifies tree-DOM (Aitkenhead-Peterson et al., 2003). It has also been suggested that sunlight driven photoreactions (Mopper et al., 2015) may act as a sink for tree-DOM (Aitkenhead-Peterson et al., 2003). Photoreactions would be expected to preferentially

remove aromatics, including black carbon (Stubbins et al., 2012b), while preserving and producing high H/C compounds such as aliphatics (Stubbins et al., 2010; Stubbins and Dittmar, 2015).

The degree to which tree-DOM is lost and altered by these processes before reaching a downstream aquatic ecosystem also depends upon the flow path traveled down the tree and from there to an inland water body (Inamdar et al., 2012). Direct input of stemflow or throughfall into a stream or lake will presumably result in negligible alteration prior to delivery. High levels of minimally modified tree-DOM may also reach inland waters during periods of heavy rainfall and resultant high flow when residence times within modifying ecosystems (e.g., soils) are reduced or bypassed completely in the case of overland flow. High flow pulses driven by heavy precipitation shunt reactive DOM downstream through river networks and are increasingly recognized as significant components of the fluvial carbon cycle (Raymond et al., 2016). These extreme, short lived pulses can account for the majority of annual river DOC loads being exported in just a few days per year (Raymond and Saiers, 2010). It remains unclear whether tree-DOM is delivered efficiently to fluvial systems during these pulse-shunt events.

To further understand the quality of tree-DOM, we collected stemflow and throughfall samples from broadleaved (oak) and needleleaved (cedar) trees with or without epiphytes during two storm events. The concentrations, optical properties and molecular signatures of the sampled tree-DOM are presented.

METHODS
Sample Site

Samples were collected on the Skidaway Institute of Oceanography (SkIO) campus, Georgia, USA (31.9885°N, 81.0212°W) (**Figure 1a**) during two rain events: Storm A on June 27th 2015 and Strom B on 28th June 2015 (**Table 1**). SkIO is in a subtropical climate zone (Köppen *Cfa*), with 30-year mean annual precipitation ranging from 750 to 1,200 mm that occurs as rainfall and mostly during the summer months (GA Office of the State Climatologist). Average daily temperatures in summer range between 30 and 35°C (Georgia Office of the State Climatologist, 2012). The elevation of SkIO ranges from 0 to 10 m above mean sea level. The sampling sites were flat (0–5% slopes) and underlain by Chipley fine sandy soils (https://websoilsurvey. sc.egov.usda.gov). Two species of tree were sampled: *Quercus virginiana* Mill. (southern live oak) referred to here as oak for brevity; and, *Juniperus virginiana* L. (eastern red cedar) referred to as cedar. The epiphytes *Tillandsia usneoides* L. (Spanish moss) and *Pleopeltis polypodioides* (resurrection fern) can be found to

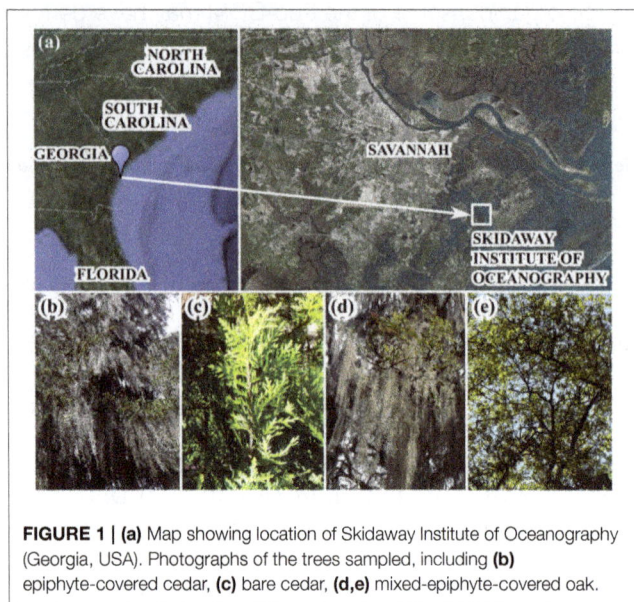

FIGURE 1 | (a) Map showing location of Skidaway Institute of Oceanography (Georgia, USA). Photographs of the trees sampled, including **(b)** epiphyte-covered cedar, **(c)** bare cedar, **(d,e)** mixed-epiphyte-covered oak.

cover these trees at high densities on SkIO (**Figures 1b–e**). A Spanish moss-covered cedar tree (cedar moss, **Figure 1b**) and an epiphyte-free cedar tree (bare cedar, **Figure 1c**) were chosen to assess whether the influence of epiphytes were apparent in the concentrations or quality of tree-DOM. Four oak trees were sampled, each with highly variable epiphyte coverage including, both resurrection ferns and Spanish moss (**Figures 1d,e**), which is typical of live oaks in the maritime southeastern US. Both stemflow and throughfall samples were collected for each tree type.

Sample Collection and Processing

All plastic and glassware were pre-cleaned by rinsing five times with ultrapure water (MilliQ), soaking in pH 2 ultrapure water (2 ppt 6N hydrochloric acid), re-rinsing five times with ultrapure water, and dried. Once dry, glassware was further baked at 450°C for 8 h. Twenty throughfall samplers (0.18 m^2, 0.5 m height, high density polyethylene (HDPE) bins) were deployed for each storm, three beneath each of four oaks (TF Oak 1–4), and four beneath the bare (TF Bare Cedar) and four beneath the epiphyte-covered cedars (TF Cedar Moss). An additional four throughfall samplers were placed upon open ground to sample rainwater (Rain). Stemflow samplers, which consisted of collars cut from polyethylene tubing wrapped about the trunk at 1.4 m height and connected to 10 L HDPE carboys, were installed on four oaks (SF Oak 1–4), the bare cedar (SF Bare Cedar) and the epiphyte-covered cedar (SF Cedar Moss). Throughfall and stemflow collectors were deployed approximately 1 h before rainfall commenced and collected within 1 h of rainfall ceasing. All sampling sites are within 10 min walk of Stubbins' laboratory at SkIO. Samples were rapidly returned to the laboratory and 0.2 μm filtered within 4 h of collection. Sample volumes were measured. Throughfall and rainwater volume fluxes (mm) were calculated by dividing the sample volumes by the surface area of the samplers (0.18 m^2).

Dissolved Organic Carbon Concentrations

After filtration, aliquots of sample were transferred to pre-combusted 40 mL glass vials, acidified to pH 2 (hydrochloric acid), and analyzed for non-purgable organic carbon using a Shimadzu TOC-VCPH analyzer fitted with a Shimadzu ASI-V autosampler. In addition to potassium hydrogen phthalate standards, aliquots of deep seawater reference material, Batch 10, Lot# 05-10, from the Consensus Reference Material Project (CRM) were analyzed to check the precision and accuracy of the DOC analyses. Analyses of the CRM deviated by <5% from the reported value for these standards (41–44 μM-DOC). Routine minimum detection limits in the investigator's laboratory using the above configuration are 2.8 ± 0.3 μM-C and standard errors are typically 1.7 ± 0.5% of the DOC concentration (Stubbins and Dittmar, 2012).

Colored Dissolved Organic Matter

Filtered sample (non-acidified) was placed in a 1 cm quartz absorbance cell situated in the light path of an Agilent 8453 ultraviolet-visible spectrophotometer and CDOM absorbance spectra were recorded from 190 to 800 nm. Ultrapure water provided a blank. Blank corrected absorbance spectra were corrected for offsets due to scattering and instrument drift by subtraction of the average absorbance between 700 and 800 nm (Stubbins et al., 2011). Data output from the spectrophotometer were in the form of dimensionless absorbance (i.e., optical density, OD) and were subsequently converted to the Napierian absorption coefficient, a (m^{-1}) (Hu et al., 2002). If sample absorbance (OD) exceeded 2 at 250 nm, samples were diluted 10-fold with ultrapure water and reanalyzed. Specific UV absorbance at 254 nm (SUVA$_{254}$; L mg-C^{-1} m^{-1}), an indicator of DOM aromaticity defined as the Decadic absorption coefficient at 254 nm (m^{-1}) normalized to DOC (mg-C L^{-1}) (Weishaar et al., 2003) was calculated along with spectral slope over the range 275–295 nm ($S_{275-295}$) (Helms et al., 2008). Spectral slope values are reported as positive values.

Fourier Transform Ion Cyclotron Resonance Mass Spectrometry

In the current study, whole water samples were analyzed without extraction or purification to provide the broadest analytical window prior to mass spectral analysis. Each stemflow sample (SF Oak 1–4; SF Cedar Moss; SF Bare Cedar) was analyzed. For throughfall samples, carbon-weighted composite samples were generated for each rainfall and throughfall sample by combining carbon-dependent volumes of sample (i.e., all four aliquots were combined for each of the rain samples and each cedar sampled for throughfall; three aliquots were combined for each of the four oak trees sampled for throughfall; the volume that each aliquot contributed to a composite sample was adjusted in order that the final composite sample contained an equal fraction of carbon from each aliquot). To generate consistent FT-ICR mass spectra all samples were analyzed under the same conditions, including DOC concentration. Therefore, all tree-DOM samples were diluted to the identical DOC concentration with ultrapure water (10 mg-C L^{-1}) and then further diluted 1:1 with methanol. However, as rainwater DOC (1–2 mg-C

TABLE 1 | Sample numbers, volumes, hydrological fluxes (calculated based upon the 0.18 m^2 surface area of rain and throughfall collectors), dissolved organic carbon concentrations (DOC), colored dissolved organic matter Napierian absorption coefficients at 300 nm (CDOM a_{300}), CDOM spectral slope values for the range 275–295 nm (S$_{275-295}$), and specific ultraviolet absorbance at 254 nm (SUVA$_{254}$) for rainwater and each of the stemflow (SF) and throughfall (TF) sample types collected during two storms.

Event	Sample name	N	Volume (mL)	Flux (mm)	DOC (mg-C L^{-1})	CDOM a_{300} (m^{-1})	S$_{275-295}$ (nm^{-1})	SUVA$_{254}$ (L mg-C^{-1} m^{-1})
Storm A 27/06/15	Rain	4	2778 ± 125	15 ± 1	2.3 ± 0.1	1.9 ± 0.9	0.0208 ± 0.0037	0.8 ± 0.2
	SF Oak	4	246 ± 140	N/A	46 ± 9	180 ± 83	0.0146 ± 0.0002	3.0 ± 0.9
	TF Oak	12	2172 ± 439	12 ± 2	15 ± 8	43 ± 22	0.0158 ± 0.0008	2.2 ± 0.3
	SF Cedar Moss	1	>10 L	N/A	52	321	0.0154	5.1
	TF Cedar Moss	4	1855 ± 355	10 ± 2	52 ± 9	153 ± 28	0.0165 ± 0.0001	2.4 ± 0.1
	SF Bare Cedar	1	>10 L	N/A	30	159	0.0145	4.1
	TF Bare Cedar	4	2996 ± 197	16.6 ± 1.1	13 ± 1	40 ± 4	0.016 ± 0.0002	2.5 ± 0.1
Storm B 28/06/15	Rain	4	3600 ± 91	20.0 ± 0.5	1.2 ± 0.1	0.7 ± 0.7	0.0457 ± 0.0237	0.8 ± 0.4
	SF Oak	4	1140 ± 674	N/A	78 ± 17	418 ± 99	0.0144 ± 0.0002	4.1 ± 0.1
	TF Oak	12	3488 ± 610	19.4 ± 3.4	10 ± 7	31 ± 19	0.0158 ± 0.0009	2.6 ± 0.4
	SF Cedar Moss	1	9583	N/A	71	378	0.0145	4.0
	TF Cedar Moss	4	2945 ± 200	16.4 ± 1.1	36 ± 9	113 ± 24	0.016 ± 0.0002	2.5 ± 0.2
	SF Bare Cedar	1	>10 L	N/A	25	129	0.0155	4.0
	TF Bare Cedar	4	3358 ± 294	18.7 ± 1.6	13 ± 2	48 ± 13	0.0157 ± 0.0006	2.9 ± 0.3

Values present are means ± one standard deviation.

L^{-1}; **Table 1**) was significantly lower than for tree-DOM and reducing all sample DOC concentrations to <1 mg-C L^{-1} would have impaired FT-ICR MS performance, rainwater samples were diluted 1:1 with methanol and run at their resulting DOC concentrations (Storm A: 1.1 mg-C L^{-1}; Storm B: 0.6 mg-C L^{-1}). As this impaired the quality of the rainwater FT-ICR MS data, this data is only used to contrast with the tree-DOM data in a cluster analysis and the molecular quality of rainwater DOM is not presented. In order to compare rainwater DOM to tree-DOM directly, the study design would have needed to include a DOM isolation and concentration step in order to allow all samples, rainwater included, to be analyzed by FT-ICR MS at the same concentrations. This option was not chosen as it would have reduced the analytical window for our focus of study: tree-DOM.

Once mixed 1:1 with methanol, samples were analyzed in negative mode electrospray ionization using a 15 Tesla FT-ICRMS (Bruker Solarix) at the University of Oldenburg, Germany. 500 broadband scans were accumulated for the mass spectra. After internal calibration, mass accuracies were within an error of <0.2 ppm. Elemental formulas were assigned to peaks with signal to noise ratios greater than five based on published rules (Koch et al., 2007; Stubbins et al., 2010; Singer et al., 2012). Peaks detected in the procedural blank (PPL extracted ultrapure water) were removed. Peak detection limits were standardized between samples by adjusting the dynamic range of each sample to that of the sample with the lowest dynamic range (dynamic range = average of the largest 20% of peaks assigned a formula divided by the signal to noise threshold intensity; standardized detection limit = average of largest 20% of peaks assigned a formula within a sample divided by the lowest dynamic range within the sample set; Spencer et al., 2014; Stubbins et al., 2014). Peaks below the standardized detection limit were removed.

These peaks were removed in order to prevent false negatives within samples with low dynamic range.

For each elemental formula, we calculated the modified Aromaticity Index (AImod) (Koch and Dittmar, 2006, 2016), which indicates the likelihood of an elemental formula representing aromatic structures, from an AImod of zero, where formulas are aliphatic, through an intermediate range, where an elemental formula could indicate aromatic or non-aromatic isomers, to AImod values above 0.5, where an elemental formula is highly likely to represent aromatic isomers (Koch and Dittmar, 2006). These AImod values were calculated as:

$$\text{AImod} = (1 + C - 0.5O - S - 0.5(N + P + H)$$
$$/(C - 0.5O - S - N - P) \quad (1)$$

AImod values 0.5–0.67 and >0.67 were assigned as aromatic and condensed aromatic structures, respectively (Koch and Dittmar, 2006). Compound classes were further defined as highly unsaturated (AImod < 0.5, H/C < 1.5, O/C < 0.9), unsaturated aliphatics (1.5 ≤ H/C < 2, O/C < 0.9, N = 0), saturated fatty acids (H/C ≥ 2, (O/C < 0.9), sugars (O/C ≥ 0.9) and peptides (1.5 ≤ H/C < 2, O/C < 0.9, N < 0). Since an individual formula could occur as multiple isomeric structures, these classifications only serve as a guide to the structures present within DOM. For instance, "peptides" have the elemental formulas of peptides, but their actual structure may differ.

Standardized peak intensities (z) within a sample were calculated following:

$$z = \frac{x - \mu}{\sigma} \quad (2)$$

where, x is the measured peak intensity, μ is mean peak intensity within the sample, and σ is the standard deviation in peak intensity within the sample (Spencer et al., 2014). Cluster analysis of the standardized peak intensities of assigned formulas (Ward clustering in JMP®) was then performed (Spencer et al., 2014).

RESULTS

Dissolved Organic Matter Concentrations and Optical Properties

DOM concentrations, quantified as DOC, ranged from 1.1 to 2.4 mg-C L^{-1} in rainwater, 40 to 95 mg-C L^{-1} in oak stemflow, 2.2 to 28 mg-C L^{-1} in oak throughfall, 52 to 71 mg-C L^{-1} in epiphyte-covered cedar stemflow, 31 to 62 mg-C L^{-1} in epiphyte-covered cedar throughfall, 25 to 30 mg-C L^{-1} in bare cedar stemflow, and 12 to 16 mg-C L^{-1} in bare cedar throughfall across the two storms sampled.

Absorbance spectra decayed exponentially with increasing wavelength (**Figure 2**). Napierian absorption coefficients for CDOM at 300 nm ranged from 0.1 to 3.2 m^{-1} in rainwater, 101 to 526 m^{-1} in oak stemflow, 6 to 81 m^{-1} in oak throughfall, 321 to 378 m^{-1} in epiphyte-covered cedar stemflow, 95 to 148 m^{-1} in epiphyte-covered cedar throughfall, 129 to 159 m^{-1} in bare cedar stemflow, and 36 to 67 m^{-1} in bare cedar throughfall across the two storms sampled.

Spectral slope values for the range 275–295 nm for CDOM ranged from 0.0156 to 0.0673 nm^{-1} in rainwater, 0.0142 to 0.0147 nm^{-1} in oak stemflow, 0.0139 to 0.0169 nm^{-1} in oak throughfall, 0.0145 to 0.0154 nm^{-1} in epiphyte-covered cedar stemflow, 0.0157 to 0.0166 nm^{-1} in epiphyte-covered cedar throughfall, 0.0145 to 0.0155 nm^{-1} in bare cedar stemflow, and 0.0152 to 0.0165 nm^{-1} in bare cedar throughfall across the two storms sampled.

$SUVA_{254}$ values ranged from 0.5 to 1.3 L mg-C^{-1} m^{-1} in rainwater, 1.9 to 4.2 L mg-C^{-1} m^{-1} in oak stemflow, 1.6 to 2.8 L mg-C^{-1} m^{-1} in oak throughfall, 4.0 to 5.1 L mg-C^{-1} m^{-1} in epiphyte-covered cedar stemflow, 2.3 to 2.8 L mg-C^{-1} m^{-1} in epiphyte-covered cedar throughfall, 4.0 to 4.1 L mg-C^{-1} m^{-1} in

bare cedar stemflow, and 2.4 to 3.3 L mg-C^{-1} m^{-1} in bare cedar throughfall across the two storms sampled.

The full dataset is presented in Table S1. Means and standard deviations for each data and sample type per storm are presented in **Table 1** and summarized in **Figure 3**. Patterns in DOC concentrations and DOM optical properties were similar between storms (**Figure 3**). Rainwater DOM had much lower values of DOC, CDOM, and SUVA, and much steeper spectral slope values than the tree-DOM samples. In general, the stemflow samples exhibited higher DOC, CDOM, and SUVA, and shallower spectral slopes, than their respective throughfall samples (**Figure 3**). The one exception being DOC for the epiphyte-covered cedar during storm A, when both stemflow and throughfall DOC concentrations were similar (**Figure 3A**).

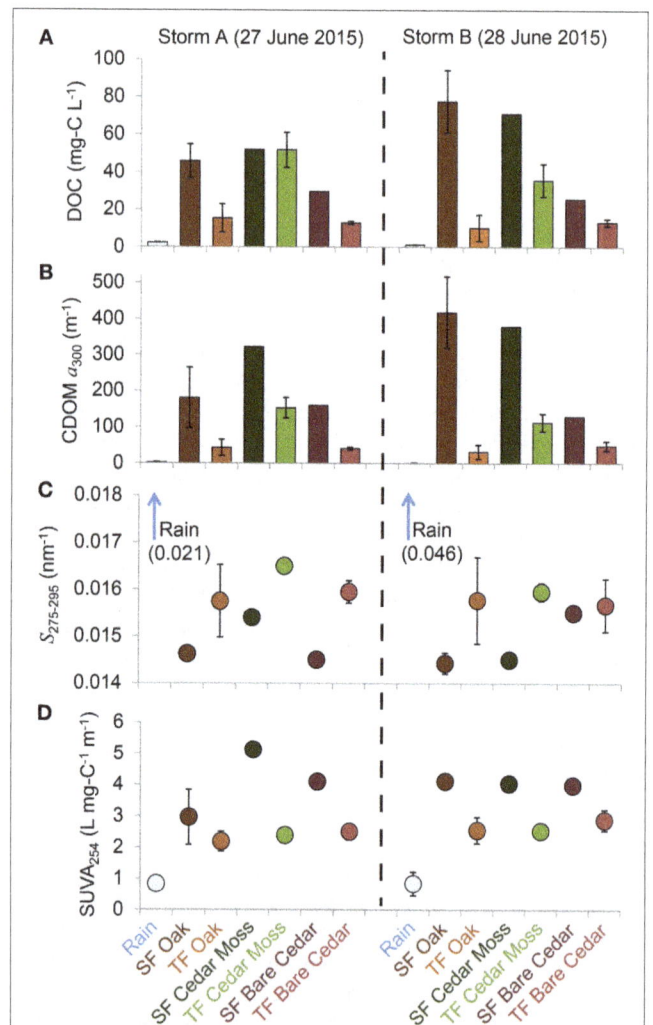

FIGURE 3 | Mean values for (A) dissolved organic carbon (DOC) concentration, (B) colored dissolved organic matter (CDOM) Napierian absorption coefficient at 300 nm (a_{300}), (C) spectral slope from 275 to 295 nm ($S_{275-295}$), and (D) specific ultraviolet absorbance at 254 nm ($SUVA_{254}$) for rainwater and each of the stemflow (SF) and throughfall (TF) sample types collected (see legend). Error bars represent 1 standard deviation and are not shown when they were narrower than the symbol.

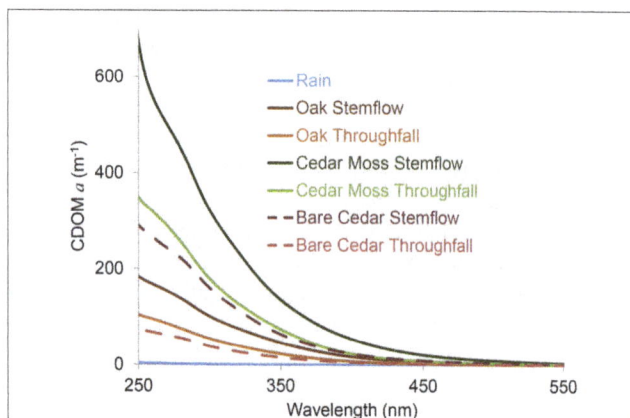

FIGURE 2 | Exemplary colored dissolved organic matter (CDOM) Napierian absorption coefficient (*a*) spectra for rainwater and each of the stemflow (SF) and throughfall (TF) sample types collected.

Molecular Signatures of Tree-Derived Dissolved Organic Matter

Whole water samples mixed 1:1 with methanol yielded mass spectra (**Figure 4**) with sufficient resolution and signal to enable the assignment of 5,852 elemental formulas to tree-DOM. The raw mass spectra for oak and cedar stemflow displayed molecular signatures consistent with those of whole river water DOM run on the same instrument, under the same conditions, during the same month (Kolyma River data in **Figure 4**; Stubbins et al., 2017). Looking at one representative mass to charge (343 m/z; **Figure 4**), DOM in oak stemflow and cedar stemflow has similar molecular diversities (i.e., there are a similar number of peaks). However, the relative abundance of some peaks varies between samples, with some peaks being below detection in one sample and present in the other.

To further explore the diversity of tree-DOM and how it varied among the samples a cluster analysis of the standardized peak intensities of assigned formulas (Ward clustering in JMP®) was performed (Spencer et al., 2014). The distance graph for this cluster analysis revealed a sharp slope break at 4 clusters (**Figure 5**; lower panel). These four clusters were: rainwater, stemflow oak, throughfall oak, cedar, the latter including stemflow and throughfall from both the epiphyte-covered and bare cedar trees (**Figure 5**). Oak DOM, including both stemflow and throughfall, and cedar DOM are clearly separated (distance between clusters, $d = 17$). Oak stemflow and throughfall are also separated from one another (distance between clusters, $d = 12$). Although other clusters are formed, the distances between them are smaller ($d < 6$). For instance, the epiphyte-covered and bare cedar trees form distinct clusters, but with a cluster distance of <2 these samples have limited molecular differences. Based upon the cluster analysis, the molecular properties of three molecularly distinct types of tree-DOM are presented: oak throughfall, oak

stemflow and cedar (**Table 2**). Data for the rainwater cluster are not presented.

The oak stemflow cluster comprised 4,765, oak throughfall 3,565, and cedar 3,643 formulas. All samples contained high proportions of CHO-only formulas (65–71%; **Table 2**). The oak stemflow was enriched in nitrogen compared to oak throughfall, and both were enriched in nitrogen compared to cedar DOM. Oak throughfall was enriched in sulfur and depleted in phosphorous compared to the other forms of tree-DOM, with cedar DOM being the most enriched in phosphorous.

The elemental formulas for each tree-DOM type were plotted in van Krevelen space (**Figures 6A–C**). Oak DOM spanned a broader range of van Krevelen space than cedar DOM. All types of tree-DOM contained high intensity elemental formulas in the region bounded by approximately H/C 1.1–1.6 and O/C 0.15 and 0.35. Oak DOM also had similarly high intensity elemental formulas in the approximate region H/C 0.5–1.0 by O/C 0.25–0.7.

All forms of tree-DOM covered a wide area of van Krevelen space and correspondingly included elemental formulas with a diverse range of possible structural properties (**Table 2**). All tree-DOM clusters had similar contributions from aromatic compounds (14–15%), but oak DOM was enriched in condensed aromatics (12–13%) compared to cedar DOM (10%). Highly unsaturated formulas were the most prominent molecular class in each type of tree-DOM and were enriched in cedar DOM (47%) compared to oak DOM (38–40%; **Table 2**). Combined aliphatics, including both unsaturated aliphatics and saturated fatty acids, constituted approximately 25% of peaks in all tree-DOM types. Sugar and peptide contributions were low across tree-DOM types, but were elevated in oak DOM compared to cedar DOM. Oak throughfall DOM had a lower average molecular mass (359 g mol^{-1}) compared to both oak stemflow and cedar DOM (382–383 g mol^{-1}). Average H/C and AImod

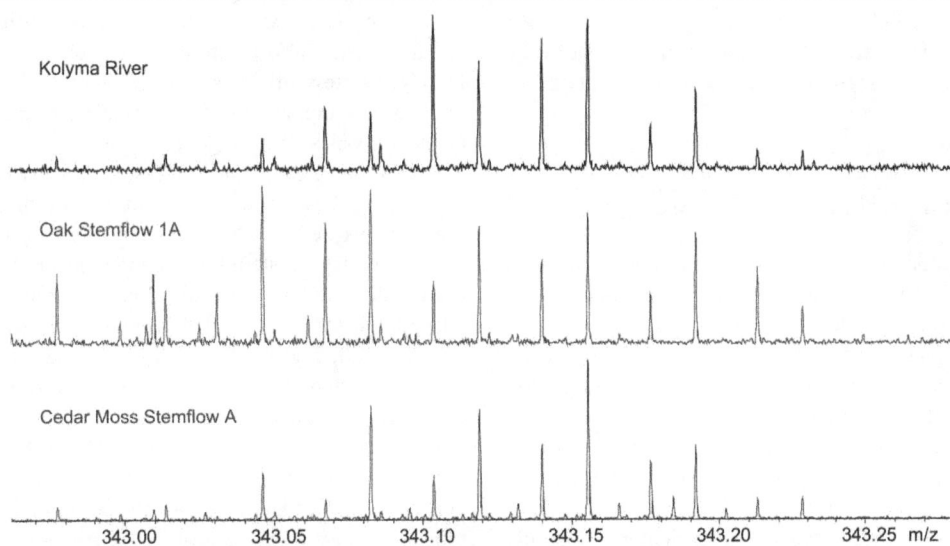

FIGURE 4 | Raw electrospray ionization Fourier transform ion cyclotron mass spectra in the 343 mass to charge (m/z) region for a representative river water sample (Kolyma River; Spencer et al., 2015), a live oak stemflow sample and a cedar stemflow sample.

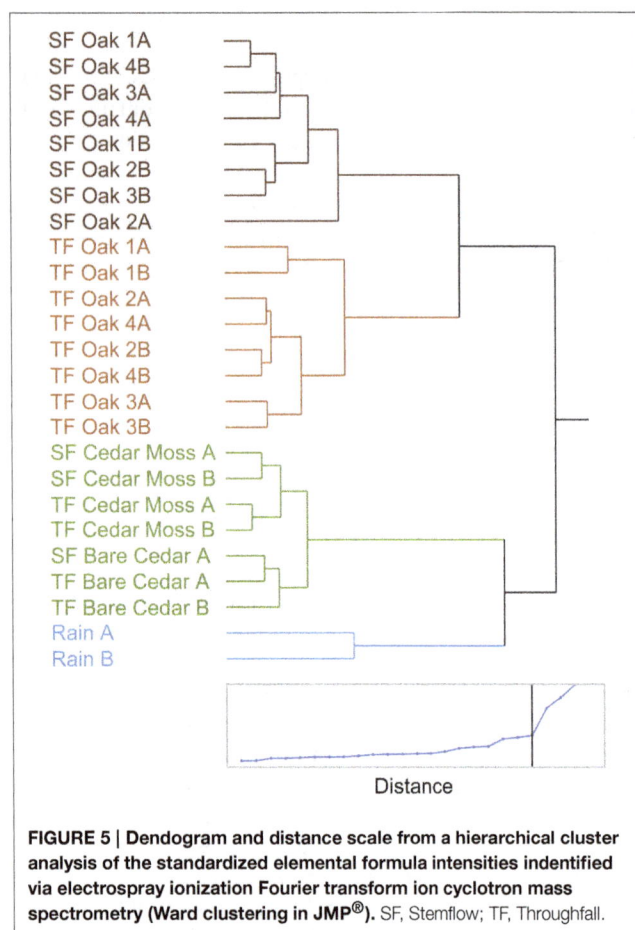

FIGURE 5 | Dendogram and distance scale from a hierarchical cluster analysis of the standardized elemental formula intensities indentified via electrospray ionization Fourier transform ion cyclotron mass spectrometry (Ward clustering in JMP®). SF, Stemflow; TF, Throughfall.

values for the clusters were similar. However, O/C decreased from oak stemflow > oak throughfall > cedar.

To reveal the quintessential molecular signatures of each tree-DOM type, the degree to which each elemental formula was enriched in a tree-DOM cluster relative to the mean for all tree-DOM types (i.e., mean intensity for a molecular formula in the whole dataset, excluding the two rain samples) was calculated as:

$$\text{Enrichment Factor} = \frac{\text{Mean Intensity in Clustered Samples}}{\text{Mean Intrensity in All Tree DOM Samples}} \quad (3)$$

An elemental formula was then classified as being enriched within a tree-DOM type if it exhibited an enrichment factor >1 and also had relatively low variations in intensity across the clustered samples (standard deviation <50% of mean intensity). The standard deviation term was included in the classification in order to exclude formulas which were not routinely enriched within the samples of a cluster. The results of this classification are presented in **Table 2** and in van Krevelen diagrams (**Figures 6D–F**). Dot size in the van Krevelen diagrams represents the mean intensity of an elemental formula in the mass spectrum, while dot color represents the degree of enrichment ranging from just above 1 (yellow) to 3 or higher (dark blue). Oak stemflow was enriched in a large

number (**Table 2**) and wide variety of molecules (**Figure 6D**) compared to oak throughfall and cedar DOM (**Figures 6E,F**). However, many of the elemental formulas that were highly enriched in oak stemflow (enrichment factor >2; darker blues) were present at relatively low intensity (small size of the dots). Oak throughfall was also enriched in a wide variety of DOM types (**Figure 6E**), but had highest enrichment in the low H/C, low O/C region typical of condensed aromatics and within the region bounded by approximately H/C 1.1–1.6 and O/C 0.15 and 0.35 where the original van Krevelens (**Figures 6A–C**) revealed high abundance within all tree-DOM types. Finally, cedar DOM was enriched in elemental formulas with H/C values from approximately 1.0–1.5 and O/C 0.1–0.45 (**Figure 6F**).

Quintessential cedar DOM formulas (i.e., formulas that were consistently enriched in cedar DOM; **Table 2** right side) were enriched in CHO-only formulas (93%) compared to oak DOM (83–84%). The number of nitrogen containing quintessential formulas decreased in the order oak stemflow > oak throughfall > cedar, while sulfur containing quintessential formulas decreased in the order oak throughfall > cedar > oak stemflow, and phosphorous containing quintessential formulas decreased in the order cedar > oak stemflow > oak throughfall (**Table 2**). The average molecular mass, O/C, and AImod of quintessential oak stemflow formulas was higher than that for cedar and oak throughfall, while quintessential cedar formulas had the highest average H/C and lowest average O/C and AImod. Quintessential oak formulas were enriched in condensed aromatics (17–21%) and contained a small proportion of peptide (0.3%) and sugar (2.0–4.3%) formulas, while the quintessential cedar formulas included zero condensed aromatic, peptide or sugar formulas (**Table 2**). Quintessential oak stemflow formulas included approximately twice the percentage of aromatic formulas (23%) when compared to the other tree DOM types (12–13%). Quintessential cedar formulas were highly enriched in highly unsaturated formulas (71%), which represented approximately half of the quintessential oak stemflow formulas (53%), and about a third (30%) of quintessential oak throughfall formulas. Finally, quintessential oak throughfall formulas were enriched in unsaturated aliphatics (30%) compared to quintessential cedar (16%) and oak stemflow (5%).

Distribution plots further resolved variations in molecular mass, H/C, O/C, and AImod of the quintessential formulas that are enriched were the different tree-DOM types (**Figure 7**). Quintessential oak stemflow formulas covered a broad, relatively evenly distributed range in molecular mass, O/C, H/C and AImod, with the H/C distribution skewed toward lower values (center ∼1.0; **Figure 7C**). Those formulas that were enriched in oak throughfall DOM exhibited pronounced peaks in abundance at low molecular mass (∼260 g mol^{-1}; **Figure 7A**), low O/C (∼0.24; **Figure 7B**), high H/C (∼1.6; **Figure 7C**), and contained three spikes in AImod (0, ∼0.2 and ∼0.8; **Figure 7D**). Quintessential cedar formulas also exhibited marked peaks in molecular mass (∼310 g mol^{-1}; **Figure 7A**), midranges in H/C (∼1.0–1.2; **Figure 7C**) and AImod (∼0.4; **Figure 7D**) and a peak in O/C that reached a maximum at ∼0.36, but exhibited two smaller shoulders at 0.21 and 0.11 (**Figure 7B**).

TABLE 2 | Molecular signatures of tree-derived dissolved organic matter (DOM) within live oak stemflow, live oak throughfall and cedar (stemflow plus throughfall).

	Formulas within each cluster			Formulas enriched within each cluster		
	SF Oak	TF Oak	Cedar	SF Oak	TF Oak	Cedar
Total Formulas	4765	3565	3643	1887	859	322
CHO	3116 (65%)	2338 (66%)	2581 (71%)	1576 (84%)	714 (83%)	301 (93%)
With N	869 (18%)	552 (15%)	420 (12%)	254 (13%)	51 (5.9%)	0 (0%)
With S	574 (12%)	568 (16%)	403 (11%)	36 (1.9%)	92 (11%)	15 (4.7%)
With P	206 (4.3%)	107 (3%)	239 (6.6%)	21 (1.1%)	2 (0.2%)	6 (1.9%)
Condensed Aromatics	590 (12%)	471 (13%)	374 (10%)	330 (17%)	184 (21%)	0 (0%)
Aromatics	676 (14%)	549 (15%)	554 (15%)	431 (23%)	115 (13%)	39 (12%)
Highly Unsaturated	1922 (40%)	1340 (38%)	1706 (47%)	992 (53%)	258 (30%)	230 (71%)
Unsaturated Aliphatics	1134 (24%)	869 (24%)	809 (22%)	95 (5.0%)	261 (30%)	52 (16%)
Saturated Fatty Acids	43 (0.9%)	40 (1.1%)	38 (1%)	0 (0%)	1 (0.1%)	1 (0.3%)
Sugars	162 (3.4%)	135 (3.8%)	82 (2.3%)	37 (2.0%)	37 (4.3%)	0 (0%)
Peptides	244 (5.1%)	162 (4.5%)	80 (2.2%)	6 (0.3%)	3 (0.3%)	0 (0%)
Average Molecular Mass (g mol^{-1})	383	359	382	370	311	330
Average H/C	1.23	1.24	1.22	1.01	1.19	1.24
Average O/C	0.45	0.42	0.39	0.46	0.39	0.27
Average Almod	0.31	0.32	0.32	0.42	0.36	0.35

Percentages represent the relative contributions of each molecular class within each type of tree-DOM. Left side: values for all formulas within each type of tree-DOM. Right side: values for formulas that were enriched within each type of tree-DOM.

DISCUSSION

Concentrations and Optical Signatures of Tree-DOM

The high levels of DOC and CDOM in stemflow and throughfall relative to rainwater samples indicate that rainwater DOM was a very minor component of stemflow and throughfall DOM, with the majority of tree-DOM being entrained during interaction with the tree canopy and stem.

Mean DOC concentrations in throughfall (10–52 mg-C L^{-1}) and stemflow (25–78 mg-C L^{-1}) were both within the range of values reported by previous throughfall (1–100 mg-C L^{-1}) (Michalzik et al., 2001; Neff and Asner, 2001; Le Mellec et al., 2010; Inamdar et al., 2012) and stemflow (5–200 mg-C L^{-1}) (Moore, 2003; Tobón et al., 2004; Levia et al., 2011) studies. The observed range in tree-DOC concentrations is at the higher end or exceeds the mean DOC concentrations in major US rivers (1–12 mg-C L^{-1}; Spencer et al., 2012), but is consistent with the high DOC values observed in black water rivers draining swamps (e.g., St. Marys = 42 mg-C L^{-1}; Spencer et al., 2012). Given the small size of the current dataset (two storms, six trees), no estimates of DOC fluxes are made.

The optical properties of tree-DOM are broadly consistent with those for CDOM in other aquatic environments. Tree-DOM CDOM spectra exhibit an exponential increase in absorbance with decreasing wavelength (**Figure 2**). The range in mean spectral slope values ($S_{275-295}$) for stemflow (0.0144 nm^{-1}) and throughfall (0.0157–0.0165 nm^{-1}) are consistent with values for US rivers (0.012–0.023 nm^{-1}) (Spencer et al., 2012). Literature values for stemflow and throughfall spectral slope were not found for comparison. SUVA$_{254}$ values for stemflow (means 3.0–5.1 L

mg-C^{-1} m^{-1}) from our oaks and cedars compare with ranges of 2.5–4.9 L mg-C^{-1} m^{-1} for American beech (*Fagus grandifolia*) and 3.7–6.2 L mg-C^{-1} m^{-1} for yellow poplar (*Liriodendron tulipfera*) (Levia et al., 2011). These values are all at the higher end or exceeding the range in mean SUVA$_{254}$ values reported for US rivers (1.3–4.6 L mg-C^{-1} m^{-1}) (Spencer et al., 2012) and are consistent with highly colored, aromatic-rich DOM (Weishaar et al., 2003). The mean SUVA$_{254}$ values for throughfall (2.2–2.9 L mg-C^{-1} m^{-1}) compare to previous literature values for throughfall of 1.8–4.7 L mg-C^{-1} m^{-1} (Inamdar et al., 2012) and also indicate a significant contribution of aromatics to throughfall DOM. All tree-DOM SUVA values were higher than for rainwater CDOM and all tree-DOM spectral slopes steeper than for rainwater DOM (**Figures 3C,D**), indicating tree-DOM to be more aromatic than the trace amounts of DOM in rainwater. Hydrological fluxes become enriched with aromatic compounds, including lignin degradation products, as contact time with bark increases (Guggenberger et al., 1994). Therefore, the enrichment of stemflow in highly aromatic, high SUVA$_{254}$ DOM compared to throughfall is likely due to the high hydrological connectivity of stemflow with tree bark and other sources of tree-derived organics. The aromatics in tree-DOM are likely dominated by autochthonous, tree-produced aromatics, such as lignin, and their degradation products (Guggenberger et al., 1994) that accumulate and are then washed off the tree surface during rain events. In addition, allochthonous organics delivered to the tree via atmospheric deposition (Guggenberger and Zech, 1994) could include aromatics derived from soils, combustion sources, or distal vegetation.

The quantity, but not the optical quality of tree-DOM exported during both storms varied with epiphyte cover. DOC

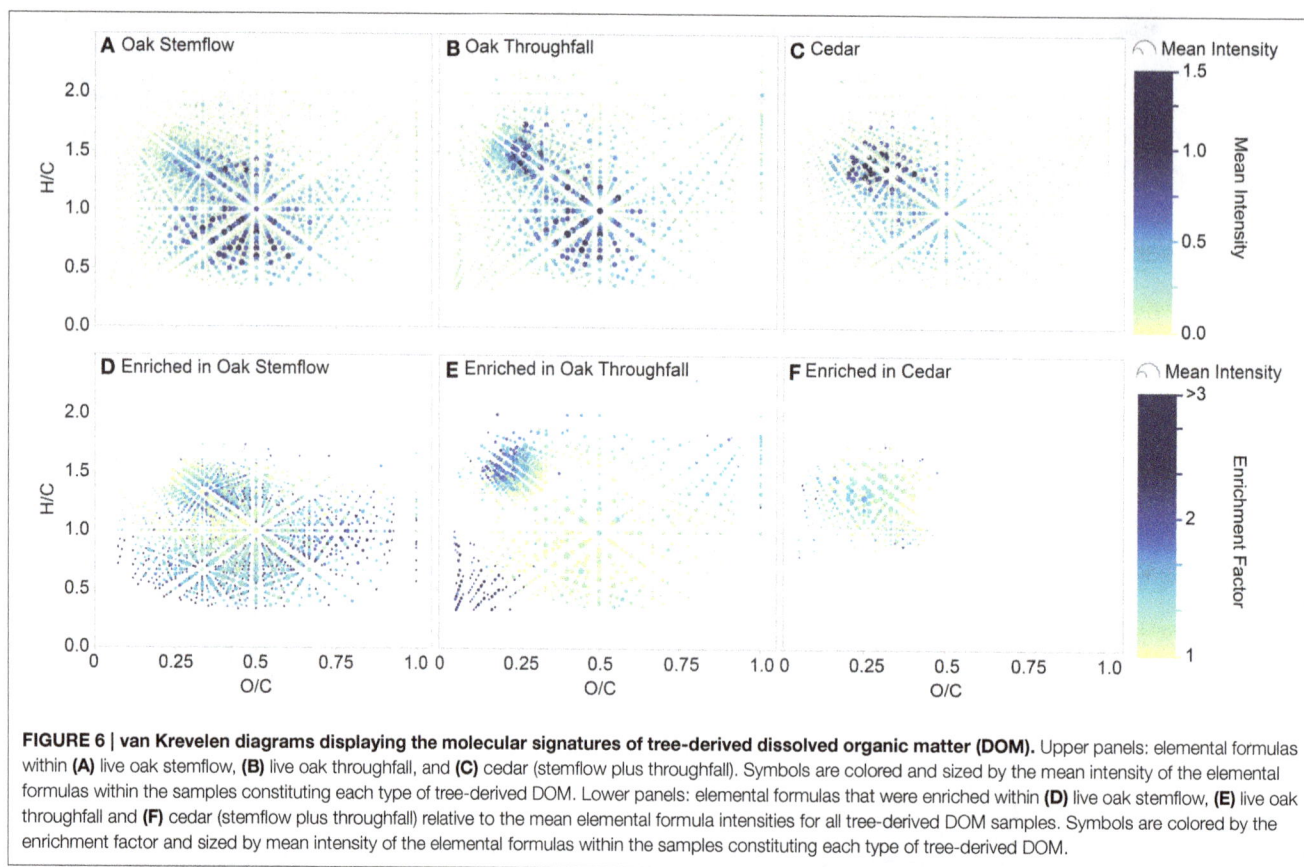

FIGURE 6 | van Krevelen diagrams displaying the molecular signatures of tree-derived dissolved organic matter (DOM). Upper panels: elemental formulas within **(A)** live oak stemflow, **(B)** live oak throughfall, and **(C)** cedar (stemflow plus throughfall). Symbols are colored and sized by the mean intensity of the elemental formulas within the samples constituting each type of tree-derived DOM. Lower panels: elemental formulas that were enriched within **(D)** live oak stemflow, **(E)** live oak throughfall and **(F)** cedar (stemflow plus throughfall) relative to the mean elemental formula intensities for all tree-derived DOM samples. Symbols are colored by the enrichment factor and sized by mean intensity of the elemental formulas within the samples constituting each type of tree-derived DOM.

concentration and CDOM were consistently lower in both stemflow and throughfall from the bare cedar than from the epiphyte covered cedar and the four oaks, which all had mixed epiphyte cover (**Table 1**; **Figures 1**, **3**). These results suggest that epiphytes were either a direct source of autochthonous, epiphyte organics or an intermediate accumulator of organics derived from the tree, fauna or atmospheric deposition. Research into the fluxes and quality of epiphyte DOM release is scarce. With respect to the dominant epiphytes encountered in our study, no data is available for resurrection ferns; while Spanish moss collected from nearby sites in coastal Georgia leached DOM with significantly lower $SUVA_{254}$ values (Van Stan et al., 2015) than for the current tree-DOM samples (**Table 1**). The Spanish moss samples leached in Van Stan et al. (2015) were cleaned of all canopy soil and any decaying or damaged moss. As such, these leachates contained organics derived directly from Spanish moss rather than from the more diverse canopy ecosystem and potential organic sources that the presence of Spanish moss in a tree cultivates. For the limited data collected in our study (two storms, six trees, one of which has no epiphyte cover), the presence of epiphytes increased DOC concentrations, but did not reduce the $SUVA_{254}$ values as would be expected if the additional DOC leached directly from healthy Spanish moss. Therefore, the DOM enrichment within stemflow and throughfall exported by epiphyte covered trees is likely due to the increase in hydrological contact time and flow path (Levia and Frost, 2003), the accumulation of organic matter from bark, litter

and fauna facilitated by the presence of epiphytes in the canopy ecosystem (Hietz et al., 2002; Woods et al., 2012), and potentially, the increase in canopy surface area for atmospheric deposition (Rodrigo et al., 1999; Woods et al., 2012).

Molecular Signatures of Tree-DOM

The molecular signatures of whole water tree-DOM as revealed by negative mode electrospray ionization FT-ICR MS (**Figure 2**) share many features of whole water and extracted DOM from other aquatic environments (Mopper et al., 2007; Singer et al., 2012; Chen et al., 2014; Dittmar and Stubbins, 2014; Cawley et al., 2016). Shared features include a high diversity of elemental formulas distributed broadly in van Krevelen space (**Figures 6A–C**) and spanning a range of molecular classes (**Table 2**). The average molecular mass of tree-DOM (359–382 g mol^{-1}; **Table 2**) was greater than for Kolyma River whole water DOM run on the same mass spectrometer (336 g mol^{-1}) (Spencer et al., 2015), but lower than for Congo River whole water DOM run on a different mass spectrometer (424 g mol^{-1}) (Stubbins et al., 2010). Tree-DOM was also H-poor (H/C 1.22–1.24) and O-rich (O/C 0.39–0.45) compared to Kolyma River DOM (H/C 1.27; O/C 0.39) (Spencer et al., 2015). Average H/C and O/C were not reported for Congo River DOM (Stubbins et al., 2010). Other reports of elemental formulas for non-extracted river water DOM are scarce.

As for other whole water FT-ICR mass spectra for terrigenous DOM from freshwater environments (Stubbins et al., 2010,

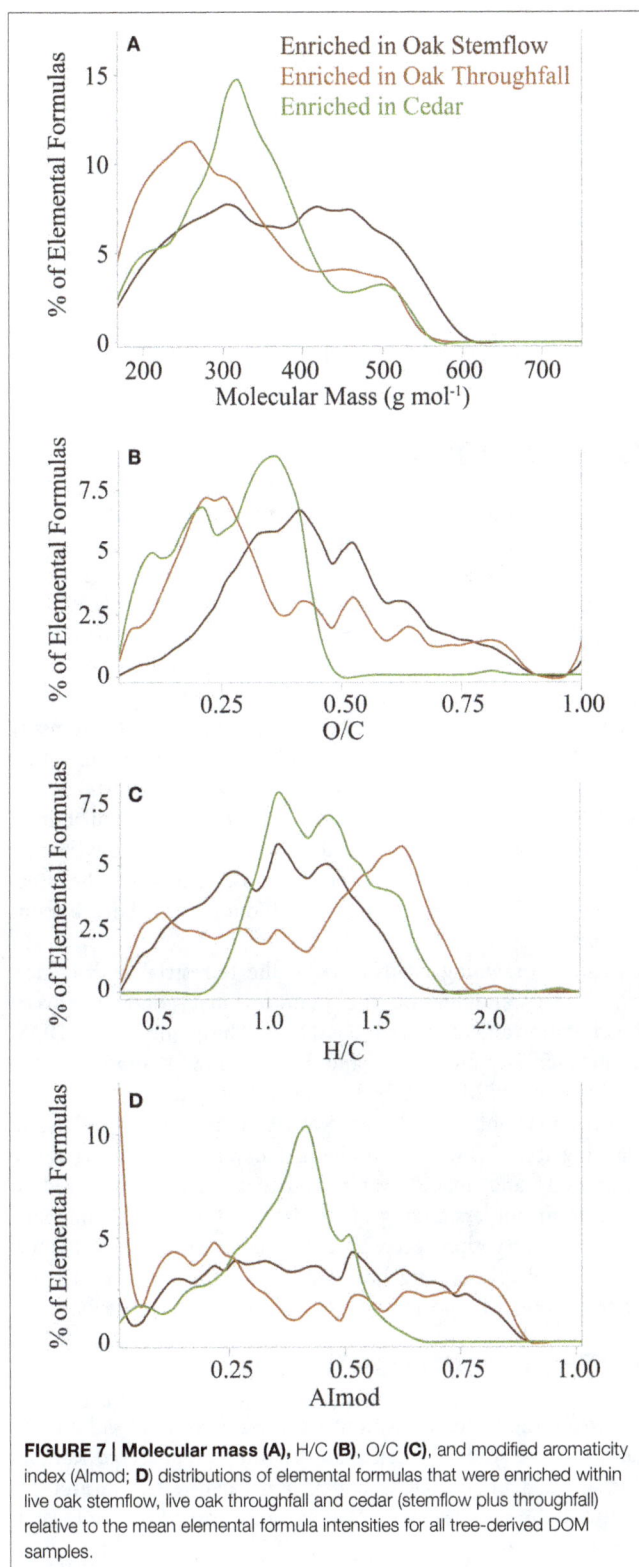

FIGURE 7 | Molecular mass (A), H/C **(B)**, O/C **(C)**, and modified aromaticity index (Almod; **D**) distributions of elemental formulas that were enriched within live oak stemflow, live oak throughfall and cedar (stemflow plus throughfall) relative to the mean elemental formula intensities for all tree-derived DOM samples.

diverse array of structures of correspondingly diverse potential biogeochemical sources and functions (Stubbins et al., 2010). For instance, the possible isomers of any highly unsaturated formula include aromatic ring containing lignin degradation products derived from vascular land plants (Stubbins et al., 2010) and carboxylic-rich alicyclic molecules (Hertkorn et al., 2006) of indeterminate, potentially microbial origin. Due the high $SUVA_{254}$ values of tree-DOM and particularly stemflow DOM, it is likely that a significant proportion of the unsaturated formulas within tree-DOM represent vascular plant derived molecules that contain aromatic rings.

The modified aromaticity index classifies formulas as either aromatic or condensed aromatic (Koch and Dittmar, 2006, 2016). Tree-DOM was enriched in aromatic formulas (14–15%; **Table 2**) compared to Congo River whole water analyzed on a different FT-ICR mass spectrometer (9%) (Stubbins et al., 2010), but similar in aromatic content to Kolyma River whole water analyzed on the FT-ICR mass spectrometer used in the current study (13%) (Spencer et al., 2015) consistent with the view that river DOM is derived predominantly from the degradation products of vascular plants (Ertel et al., 1984) and the enrichment of similar compounds within tree-DOM.

Tree-DOM was also enriched in condensed aromatics (10–12%; **Table 2**) compared to both Congo (1%) (Stubbins et al., 2010) and Kolyma River (6%) (Spencer et al., 2015) samples. Condensed aromatics are known to form during the incomplete combustion of organics (Goldberg, 1985) and when observed in DOM are usually termed dissolved black carbon and ascribed a thermogenic source (Kim et al., 2004; Hockaday et al., 2006; Ziolkowski and Druffel, 2010). The ubiquity of dissolved black carbon in river waters (Dittmar et al., 2012; Jaffé et al., 2013; Stubbins et al., 2015; Wagner et al., 2015) is explained as resulting from the ubiquity of refractory, apparently thermogenic black carbon in soils (Forbes et al., 2006; Guggenberger et al., 2008; Schmidt et al., 2011). Other sources of black carbon to natural waters and landscapes include direct input from local combustion sources and atmospheric deposition from distant combustion sources. Atmospheric deposition has been posited as a source of organics to remote regions of the earth (Stubbins et al., 2012a; Spencer et al., 2014) and organics transported from global and regional sources of combustion (e.g., automobile, industrial, domestic, agricultural, wildfire, and biomass burning), as well as produced locally on the Skidaway Institute of Oceanography campus, could all produce black carbon for deposition to the trees sampled.

Aliphatic formulas (sum of unsaturated aliphatics and saturated fatty acids) were slightly enriched in tree-DOM (23–25%; **Table 2**) relative to Congo (19%) (Stubbins et al., 2010) and Kolyma River (22%) (Spencer et al., 2015) DOM. Tree-DOM was also enriched in sugar (2.3–3.8%) and peptide (2.2–5.1%) formulas compared to Kolyma River DOM (0.6% sugar; 2.2% peptide) (Spencer et al., 2015). Sugar and peptide formulas were not reported for the Congo River (Stubbins et al., 2010). All of these compound classes likely derive directly from foliar leachates, foliar washoff, and their breakdown products (Guggenberger and Zech, 1994; Michalzik et al., 2001; Kalbitz et al., 2007). However, similar molecular formulas

2012a; Spencer et al., 2015), tree-DOM was dominated by CHO-only formulas and was rich in highly unsaturated molecules (**Table 2**). As they have a high degree of isomeric freedom, any single highly unsaturated elemental formula may represent a

are also observed in atmospheric aerosols (Wozniak et al., 2008).

Quintessential Signatures of Oak Stemflow, Oak Throughfall and Cedar DOM

Tree-DOM molecular signatures were significantly different from rainwater DOM and clustered in three groups: oak stemflow, oak throughfall and cedar, the latter including stemflow and throughfall samples from both the epiphyte covered and bare cedars (**Figure 5**). Distinct differences between the molecular signatures of each class of tree-DOM were evident in van Krevelen plots (**Figures 6A–C**). In order to determine the elemental formulas that distinguished these three classes of tree-DOM from one another, the quintessential formulas associated with each type of tree-DOM were classified as those formulas that are consistently enriched across the samples within a cluster (Equation 3). Plotting the resultant data in van Krevelen space revealed that the above data treatment accentuated differences in the molecular signatures of oak stemflow, oak throughfall, and cedar DOM (**Figures 6D–F**). As noted in the results, dot size in the van Krevelen diagrams (**Figures 6D–F**) represents the mean intensity of an elemental formula in the mass spectrum, while dot color represents the degree of enrichment ranging from just above 1 (yellow) to 3 or higher (dark blue).

Quintessential oak stemflow formulas (**Figure 6D**) occupied much of the van Krevelen space typically populated by riverine DOM (e.g., Spencer et al., 2015 in which data for a Kolyma River whole water sample run on the same mass spectrometer is displayed). The van Krevelen is densely populated, as is also typical of riverine DOM samples. The enrichment of stemflow DOM in this diverse population of formulas indicates that the organics within oak stemflow are of more diverse origin or have undergone more extensive processing than the organics within oak throughfall and cedar DOM. The canopy structure and coarse bark of live oaks makes them excellent habitats for colonization by epiphytes, such as the Spanish moss and resurrection ferns observed on our sampled trees (**Figure 1d**), the accumulation of organic debris, and utilization by fauna. The development of these canopy microhabitats, replete with observable canopy soils within the trees on SkIO campus, may have led to the development of a rich molecular mix of organics for export. While of diverse stoichiometric composition, the quintessential oak stemflow elemental formulas were of higher average molecular mass and AImod (**Table 2**; **Figure 7A**), and enriched in aromatics, compared to quintessential oak throughfall and cedar formulas, suggestive of the greater interaction of stemflow with the oak bark, debris and canopy soils.

Quintessential oak throughfall formulas (**Figure 6E**) were not as evenly distributed in van Krevelen space as the quintessential formulas of oak stemflow suggesting formulas enriched in throughfall derive from more distinct sources and have undergone less processing. Compared to the quintessential formulas associated with oak stemflow and cedar, quintessential oak throughfall formulas were of low average molecular weight

($311\,\text{g mol}^{-1}$) and were enriched in unsaturated aliphatics (30%) and sugars (4.3%; **Table 2**), suggesting significant inputs from the direct leaching or washing of foliar surfaces (Guggenberger and Zech, 1994; Michalzik et al., 2001; Kalbitz et al., 2007). Oak throughfall also contained elevated levels of condensed aromatics (21%; **Table 2**; **Figure 7B**) compared to the other tree-DOM classes, suggesting potential wash off of deposited combustion products from leaf surfaces.

By comparison to oak DOM, cedar DOM was enriched in CHO-only and highly unsaturated formulas (**Table 2**) of limited molecular diversity (**Figure 6F**; **Figure 7C**), suggesting cedar DOM is enriched in minimally processed non-descript tree-DOM.

CONCLUSIONS

The relatively high $SUVA_{254}$ values and abundance of aliphatics and aromatic formulas within tree-DOM are consistent with autochthonous (i.e., tree-derived) organics. However, the presence of condensed aromatics within tree-DOM also suggests that some of the DOM exported from trees derives from the atmospheric deposition of allochthonous organics. As electrospray ionization efficiency varies with analyte chemistry, the current dataset cannot be used to robustly quantify the contribution of condensed aromatics or other forms of deposited organics to tree-DOM export. Future work should therefore seek to quantify the contribution of condensed aromatics and other allochthonous forms of DOM to tree-DOM to assess how much of the tree-DOM flux is derived from autochthonous vs. allochthonous sources.

As the crowning headwaters of the terrestrial hydrological cycle, tree canopies are the point of first contact between precipitation and terrestrial ecosystems. The quality of tree-DOM as detailed by absorbance and FT-ICR MS is similar to the terrigenous DOM described in inland waters, but sufficiently distinct that optical and chemical signatures may be of use in tracking tree-DOM into receiving ecosystems including forest floor soils and inland waters. Further study is required to develop an understanding of the fate and ecological functions of tree-DOM within receiving ecosystems. Such knowledge will be essential in assessing how ongoing changes to forest cover distributions will impact both soil and aquatic ecosystems.

AUTHOR CONTRIBUTIONS

AS wrote the paper with input from other authors. AS and JV designed the study. LS collected the samples, analyzed samples and worked up the DOC and CDOM data sets. AS and TD analyzed the samples by FT-ICR MS and worked up the FT-ICR MS data.

REFERENCES

Aitkenhead-Peterson, J., McDowell, W., Neff, J., Stuart, E., and Robert, L. (2003). "Sources, production, and regulation of allochthonous dissolved organic matter inputs to surface waters," in *Aquatic Ecosystems Interactivity of Dissolved Organic Matter*, eds S. E. G. Findlay and R. L. Sinsabaugh (New York, NY: Academic Press), 26–70. doi: 10.1016/b978-012256371-3/50003-2

Algeo, T. J., Scheckler, S. E., and Maynard, J. B. (2001). "Effects of the middle to late devonian spread of vascular land plants on weathering regimes, Marine Biotas, and Global Climate," in *Plants Invade the Land: Evolutionary and Environmental Approaches*, eds P. G. Gensel and D. Edwards (New York, NY: Columbia University Press), 213–236.

Cawley, K. M., Murray, A. E., Doran, P. T., Kenig, F., Stubbins, A., Chen, H., et al. (2016). Characterization of dissolved organic material in the interstitial brine of Lake Vida, Antarctica. *Geochim. Cosmochim. Acta* 183, 63–78. doi: 10.1016/j.gca.2016.03.023

Chen, H., Stubbins, A., Perdue, E. M., Green, N. W., Helms, J. R., Mopper, K., et al. (2014). Ultrahigh resolution mass spectrometric differentiation of dissolved organic matter isolated by coupled reverse osmosis-electrodialysis from various major oceanic water masses. *Mar. Chem.* 164, 48–59. doi: 10.1016/j.marchem.2014.06.002

Crutzen, P. J. (2002). Geology of mankind. *Nature* 415, 23–23. doi: 10.1038/415023a

Dittmar, T., De Rezende, C. E., Manecki, M., Niggemann, J., Coelho Ovalle, A. R., Stubbins, A., et al. (2012). Continuous flux of dissolved black carbon from a vanished tropical forest biome. *Nat. Geosci.* 5, 618–622. doi: 10.1038/ngeo1541

Dittmar, T., and Stubbins, A. (2014). "Dissolved Organic Matter in Aquatic Systems," in *Treatise on Geochemistry, 2nd Edn.*, ed K. K. Turekian (Oxford: Elsevier), 125–156.

Ertel, J. R., Hedges, J. I., and Perdue, E. M. (1984). Lignin signature of aquatic humic substances. *Science* 223, 485–487. doi: 10.1126/science.223.4635.485

FAO (2016). "State of the World's Forests 2016," in *Forests and Agriculture: Land-Use Challenges and Opportunities* (Rome).

Forbes, M. S., Raison, R. J., and Skjemstad, J. O. (2006). Formation, transformation and transport of black carbon (charcoal) in terrestrial and aquatic ecosystems. *Sci. Total Environ.* 370, 190–206. doi: 10.1016/j.scitotenv.2006.06.007

Gensel, P. G. (2001). "Introduction," in *Plants Invade the Land: Evolutionary and Environmental Approaches*, eds P. G. Gensel and D. Edwards (New York, NY: Columbia University Press), 1–2.

Georgia Office of the State Climatologist (2012). *Office of the State Climatologist.* Available online at: https://epd.georgia.gov/office-state-climatologist (Accessed September 13, 2013).

Goldberg, E. (1985). *Black Carbon in the Environment.* New York, NY: Wiley.

Greb, S. F., DiMichele, W. A., and Gastaldo, R. A. (2006). "Evolution and importance of wetlands in earth history," in *Wetlands Through Time*, eds W. A. DiMichele and S. Greb (Boulder, CO: Geological Society of America), 1–40.

Guggenberger, G., Rodionov, A., Shibistova, O., Grabe, M., Kasansky, O. A., Fuchs, H., et al. (2008). Storage and mobility of black carbon in permafrost soils of the forest tundra ecotone in Northern Siberia. *Glob. Chang. Biol.* 14, 1367–1381. doi: 10.1111/j.1365-2486.2008.01568.x

Guggenberger, G., and Zech, W. (1994). Composition and dynamics of dissolved carbohydrates and lignin-degradation products in two coniferous forests, NE Bavaria, Germany. *Soil Biol. Biochem.* 26, 19–27. doi: 10.1016/0038-0717(94)90191-0

Guggenberger, G., Zech, W., and Schulten, H.-R. (1994). Formation and mobilization pathways of dissolved organic matter: evidence from chemical structural studies of organic matter fractions in acid forest floor solutions. *Org. Geochem.* 21, 51–66. doi: 10.1016/0146-6380(94)90087-6

Hansen, M. C., Potapov, P. V., Moore, R., Hancher, M., Turubanova, S., Tyukavina, A., et al. (2013). High-resolution global maps of 21st-century forest cover change. *Science* 342, 850–853. doi: 10.1126/science.1244693

Hansen, M. C., Stehman, S. V., and Potapov, P. V. (2010). Quantification of global gross forest cover loss. *Proc. Natl. Acad. Sci. U.S.A.* 107, 8650–8655. doi: 10.1073/pnas.0912668107

Helms, J. R., Stubbins, A., Ritchie, J. D., Minor, E. C., Kieber, D. J., and Mopper, K. (2008). Absorption spectral slopes and slope ratios as indicators of molecular weight, source, and photobleaching of chromophoric dissolved organic matter. *Limnol. Oceanogr.* 53, 955–969. doi: 10.4319/lo.2008.53.3.0955

Hertkorn, N., Benner, R., Frommberger, M., Schmitt-Kopplin, P., Witt, M., Kaiser, K., et al. (2006). Characterization of a major refractory component of marine dissolved organic matter. *Geochim. Cosmochim. Acta* 70, 2990–3010. doi: 10.1016/j.gca.2006.03.021

Hietz, P., Wanek, W., Wania, R., and Nadkarni, N. M. (2002). Nitrogen-15 natural abundance in a montane cloud forest canopy as an indicator of nitrogen cycling and epiphyte nutrition. *Oecologia* 131, 350–355. doi: 10.1007/s00442-002-0896-6

Hockaday, W. C., Grannas, A. M., Kim, S., and Hatcher, P. G. (2006). Direct molecular evidence for the degradation and mobility of black carbon in soils from ultrahigh-resolution mass spectral analysis of dissolved organic matter from a fire-impacted forest soil. *Org. Geochem.* 37, 501–510. doi: 10.1016/j.orggeochem.2005.11.003

Hu, C., Muller-Karger, F. E., and Zepp, R. G. (2002). Absorbance, absorption coefficient, and apparent quantum yield: a comment on common ambiguity in the use of these optical concepts. *Limnol. Oceanogr.* 47, 1261–1267. doi: 10.4319/lo.2002.47.4.1261

Inamdar, S., Finger, N., Singh, S., Mitchell, M., Levia, D., Bais, H., et al. (2012). Dissolved organic matter (DOM) concentration and quality in a forested mid-Atlantic watershed, USA. *Biogeochemistry* 108, 55–76. doi: 10.1007/s10533-011-9572-4

Jaffé, R., Ding, Y., Niggemann, J., Vähätalo, A. V., Stubbins, A., Spencer, R. G. M., et al. (2013). Global charcoal mobilization from soils via dissolution and riverine transport to the oceans. *Science* 340, 345–347. doi: 10.1126/science.1231476

Jardine, P., McCarthy, J., and Weber, N. (1989). Mechanisms of dissolved organic carbon adsorption on soil. *Soil Sci. Soc. Am. J.* 53, 1378–1385. doi: 10.2136/sssaj1989.03615995005300050013x

Kaiser, K., and Zech, W. (1998). Soil dissolved organic matter sorption as influenced by organic and sesquioxide coatings and sorbed sulfate. *Soil Sci. Soc. Am. J.* 62, 129–136. doi: 10.2136/sssaj1998.03615995006200010017x

Kaiser, K., and Zech, W. (2000). Sorption of dissolved organic nitrogen by acid subsoil horizons and individual mineral phases. *Eur. J. Soil Sci.* 51, 403–411. doi: 10.1046/j.1365-2389.2000.00320.x

Kalbitz, K., Meyer, A., Yang, R., and Gerstberger, P. (2007). Response of dissolved organic matter in the forest floor to long-term manipulation of litter and throughfall inputs. *Biogeochemistry* 86, 301–318. doi: 10.1007/s10533-007-9161-8

Kim, S., Kaplan, L. A., Benner, R., and Hatcher, P. G. (2004). Hydrogen-deficient molecules in natural riverine water samples - Evidence for the existence of black carbon in DOM. *Mar. Chem.* 92, 225–234. doi: 10.1016/j.marchem.2004.06.042

Koch, B. P., and Dittmar, T. (2006). From mass to structure: an aromaticity index for high-resolution mass data of natural organic matter. *Rapid Commun. Mass Spectrometry* 20, 926–932. doi: 10.1002/rcm.2386

Koch, B. P., and Dittmar, T. (2016). From mass to structure: an aromaticity index for high-resolution mass data of natural organic matter. *Rapid Commun. Mass Spectrometry* 30, 250–250. doi: 10.1002/rcm.7433

Koch, B. P., Dittmar, T., Witt, M., and Kattner, G. (2007). Fundamentals of molecular formula assignment to ultrahigh resolution mass data of natural organic matter. *Anal. Chem.* 79, 1758–1763. doi: 10.1021/ac061949s

Kolka, R. K., Nater, E., Grigal, D., and Verry, E. (1999). Atmospheric inputs of mercury and organic carbon into a forested upland/bog watershed. *Water Air Soil Pollut.* 113, 273–294. doi: 10.1023/A:1005020326683

Le Mellec, A., Meesenburg, H., and Michalzik, B. (2010). The importance of canopy-derived dissolved and particulate organic matter (DOM and POM)– comparing throughfall solutionfrom broadleaved and coniferous forests. *Ann. For. Sci.* 67:411. doi: 10.1051/forest/2009130

Levia, D. F., and Frost, E. E. (2003). A review and evaluation of stemflow literature in the hydrologic and biogeochemical cycles of forested and agricultural ecosystems. *J. Hydrol.* 274, 1–29. doi: 10.1016/S0022-1694(02)00399-2

Levia, D. F., Van Stan, I., John, T., Inamdar, S. P., Jarvis, M. T., Mitchell, M. J., et al. (2011). Stemflow and dissolved organic carbon cycling: temporal variability in concentration, flux, and UV-Vis spectral metrics in a temperate broadleaved deciduous forest in the eastern United States. *Can. J. Forest Res.* 42, 207–216. doi: 10.1139/x11-173

McClain, M. E., Boyer, E. W., Dent, C. L., Gergel, S. E., Grimm, N. B., Groffman, P. M., et al. (2003). Biogeochemical hot spots and hot moments at the interface of terrestrial and aquatic ecosystems. *Ecosystems* 6, 301–312. doi: 10.1007/s10021-003-0161-9

Michalzik, B., Kalbitz, K., Park, J.-H., Solinger, S., and Matzner, E. (2001). Fluxes and concentrations of dissolved organic carbon and nitrogen–a synthesis for temperate forests. *Biogeochemistry* 52, 173–205. doi: 10.1023/A:1006441620810

Moore, T. (2003). Dissolved organic carbon in a northern boreal landscape. *Glob. Biogeochem. Cycles* 17, 1109. doi: 10.1029/2003GB002050

Mopper, K., Kieber, D. J., and Stubbins, A. (2015). "Marine photochemistry: processes and impacts," in *Biogeochemistry of Marine Dissolved Organic Matter, 2nd Edn.*, eds D. A. Hansell and C. A. Carlson (New York, NY: Elsevier), 389–450.

Mopper, K., Stubbins, A., Ritchie, J. D., Bialk, H. M., and Hatcher, P. G. (2007). Advanced instrumental approaches for characterization of marine dissolved organic matter: extraction techniques, mass spectrometry, and nuclear magnetic resonance spectroscopy. *Chem. Rev.* 107, 419–442. doi: 10.1021/cr050359b

Neff, J. C., and Asner, G. P. (2001). Dissolved organic carbon in terrestrial ecosystems: synthesis and a model. *Ecosystems* 4, 29–48. doi: 10.1007/s100210000058

Raymond, P. A., and Saiers, J. E. (2010). Event controlled DOC export from forested watersheds. *Biogeochemistry* 100, 197–209. doi: 10.1007/s10533-010-9416-7

Raymond, P. A., Saiers, J. E., and Sobczak, W. V. (2016). Hydrological and biogeochemical controls on watershed dissolved organic matter transport: pulse-shunt concept. *Ecology* 97, 5–16. doi: 10.1890/14-1684.1

Rodrigo, A., Avila, A., and Gomez-Bolea, A. (1999). Trace metal contents in Parmelia caperata (L.) Ach. Compared to bulk deposition, throughfall and leaf-wash fluxes in two holm oak forests in Montseny (NE Spain). *Atmos. Environ.* 33, 359–367. doi: 10.1016/S1352-2310(98)00167-8

Schmidt, M. W. I., Torn, M. S., Abiven, S., Dittmar, T., Guggenberger, G., Janssens, I. A., et al. (2011). Persistence of soil organic matter as an ecosystem property. *Nature* 478, 49–56. doi: 10.1029/2008GB003327

Singer, G. A., Fasching, C., Wilhelm, L., Niggemann, J., Steier, P., Dittmar, T., et al. (2012). Biogeochemically diverse organic matter in Alpine glaciers and its downstream fate. *Nat. Geosci.* 5, 710–714. doi: 10.1038/ngeo1581

Spencer, R. G. M., Butler, K. D., and Aiken, G. R. (2012). Dissolved organic carbon and chromophoric dissolved organic matter properties of rivers in the USA. *J. Geophys. Res. Biogeosciences* 117, G03001. doi: 10.1029/2011jg001928

Spencer, R. G. M., Guo, W., Raymond, P. A., Dittmar, T., Hood, E., Fellman, J., et al. (2014). Source and biolability of ancient dissolved organic matter in glacier and lake ecosystems on the Tibetan Plateau. *Geochim. Cosmochim. Acta* 142, 64–74. doi: 10.1016/j.gca.2014.08.006

Spencer, R. G. M., Mann, P. J., Dittmar, T., Eglinton, T. I., McIntyre, C., Holmes, R. M., et al. (2015). Detecting the signature of permafrost thaw in Arctic rivers. *Geophys. Res. Lett.* 42, 2830–2835. doi: 10.1002/2015GL063498

Stein, W. E., Mannolini, F., Hernick, L. V., Landing, E., and Berry, C. M. (2007). Giant cladoxylopsid trees resolve the enigma of the Earth/'s earliest forest stumps at Gilboa. *Nature* 446, 904–907. doi: 10.1038/nature05705

Stubbins, A., and Dittmar, T. (2012). Low volume quantification of dissolved organic carbon and dissolved nitrogen. *Limnol. Oceanogr.* 10, 347–352. doi: 10.4319/lom.2012.10.347

Stubbins, A., and Dittmar, T. (2015). Illuminating the deep: molecular signatures of photochemical alteration of dissolved organic matter from North Atlantic Deep Water. *Mar. Chem.* 177, 318–324. doi: 10.1016/j.marchem.2015.06.020

Stubbins, A., Hood, E., Raymond, P. A., Aiken, G. R., Sleighter, R. L., Hernes, P. J., et al. (2012a). Anthropogenic aerosols as a source of ancient dissolved organic matter in glaciers. *Nat. Geosci.* 5, 198–201. doi: 10.1038/ngeo1403

Stubbins, A., Lapierre, J. F., Berggren, M., Prairie, Y. T., Dittmar, T., and del Giorgio, P. A. (2014). What's in an EEM? Molecular Signatures Associated with Dissolved Organic Fluorescence in Boreal Canada. *Environ. Sci. Technol.* 48, 10598–10606. doi: 10.1021/es502086e

Stubbins, A., Law, C. S., Uher, G., and Upstill-Goddard, R. C. (2011). Carbon monoxide apparent quantum yields and photoproduction in the Tyne estuary. *Biogeosciences* 8, 703–713. doi: 10.1029/2009GL041158

Stubbins, A., Mann, P. J., Powers, L., Bittar, T. B., Dittmar, T., McIntyre, C. P., et al. (2017). Low photolability of yedoma permafrost dissolved organic carbon. *J. Geophys. Res.* 122, 200–211. doi: 10.1002/2016JG003688

Stubbins, A., Niggemann, J., and Dittmar, T. (2012b). Photo-lability of deep ocean dissolved black carbon. *Biogeosciences* 9, 1661–1670. doi: 10.1029/2008GL036169

Stubbins, A., Spencer, R. G. M., Chen, H., Hatcher, P. G., Mopper, K., Hernes, P. J., et al. (2010). Illuminated darkness: molecular signatures of Congo River dissolved organic matter and its photochemical alteration as revealed by ultrahigh precision mass spectrometry. *Limnol. Oceanogr.* 55, 1467–1477. doi: 10.1890/0012-09658

Stubbins, A., Spencer, R., Mann, P. J., Holmes, R. M., McClelland, J., Niggemann, J., et al. (2015). Utilizing colored dissolved organic matter to derive dissolved black carbon export by Arctic Rivers. *Front. Earth Sci.* 3:63. doi: 10.3389/feart.2015.00063

Tobón, C., Sevink, J., and Verstraten, J. M. (2004). Solute fluxes in throughfall and stemflow in four forest ecosystems in northwest Amazonia. *Biogeochemistry* 70, 1–25. doi: 10.1023/B:BIOG.0000049334.10381.f8

Van Stan, J. T., Stubbins, A., Bittar, T., Reichard, J. S., Wright, K. A., and Jenkins, R. B. (2015). Tillandsia usneoides (L.) L.(Spanish moss) water storage and leachate characteristics from two maritime oak forest settings. *Ecohydrology* 8, 988–1004. doi: 10.1002/eco.1549

Vidon, P., Allan, C., Burns, D., Duval, T. P., Gurwick, N., Inamdar, S., et al. (2010). Hot spots and hot moments in riparian zones: potential for improved water quality management. *J. Am. Water Resour. Assoc.* 46, 278–298. doi: 10.1111/j.1752-1688.2010.00420.x

Wagner, S., Cawley, K. M., Rosario-Ortiz, F. L., and Jaffé, R. (2015). In-stream sources and links between particulate and dissolved black carbon following a wildfire. *Biogeochemistry* 124, 145–161. doi: 10.1007/s10533-015-0088-1

Weishaar, J. L., Aiken, G. R., Bergamaschi, B. A., Fram, M. S., Fujii, R., and Mopper, K. (2003). Evaluation of specific ultraviolet absorbance as an indicator of the chemical composition and reactivity of dissolved organic carbon. *Environ. Sci. Technol.* 37, 4702–4708. doi: 10.1021/es030360x

Willey, J. D., Kieber, R. J., Eyman, M. S., and Avery, G. B. (2000). Rainwater dissolved organic carbon: concentrations and global flux. *Glob. Biogeochem. Cycles* 14, 139–148. doi: 10.1029/1999gb900036

Woods, C. L., Hunt, S. L., Morris, D. M., and Gordon, A. M. (2012). Epiphytes influence the transformation of nitrogen in coniferous forest canopies. *Boreal Environ. Res.* 17, 411–425.

Wozniak, A., Bauer, J., Sleighter, R., Dickhut, R., and Hatcher, P. (2008). Technical Note: molecular characterization of aerosol-derived water soluble organic carbon using ultrahigh resolution electrospray ionization Fourier transform ion cyclotron resonance mass spectrometry. *Atmosp. Chem. Phys.* 8, 5099–5111. doi: 10.5194/acp-8-5099-2008

Ziolkowski, L. A., and Druffel, E. R. M. (2010). Aged black carbon identified in marine dissolved organic carbon. *Geophys. Res. Lett.* 37, L16601. doi: 10.1016/j.drs.2009.05.008

Conflict of Interest Statement: The authors declare that the research was conducted in the absence of any commercial or financial relationships that could be construed as a potential conflict of interest.

16

Determining the Stress Field in Active Volcanoes Using Focal Mechanisms

Bruno Massa [1,2]*, Luca D'Auria [2,3], Elena Cristiano [2] and Ada De Matteo [1]

[1] Dipartimento di Scienze e Tecnologie, Universitá degli Studi del Sannio, Benevento, Italy, [2] Istituto Nazionale di Geofisica e Vulcanologia, Sezione di Napoli, Osservatorio Vesuviano, Napoli, Italy, [3] Istituto per il Rilevamento Elettromagnetico dell'Ambiente, Consiglio Nazionale delle Ricerche, Napoli, Italy

Stress inversion of seismological datasets became an essential tool to retrieve the stress field of active tectonics and volcanic areas. In particular, in volcanic areas, it is able to put constrains on volcano-tectonics and in general in a better understanding of the volcano dynamics. During the last decades, a wide range of stress inversion techniques has been proposed, some of them specifically conceived to manage seismological datasets. A modern technique of stress inversion, the BRTM, has been applied to seismological datasets available at three different regions of active volcanism: Mt. Somma-Vesuvius (197 Fault Plane Solutions, FPSs), Campi Flegrei (217 FPSs) and Long Valley Caldera (38,000 FPSs). The key role of stress inversion techniques in the analysis of the volcano dynamics has been critically discussed. A particular emphasis was devoted to performances of the BRTM applied to volcanic areas.

Keywords: stress field, focal mechanism, BRTM, volcano-tectonics, monitoring

Edited by:
Antonio Costa,
National Institute of Geophysics and Volcanology, Bologna, Italy

Reviewed by:
Eisuke Fujita,
National Research Institute for Earth Science and Disaster Prevention,
Japan
Mimmo Palano,
National Institute of Geophysics and Volcanology, Roma, Italy

*Correspondence:
Bruno Massa
massa@unisannio.it

Specialty section:
This article was submitted to Volcanology,
a section of the journal Frontiers in Earth Science

INTRODUCTION

Theoretical Background to Stress Field Inversion

In last decades focal mechanisms have shown to be very useful to infer about the stress field within the Earth. Being seismicity a common feature of both quiescent and erupting volcanoes, the study of volcano-tectonic earthquakes is an effective tool to retrieve spatial and temporal patterns in the pre-, syn-, and inter-eruptive stress fields. Stress changes associated with magma migration and more in general with the dynamics of magma chambers and hydrothermal systems are causally linked to these earthquakes. Hence retrieving the stress field pattern in volcanoes has important implications in studying their dynamics and the interaction with the regional tectonic stress field (Umakoshi et al., 2001; Pedersen and Sigmundsson, 2004; Segall, 2013; Cannavò et al., 2014; D'Auria et al., 2014a,b).

Rocks in the Earth lithosphere possess a specific state of stress strictly related to the tectonic setting of the region. The state of stress at any given point can be represented geometrically as a Stress Ellipsoid, a tri-axial solid built with the Principal Stress Components (PSC) $\sigma_1 > \sigma_2 > \sigma_3$ as main axes. The shape of this ellipsoid can be expressed using the Bishop's ratio $\Phi_B = (\sigma_2 - \sigma_3) / (\sigma_1 - \sigma_3)$ (Bishop, 1966). It expresses the relationship between the intermediate and maximum PSC respect to the minimum one and can assume values varying from 0 to 1. In a general tri-axial state of stress $\sigma_1 > \sigma_2 > \sigma_3 \neq 0$, and the $\Phi_B \approx 0.5$. A slip event along a fault surface occurs when the accumulated energy exceeds the internal strength of the rock (Reid, 1910). The rupture starts at the hypocenter and then propagates along the whole fault plane. Analysis of seismograms allows the retrieving of information about the seismogenic structure responsible for an event. Byerly (1928) proposed a method that allows the reconstruction of the Fault Plane Solution

(FPS) of an earthquake, which is a stereographic projection representing the geometry of a seismogenic structure. It can be easily explained picturing a focal sphere ideally located around the earthquake hypocenter. The focal sphere can be separated into four dihedra, two contractional, and two dilatational, delimited by two planes known as Nodal Planes (the main and the auxiliary planes). The main nodal plane represents the actual fault, with the slip occurring during the earthquake being along the direction orthogonal to the auxiliary plane (**Figure 1**). Therefore the geometry of the FPS is tightly linked to the fault kinematics. At the center of the contractional dihedra is located the P-axis (compressive) while the center of extensional dihedra (dilatational) hosts the T-axis. The neutral B-axis corresponds to the intersection of nodal planes (**Figure 1**). P, T, B axes are mutually normal. The relationship between P and T axes of a FPS and the actual σ_1 and σ_3 stress directions is not straightforward. The aim of stress inversion techniques is precisely to determine it. A FPS can be reconstructed a posteriori, starting from the analysis of the first arrival of P-waves at an adequate number of seismographs arranged around the hypocenter. Using the equal-area stereographic projection technique (lower hemisphere), each station can be plotted using the angular coordinates of the direct seismic ray leaving the hypocenter. Stations falling in a contractional quadrant will record a downward first P-wave arrival (void circle of **Figure 1A**). Conversely, stations falling in a dilatational quadrant will register an upward first P-wave arrival (solid circle of **Figure 1A**). Plotting this information on an equal-area stereographic projection net it is possible to find a couple of nodal planes that is able to divide up the plot into four quadrants, corresponding to the four dihedra of the focal sphere (**Figure 1**). Nowadays, these procedures are implemented in a wide range of algorithms performing the determination of the FPS automatically. One of the most commonly used

algorithms is FPFIT (Reasenberg and Oppenheimer, 1985). It is a suite of Fortran codes for calculating and displaying earthquake fault-plane solutions. FPFIT finds the double-couple fault plane solution (source model) that best fits a given set of observed first motion polarities for an earthquake.

Starting from observed fault kinematics or focal mechanisms, various stress inversion procedures can be implemented. These procedures allow the reconstruction of the reduced stress field that carries information about the stress orientation and the ratios between principal stresses. Inversion of focal mechanism datasets became a common task in studies aimed at constrain seismotectonic stress field in active tectonic areas (Carey-Gailhardis and Mercier, 1992; Gillard and Wyss, 1995; Hardebeck and Hauksson, 2001; Massa, 2003; Angelier and Baruah, 2009; De Matteis et al., 2012) and more recently also in volcanic environments (Wyss et al., 1992; Cocina et al., 1997; D'Auria et al., 2014a,b; Plateaux et al., 2014). A wide range of graphical and analytical methods were proposed by authors in last decades (e.g., Angelier and Mechler, 1977; Gephart and Forsyth, 1984; Carey-Gailhardis and Mercier, 1987; Lisle, 1987, 1988; Michael, 1987; Angelier, 1990, 2002; Rivera and Cisternas, 1990; D'Auria and Massa, 2015). Stress inversion procedures are based on the Wallace-Bott hypothesis: shear traction acting on a fault plane (irrespective of newly formed structure or a reactivated preexisting one) causes a slip event in the direction and sense of the shear traction itself (Wallace, 1951; Bott, 1959; Angelier and Mechler, 1977; Yamaji, 2007). Analytical techniques of stress inversion are based on the systematic comparison of the actual slip vector with respect to the theoretical (modeled) maximum shear stress acting on the same surface in response to the active stress field (**Figure 2**). Graphical approaches are based on the definition of a probability function (or an equivalent formulation) on the whole focal sphere. They allow the

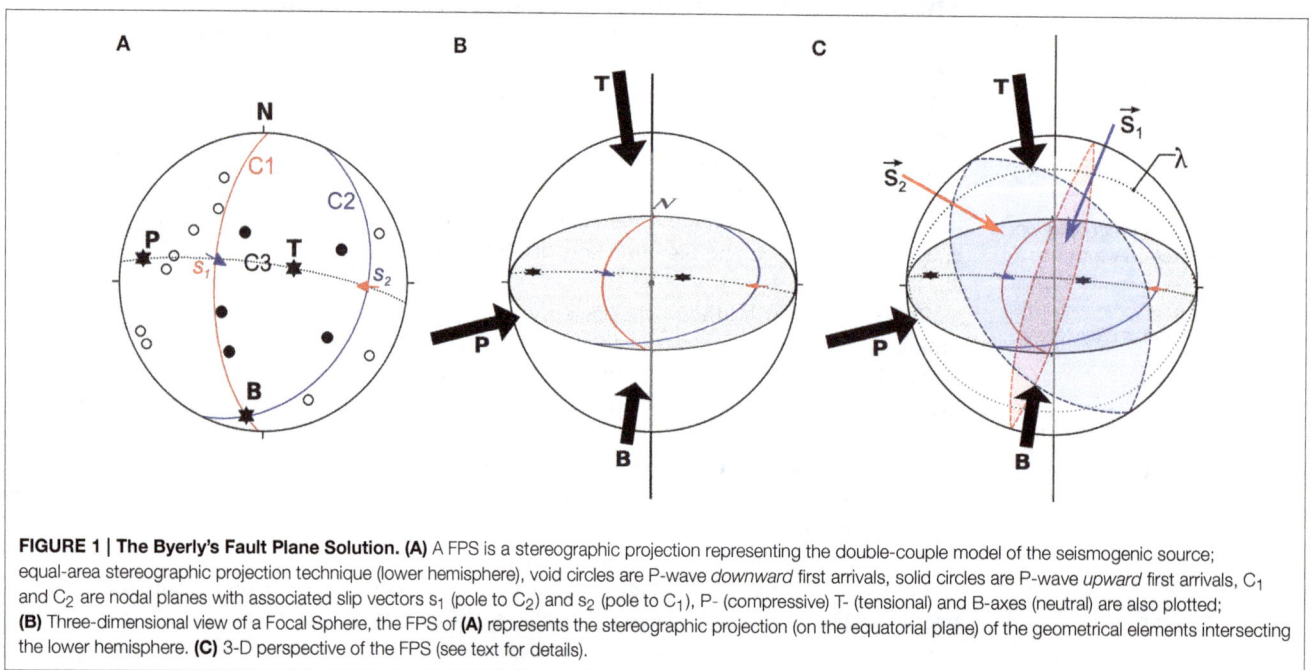

FIGURE 1 | The Byerly's Fault Plane Solution. (A) A FPS is a stereographic projection representing the double-couple model of the seismogenic source; equal-area stereographic projection technique (lower hemisphere), void circles are P-wave *downward* first arrivals, solid circles are P-wave *upward* first arrivals, C_1 and C_2 are nodal planes with associated slip vectors s_1 (pole to C_2) and s_2 (pole to C_1), P- (compressive) T- (tensional) and B-axes (neutral) are also plotted; **(B)** Three-dimensional view of a Focal Sphere, the FPS of **(A)** represents the stereographic projection (on the equatorial plane) of the geometrical elements intersecting the lower hemisphere. **(C)** 3-D perspective of the FPS (see text for details).

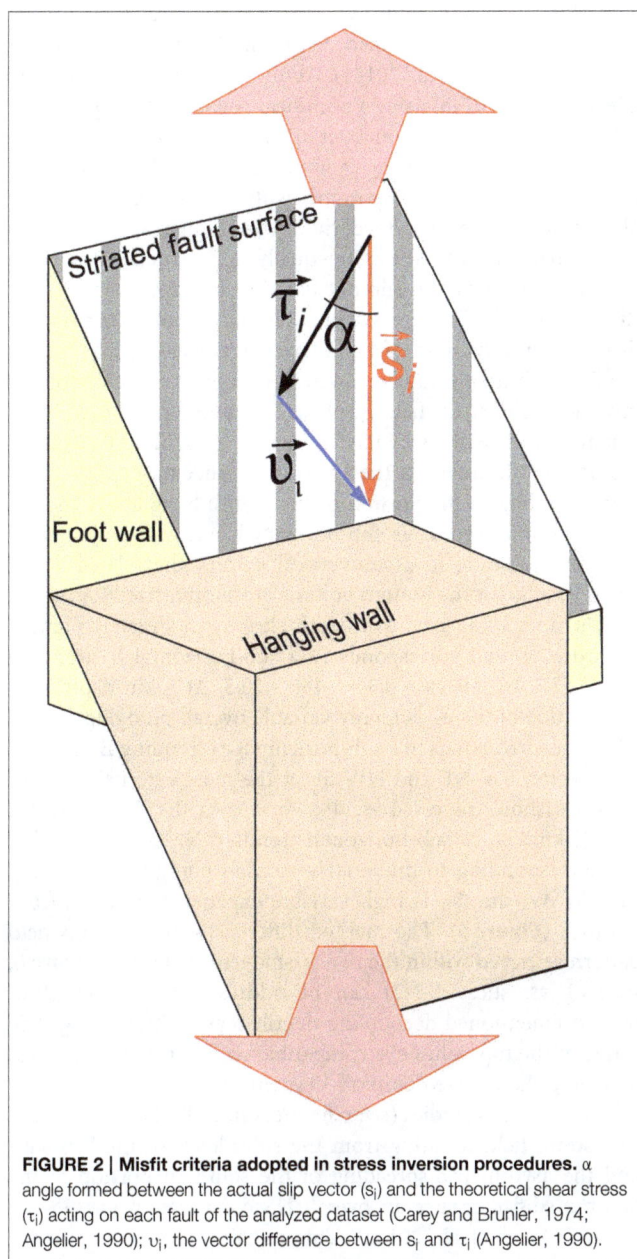

FIGURE 2 | Misfit criteria adopted in stress inversion procedures. α angle formed between the actual slip vector (s_i) and the theoretical shear stress (τ_i) acting on each fault of the analyzed dataset (Carey and Brunier, 1974; Angelier, 1990); υ_i, the vector difference between s_i and τ_i (Angelier, 1990).

determination of a range of possible attitudes for principal stress axes using hand stereo-plot. Right Dihedra Method (RDM) represents the simplest of graphical stress inversions (Angelier and Mechler, 1977). It basically consists of averaging the focal spheres of a focal mechanisms dataset. In the late 80's, RDM was improved adding an additional geometrical constrain and was re-proposed as Right Trihedra Method (Lisle, 1987, 1988). Both RDM and RTM are able to manage both planes of focal mechanism regardless the pre-selection of the actual fault plane. D'Auria and Massa (2015) proposed a novel Bayesian approach for the determination of the stress field from focal mechanism datasets (BRTM). This method can be regarded as a re-visitation of the RTM in a sound probabilistic framework. It provides a probability function over the focal sphere for both σ_1 and σ_3 principal stress directions (**Figure 3**). BRTM

algorithm has shown to be robust and efficiently enough to manage all kind of kinematics and nodal plane attitudes. A comparison of the BRTM performances to respect a classical Direct Inversion Method (Angelier, 1990, 2002) and the Multiple Inverse Method (Yamaji, 2000), confirms that BRTM is able to successfully manage both homogeneous and heterogeneous datasets (see Figure 2 in D'Auria and Massa, 2015). Furthermore BRTM allows both a reliable determination of the principal stress axes attitude and a quantitative estimation of the corresponding confidence regions (D'Auria and Massa, 2015). Finally, to give a complete and accurate inversion the corresponding Bishop's ratio Φ_B can be determined, exploiting the BRTM results, following the approach proposed by Angelier (1990; **Figure 3**). In case of heterogeneous datasets result of stress field inversion could be misleading. The problem can be approached using different un-supervisioned techniques (e.g., Angelier and Manoussis, 1980; Yamaji, 2000). Among those, the Multiple Inverse Method (Yamaji, 2000), makes use of a resampling technique of the analyzed dataset, through the construction of many data subsets, inverted separately using an analytical approach (Otsubo et al., 2008). Results can be plotted on a stereographic projection to outline graphically different clusters of stress tensors, corresponding to the different components of the stress field acting in the lithospheric volume associated to the dataset. The major drawback of this technique is the difficulty in attributing to the retrieved stress field components a proper spatio-temporal collocation. Therefore heterogeneous datasets can be approached, more properly, taking into account spatial and possibly temporal variations of the stress pattern (Wyss et al., 1992; Hardebeck and Michael, 2006; D'Auria et al., 2014b). In other words, these methods take into account the spatio-temporal distribution of the data, providing as results not a single stress tensor, but a spatially (and possibly temporally) variable stress field. Applying the BRTM to subsets of FPS corresponding to a given sub-volume of the area under investigation allows retrieving the spatial pattern of the stress field. This is an essential approach in the analysis of seismological dataset of volcanic districts. In the next section we show the application of stress inversion procedures to three different volcanic areas. Analyses were carried out on datasets available from previous studies. For Mt.Somma-Vesuvius and Campi Flegrei we used data from D'Auria et al. (2014a,b). For the Long Valley Caldera, data are freely accessible to users via the Internet (NCEDC, 2014). We preferred to invert published datasets to novel ones as this allows a more efficient comparison of obtained results got with different processing approaches.

This study aims at highlighting the key role played by stress inversion procedures in the analysis of the volcano dynamics. In detail, we show the methodological potential offered by the BRTM here tested on seismological datasets collected at three different volcanoes.

CASE STUDIES

In order to show the capability of the stress inversion procedures on volcanoes, we applied the BRTM to determine the spatial variations of the stress field in three different volcano-tectonic environments: the Mt. Somma-Vesuvius Volcano (Southern

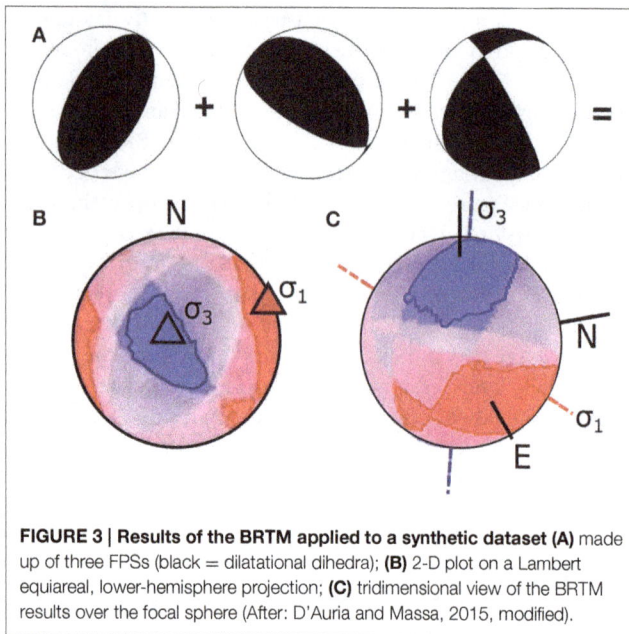

FIGURE 3 | Results of the BRTM applied to a synthetic dataset (A) made up of three FPSs (black = dilatational dihedra); **(B)** 2-D plot on a Lambert equiareal, lower-hemisphere projection; **(C)** tridimensional view of the BRTM results over the focal sphere (After: D'Auria and Massa, 2015, modified).

Italy; **Figure 4**), the Campi Flegrei Caldera (Southern Italy; **Figure 4**) and the Long Valley Caldera (California-Sierra Nevada border, USA; **Figure 9**). In the following we report a synthesis of the key results, focusing on the relationship between the local active stress fields and the background regional ones.

Mt. Somma-Vesuvius

The Somma-Vesuvius (**Figures 4**, **5**) is the youngest volcano of the Neapolitan district, it is characterized both by explosive and effusive activity (Santacroce, 1987). Its oldest products are dated 0.369 ± 0.028 Ma ($^{40}Ar/^{39}Ar$ age from Brocchini et al., 2001). The last eruption of Vesuvius occurred in 1944 and after this event it has become quiescent, showing only fumarolic activity and low seismicity ($M \leq 3.6$). Since 1964 to nowadays the seismicity started to increase in the occurrence and in the magnitude (Giudicepietro et al., 2010), with four episodes (1978–1980, 1989–1990, 1995–1996, 1999–2000) of strong increased strain release rate, occurrence and magnitude rate (D'Auria et al., 2013). The Mesozoic basement of the volcano is displaced by both SW- and NW-dipping normal fault systems and secondarily by NE-SW and E-W faults (Bianco et al., 1998; Ventura and Vilardo, 1999; Zollo et al., 2002; Acocella and Funiciello, 2006; **Figure 4**). Mesoscale faults and eruptive fractures striking NW-SE, NE-SW and ENE-WSW affect volcanic units outside the Somma caldera (Rosi et al., 1987; Andronico et al., 1995; Ventura and Vilardo, 1999; D'Auria et al., 2014a, **Figure 1** for a review). Seismicity at Mt. Somma-Vesuvius shows the presence of two different seismogenic volumes: a top volume (above sea level) and a bottom one (1–5 km depth). These two volumes appear to be separated by a volume with a markedly reduced seismic strain release, possibly corresponding to a ductile sedimentary layer buried at about 1000 m b.s.l. (Bernasconi et al., 1981; Borgia et al., 2005; D'Auria et al., 2014a).

At Mt. Somma-Vesuvius we analyzed a dataset consisting of 197 FPSs of earthquakes recorded from 1999 to 2012 with $0 < M < 3.6$ (D'Auria et al., 2014a). All the events were relocated by D'Auria et al. (2014a) using a nonlinear probabilistic approach in a 3-D velocity model (Lomax et al., 2001; D'Auria et al., 2008). FPSs were computed using P wave polarities (Reasenberg and Oppenheimer, 1985). Only events with at least six first motion observations were used to derive FPSs (**Figure 5**). Kinematics of the analyzed FPSs are quite equally represented in the three extreme categories, dip-slip normal, dip-slip reverse and strike-slip (**Figure 5D**). The stress field has been computed on a regular three-dimensional grid of 1-km spacing. For each grid node all the FPSs within a radius of 1-km have been considered for the inversion. The stress field has been computed only at volumes containing at least five FPSs. A synthesis of BRTM results is reported in **Figure 6**. For the top-volume (slice for depth −1 km, corresponding to the portion of the edifice lying above the sea level) the retrieved σ_1 is sub-horizontal overall trending NW-SE, corresponding to a sub-vertical σ_3 and the related Φ_B is <0.5. Results for the bottom volume are summarized for depths of 1, 3, and 5 km. At 1 km depth slice σ_1 moderately plunges toward ENE and corresponds to a sub-horizontal σ_3 trending NNW-SSE. In this case $0.3 < \Phi_B < 0.5$. At 3 km depth slice the attitude of the σ_1 is highly variable overall plunging toward NNW. It corresponds to a sub-horizontal σ_3 trending E-W in the axial sector, SW-NE and NW-SE at the most external volumes with a Bishop's ratio $0.4 < \Phi_B < 0.5$. At the deepest depth slice (5 km), σ_1 is sub-horizontal trending NNE-SSW to NW-SE, corresponding to moderately to high plunging σ_3 trending ENE-WSW, the Φ_B is high variable experiencing values from 0.2 to 1 (**Figure 6**). The marked differences in the stress field pattern retrieved within the two seismogenic volumes (**Figure 6**, slice -1 vs. slices 1-3-5) can be addressed to the effect of the aforementioned decoupling ductile layer. The faulting style active in the top-volume is compatible with a spreading process involving the exposed Somma-Vesuvius edifice, as proposed by various previous studies (see references in D'Auria et al., 2014a). The strain field resulting from the subsidence of the Vesuvius and the asymmetric spreading of the southern portion of the Somma edifice creates an overall NS compression (see Figure 17 in D'Auria et al., 2014a). This ongoing process is also the source of the persistent seismicity located within the top-volume (Borgia et al., 2005; D'Auria et al., 2013, 2014a; **Figures 5, 6**). Additionally, the low Φ_B values associated to stress inversion in the top volume confirm that the retrieved fields are strongly driven by the σ_1. According to previous studies, the bottom-volume seems to be deeply related to a regional background stress field slightly perturbed by local heterogeneities in the volcanic structure and by the complex topography of the volcano. The intermittent behavior of the seismic activity in the deepest volume is probably due to a relation with the dynamics of the hydrothermal system: the pore pressure within the hydrothermal system can be perturbed by episodes of fluid injection that essentially influence the stress field pattern in the bottom-volume (**Figures 5, 6**; Chiodini et al., 2001; D'Auria et al., 2014a). The Φ_B values associated to stress inversion in the bottom volume show that for the 1- and 3-km depth-slices the three principal stresses

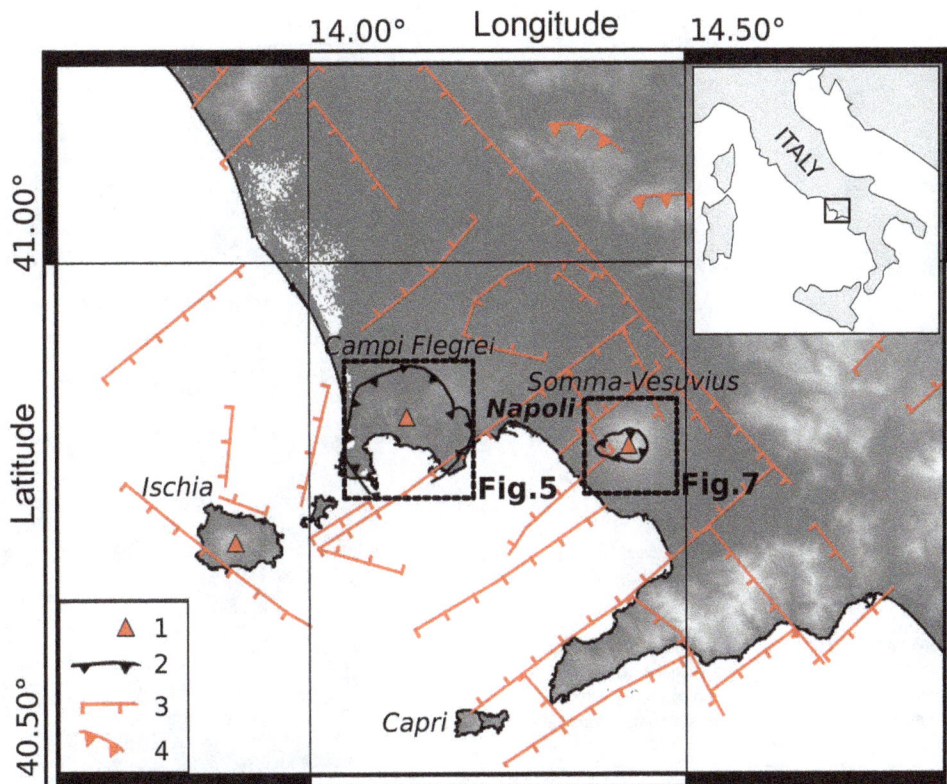

FIGURE 4 | Tectonic setting of the Campi Flegrei-Vesuvius area. (1) Main volcanoes; (2) Campi Flegrei (Campanian Ignimbrite) and Mt. Somma caldera rims; (3) main Plio-Quaternary faults; (4) thrust faults. Dashed squares identify the location of **Figures 5, 7**. Data from: Ippolito et al., 1973; Di Vito et al., 1999; Lavecchia et al., 2003; Acocella and Funiciello, 2006; Milia et al., 2013; Vitale and Isaia, 2013; D'Auria et al., 2014a,b, and referencer therein.

are well defined (overall, $0.3 < \Phi_B < 0.5$). A quite different result was obtained for the deepest depth-slice, the 5-km one: here Φ_B values show an high variability with values ranging from 0.5 to 1 and a σ_3 overall trending NE-SW. This last result appears in accordance with the local minimum horizontal principal stress component (S_h) measured within the Trecase 1 well (located at the SE slope of the Mt. Somma-Vesuvius) as borehole breakout, showing an elongation in ENE-WSW direction (Montone et al., 2012).

Campi Flegrei

Campi Flegrei caldera (CFc) is located west of the city of Naples (**Figures 4, 7**). It is a partially submerged collapse caldera shaped by two main eruptive episodes: the Campanian Ignimbrite eruption (40.6 ka, Gebauer et al., 2014) and the Neapolitan Yellow Tuff eruption (14.9 ka, Deino et al., 2004). Between 12 and 3.8 ka there were three eruptive episodes followed by a long period of quiescence until the last eruption of Monte Nuovo in 1538 CE (Di Vito et al., 1999). During the last decades CFc is subjected to seismic activity, gas emissions and intense ground deformations (Chiodini et al., 2001; D'Auria et al., 2011). Recent events of uplift occurred in 1950–1952, 1969–1972, and in 1982–1984 (Del Gaudio et al., 2010). During the 1982–1984 episode there was a strong increase of the seismicity (D'Auria et al., 2011). D'Auria et al. (2014b) performed a detailed analysis of the

events occurred during this last crisis in order to determine the spatial-temporal variations of the stress field within the CFc. The joint inversion of seismological and geodetic datasets evidences the presence of a weak NNE-SSW extensional stress field that is progressively overcome by a local volcanic one, active during the 1982–1985 unrest episode (D'Auria et al., 2014b).

The FPS dataset analyzed in this research consists of 217 events with $0 < M < 4$ occurred between 1983 and 1984 supplied by D'Auria et al. (2014b; **Figure 7**). Epicenters are mainly concentrated at the axial sector of the CFc. Hypocentral depths are up to 6 km. The three extreme slip classes are well represented, with a slight prevalence of the dip-slip normal solutions (**Figure 7D**). The stress field has been computed applying the BRTM, using a three-dimensional grid similar to that adopted for Vesuvius: 1-km spaced nodes, for each grid node all the FPSs within a radius of 1-km have been considered for the subset inversions. Only subsets related to volumes containing at least five FPSs have been inverted (**Figure 7**). Results of the BRTM stress inversion procedures are shown in **Figure 8** and summarized for 1, 2, and 3 km depth-slices. Only a few volumes have been inverted for the 1 km depth slice. The resulting σ_1 is high to moderately plunging and corresponds to a sub-horizontal σ_3 trending E-W. The related Φ_B values are in the range 0.5–0.8. At 2 km depth slice the attitude of the inverted σ_1 is variable, with a prevalence of sub-vertical plunges while the σ_3 trends

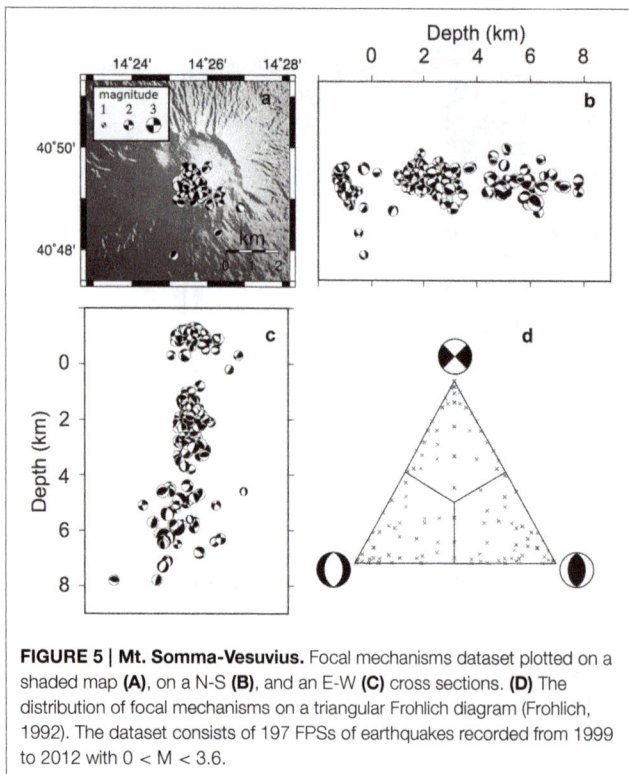

FIGURE 5 | Mt. Somma-Vesuvius. Focal mechanisms dataset plotted on a shaded map **(A)**, on a N-S **(B)**, and an E-W **(C)** cross sections. **(D)** The distribution of focal mechanisms on a triangular Frohlich diagram (Frohlich, 1992). The dataset consists of 197 FPSs of earthquakes recorded from 1999 to 2012 with $0 < M < 3.6$.

mainly NNE-SSW. The corresponding Bishop's ratio is mainly $0.2 < \Phi_B < 0.6$. At 3 km depth slice the attitude of the principal stress axes is variable and the corresponding Bishop's ratio ranges from 0 to 0.8. Nevertheless at this depth too there is a prevalence of sub-vertical plunging σ_1 and a sub-horizontal NNE-SSW trending σ_3. Overall, the key features of the stress field in the area are: a nearly sub-vertical σ_1 at the center of the CFc and a low-plunging σ_1 trending radially in the surrounding areas. The corresponding σ_3 has a roughly horizontal NNE-SSW trend corresponding to the regional extensional stress field. Related Φ_B values vary between 0.3 and 0.8 (**Figure 8**). According to D'Auria et al. (2014b), this result is compatible with the presence of a varying stress field related to a source of deformation, located at about 2.7 km depth, that during inflation and deflation episodes (associated to an increased seismicity) is able to overcome the weak regional extensional stress field having a NNE-SSW trend.

Long Valley

Long Valley caldera (LVc) is located in eastern California at the boundary between the Sierra Nevada block and the westernmost extensional Basin And Range Province (**Figure 9**). LVc develops inside a transfer zone hosted between NNW-SSE trending normal faults (e.g., Dickinson, 1979; Bosworth et al., 2003). This caldera is one of the youngest volcanic systems active in California. LVc has a surface of about 500 km² and is surrounded by numerous basins and ranges (like Mammoth Mountain at SW, Glass Mountain at NE, Laurel Mountain at S). It is located near some important ENE dipping normal fault

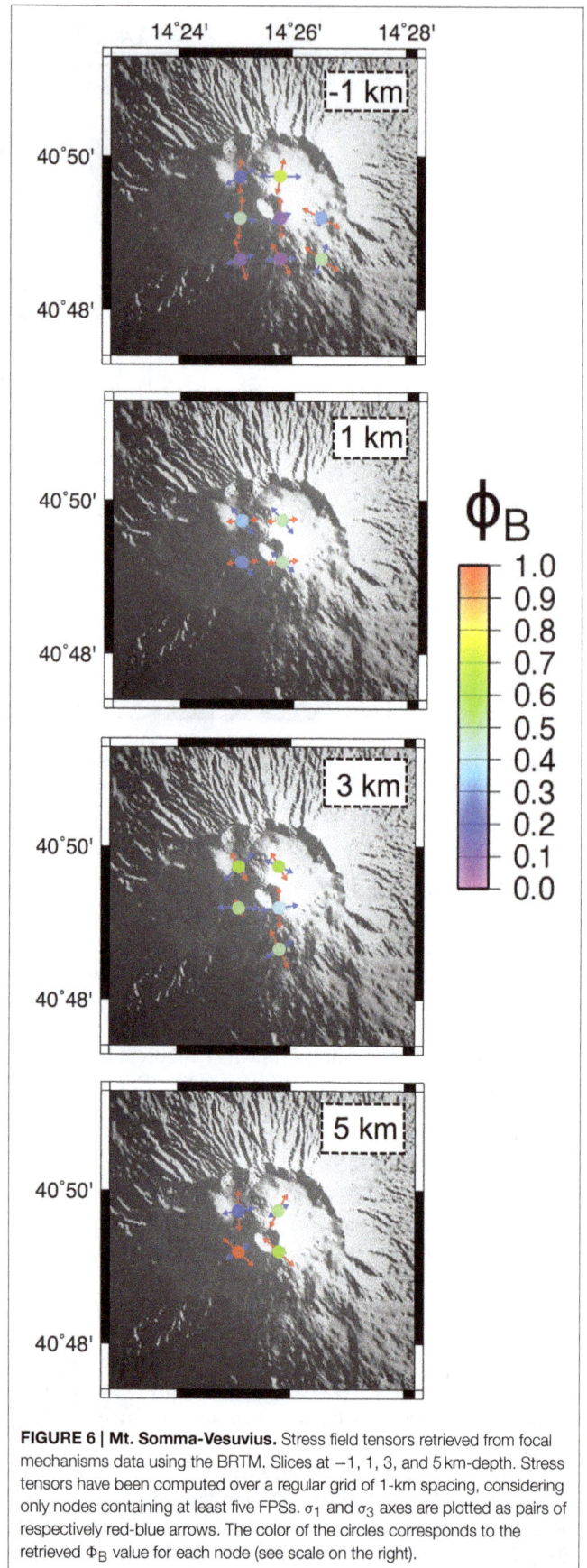

FIGURE 6 | Mt. Somma-Vesuvius. Stress field tensors retrieved from focal mechanisms data using the BRTM. Slices at −1, 1, 3, and 5 km-depth. Stress tensors have been computed over a regular grid of 1-km spacing, considering only nodes containing at least five FPSs. σ_1 and σ_3 axes are plotted as pairs of respectively red-blue arrows. The color of the circles corresponds to the retrieved Φ_B value for each node (see scale on the right).

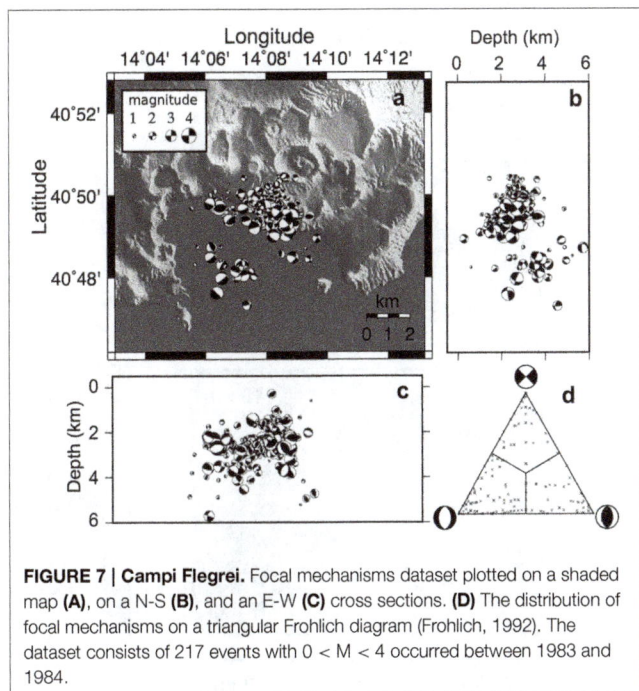

FIGURE 7 | Campi Flegrei. Focal mechanisms dataset plotted on a shaded map **(A)**, on a N-S **(B)**, and an E-W **(C)** cross sections. **(D)** The distribution of focal mechanisms on a triangular Frohlich diagram (Frohlich, 1992). The dataset consists of 217 events with 0 < M < 4 occurred between 1983 and 1984.

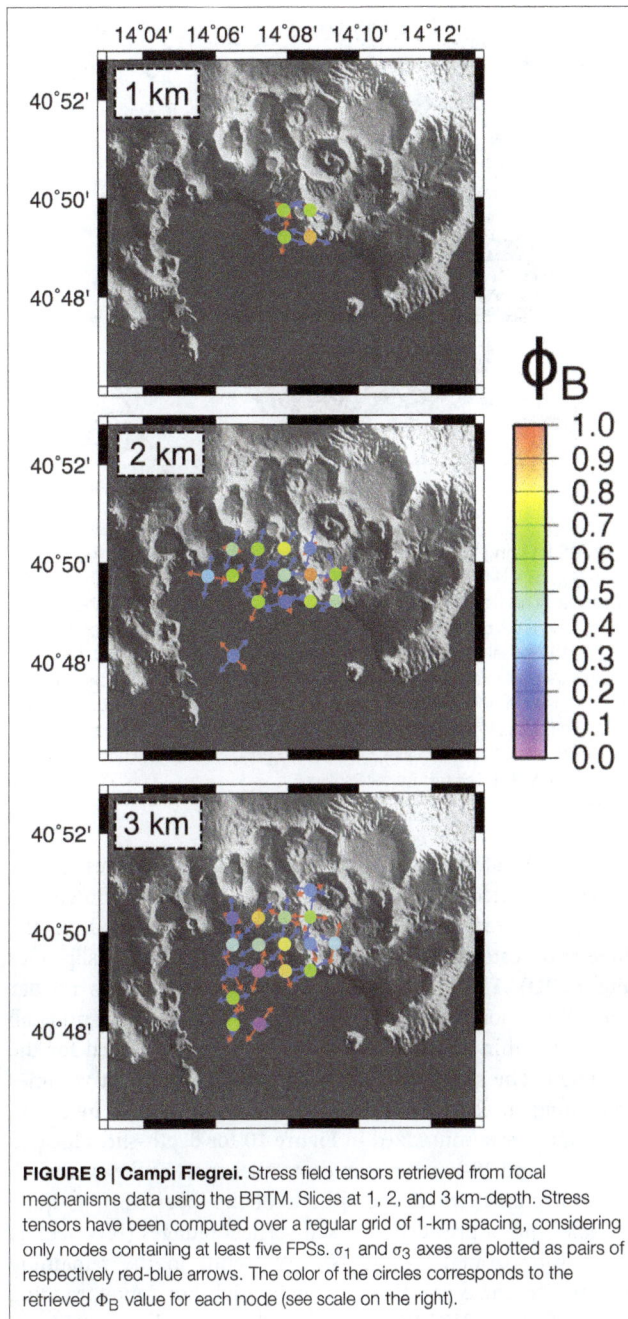

FIGURE 8 | Campi Flegrei. Stress field tensors retrieved from focal mechanisms data using the BRTM. Slices at 1, 2, and 3 km-depth. Stress tensors have been computed over a regular grid of 1-km spacing, considering only nodes containing at least five FPSs. σ_1 and σ_3 axes are plotted as pairs of respectively red-blue arrows. The color of the circles corresponds to the retrieved Φ_B value for each node (see scale on the right).

systems: Hartley Springs-Silver Lake Fault system to the North (HS, SL, in **Figure 9**), Hilton Creek and Round Valley Fault systems to the South (e.g., Prejean et al., 2002; Sorey et al., 2003; Bursik, 2009; HC an RV in **Figure 9**). Additionally, the LVc area is bounded to the east by the White Mountain Range and the related SW-dipping border faults (WMt in **Figure 9**). First volcanic activity can be dated approximately 760 ka ago as result of a large explosive eruption, during which more than 600 km^3 of pyroclastics and ash (Bishop Tuff) have been erupted (Bailey et al., 1976). The emplacement of a resurgent dome (RD in **Figure 9**) located almost at the center-western sector of the caldera (started about 600 ka ago) has been causing an intense localized uplift (Hill et al., 1991; Langbein et al., 1993, 1995; Tizzani et al., 2009). Recent activity in the area consists of small eruptions and phreatic explosions at Inyo-Mono Chain (700–550 years ago; IC in **Figure 9**) and at Mono Lake, at NW of the LVc (about 200 years ago; Miller, 1985). Since 1978 seismicity and surface deformations have been continuously recorded in the area of the resurgent dome and in an active seismic zone along the southern margin of the caldera (Langbein, 2003). Location, size and geometry of magma bodies at LVc are still matter of debates (Carle, 1988; Battaglia et al., 1999, 2003a,b; Langbein, 2003; Tizzani et al., 2009; Guoqing, 2015). The regional active stress field in the western Basin and Range Province shows a minimum horizontal principal stress component (S$_h$) trending ESE-WNW, it rotates roughly to ENE-WSW at the border with the Sierra Nevada Range (Zoback, 1989, 1992; Heidbach et al., 2009). Breakout and seismic data depict a more complex pattern for the S$_h$ across the LVc. NE-SW-trending S$_h$ in the Resurgent Dome and at the South Moat Range, NW-SE-trending S$_h$ in the West Moat and at Mammoth Mts. (**Figure 9**; Moos and Zoback, 1993; Bosworth et al., 2003, and references therein).

For the Long Valley Caldera we used a dataset made up of about 38,000 FPSs of earthquakes occurred from 1978 to 2015 (0 < M < 6.4; Dataset by: NCEDC, 2014). These data were used to determine, with high resolution, the spatial variations of the stress field using the BRTM approach. In **Figure 9** we plotted only the most energetic 3000 events (M > 3) in order to avoid an excessive overlapping of data. Epicenters are densely clustered at the southern sector of the LVc rim at the West Moat-Mammoth Mts.-South Moat Range sector. Additionally, many events are located along key tectonic lineaments of Hilton Creek and Round Valley to the south and along the western border fault of the White Mountains (**Figure 9A**). Hypocenters are

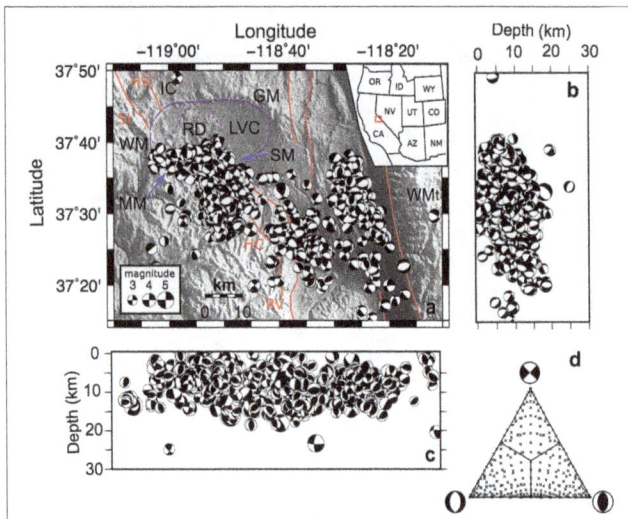

FIGURE 9 | Long Valley. Focal mechanisms dataset plotted on a shaded map (A);, on a N-S (B), and an E-W (C) cross sections. (D) The distribution of focal mechanisms on a triangular Frohlich diagram (Frohlich, 1992). The dataset consists of 38,000 events with 0 < M < 6.4 occurred between 1978 and 2015. For sake of clarity only the events with m >= 3 are plotted. LVC, Long Valley Caldera (purple dashed line); RD, Resurgent Dome (purple dotted line); IC, Inyo Craters; WM, West Moat; MM, Mammoth Mts.; SM, South Moat; GM, Glass Mts.; WMt, White Mts. Red lines represent the main fault systems of the LVc area: SL, Silver Lake; HS, Hartley Springs Fault; HC, Hilton Creek; RV, Round Valley. See text for data sources.

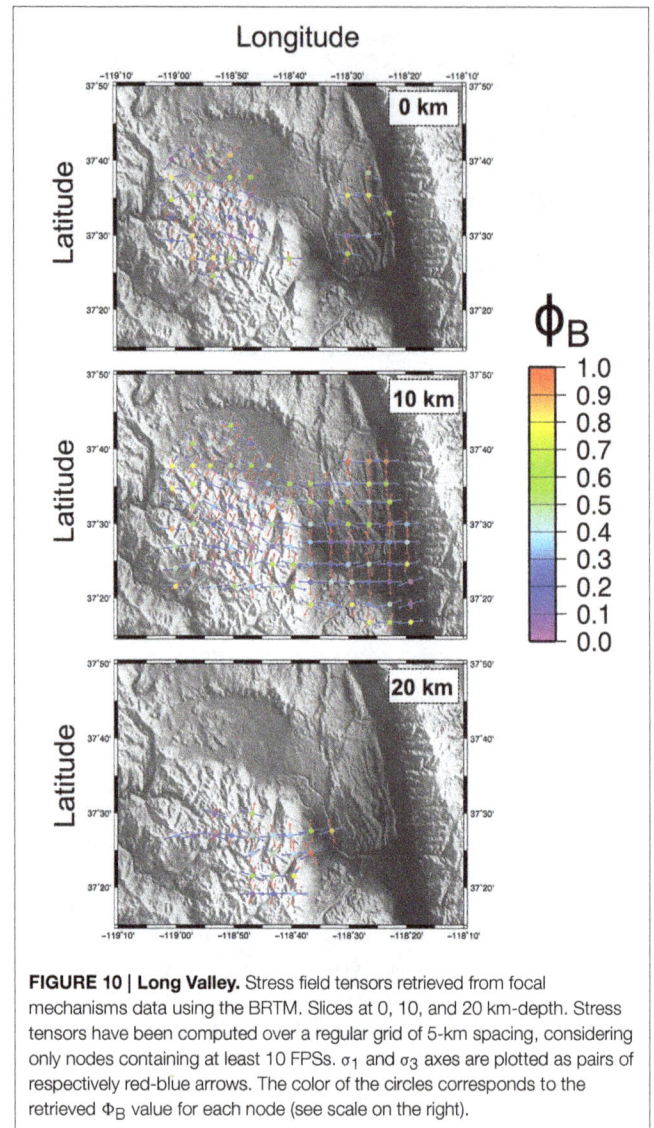

FIGURE 10 | Long Valley. Stress field tensors retrieved from focal mechanisms data using the BRTM. Slices at 0, 10, and 20 km-depth. Stress tensors have been computed over a regular grid of 5-km spacing, considering only nodes containing at least 10 FPSs. σ_1 and σ_3 axes are plotted as pairs of respectively red-blue arrows. The color of the circles corresponds to the retrieved Φ_B value for each node (see scale on the right).

essentially located above 20 km of depth. The shallowest events are concentrated at LVC and RD (**Figures 9B,C**). Kinematics of the analyzed FPSs are quite equally represented in the three main categories, with a slight prevalence of dip-slip ones (**Figure 9D**). The stress field has been computed on a regular three-dimensional grid of 5-km spacing. For each grid node all the FPSs within a radius of 5-km have been considered for the inversion. The stress field has been computed only at volumes containing at least 10 FPSs. Results of the stress inversion procedure are summarized in **Figure 10** for depth-slices located at 0-, 10-, and 20-km. At 0 km depth slice most of the events fall in the LVc area, where the attitude of the inverted σ_1 is quite variable, with a prevalence of sub-vertical plunges (NW sector) and sub-horizontal N-S to NE-SW trends in the remaining sectors. The corresponding σ_3 is mainly sub-horizontal trending mainly E-W to NW-SE. The Bishop's ratio values are highly variable, ranging from 0.1 to 0.9. The 10 km depth slice samples most of the events of the analyzed dataset. Results can be divided up in two groups, the eastern one is dominated by a sub-horizontal σ_1 trending N-S associated to a sub-horizontal E-W trending σ_3. For the western sector, the retrieved stress field appears very similar to the one obtained for the shallowest 0-km slice; a sub-horizontal σ_1 overall trending NE-SW associated to a sub-horizontal NW-SE trending σ_3. Bishop's ratio values are highly variable ranging between 0.1 and 0.9, without a clear pattern (**Figure 10**). At the 20-km depth slice the stress inversion procedure highlights the presence of an overall stress field comparable to the corresponding nodes of the 10-km depth slice: in the easternmost region, a sub-horizontal NNW-SSE

trending σ_1, corresponding to a sub-horizontal ENE-WSW trending σ_3. Bishop's ratios show a slight prevalence of lower values $\Phi_B < 0.5$. Summing up, a strong background stress field seems to dominate the investigated volume: it is characterized by a sub-horizontal NNW-SSE-trending σ_1 and a sub-horizontal ENE-WSW-trending σ_3. This field is clearly evidenced by the inversion results at the deepest slices in particular at 20-km depth slice (**Figure 10**). The region at the south of the caldera rim experiences a clock-wise rotation of the $\sigma_1 - \sigma_3$ axes up to $45°$, as shown at 0- and 10-km depth slices (**Figure 10**). Our results can be interpreted as the effect of the interaction between a background regional stress field with the local volcano-tectonic structures. The presence of two distinct stress patterns at 10-km depth slice clues the presence of a regional stress field, characterized by a sub-horizontal σ_1 roughly trending from N-S to NNW-SSE associated with a sub-horizontal σ_3 trending from E-W to NNE-SSW. The retrieved background stress field drives the evolution of the main tectonic trends of the area (**Figure 8A**), overcome in the western sector by a local volcano-tectonic regime, involving mainly the shallowest lithospheric portion. This last allows the above mentioned clock-wise rotation of the $\sigma_1 - \sigma_3$ axes well documented at the -10-km slice (**Figure 10**). Our results are in agreement with S_h found in previous studies, derived by the analyses of earthquake focal mechanisms, borehole breakouts, fault offsets, hydraulic fracturing, and alignment of young volcanic vents. In detail, the $\sigma_1 - \sigma_3$ axes attitude derived for the easternmost subset inverted at -20-km slice is in good agreement with the ENE-WSW-trending S_h retrieved at the border with the Basin and Range and Sierra Nevada Range (Moos and Zoback, 1993; Prejean et al., 2002; Bosworth et al., 2003, and references therein).

DISCUSSION AND CONCLUSIONS

In volcanic environments, a spatio-temporal analysis of the stress fields represents a valuable approach to infer the volcano dynamics (Wyss et al., 1992; Hardebeck and Michael, 2006; Gudmundsson et al., 2009; Plateaux et al., 2014; Costa and Marti, 2016). Stress field variations play a key role in driving magmas and/or fluids migration. On the other hand, injection of fluids can be responsible for local variations of the stress field allowing the reactivation of locked faults. Stress changes may influence fluid circulation within the shallow crust and they can be also responsible for the triggering of tectonic earthquakes associated to strong variations in the dynamics of volcanic eruption and behavior of hydrothermal phenomena (Linde and Sacks, 1998; Hill et al., 2002; Walter et al., 2007).

During the last decades, a wide range of stress inversion techniques has been proposed, some of them specifically conceived to manage seismological datasets. Several key aspects, that could impact on the reliability of retrieved results, require a short discussion. First of all, the discernment of the actual fault plane among the nodal planes of a FPS represents a known critical step in many analytical techniques of stress inversion. (e.g., Gephart and Forsyth, 1984; Michael, 1987; Angelier, 1990). A good solution can be an a-posteriori approach that pick the actual fault following a best-fit procedure or a massive computational approach on small datasets (Michael, 1987). Of course, it implies a complication in the processing procedure that can hide pitfalls. Only a few analytical methods do not require this choice, for instance Angelier (2002) based on the slip shear component criterion. Additionally, all graphical (e.g., RDM) or graphical-derived (e.g., RTM, BRTM) techniques do not require this pre-selection; this is a valuable asset that makes these processing approaches as robust as they are simple. As premised, graphical-derived methods are based on the implementation of an algorithm that is able to figure a probability function over the focal sphere for both σ_1 and σ_3 principal stress attitude. The analytical implementation of the classical graphical methods allows the efficient managing of large datasets following a proper statistical approach (e.g., RDTM; Ramsay and Lisle, 2000). Generally, the output of these procedures needs a strong graphical post-processing in order to obtain an adequate representation of results (e.g., MORE;

Massa, 2003). BRTM proposes a solution to this limitation offering a standard procedure of data ingestion, processing and graphical post-processing in both 2- and 3-Dimensional rendering. All these task can be performed through the use of editable scripts run in the same computational environment. A known drawback of stress inversion procedures (for both graphical and analytical) is that they generally lack of a robust statistical estimation of the uncertainty on the retrieved tensors. To overcome this lack, BRTM performs an estimation of solution uncertainty and trough the implementation of the approach of Jackson and Matsu'ura (1985), BRTM is able to delimit the confidence intervals around the σ_1 and σ_3 retrieved axes (D'Auria and Massa, 2015). Another key aspect to consider is that very frequently datasets to be processed are large and heterogeneous. Without an a-priori selection, many datasets collect faults/FPS related to different stress fields. For instance, seismological datasets collected across wide lithospheric volume, mesoscale faults hosted in a limited rock volume but generated during different tectonic events, etc. The solution to identify the different contributions from multiple sources can be a clustering approach as the MIM (Yamaji, 2000; Otsubo et al., 2008) or the application of other automatic analytical procedures (Angelier and Manoussis, 1980). These approaches allows the identification of the different stress fields but they are not designed to give a quantitative assessment of the solution quality; additionally they are not able to locate the spatial distribution of the retrieved tensors (Yamaji, 2000; D'Auria and Massa, 2015). The relevance of this point becomes larger considering the seismological datasets collected in volcanic areas. Here, the retrieving of stress field spatial variations represents a crucial task: sensible variation in stress axes attitude can characterize stress field active in contiguous volumes (e.g., D'Auria et al., 2014a,b; Plateaux et al., 2014). For this reason, unsupervisioned stress inversions performed on an unselected dataset can be quite useless. A similar discussion can be done for the analysis of stress field temporal variations. The ideal way to obtain a reliable stress field inversion would be the analysis of a dataset covering a time interval as short as possible, in order to consider the responsible stress field quite constant in its parameters (i.e., the attitude of principal stress axes and the related Φ_B). The preliminary temporal selection of data can be unfeasible in case of mesoscale faults also when collected in a limited volume of rock: several tectonic "phases" could have superimposed in the geological record. Conversely, the temporal selection of data can be conveniently done in the analysis of seismological datasets where the focal parameters are well known. In volcanic environments, a spatio-temporal analysis of the stress fields represents the best way to infer about the volcano dynamics (Wyss et al., 1992; Hardebeck and Michael, 2006; D'Auria et al., 2014b; Plateaux et al., 2014). In this perspective, a supervisioned splitting of data in 3D sub-volumes represents the best solution to retrieve the spatial distribution of active coeval stress fields. As a final consideration about the stress inversion procedures, it should be noted that the publication of a rigorous formulation of a technique is not enough to determine its success. A strong limitation to the dissemination of a processing procedure consists of the availability of a free or open-source platform of implementation. Available methods are frequently implemented by authors and freely distributed to the community only as PC programs. For instance, the Direct Inversion algorithm was implemented by the author in the executing software TENSOR (Angelier, 1990), the Multiple Inverse Method can be performed through a set of executable supplied by the author (Yamaji, 2000), RDM and RTM can be applied using RDTM, supplied by authors both as executing program than as code (Ramsay and Lisle, 2000). The availability of an editable code represents a key point to allow a flexible use of a technique, allowing, in addition, a facilitation in its spreading over the community. A BRTM implementation in MATLAB® was proposed by the Authors, allowing a basic stress-inversion procedure and an advanced plot of results (**Figure 3**; D'Auria and Massa, 2015). The implementation of the basic inversion algorithm can be easily exported to open-source and free platform. Results can be plotted using the preferred platform. The graphical post-processing proposed by D'Auria and Massa (2015) was implemented in MATLAB®. It would be useful a revision in order to make it available to free platforms.

According to previous researches, the new stress-inversions discussed in this paper (**Figures 6**, **8**, **10**), have shown that the stress field in the studied volcanoes results from the interaction of a regional background field with local volcanic structures and dynamics. In detail, during the 1983–1984 seismic crisis in the Campi Flegrei area, a sub-vertical σ_1 dominates the axial volume of the caldera. Conversely, a low-plunging σ_1 appears trending radially in the surrounding volumes. The corresponding σ_3 attitudes are roughly sub-horizontal NNE-SSW trending, in accordance with the regional S_h retrieved by breakout data (Montone et al., 2012). The retrieved stress fields are in accordance with Zuppetta and Sava (1991). The background regional stress field has been also modeled by D'Auria et al. (2014b) using a joint inversion of a geodetical-seismological dataset, obtaining a sub-horizontal σ_3 trending N-S. This result is in good accordance with the Mt. Somma-Vesuvius where D'Auria et al. (2014a) found a very similar configuration for the regional stress field. The retrieved variability of the stress field can be related to fluids migration within a planar crack probably located at shallow depth and possibly responsible for unrest episodes (D'Auria et al., 2014b; Macedonio et al., 2014). At Mt. Vesuvius, rheological variations of the involved geological units and the action of key volcano-tectonic structures seem to play a significant role in determining the local stress fields (**Figures 4–6**; D'Auria et al., 2014a). The configuration of the retrieved stress fields shows the superposition of two volumes with a marked difference in their evolution history. A top volume dominated by a gravitational volcanic spreading that allows the setting of extensional stress fields (low Φ_B values and sub-vertical σ_1) active in the analyzed sub-volumes. A bottom volume strictly related to a regional extensional background stress field in accordance with the S_h derived by breakout data (**Figure 6**; Borgia et al., 2005; Montone et al., 2012; D'Auria et al., 2014a). The computation of the stress field using the BRTM on a regular three-dimensional grid allows the 3-D figuration of the stress fields "simultaneously" acting from the surface to about 8-km depth (**Figures 5**, **6**). This approach appears very useful in supporting studies on the volcano-dynamics of complex system as the Mt. Somma-Vesuvius. Finally, analyses performed at Long Valley Caldera show a complex configuration of the stress

field pattern associated to a volcanic district developed inside a transfer zone hosted between NNW-SSE trending normal faults (e.g., Dickinson, 1979; Bosworth et al., 2003). A dataset of 38,000 FPSs allows to retrieve with high resolution, the spatial variations of the stress field active in the first 20 km of depth. A regional background stress field dominates the deepest portion of the investigated volume accordingly to the ENE-WSW-trending S_h retrieved at the border with the Basin and Range and Sierra Nevada Range (Moos and Zoback, 1993; Prejean et al., 2002; Bosworth et al., 2003). In the shallowest lithospheric portions this regional stress field has been overcome by the volcano-tectonic regime; this is clear in particular at the 10 km depth-slice (**Figures 9**, **10**). The managing of the very large dataset has been approached using a regular sampling of data inverted using the BRTM. The result is a very detailed figuration of the spatial variation of the active stress fields in the 500 km^2 Long Valley Caldera, one of the youngest volcanic systems active in California.

The above results confirm that the application of stress inversion procedures to volcanic environments provides crucial information about the volcano dynamics with particular care to its interaction with regional tectonics. In detail, BRTM has shown to be very efficient in managing heterogeneous datasets in a user-friendly processing environment, with a clear graphic output and an efficient evaluation of the regions of confidence around the retrieved principal stress axes. The proposed processing approach of FPS datasets in active volcanic areas is suitable to be applied to a wide range of contexts. Finally, the results presented in this work suggest that these methods could be useful also as a near real-time monitoring tool to characterize spatial and temporal variations in the stress field linked to a volcanic unrest (e.g., Toda et al., 2002; Plateaux et al., 2014). A final consideration concerns the relevance of the stress-inversion approach based on the integration and/or the joint inversion of seismological and geodetical datasets. The joint inversion of ground deformation and focal mechanism is more efficient than the mere comparison of the results obtained by the separate inversion (Segall, 2013; D'Auria et al., 2014b; Viccaro et al., 2016). Spatial-temporal analysis of the stress field derived from seismological datasets associated to a continuous recording of ground deformations (remote and/or field classical techniques) could be a reliable tool to monitor volcano dynamics and infer about their evolution.

AUTHOR CONTRIBUTIONS

BM, LD, EC, and AD: rock mechanics, stress inversion, join inversion, and focal mechanisms.

FUNDING

This research is financially supported by: MED-SUV project (European Union's Seventh Programme for research, technological development and demonstration, Grant Agreement Number 308665); Università degli Studi del Sannio FRA 2014–2015 "Modeling of geological processes" (P.I. B. Massa).

ACKNOWLEDGMENTS

Thanks are due to the two Referees, Eisuke Fujita and Mimmo Palano, to the Associate Editor Antonio Costa and to the Chief Editor Valerio Acocella for the constructive and valuable review that greatly improved the manuscript. BRTM can be accessed as Electronic Supplement to D'Auria and Massa (2015) following the link: http://srl.geoscienceworld.org/content/early/2015/03/17/0220140153.

REFERENCES

Acocella, V., and Funiciello, R. (2006). Transverse systems along the extensional Tyrrhenian margin of central Italy and their influence on volcanism. *Tectonics*, 25:TC2003. doi: 10.1029/2005TC001845

Andronico, D., Calderoni, G., Cioni, R., Sbrana, A., Sulpizio, R., and Santacroce, R. (1995). Geological map of Somma-Vesuvius volcano. *Period. Mineral.* 64, 77–78.

Angelier, J. (1990). Inversion of field data in fault tectonics to obtain the regional stress. III—A new rapid direct inversion method by analytical means. *Geophys. J. Int.* 103, 363–376. doi: 10.1111/j.1365-246X.1990.tb01777.x

Angelier, J. (2002). Inversion of earthquake focal mechanisms to obtain the seismotectonic stress IV—A new method free of choice among nodal planes. *Geophys. J. Int.* 150, 588–609. doi: 10.1046/j.1365-246X.2002.01713.x

Angelier, J., and Baruah, S. (2009). Seismotectonics in Northeast India: a stress analysis of focal mechanism solutions of earthquakes and its kinematic implications. *Geophys. J. Int.* 178, 303–326. doi: 10.1111/j.1365-246X.2009.04107.x.

Angelier, J., and Manoussis, S. (1980). Classification automatique et distinction de phases superposées en tectonique cassante. *C. R. Acad. Sci. Paris* 290, 651–654.

Angelier, J., and Mechler, P. (1977). Sur une méthode graphique de recherche des contraintes principales également utilisable en tectonique et en séismologie: La méthode des diédres droits. *Bulletin de la SocielÂtelÂ GelÂologique de France* 19, 1309–1318.

Bailey, R. A., Dalrymple, G. B., and Lanphere, M. A. (1976). Volcanism, structure and geochronology of Long Valley caldera, Mono County, California. *J. Geophys. Res.* 81, 725– 744. doi: 10.1029/JB081i005p00725

Battaglia, M., Roberts, C., and Segall, P. (1999). Magma intrusion beneath Long Valley caldera confirmed by temporal changes in gravity. *Science* 285, 2119–2122. doi: 10.1126/science.285.5436.2119

Battaglia, M., Segall, P., Murray, J., Cervell, P., and Langbein, J. (2003a). The mechanics of unrest at Long Valley caldera, California: 1. Modeling the geometry of the source using GPS, leveling and two-color EDM data. *J. Volcanol. Geother. Res.* 127, 195–217. doi: 10.1016/S0377-0273(03)00170-7

Battaglia, M., Segall, P., and Roberts, C. (2003b). The mechanics of unrest at Long Valley caldera, California. 2. Constraining the nature of the source using geodetic and micro-gravity data. *J. Volcanol. Geother. Res.* 127, 219–245. doi: 10.1016/S0377-0273(03)00171-9

Bernasconi, A., Bruni, P., Gorla, L., Principe, C., and Sbrana, A. (1981). Risultati preliminari dell'esplorazione geotermica profonda nell'area vulcanica del Somma-Vesuvio. *Rend. Soc. Geol.* 4, 237–240.

Bianco, F., Castellano, M., Milano, G., Ventura, G., and Vilardo, G. (1998). The Somma-Vesuvius stress field induced by regional tectonics: evidences from seismological and mesostructural data. *J. Volcanol. Geother. Res.* 82, 119–218. doi: 10.1016/S0377-0273(97)00065-6

Bishop, A. (1966). The strength of solids as engineering materials. *Geotechnique* 16, 91–130. doi: 10.1680/geot.1966.16.2.91

Borgia, A., Tizzani, P., Solaro, G., Manzo, M., Casu, F., Luongo, et al. (2005). Volcanic spreading of Vesuvius, a new paradigm for interpreting its volcanic activity. *Geophys. Res. Lett.* 32:L03303, doi: 10.1029/2004GL022155

Bosworth, W., Burke, K., and Strecker, M. (2003). Effect of stress fields on magma chamber stability and the formation of collapse calderas. *Tectonics* 22:1042. doi: 10.1029/2002TC001369

Bott, M. (1959). The mechanisms of oblique slip faulting. *Geol. Mag.* 96, 109–117.

Brocchini, D., Principe, C., Castradori, D., Laurenzi, M. A., and Gorla, L. (2001). Quaternary evolution of the southern sector of the Campania Plain and early Somma-Vesuvius activity: insights from the Trecase 1 well. *Mineral. Petrol.* 73, 67–91. doi: 10.1007/s007100170011

Bursik, M. (2009). A general model for tectonic control of magmatism: Examples from Long Valley Caldera (USA) and El Chichón (México). *Geofísica Int.* 48, 171–183. Available online at: http://www.scielo.org.mx/pdf/geoint/v48n1/v48n1a12.pdf

Byerly, P. (1928). The nature of the first motion in the Chilean earthquake of November 11, 1922. *Am. J. Sci.* 16, 232–236. doi: 10.2475/ajs.s5-16.93.232

Cannavò, F., Scandura, D., Palano, M., and Musumeci, C. (2014). A joint inversion of ground deformation and focal mechanisms data for magmatic source modelling. *Pure Appl. Geophys.* 171, 1695–1704. doi: 10.1007/s00024-013-0771-x

Carey, E., and Brunier, B. (1974). Analyse théorique et numérique d'un modéle mécanique élémentaire appliquéáá l'étude d'une population de failles. *C. R. Acad. Sci.* D179, 891–894.

Carey-Gailhardis, E., and Mercier, J. (1987). A numerical method for determining the state of stress using focal mechanisms of earthquake populations: application to Tibetan teleseisms and microseismicity of southern Peru. *Earth Planet. Sci. Lett.* 82, 165–179. doi: 10.1016/0012-821X(87)90117-8

Carey-Gailhardis, E., and Mercier, J. L. (1992). Regional state of stress, fault kinematics and adjustments of blocks in a fractured body of rock: application to the micro-seismicity of the Rhine graben. *J. Struct. Geol.* 14, 1007–1017. doi: 10.1016/0191-8141(92)90032-R

Carle, S. F. (1988). Three-dimensional gravity modeling of the geologic structure of Long Valley Caldera. *J. Geophys. Res.* 93, 13237–13250. doi: 10.1029/jb093ib11p13237

Chiodini, G. L., and Marini, and, M., Russo (2001). Geochemical evidence for the existence of high-temperature hydrothermal brines at Vesuvio volcano, Italy. *Geochim. Cosmochim. Acta* 65, 2129–2147. doi: 10.1016/S0016-7037(01)00583-X

Cocina, O., Neri, G., Privitera, E., and Spampinato, S. (1997). Stress tensor computations in the Mount Etna area (Southern Italy) and tectonic implications. *J. Geodynamics* 23, 109–127. doi: 10.1016/S0264-3707(96)00027-0

Costa, A., and Marti, J. (2016). Stress field control during large caldera-forming eruptions. *Front. Earth Sci.* 4:92, doi: 10.3389/feart.2016.00092

D'Auria, L., Esposito, A. M., Lo Bascio, D., Ricciolino, P., Giudicepietro, F., Martini, M., et al. (2013). The recent seismicity of Mt. Vesuvius: inference on seismogenic processes. *Ann. Geophys.* 56:S0442, doi: 10.4401/ag-6448

D'Auria, L. F., Giudicepietro, I., Aquino, G., Borriello, C., Del Gaudio, D., Lo Bascio, M., et al. (2011). Repeated fluid-transfer episodes as a mechanism for the recent dynamics of Campi Flegrei caldera (1989–2010). *J. Geophys. Res.* 116:B04313, doi: 10.1029/2010JB007837

D'Auria, L., Martini, M., Esposito, A., Ricciolino, P., and Giudicepietro, F. (2008). "A unified 3D velocity model for the Neapolitan volcanic areas," in *Conception, Verification and Application of Innovative Techniques to Study Active Volcanoes*, eds W. Marzocchi and A. Zollo (Napoli: INGV-DPC), 375–390.

D'Auria, L., and Massa, B. (2015). Stress inversion of focal mechanism data using a bayesian approach: a novel formulation of the right trihedra method. *Seism. Res. Lett.* 86, 968–977. doi: 10.1785/0220140153

D'Auria, L., Massa, B., Cristiano, E., Del Gaudio, C., Giudicepietro, F., Ricciardi, G., et al. (2014b). Retrieving the stress field within the campi flegrei caldera (southern italy) through an integrated geodetical and seismological approach. *Pure Appl. Geophy.* 172, 3247–3263. doi: 10.1007/s00024-014-1004-7

D'Auria, L., Massa, B., and De Matteo, A. (2014a). The stress field beneath a quiescent stratovolcano: the case of mount Vesuvius. *J. Geophys. Res.* 119, 1181–1199. doi: 10.1002/2013JB010792

Deino, A. L., Orsi, G., Piochi, M., and De Vita, S. (2004). The age of the neapolitan Yellow Tuff caldera-forming eruption (Campi Flegrei caldera–Italy) assessed by 40ar/39ar dating method. *J. Volcanol. Geother. Res.* 133, 157–170. doi: 10.1016/S0377-0273(03)00396-2

Del Gaudio, C., Aquino, I., Ricciardi, G. P., Ricco, C., and Scandone, R. (2010). Unrest episodes at Campi Flegrei: a reconstruction of vertical ground movements during 1905–2009. *J. Volcanol. Geother. Res.* 195, 48–56. doi: 10.1016/j.jvolgeores.2010.05.014

De Matteis, R., Matrullo, E., Rivera, L., Stabile, T. A., Pasquale,G., and Zollo, A. (2012). Fault Delineation and Regional Stress Direction from the Analysis of Background Microseismicity in the southern Apennines, Italy. *Bull. Seismol. Soc. Am.* 102, 1899–1907. doi: 10.1785/0120110225

Dickinson, W. R. (1979). "Cenozoic plate tectonic setting of the Cordilleran region in the United States," in *Pacific Coast Paleogeography Symposium 3; Cenozoic Paleogeography of the Western United States*, eds J. M. Armentrout, M. R. Cole, and H. TerBest Jr. (Anaheim, CA: Pac. Sect., Soc. of Econ, Paleontol. and Mineral), 1–13.

Di Vito, M. A., Isaia, R., Orsi, G., Southon, J., De Vita, S., D'antonio, M., et al. (1999). Volcanism and deformation since 12,000 years at the Campi Flegrei caldera (Italy). *J. Volcanol. Geother. Res.* 91, 221–246. doi: 10.1016/S0377-0273(99)00037-2

Frohlich, C. (1992). Triangle diagrams: ternary graphs to display similarity and diversity of earthquake focal mechanisms. *Phys. Earth. Planet. Inter.* 75, 193–198. doi: 10.1016/0031-9201(92)90130-N

Gebauer, S., Schmitt, A. K., Pappalardo, L., Stockli, D. F., and Lovera, O. M. (2014). Crystallization and eruption ages of Breccia Museo (Campi Flegrei caldera, Italy) plutonic clasts and their relation to the Campanian ignimbrite. *Contrib. Mineral. Petrol.* 167:953. doi: 10.1007/s00410-013-0953-7

Gephart, J., and Forsyth, D. (1984). An improved method for determining the regional stress tensor using earthquake focal mechanism data: application to the San Fernando earthquake sequence. *J. Geophys. Res.* 89, 9305–9320. doi: 10.1029/JB089iB11p09305

Gillard, D., and Wyss, M. (1995). Comparison of strain and stress tensor orientation: application to Iran and southern California. *J. Geophys. Res.* 100, 22197–22213. doi: 10.1029/95JB01871

Giudicepietro, F., Orazi, M., Scarpato, G., Peluso, R., D'Auria, L., Ricciolino, P., et al. (2010). Seismological monitoring of Mount Vesuvius (Italy): more than a century of observations. *Seism. Res. Lett.* 81, 625–634, doi: 10.1785/gssrl.81.4.625

Guoqing, L. (2015). Seismic velocity structure and earthquake relocation for the magmatic system beneath Long Valley Caldera, eastern California. *J. Volcanol. Geother. Res.* 296, 19–30. doi: 10.1016/j.jvolgeores.2015.03.007

Gudmundsson, A., Acocella, V., and Vinciguerra, S. (2009). Understanding stress and deformation in active volcanoes. *Tectonophysics* 471, 1–3. doi: 10.1016/j.tecto.2009.04.014

Hardebeck, J. L., and Hauksson, E. (2001). Stress orientations obtained from earthquake focal mechanisms: what are appropriate uncertainty estimates? *Bull. Seismol. Soc. Am.* 91, 250–252. doi: 10.1785/0120000032

Hardebeck, J. L., and Michael, A. J. (2006). Damped regional-scale stress inversions: methodology and examples for southern California and the Coalinga aftershock sequence. *J. Geophys. Res.* 111:B11310, doi: 10.1029/2005JB004144

Heidbach, O., Tingay, M., Barth, A., Reinecker, J., Kurfeß, D. and Müller, B. (2009). *The World Stress Map based on the database release 2008, equatorial scale 1:46,000,000.* Paris: Commission for the Geological Map of the World.

Hill, D. P., Johnston, M. J. S., Langbein, J. O., McNutt, S. R., Miller, C. D., Mortensen, C. E., et al. (1991). *Response Plans for Volcanic Hazards in the Long Valley Caldera and Mono Craters area, California.* Open-File Report, U.S. Geology Survey.

Hill, D. P., Pollitz, F., and Newhall, C. (2002). Earthquake-volcano interactions. *Phys. Today* 55, 41–47. doi: 10.1063/1.1535006

Ippolito, F., Ortolani, F., and Russo, M. (1973). Struttura marginale tirrenica dell'Appennino campano: reinterpretazione di dati di antiche ricerche di idrocarburi. *Mem. Soc. Geol.* 12, 227–250.

Jackson, D. D., and Matsu'ura, M. (1985). A Bayesian approach to nonlinear inversion. *J. Geophys. Res.* 90, 581–591. doi: 10.1029/JB090iB01p00581

Langbein, J. O. (2003). Deformation of the Long Valley caldera, California: inferences from measurements from 1988 to 2001. *J. Volcanol. Geother. Res.* 127, 247–267. doi: 10.1016/S0377-0273(03)00172-0

Langbein, J. O., Dzurisin, D., Marshall, G., Stein, R., and Rundle, J. (1995). Shallow and peripheral volcanic sources of inflativo revealed by modeling two-color

geodimeter and leveling data from Long Valley caldera, California, 1988–1992. *J. Geophys. Res.* 100, 12487–12495. doi: 10.1029/95JB01052

Langbein, J. O., Hill, D. P., Parker, T. N., and Wilkinson, S. K. (1993). An episode of re-inflation of the Long Valley caldera, eastern California, 1989–1991. *J. Geophys. Res.* 98, 15851–15870. doi: 10.1029/93JB00558

Lavecchia, G., Boncio, P., Creati, N., and Brozzetti, F. (2003). Some aspects of the Italian geology not fitting with a subduction scenario. *J. Virtual Explor.* 10, 1–42. doi: 10.3809/jvirtex.2003.00064

Linde, A. T., and Sacks, I. S. (1998). Triggering of volcanic eruptions. *Nature* 395, 888–890. doi: 10.1038/27650

Lisle, R. (1988). ROMSA: a basic program for paleostress analysis using fault striation data. *Comput. Geosci.* 14, 255–259. doi: 10.1016/0098-3004(88)90007-6

Lisle, R. J. (1987). Principal stress orientations from faults: an additional constraint. *Annales Tectonicae* 1, 155–158.

Lomax, A., Zollo, A., Capuano, P., and Virieux, J. (2001). Precise, absolute earthquake location under Somma-Vesuvius volcano using a new three-dimensional velocity model. *Geophys. J. Int.* 146, 313–331. doi: 10.1046/j.0956-540x.2001.01444.x

Macedonio, G., Giudicepietro, F., D'auria, L., and Martini, M. (2014). Sill intrusion as a source mechanism of unrest at volcanic calderas. *J. Geophys. Res. Solid Earth* 119, 3986–4000. doi: 10.1002/ 2013JB010868

Massa, B. (2003). *Relazione Tra Faglie Quaternarie e Sismicita NellŠarea Sannita*, Printed, Dottorato di Ricerca in Scienze della Terra e della Vita, XV ciclo., Universita degli Studi del Sannio, Benevento, Italy.

Michael, A. J. (1987). Use of Focal Mechanisms to Determine Stress: a control study. *J. Geophys. Res.* 92, 357–368. doi: 10.1029/JB092iB01p00357

Milia, A., Torrente, M. M., Massa, B., and Iannace, P. (2013). Progressive changes in rifting directions in the Campania margin (Italy): New constrains for the Tyrrhenian Sea opening. *Global Planetary Change* 109, 3–17. doi: 10.1016/j.gloplacha.2013.07.003

Miller, C. D. (1985). Holocene eruptions at the Inyo volcanic chain, California: implications for possible eruptions in Long Valley caldera. *Geology* 13, 14–17. doi: 10.1130/0091-7613(1985)13<14:HEATIV>2.0.CO;2

Montone, P., Mariucci, M. T., and Pierdominici, S. (2012). The Italian present-day stress map. *Geophys. J. Int.* 189, 705–716, doi: 10.1111/j.1365-246X.2012.05391.x

Moos, D., and Zoback, M. D. (1993). State of stress in the Long Valley caldera, California. *Geology* 21, 837–840. doi: 10.1130/0091-7613(1993)021<0837:SOSITL>2.3.CO;2

NCEDC (2014). *Northern California Earthquake Data Center.* Berkeley: UC Berkeley Seismological Laboratory. Dataset.

Otsubo, M., Yamaji, A., and Kubo, A. (2008). Determination of stresses from heterogeneous focal mechanism data: an adaptation of the multiple inverse method. *Tectonophysics* 475, 150–160. doi: 10.1016/j.tecto.2008.06.012

Pedersen, R., and Sigmundsson, F. (2004). InSAR based sill model links spatially offset areas of deformation and seismicity for the 1994 unrest episode at Eyjafjallajökull volcano, Iceland. *Geophys. Res. Lett.* 31:L14610. doi: 10.1029/2004GL020368

Plateaux, R., Béthoux, N., Bergerat, F., and Mercier de Lépinay, B. (2014). Volcano-tectonic interactions revealed by inversion of focal mechanisms: stress field insight around and beneath the Vatnajökull ice cap in Iceland. *Front. Earth Sci.* 2:9, 1–21. doi: 10.3389/feart.2014.00009

Prejean, S., Ellsworth, W., Zoback, M., and Walhouser, F. (2002). Fault structure and kinematics of the Long Valley Caldera region, California, revealed by high-accuracy earthquake hypocenters and focal mechanism stress inversion. *J. Geophys. Res.* 107:2355, doi: 10.1029/2001JB001168

Ramsay, J., and Lisle, R. (2000). *The Techniques of Modern Structural Geology, Volume 3: Applications of Continuum Mechanics in Structural Geology.* London: Academic Press.

Reasenberg, P., and Oppenheimer, D. (1985). *FPFIT, FPPLOT and FPPAGE: Fortran Computer Programs for Calculating and Displaying Earthquake Fault Plane Solutions.* Open File Report U.S. Geological Survey.

Reid, H. F. (1910). "The mechanism of the earthquake," in *The California Earth-quake of April 18, 1906. Report of the State Earthquake Investigation Commission*, Vol. 2 (Washington, DC: Carnegie Institution for Science), 1–192.

Rivera, L., and Cisternas, A. (1990). Stress tensor and fault plane solutions for a population of earthquakes. *Bull. Seismol. Soc. Am.* 80, 600–614.

Rosi, M., Santacroce, R., and Sbrana, A. (1987). *Geological Map of the Somma-Vesuvius Volcanic Complex (Scale 1:25000)*. Roma: CNR,PF Geodinamica, L Salomone.

Santacroce, R. (ed.). (1987). *Somma-Vesuvius, Vol. 114*, Rome: CNR.

Segall, P. (2013). *Volcano Deformation and Eruption Forecasting*. London: Geological Society; Special Publications.

Sorey, M. L., McConnell, V. S., and Roeloffs, E. (2003). Summary of recent research in Long Valley caldera, California. *J. Volcanol. Geother. Res.* 127, 165–173. doi: 10.1016/S0377-0273(03)00168-9

Tizzani, P., Battaglia, M., Zeni, G., Atzori, S., Berardino, P., and Lanari, R. (2009). Uplift and magma intrusion at Long Valley caldera from InSAR and gravity measurements. *Geology* 37, 63–66. doi: 10.1130/G25318A.1

Toda, S., Stein, R. S., and Sagiya, T. (2002). Evidence from the AD 2000 Izu islands earthquake swarm that stressing rate governs seismicity. *Nature* 419, 58–61. doi: 10.1038/nature00997

Umakoshi, K., Shimizu, H., and And Matsuwo, N. (2001). Volcano-tectonic seismicity at Unzen Volcano, Japan, 1985–1999. *J. Volcanol. Geother. Res.* 112, 117–131. doi: 10.1016/S0377-0273(01)00238-4

Ventura, G., and Vilardo, G. (1999). Slip tendency analysis of the Vesuvius faults: implication for the seismotectonic and volcanic hazard assessment. *Geophys. Res. Lett.* 26, 3229–3232. doi: 10.1029/1999GL005393

Viccaro, M., Zuccarello, F., Cannata, A., Palano, M., and Gresta, S. (2016). How a complex basaltic volcanic system works: constraints from integrating seismic, geodetic, and petrological data at Mount Etna volcano during the July-August 2014 eruption. *J. Geophys. Res. Solid Earth* 121, 5659–5678. doi: 10.1002/2016JB013164

Vitale, S., and Isaia, R. (2013). Fractures and faults in volcanic rocks (Campi Flegrei, southern Italy): insight into volcano-tectonic processes. *Int. J. Earth Sci.* 103, 801–819. doi: 10.1007/s00531-013-0979-0

Wallace, R. E. (1951). Geometry of shearing stress and relation to faulting. *J. Geol.* 22, 118–130. doi: 10.1086/625831

Walter, T. R., Wang, R., Zimmer, M., Grosser, H., Luhr, B., and Ratdomopurbo, A. (2007). Volcanic activity influenced by tectonic earthquakes: static and dynamic stress triggering at Mt. Merapi. *Geophys. Res. Lett.* 34:L05304. doi: 10.1029/2006gl028710

Wyss, M., Liang, B., Tanigawa, W. R., and Xiaoping, W. (1992). Comparison of orientations of stress and strain tensor based on fault plane solutions in Kaoiki, Hawaii. *J. Geophys. Res.* 97, 4769–4790. doi: 10.1029/91JB02968

Yamaji, A. (2000). The multiple inverse method: a new technique to separate stresses from heterogeneous fault-slip data. *J. Struct. Geol.* 22, 441–452. doi: 10.1016/S0191-8141(99)00163-7

Yamaji, A. (2007). *An Introduction to Tectonophysics: Theoretical Aspects of Structural Geology*. Tokyo: TERRAPUB.

Zoback, M. L. (1989). State of stress and modern deformation of the northern basin and rangeprovince. *J. Geophys. Res.* 94, 7105–7128. doi: 10.1029/JB094iB06p07105

Zoback, M. L. (1992). First and second-order patterns of stress in the lithosphere: the world stress map project. *J. Geophys. Res.* 97, 11703–11728. doi: 10.1029/92JB00132

Zollo, A., Marzocchi, W., Capuano, P., Lomax, A., and Iannaccone, G. (2002). Space and time behaviour of seismic activity and Mt. Vesuvius volcano, Southern Italy. *Bull. Seismol. Soc. Am.* 92, 625–640. doi: 10.1785/0120000287

Zuppetta, A., and Sava, A. (1991). Stress pattern at Campi Flegrei from focal mechanisms of the 1982–1984 earthquakes (Southern Italy). *J. Volcanol. Geother. Res.* 48, 127–137. doi: 10.1016/0377-0273(91)90038-2

Conflict of Interest Statement: The authors declare that the research was conducted in the absence of any commercial or financial relationships that could be construed as a potential conflict of interest.

The handling Editor declared a shared affiliation, though no other collaboration, with the authors BM, LD, and EC and states that the process nevertheless met the standards of a fair and objective review.

17

The Stoichiometry of Nutrient Release by Terrestrial Herbivores and Its Ecosystem Consequences

Judith Sitters [1,2,3]*, Elisabeth S. Bakker [2], Michiel P. Veldhuis [4], G. F. Veen [3], Harry Olde Venterink [1] and Michael J. Vanni [5]

[1] Ecology and Biodiversity, Department Biology, Vrije Universiteit Brussel, Brussels, Belgium, [2] Department of Aquatic Ecology, Netherlands Institute of Ecology (NIOO-KNAW), Wageningen, Netherlands, [3] Department of Terrestrial Ecology, Netherlands Institute of Ecology (NIOO-KNAW), Wageningen, Netherlands, [4] Faculty of Science and Engineering, Groningen Institute for Evolutionary Life Sciences, University of Groningen, Groningen, Netherlands, [5] Department of Biology, Miami University, Oxford, OH, USA

Edited by:
James Joseph Elser,
University of Montana, USA

Reviewed by:
John Pastor,
University of Minnesota, USA
Jacob Edward Allgeier,
University of California, Santa Barbara,
USA
Angelica L. Gonzalez,
Rutgers University, USA

*Correspondence:
Judith Sitters
judith.sitters@vub.be

Specialty section:
This article was submitted to
Biogeoscience,
a section of the journal
Frontiers in Earth Science

It is widely recognized that the release of nutrients by herbivores via their waste products strongly impacts nutrient availability for autotrophs. The ratios of nitrogen (N) and phosphorus (P) recycled through herbivore release (i.e., waste N:P) are mainly determined by the stoichiometric composition of the herbivore's food (food N:P) and its body nutrient content (body N:P). Waste N:P can in turn impact autotroph nutrient limitation and productivity. Herbivore-driven nutrient recycling based on stoichiometric principles is dominated by theoretical and experimental research in freshwater systems, in particular interactions between algae and invertebrate herbivores. In terrestrial ecosystems, the impact of herbivores on nutrient cycling and availability is often limited to studying carbon (C):N and C:P ratios, while the role of terrestrial herbivores in mediating N:P ratios is also likely to influence herbivore-driven nutrient recycling. In this review, we use rules and predictions on the stoichiometry of nutrient release originating from algal-based aquatic systems to identify the factors that determine the stoichiometry of nutrient release by herbivores. We then explore how these rules can be used to understand the stoichiometry of nutrient release by terrestrial herbivores, ranging from invertebrates to mammals, and its impact on plant nutrient limitation and productivity. Future studies should focus on measuring both N and P when investigating herbivore-driven nutrient recycling in terrestrial ecosystems, while also taking the form of waste product (urine or feces) and other pathways by which herbivores change nutrients into account, to be able to quantify the impact of waste stoichiometry on plant communities.

Keywords: autotroph productivity, aquatic ecosystems, C:N:P ratios, excretion, feces, herbivore-driven nutrient recycling, nitrogen, phosphorus

INTRODUCTION

Herbivores are a major component of most ecosystems, ranging in size from zooplankton to elephants. All herbivores consume and digest autotroph biomass, and release nutrients, e.g., nitrogen (N) and phosphorus (P), in wastes through excretion (urine) or egestion (feces). Nutrient release by herbivores can strongly impact nutrient availability for autotrophs in terrestrial, marine, and freshwater ecosystems (Pastor et al., 1993; McNaughton et al., 1997; Covich et al., 1999; Sirotnak and Huntly, 2000; Hunter, 2001; Vanni, 2002; Bardgett and Wardle, 2003; McIntyre et al., 2007; Cech et al., 2008; Roman and McCarthy, 2010; Metcalfe et al., 2014; Turner, 2015; Doughty et al., 2016). The ratio of N to P released (i.e., waste N:P) may be crucial for mediating ecosystem impacts of herbivore-driven nutrient recycling (Sterner, 1990; Urabe et al., 1995; Elser and Urabe, 1999). Two basic "stoichiometric rules" have been formulated, one based on how food and consumer body N:P determine waste N:P (rule 1), and the other on how waste N:P affects autotroph nutrient limitation and productivity (rule 2). Both rules allow for explicit predictions about the N:P stoichiometry of nutrient release and its ecosystem consequence (**Table 1**).

Thus far, evidence for these rules is mainly restricted to interactions between freshwater (pelagic) algae and invertebrate herbivores (Elser et al., 1988; Sterner, 1990; Sterner et al., 1992; Elser and Urabe, 1999; Sterner and Elser, 2002; Vanni, 2002), and to a lesser extent herbivorous fish (Schindler and Eby, 1997; Hood et al., 2005). However, herbivore-driven nutrient recycling also likely plays a major role in terrestrial ecosystems (Pastor et al., 1993; McNaughton et al., 1997; Hunter, 2001; Wardle et al., 2004; Metcalfe et al., 2014; Doughty et al., 2016). Indeed, the ratio of carbon (C) to nutrient (N and/or P) in plant tissues has long been recognized as an important determinant of herbivore feeding selectivity and subsequent nutrient cycling and availability in terrestrial ecosystems (Ritchie et al., 1998; Pastor et al., 2006; Bakker et al., 2009b). However, compared to aquatic systems, the role of terrestrial herbivores in mediating N:P ratios has received little attention so far. Because the ratio of N:P availability influences the type of growth limitation and the functional composition of terrestrial plant communities (Elser et al., 2007; Fujita et al., 2014), we hypothesize that the impact of terrestrial herbivores on this ratio has potentially strong ecosystem consequences.

In this review, we explore how we can apply these stoichiometric rules to terrestrial ecosystems, focusing on N:P ratios. We first explain the two rules derived from algae-invertebrate interactions in more detail. We then synthesize studies that applied these rules to terrestrial herbivores and ecosystems, identify research gaps, and suggest perspectives for future research.

RULE 1 (INDIVIDUAL CONSUMER LEVEL)—RELATIONSHIPS BETWEEN FOOD, BODY AND WASTE N:P

Rule 1 is based on mass balance and the assumption that consumers maintain elemental homeostasis in their tissues by differential release of N and P. Stoichiometry theory predicts a positive relationship between food N:P and waste N:P, assuming constant consumer body N:P (Sterner and Elser, 2002). Second, waste N:P is predicted to be negatively related to body N:P, if food N:P is constant (**Figure 1**). These predictions have been supported in lab studies using aquatic invertebrate herbivores (*Daphnia*) feeding on phytoplankton (Elser and Urabe, 1999;

TABLE 1 | Overview of the findings of the two stoichiometric rules, derived from interactions between algae and invertebrate herbivores in freshwater ecosystems, applied to terrestrial invertebrate (IV) and vertebrate (V) herbivores and ecosystems, including future research perspectives on each rule/prediction.

Stoichiometric rules	Terrestrial findings	Future research perspectives
Rule 1: To maintain homeostasis, a consumer will retain the element that most limits its growth, while releasing relatively large amounts of the element in excess relative to its needs.	See findings under prediction 1 and 2.	In general for rule 1 we are in need of studies that: • Measure both N and P in food, body and waste. • Quantify the relative degree of homeostasis. • Quantify total nutrient release, not only excretion or egestion. • Investigate the role of body size in determining body N:P both within and between invertebrate and vertebrate herbivore species.
Prediction 1: There is a positive relationship between food N:P and waste N:P (assuming body N:P is constant).	IV: Mixed findings for only N and only P (positive or unrelated), no correlation for N:P. V: True for only N and only P, no studies on N:P.	Specifically for prediction 1 of rule 1 we are in need of: • Experimental and field studies that measure both N and P in food and waste, preferably with single herbivore species (with constant body N:P).
Prediction 2: There is a negative relationship between herbivore body N:P and waste N:P (assuming food N:P is constant).	IV: No correlation for N:P. V: No studies on N:P.	Specifically for prediction 2 of rule 1 we are in need of: • Studies that measure both N and P in herbivore body and waste, while keeping food N:P equal.
Rule 2: The stoichiometry of nutrient release by herbivores can strongly affect autotroph nutrient limitation and primary production in ecosystems.	IV: True, but only on N. V: True, but often due to overall net effect of herbivores.	• Studies that focus on both N and P, and not only N. • Studies that quantify the different pathways by which herbivores impact the stoichiometry of plants. • Studies that compare the consequences of nutrient return through excretion (urine) or egestion (feces) for the stoichiometry of nutrient availability to plants. • Studies that incorporate the spatial return of N and P.

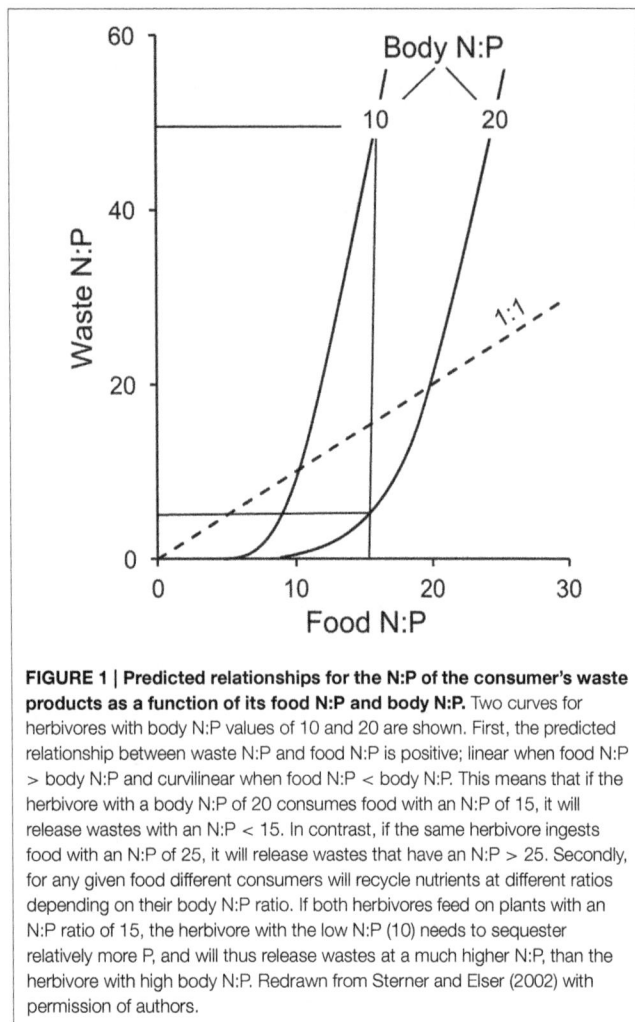

FIGURE 1 | Predicted relationships for the N:P of the consumer's waste products as a function of its food N:P and body N:P. Two curves for herbivores with body N:P values of 10 and 20 are shown. First, the predicted relationship between waste N:P and food N:P is positive; linear when food N:P > body N:P and curvilinear when food N:P < body N:P. This means that if the herbivore with a body N:P of 20 consumes food with an N:P of 15, it will release wastes with an N:P < 15. In contrast, if the same herbivore ingests food with an N:P of 25, it will release wastes that have an N:P > 25. Secondly, for any given food different consumers will recycle nutrients at different ratios depending on their body N:P ratio. If both herbivores feed on plants with an N:P ratio of 15, the herbivore with the low N:P (10) needs to sequester relatively more P, and will thus release wastes at a much higher N:P, than the herbivore with high body N:P. Redrawn from Sterner and Elser (2002) with permission of authors.

Sterner and Elser, 2002). However, support for these predictions from field data is mixed. For example, a strong negative correlation between consumer body N:P and waste N:P was found in systems with large variation among animal species in body N:P (e.g., Vanni et al., 2002; McManamay et al., 2011), while in other systems food N:P was more important in predicting waste N:P (e.g., Urabe, 1993; Torres and Vanni, 2007). Recent syntheses suggest that body size and temperature have much more influence than body nutrients on excretion rates and ratios (Allgeier et al., 2015; Vanni and McIntyre, 2016).

RULE 2 (ECOSYSTEM LEVEL)—IMPACT OF N:P RELEASE BY HERBIVORES ON PLANTS

Rule 2 states that the stoichiometry of nutrient release by herbivores strongly affects autotroph nutrient limitation and primary production (Elser and Urabe, 1999). For example, if a consumer feeding on N-limited plants (low plant N:P) releases waste products with an even lower N:P than that in plant tissue (following rule 1), the herbivore could render the

plant community even more N-limited, which can then impact competitive interactions between plants and plant community composition (Sterner, 1990; Fujita et al., 2014). However, the impact on plant communities will depend on the proportion of nutrient demand met by consumer-driven recycling. In freshwater systems, for instance, it sustains anywhere from <5 to >80% of algal nutrient uptake (Taylor et al., 2015). So far, field evidence that nutrient recycling by herbivores can shift autotroph assemblages between N- and P-limitation is scant (e.g., Sterner et al., 1992; Knoll et al., 2009), and hence a general underpinning of rule 2 is still lacking.

Even though tests of these rules are still scarce, especially under field conditions, according to mass-balance principles the relative ratios of nutrient release by herbivores should be influenced by stoichiometric balance. In the following sections, we take on the challenge of applying this stoichiometric view to herbivore-driven nutrient recycling in terrestrial ecosystems.

APPLYING RULE 1 TO TERRESTRIAL HERBIVORES

Although rule 1 predicts a positive relationship between food N:P and waste N:P (**Figure 1**, **Table 1**), most studies with terrestrial herbivores focused on single nutrients. Mixed results are found for invertebrate herbivores; food N and waste N can be positively related (lepidopterans; Kagata and Ohgushi, 2012), or unrelated (grasshopper; Zhang et al., 2014). Similarly, food P and waste P were unrelated for caterpillars (Meehan and Lindroth, 2009), but positively related for a grasshopper species (Zhang et al., 2014). The latter study also investigated the N:P ratio of food and waste simultaneously (the only terrestrial study we know of), and the ratios were not correlated. The authors suggest that the lack of relationship between food and waste nutrients is likely because mechanisms other than excretion maintain N:P homeostasis, such as pre-ingestive regulation of the nutrient balance through food selection (Zhang et al., 2014) or compensatory feeding (Meehan and Lindroth, 2009). Work on large vertebrate herbivores is much more extensive—but again, measurements were often done on either N or P and not both—and generally finds positive correlations between food and waste nutrient contents. For example, diet N and fecal N were positively correlated for rabbits (Gil-Jimenez et al., 2015), roe deer (Verheyden et al., 2011), white-tailed deer (Osborn and Ginnett, 2001), blackbuck antelope (Jhala, 1997), and several African herbivores (Wrench et al., 1997). This positive relationship was also found for P for cattle (Zhang et al., 2016) and several African herbivores (Wrench et al., 1997). Furthermore, urinary N (excretion) of large ungulates increases with plant N concentration (Hobbs, 1996). These relationships seem so reliable that fecal N and P contents are used to predict food N and P contents (Wrench et al., 1997; Verheyden et al., 2011; Gil-Jimenez et al., 2015), although corrections for the presence of indigestible forms of N, e.g., tannins, are needed (Verheyden et al., 2011; Steuer et al., 2014).

The second prediction derived from rule 1 is a negative relationship between consumer body N:P and waste N:P

(Table 1). The only published study of terrestrial herbivores found no clear relationship between body and waste N:P for an invertebrate herbivore (grasshopper; Zhang et al., 2014); we found no studies on vertebrate herbivores.

This synthesis reveals the lack of studies examining relationships between food, body, and waste N:P (and not just N or P) in terrestrial herbivores, making it impossible to draw any general conclusions. Future work should include simultaneous measurements of N and P in food, bodies, and waste (Table 1). Controlled experiments where single herbivore species (constant body N:P) feed on food sources varying in N:P will provide a good test of the first prediction. Additionally, field studies on single species experiencing seasonal changes in food quality (food N:P) will be important, and again waste products and food sources of these herbivores need to be analyzed for both N and P. To test the second prediction, studies are needed where herbivores with a range of body N:P are fed a constant food N:P, and waste N:P is measured. Importantly, the predicted stoichiometric relationships between food, body and waste N:P might be impacted by several mitigating factors. These include the degree to which animals maintain homeostasis, which is variable and perhaps related to growth rate (Hood and Sterner, 2010; Downs et al., 2016), and which mechanisms they use to regulate this homeostasis (i.e., pre- or post-ingestive). Furthermore, the type of waste product (excretion or egestion; Halvorson et al., 2015) and the relationship between body size and body N:P are important mitigating factors, which should be taken into account and discussed below.

Animals that produce two types of wastes (feces and urine) can regulate their body composition pre-assimilation (by preferentially assimilating elements in short supply and releasing excess nutrients in feces) or post-assimilation (by excreting excess metabolic nutrients in urine before they reach toxic levels in the blood). Interestingly, the N:P stoichiometry of these two forms of nutrient release differ for terrestrial vertebrate herbivores; i.e., urine contains hardly any P but a high concentration of N in soluble form, while feces contain most of the P and some N (Morse et al., 1992; Hobbs, 1996). Relative concentrations of N in urine and feces depend on forage N, whereby herbivores that consume plants of high N (e.g., ungulate grazers consuming green grasses) return N to the soil mainly in the form of urine, while herbivores consuming plants of low N (e.g., ungulate browsers consuming tree twigs) need to extract as much N as possible and mainly produce feces of very low N (Pastor et al., 2006). However, very few studies quantify total nutrient release, instead of only excretion (often the case for aquatic animals) or egestion (often the case for terrestrial vertebrates) to test predictions of stoichiometry theory. This needs to be addressed in future studies, as the theory is based on mass-balance, and tests must therefore include all fluxes mediated by animal physiology (Table 1).

Generally, differences in body N:P are driven by patterns of investment in P-rich materials such as RNA and bone (Gillooly et al., 2005). Investments in RNA decrease significantly with body size (small organisms generally have higher growth rates), suggesting an increase in body N:P with increasing body size for invertebrates (Sterner and Elser, 2002; Back and King,

2013). However, more P is sequestered into supportive tissue like bones, suggesting that for vertebrates body N:P decreases with body size (Sterner and Elser, 2002) since skeleton mass scales allometrically (more than proportionally) with body mass (Anderson et al., 1979; Prange et al., 1979). This uncovers an important difference between aquatic and terrestrial systems, where terrestrial herbivores, especially larger individuals, need to invest more in P-rich structural tissue to counterbalance gravity. Hence, more studies investigating the role of body size in determining body N:P both within and between invertebrate and vertebrate herbivore species are needed (Table 1).

APPLYING RULE 2 TO TERRESTRIAL ECOSYSTEMS

Rule 2 states that the stoichiometry of nutrient release by herbivores can strongly affect autotroph nutrient limitation and primary production (Elser and Urabe, 1999). However, in terrestrial ecosystems it is hard to isolate the effect of herbivore-driven nutrient recycling, as the "net effect" of herbivores on nutrient cycling depends not only on direct effects of nutrient release through waste products, but also on indirect effects through modification of plant litter quantity and quality, and in the case of vertebrates, by alteration of soil physical properties through trampling (Ritchie et al., 1998; Belovsky and Slade, 2000; Hunter, 2001; Bardgett and Wardle, 2003; Schrama et al., 2013). Therefore, most empirical studies addressing how terrestrial herbivores shift plant assemblages between N- or P-limitation examined the "net effect" of herbivores (e.g., Carline et al., 2005; Frank, 2008; Zhang et al., 2011; Bai et al., 2012; Nitschke et al., 2015; Sitters et al., 2017) and not on the effects of nutrient release *per se.*

Many studies on herbivore-driven nutrient recycling in terrestrial ecosystems have focused on N, both for invertebrates (e.g., Seastedt and Crossley, 1984; Lovett and Ruesink, 1995; Belovsky and Slade, 2000; Reynolds and Hunter, 2001; Hunter et al., 2003; Metcalfe et al., 2014) and vertebrates (e.g., Pastor et al., 1988, 1993, 2006; McNaughton et al., 1988; Hobbs et al., 1991; Frank and McNaughton, 1993; Frank and Evans, 1997; McNaughton et al., 1997; Ritchie et al., 1998; Sirotnak and Huntly, 2000; Olofsson et al., 2001; Stark et al., 2003; Fornara and Du Toit, 2008). For invertebrate herbivores, the general view is that they speed up nutrient cycling in terrestrial systems by changing litter quantity and quality, modifying the nutrient content of throughfall, and releasing easily-available nutrients in frass and cadavers (Hunter, 2001). The direction of the impact of vertebrate herbivores on N cycling has traditionally been considered to depend on system fertility and corresponding plant N content; vertebrates have a positive effect on N availability and primary production in systems of high fertility, and a negative effect in low fertility systems (Hobbs, 1996; Bardgett and Wardle, 2003; Pastor et al., 2006). This is partly based on the proportion of N released, which is higher and mainly through urine when the nutrient content of plants is higher (Hobbs, 1996), but also on changes in plant litter quality, as herbivores feeding in systems with low plant N facilitate a shift toward litter of low N by selectively consuming high-N plant tissue (Bardgett and Wardle, 2003).

Stoichiometry theory has challenged the traditional views for impacts of herbivores on nutrient cycling. Modeling results demonstrate that if herbivores promote microbial C-limitation through vegetation consumption and respiration, they will have a positive effect on N availability in sites with low plant N by decreasing microbial immobilization rates, and a positive effect in sites with high plant N by decreasing mineralization rates (Cherif and Loreau, 2013). For vertebrate herbivores these results are party supported by field data (Bakker et al., 2009a; Sitters et al., 2017). Also, for invertebrates, labile C in excreta can result in N immobilization and lower availability (Lovett and Ruesink, 1995). These studies again show the need for a more integrative framework to understand and quantify the different pathways by which vertebrate (Sitters and Olde Venterink, 2015) and invertebrate herbivores (Hunter, 2001) impact nutrient cycling and availability to plants.

To predict how the stoichiometry of nutrient release by herbivores affects autotroph nutrient limitation and primary production in terrestrial systems, we must expand studies on the role of P, because N- and P-limitation are both prevalent in terrestrial ecosystems (Elser et al., 2007). The effect of herbivores on N- and P-limitation depends first on the form in which N and P are returned to the soil (urine or feces). N in urine is soluble and directly available to plants, while feces contain a substantial amount of organic matter, which needs to be decomposed and mineralized to render the N and P available to plants (Hobbs, 1996). Furthermore, N in feces and urine is subject to a significant loss from the system via ammonia volatilization and leaching (Ruess and McNaughton, 1988; Frank and Zhang, 1997; Augustine, 2003), suggesting that terrestrial herbivores (that produce both types of waste products) may drive ecosystems to N-limitation through nutrient release (e.g., Cech et al., 2008). Very little data however exist, comparing the consequences of nutrient return through urine or feces for the stoichiometry of nutrient availability to plants.

Additionally, nutrient release by herbivores strongly increases the spatial heterogeneity of N:P availability across the landscape. Terrestrial herbivores typically do not graze and excrete randomly, but are attracted to landscape features, such as nutrient-rich areas with high food quality, resulting in a net import of nutrients and creating nutrient hotspots in the landscape. At the same time, large parts of the landscape with poorer quality vegetation experience a net removal of nutrients (McNaughton et al., 1997; Augustine et al., 2003; van der Waal et al., 2011). Water bodies may also induce spatial patterns; semi-aquatic herbivores such as hippopotamus, beaver or water birds can transport nutrients across ecosystem boundaries and thus strongly impact nutrient redistribution (Sitters et al., 2015; Subalusky et al., 2015; Bakker et al., 2016). Furthermore, social behavior may affect nutrient release. When herbivores defecate in common latrines, they concentrate nutrients in the landscape (e.g., rhinos, rabbits, horses), which function as hotspots of nutrients, possibly with a low N:P ratio, though data are scarce and not fully consistent regarding the effect on soil P (see Edwards and Hollis, 1982; Willott et al., 2000; Jewell et al., 2007). In this respect, there are similarities and differences between terrestrial and aquatic habitats. In both, animals can mediate a net translocation of nutrients across habitats and ecosystems (Vanni et al., 2001; Flecker et al., 2010; Ebel et al., 2015; Sitters et al., 2015). However, in aquatic habitats excreted nutrients may easily mix in the water, whereas in terrestrial habitats patches of released nutrients are much more spatially disconnected; this suggests that terrestrial animals may induce spatial variation in nutrient supply and stoichiometry more so than aquatic animals.

CONCLUDING REMARKS

The stoichiometric view of herbivore-driven nutrient recycling in terrestrial ecosystems has not yet received the attention it deserves. We were unable to find firm evidence for rule 1, as most studies investigating relationships between food, herbivore bodies, and wastes focused on single nutrients. At the same time, many studies consider the "net" effect of herbivores on nutrient cycling, making it impossible to determine the impact of waste stoichiometry on plant communities *per se* (rule 2). We therefore suggest several perspectives on future research, to increase our understanding of the stoichiometry of nutrient release by terrestrial herbivores, ranging from invertebrates to mammals, and its impact on ecosystem stoichiometry, plant nutrient limitation, and productivity (**Table 1**).

AUTHOR CONTRIBUTIONS

All authors listed, have made substantial, direct and intellectual contribution to the work, and approved it for publication.

FUNDING

JS was financially supported by a grant of the Research Foundation Flanders (FWO), grant 12N2615N. MPV has been financially supported by the AfricanBioServices project which received funding from the European Union's Horizon 2020 research and innovation programme under grant agreement No 641918. MJV was supported by US National Science Foundation grants 0918993 and 1255159. Additional funding was provided by Strategic Resources of the Netherlands Institute of Ecology (NIOO-KNAW); this is publication number 6272.

REFERENCES

Allgeier, J. E., Wenger, S. J., Rosemond, A. D., Schindler, D. E., and Layman, C. A. (2015). Metabolic theory and taxonomic identity predict nutrient recycling in a diverse food web. *Proc. Natl. Acad. Sci. U.S.A.* 112, 2640–2647. doi: 10.1073/pnas.1420819112

Anderson, J. F., Rahn, H., and Prange, H. D. (1979). Scaling of supportive tissue mass. *Q. Rev. Biol.* 54, 139–148. doi: 10.1086/411153

Augustine, D. J. (2003). Long-term, livestock-mediated redistribution of nitrogen and phosphorus in an East African savanna. *J. Appl. Ecol.* 40, 137–149. doi: 10.1046/j.1365-2664.2003.00778.x

Augustine, D. J., McNaughton, S. J., and Frank, D. A. (2003). Feedbacks between soil nutrients and large herbivores in a managed savanna ecosystem. *Ecol. Appl.* 13, 1325–1337. doi: 10.1890/02-5283

Back, J. A., and King, R. S. (2013). Sex and size matter: ontogenetic patterns of nutrient content of aquatic insects. *Freshwater Science* 32, 837–848. doi: 10.1899/12-181.1

Bai, Y. F., Wu, J. G., Clark, C. M., Pan, Q. M., Zhang, L. X., Chen, S. P., et al. (2012). Grazing alters ecosystem functioning and C:N:P stoichiometry of grasslands along a regional precipitation gradient. *J. Appl. Ecol.* 49, 1204–1215. doi: 10.1111/j.1365-2664.2012.02205.x

Bakker, E. S., Knops, J. M. H., Milchunas, D. G., Ritchie, M. E., and Olff, H. (2009a). Cross-site comparison of herbivore impact on nitrogen availability in grasslands: the role of plant nitrogen concentration. *Oikos* 118, 1613–1622. doi: 10.1111/j.1600-0706.2009.17199.x

Bakker, E. S., Olff, H., and Gleichman, J. M. (2009b). Contrasting effects of large herbivore grazing on smaller herbivores. *Basic Appl. Ecol.* 10, 141–150. doi: 10.1016/j.baae.2007.10.009

Bakker, E. S., Pages, J. F., Arthur, R., and Alcoverro, T. (2016). Assessing the role of large herbivores in the structuring and functioning of freshwater and marine angiosperm ecosystems. *Ecography* 39, 162–179. doi: 10.1111/ecog.01651

Bardgett, R. D., and Wardle, D. A. (2003). Herbivore-mediated linkages between aboveground and belowground communities. *Ecology* 84, 2258–2268. doi: 10.1890/02-0274

Belovsky, G. E., and Slade, J. B. (2000). Insect herbivory accelerates nutrient cycling and increases plant production. *Proc. Natl. Acad. Sci. U.S.A.* 97, 14412–14417. doi: 10.1073/pnas.250483797

Carline, K. A., Jones, H. E., and Bardgett, R. D. (2005). Large herbivores affect the stoichiometry of nutrients in a regenerating woodland ecosystem. *Oikos* 110, 453–460. doi: 10.1111/j.0030-1299.2005.13550.x

Cech, P. G., Kuster, T., Edwards, P. J., and Olde Venterink, H. (2008). Effects of herbivory, fire and N2-fixation on nutrient limitation in a humid African savanna. *Ecosystems* 11, 991–1004. doi: 10.1007/s10021-008-9175-7

Cherif, M., and Loreau, M. (2013). Plant–herbivore–decomposer stoichiometric mismatches and nutrient cycling in ecosystems. *Proc. R. Soc. B. Biol. Sci.* 280:20122453. doi: 10.1098/rspb.2012.2453

Covich, A. P., Palmer, M. A., and Crowl, T. A. (1999). The role of benthic invertebrate species in freshwater ecosystems - Zoobenthic species influence energy flows and nutrient cycling. *Bioscience* 49, 119–127. doi: 10.2307/1313537

Doughty, C. E., Roman, J., Faurby, S., Wolf, A., Haque, A., Bakker, E. S., et al. (2016). Global nutrient transport in a world of giants. *Proc. Natl. Acad. Sci. U.S.A.* 113, 868–873. doi: 10.1073/pnas.1502549112

Downs, K. N., Hayes, N. M., Rock, A. M., Vanni, M. J., and Gonzalez, M. J. (2016). Light and nutrient supply mediate intraspecific variation in the nutrient stoichiometry of juvenile fish. *Ecosphere* 7:e01452. doi: 10.1002/ecs2.1452

Ebel, J. D., Leroux, S. J., Robertson, M. J., and Dempson, J. B. (2015). Ontogenetic differences in Atlantic salmon phosphorus concentration and its implications for cross ecosystem fluxes. *Ecosphere* 6, 1–18. doi: 10.1890/es14-00516.1

Edwards, P. J., and Hollis, S. (1982). The distribution of excreta on new forest grassland used by cattle, ponies and deer. *J. Appl. Ecol.* 19, 953–964.

Elser, J. J., Bracken, M. E. S., Cleland, E. E., Gruner, D. S., Harpole, W. S., Hillebrand, H., et al. (2007). Global analysis of nitrogen and phosphorus limitation of primary producers in freshwater, marine and terrestrial ecosystems. *Ecol. Lett.* 10, 1135–1142. doi: 10.1111/j.1461-0248.2007.01113.x

Elser, J. J., Elser, M. M., Mackay, N. A., and Carpenter, S. R. (1988). Zooplankton-mediated transitions between N-limited and P-limited algal growth. *Limnol. Oceanogr.* 33, 1–14.

Elser, J. J., and Urabe, J. (1999). The stoichiometry of consumer-driven nutrient recycling: theory, observations, and consequences. *Ecology* 80, 735–751. doi: 10.1890/0012-9658(1999)080[0735:tsocdn]2.0.co;2

Flecker, A. S., McIntyre, P. B., Moore, J. W., Anderson, J. T., Taylor, B. W., and Hall, R. O. (2010). Migratory fishes as material and process subsidies in riverine ecosystems. *Am. Fish. Soc. Symp.* 73, 559–592.

Fornara, D. A., and Du Toit, J. T. (2008). Browsing-induced effects on leaf litter quality and decomposition in a southern african savanna. *Ecosystems* 11, 238–249. doi: 10.1007/s10021-007-9119-7

Frank, D. A. (2008). Ungulate and topographic control of nitrogen: phosphorus stoichiometry in a temperate grassland; soils, plants and mineralization rates. *Oikos* 117, 591–601. doi: 10.1111/j.2007.0030-1299.16220.x

Frank, D. A., and Evans, R. D. (1997). Effects of native grazers on grassland N cycling in Yellowstone National Park. *Ecology* 78, 2238–2248. doi: 10.1890/0012-9658(1997)078[2238:eongog]2.0.co;2

Frank, D. A., and McNaughton, S. J. (1993). Evidence for the promotion of aboveground grassland production by native large herbivores in Yellowstone National Park. *Oecologia* 96, 157–161.

Frank, D. A., and Zhang, Y. M. (1997). Ammonia volatilization from a seasonally and spatially variable grazed grassland: Yellowstone National Park. *Biogeochemistry* 36, 189–203. doi: 10.1023/a:100570512 1160

Fujita, Y., Olde Venterink, H., van Bodegom, P. M., Douma, J. C., Heil, G. W., Holzel, N., et al. (2014). Low investment in sexual reproduction threatens plants adapted to phosphorus limitation. *Nature* 505, 82–86. doi: 10.1038/nature12733

Gil-Jimenez, E., Villamuelas, M., Serrano, E., Delibes, M., and Fernandez, N. (2015). Fecal nitrogen concentration as a nutritional quality indicator for european rabbit ecological studies. *PLoS ONE* 10:e0125190. doi: 10.1371/journal.pone.0125190

Gillooly, J. F., Allen, A. P., Brown, J. H., Elser, J. J., del Rio, C. M., Savage, V. M., et al. (2005). The metabolic basis of whole-organism RNA and phosphorus content. *Proc. Natl. Acad. Sci. U.S.A.* 102, 11923–11927. doi: 10.1073/pnas.0504756102

Halvorson, H. M., Fuller, C., Entrekin, S. A., and Evans-White, M. A. (2015). Dietary influences on production, stoichiometry and decomposition of particulate wastes from shredders. *Freshw. Biol.* 60, 466–478. doi: 10.1111/fwb.12462

Hobbs, N. T. (1996). Modification of ecosystems by ungulates. *J. Wildl. Manag.* 60, 695–713. doi: 10.2307/3802368

Hobbs, N. T., Schimel, D. S., Owensby, C. E., and Ojima, D. S. (1991). Fire and grazing in the tallgrass prairie - contingent effects on nitrogen budgets. *Ecology* 72, 1374–1382. doi: 10.2307/1941109

Hood, J. M., and Sterner, R. W. (2010). Diet mixing: do animals integrate growth or resources across temporal heterogeneity? *Am. Nat.* 176, 651–663. doi: 10.1086/656489

Hood, J. M., Vanni, M. J., and Flecker, A. S. (2005). Nutrient recycling by two phosphorus-rich grazing catfish: the potential for phosphorus-limitation of fish growth. *Oecologia* 146, 247–257. doi: 10.1007/s00442-005-0202-5

Hunter, M. D. (2001). Insect population dynamics meets ecosystem ecology: effects of herbivory on soil nutrient dynamics. *Agric. For. Entomol.* 3, 77–84. doi: 10.1046/j.1461-9563.2001.00100.x

Hunter, M. D., Linnen, C. R., and Reynolds, B. C. (2003). Effects of endemic densities of canopy herbivores on nutrient dynamics along a gradient in elevation in the southern Appalachians. *Pedobiologia* 47, 231–244. doi: 10.1078/0031-4056-00187

Jewell, P. L., Kauferle, D., Guesewell, S., Berry, N. R., Kreuzer, M., and Edwards, P. J. (2007). Redistribution of phosphorus by mountain pasture in cattle on a traditional the Alps. *Agric. Ecosyst. Environ.* 122, 377–386. doi: 10.1016/j.agee.2007.02.012

Jhala, Y. V. (1997). Seasonal effects on the nutritional ecology of blackbuck *Antelope cervicapra*. *J. Appl. Ecol.* 34, 1348–1358. doi: 10.2307/2405252

Kagata, H., and Ohgushi, T. (2012). Carbon to nitrogen excretion ratio in lepidopteran larvae: relative importance of ecological stoichiometry and metabolic scaling. *Oikos* 121, 1869–1877. doi: 10.1111/j.1600-0706.2012.20274.x

Knoll, L. B., McIntyre, P. B., Vanni, M. J., and Flecker, A. S. (2009). Feedbacks of consumer nutrient recycling on producer biomass and stoichiometry: separating direct and indirect effects. *Oikos* 118, 1732–1742. doi: 10.1111/j.1600-0706.2009.17367.x

Lovett, G. M., and Ruesink, A. E. (1995). Carbon and nitrogen mineralization from decomposing gypsy-moth frass. *Oecologia* 104, 133–138. doi: 10.1007/bf00328577

McIntyre, P. B., Jones, L. E., Flecker, A. S., and Vanni, M. J. (2007). Fish extinctions alter nutrient recycling in tropical freshwaters. *Proc. Natl. Acad. Sci. U.S.A.* 104, 4461–4466. doi: 10.1073/pnas.0608148104

McManamay, R. A., Webster, J. R., Valett, H. M., and Dolloff, C. A. (2011). Does diet influence consumer nutrient cycling? Macroinvertebrate and fish excretion in streams. *J. North Am. Benthol. Soc.* 30, 84–102. doi: 10.1899/09-152.1

McNaughton, S. J., Banyikwa, F. F., and McNaughton, M. M. (1997). Promotion of the cycling of diet-enhancing nutrients by African grazers. *Science* 278, 1798–1800. doi: 10.1126/science.278.5344.1798

McNaughton, S. J., Ruess, R. W., and Seagle, S. W. (1988). Large mammals and process dynamics in African ecosystems. *Bioscience* 38, 794–800. doi: 10.2307/1310789

Meehan, T. D., and Lindroth, R. L. (2009). Scaling of individual phosphorus flux by caterpillars of the whitemarked tussock moth, Orygia leucostigma. *J. Insect Sci.* 9:42. doi: 10.1673/031.009.4201

Metcalfe, D. B., Asner, G. P., Martin, R. E., Espejo, J. E. S., Huasco, W. H., Amezquita, F. F. F., et al. (2014). Herbivory makes major contributions to ecosystem carbon and nutrient cycling in tropical forests. *Ecol. Lett.* 17, 324–332. doi: 10.1111/ele.12233

Morse, D., Head, H. H., Wilcox, C. J., Vanhorn, H. H., Hissem, C. D., and Harris, B. (1992). Effects of concentration of dietary phosphorus on amount and route of excretion. *J. Dairy Sci.* 75, 3039–3049.

Nitschke, N., Wiesner, K., Hilke, I., Eisenhauer, N., Oelmann, Y., and Weisser, W. W. (2015). Increase of fast nutrient cycling in grassland microcosms through insect herbivory depends on plant functional composition and species diversity. *Oikos* 124, 161–173. doi: 10.1111/oik.01476

Olofsson, J., Kitti, H., Rautiainen, P., Stark, S., and Oksanen, L. (2001). Effects of summer grazing by reindeer on composition of vegetation, productivity and nitrogen cycling. *Ecography* 24, 13–24. doi: 10.1034/j.1600-0587.2001.240103.x

Osborn, R. G., and Ginnett, T. F. (2001). Fecal nitrogen and 2,6-diaminopimelic acid as indices to dietary nitrogen in white-tailed deer. *Wildl. Soc. Bull.* 29, 1131–1139. doi: 10.2307/3784136

Pastor, J., Cohen, Y., and Hobbs, N. T. (2006). "The role of large herbivores in ecosystem nutrient cycles," in *Large Herbivore Ecology, Ecosystem Dynamics and Conservation,* eds R. Dannell, R. Bergstrom, P. Duncan, and J. Pastor (Cambridge: Cambridge University Press), 289–319.

Pastor, J., Dewey, B., Naiman, R. J., McInnes, P. F., and Cohen, Y. (1993). Moose browsing and soil fertility in the boreal forests of Isle Royale National Park. *Ecology* 74, 467–480. doi: 10.2307/1939308

Pastor, J., Naiman, R. J., Dewey, B., and McInnes, P. (1988). Moose, microbes, and the boreal forest. *Bioscience* 38, 770–777. doi: 10.2307/1310786

Prange, H. D., Anderson, J. F., and Rahn, H. (1979). Scaling of skeletal mass to body-mass in birds and mammals. *Am. Nat.* 113, 103–122. doi: 10.1086/283367

Reynolds, B. C., and Hunter, M. D. (2001). Responses of soil respiration, soil nutrients, and litter decomposition to inputs from canopy herbivores. *Soil Biol. Biochem.* 33, 1641–1652. doi: 10.1016/s0038-0717(01)00085-2

Ritchie, M. E., Tilman, D., and Knops, J. M. H. (1998). Herbivore effects on plant and nitrogen dynamics in oak savanna. *Ecology* 79, 165–177. doi: 10.2307/176872

Roman, J., and McCarthy, J. J. (2010). The whale pump: marine mammals enhance primary productivity in a Coastal Basin. *PLoS ONE* 5:e13255. doi: 10.1371/journal.pone.0013255

Ruess, R. W., and McNaughton, S. J. (1988). Ammonia volatilization and the effects of large grazing mammals on nutrient loss from East African grasslands. *Oecologia* 77, 382–386. doi: 10.1007/bf00378047

Schindler, D. E., and Eby, L. A. (1997). Stoichiometry of fishes and their prey: implications for nutrient recycling. *Ecology* 78, 1816–1831.

Schrama, M., Veen, G. F., Bakker, E. S., Ruifrok, J. L., Bakker, J. P., and Olff, H. (2013). An integrated perspective to explain nitrogen mineralization in grazed ecosystems. *Perspect. Plant Ecol. Evolut. Syst.* 15, 32–44. doi: 10.1016/j.ppees.2012.12.001

Seastedt, T. R., and Crossley, D. A. (1984). The influence of arthropods on ecosystems. *Bioscience* 34, 157–161. doi: 10.2307/1309750

Sirotnak, J. M., and Huntly, N. J. (2000). Direct and indirect effects of herbivores on nitrogen dynamics: voles in riparian areas. *Ecology* 81, 78–87. doi: 10.1890/0012-9658(2000)081[0078:DAIEOH]2.0.CO;2

Sitters, J., Atkinson, C. L., Guelzow, N., Kelly, P., and Sullivan, L. L. (2015). Spatial stoichiometry: cross-ecosystem material flows and their impact on recipient ecosystems and organisms. *Oikos* 124, 920–930. doi: 10.1111/oik.02392

Sitters, J., te Beest, M., Cherif, M., Giesler, R., and Olofsson, J. (2017). Interactive effects between reindeer and habitat fertility drive soil nutrient availabilities in arctic tundra. *Ecosystems* 1–12. doi: 10.1007/s10021-017-0108-1

Sitters, J., and Olde Venterink, H. (2015). The need for a novel integrative theory on feedbacks between herbivores, plants and soil nutrient cycling. *Plant Soil* 396, 421–426. doi: 10.1007/s11104-015-2679-y

Stark, S., Tuomi, J., Strommer, R., and Helle, T. (2003). Non-parallel changes in soil microbial carbon and nitrogen dynamics due to reindeer grazing in northern boreal forests. *Ecography* 26, 51–59. doi: 10.1034/j.1600-0587.2003.03336.x

Sterner, R. W. (1990). The ratio of nitrogen to phosphorus resupplied by herbivores–zooplankton and the algal competitive arena. *Am. Nat.* 136, 209–229. doi: 10.1086/285092

Sterner, R. W., and Elser, J. J. (2002). *Ecological Stoichiometry.* Princeton, NJ: Princeton University Press.

Sterner, R. W., Elser, J. J., and Hessen, D. O. (1992). Stoichiometric relationships among producers, consumers and nutrient cycling in pelagic ecosystems. *Biogeochemistry* 17, 49–67.

Steuer, P., Sudekum, K. H., Tutken, T., Muller, D. W. H., Kaandorp, J., Bucher, M., et al. (2014). Does body mass convey a digestive advantage for large herbivores? *Funct. Ecol.* 28, 1127–1134. doi: 10.1111/1365-2435.12275

Subalusky, A. L., Dutton, C. L., Rosi-Marshall, E. J., and Post, D. M. (2015). The hippopotamus conveyor belt: vectors of carbon and nutrients from terrestrial grasslands to aquatic systems in sub-Saharan Africa. *Freshw. Biol.* 60, 512–525. doi: 10.1111/fwb.12474

Taylor, J. M., Vanni, M. J., and Flecker, A. S. (2015). "Top-down and bottom-up interactions in freshwater ecosystems: emerging complexities," in *Trophic Ecology: Bottom-Up and Top-Down Interactions Across Aquatic and Terrestrial Systems,* eds T. C. Hanley and K. J. La Pierre. (Cambridge: Cambridge University Press), 55–85.

Torres, L. E., and Vanni, M. J. (2007). Stoichiometry of nutrient excretion by fish: interspecific variation in a hypereutrophic lake. *Oikos* 116, 259–270. doi: 10.1111/j.2006.0030-1299.15268.x

Turner, J. T. (2015). Zooplankton fecal pellets, marine snow, phytodetritus and the ocean's biological pump. *Prog. Oceanogr.* 130, 205–248. doi: 10.1016/j.pocean.2014.08.005

Urabe, J. (1993). N-cycling and P-cycling coupled by grazer activities-food quality and nutrient release by zooplankton. *Ecology* 74, 2337–2350. doi: 10.2307/1939586

Urabe, J., Nakanishi, M., and Kawabata, K. (1995). Contribution of metazoan plankton to the cycling of nitrogen and phosphorus in Lake Biwa. *Limnol. Oceanogr.* 40, 232–241.

van der Waal, C., Kool, A., Meijer, S. S., Kohi, E., Heitkönig, I. M. A., de Boer, W. F., et al. (2011). Large herbivores may alter vegetation structure of semi-arid savannas through soil nutrient mediation. *Oecologia* 165, 1095–1107. doi: 10.1007/s00442-010-1899-3

Vanni, M. J. (2002). Nutrient cycling by animals in freshwater ecosystems. *Annu. Rev. Ecol. Syst.* 33, 341–370. doi: 10.1146/annurev.ecolysis.33.010802.150519

Vanni, M. J., Flecker, A. S., Hood, J. M., and Headworth, J. L. (2002). Stoichiometry of nutrient recycling by vertebrates in a tropical stream: linking species identity and ecosystem processes. *Ecol. Lett.* 5, 285–293. doi: 10.1046/j.1461-0248.2002.00314.x

Vanni, M. J., and McIntyre, P. B. (2016). Predicting nutrient excretion of aquatic animals with metabolic ecology and ecological stoichiometry: a global synthesis. *Ecology* 97, 3460–3471. doi: 10.1002/ecy.1582

Vanni, M. J., Renwick, W. H., Headworth, J. L., Auch, J. D., and Schaus, M. H. (2001). Dissolved and particulate nutrient flux from three adjacent agricultural watersheds: a five-year study. *Biogeochemistry* 54, 85–114. doi: 10.1023/a:1010681229460

Verheyden, H., Aubry, L., Merlet, J., Petibon, P., Chauveau-Duriot, B., Guillon, N., et al. (2011). Faecal nitrogen, an index of diet quality in roe deer *Capreolus capreolus*? *Wildlife Biol.* 17, 166–175. doi: 10.2981/10-111

Wardle, D. A., Bardgett, R. D., Klironomos, J. N., Setala, H., van der Putten, W. H., and Wall, D. H. (2004). Ecological linkages between aboveground and belowground biota. *Science* 304, 1629–1633. doi: 10.1126/Science.1094875

Willott, S. J., Miller, A. J., Incoll, L. D., and Compton, S. G. (2000). The contribution of rabbits (*Oryctolagus cuniculus* L.) to soil fertility in semi-arid Spain. *Biol. Fertil. Soils* 31, 379–384. doi: 10.1007/s003749900183

Wrench, J. M., Meissner, H. H., and Grant, C. C. (1997). Assessing diet quality of African ungulates from faecal analyses: the effect of forage quality, intake and herbivore species. *Koedoe* 40, 125–136. doi: 10.4102/koedoe.v40i1.268

Zhang, B., Wang, C., Wei, Z. H., Sun, H. Z., Xu, G. Z., Liu, J. X., et al. (2016). The effects of dietary phosphorus on the growth performance and phosphorus excretion of dairy heifers. Asian-*Australas. J. Anim. Sci.* 29, 960–964. doi: 10.5713/ajas.15.0548

Zhang, G. M., Han, X. G., and Elser, J. J. (2011). Rapid top-down regulation of plant C:N:P stoichiometry by grasshoppers in an Inner Mongolia grassland ecosystem. *Oecologia* 166, 253–264. doi: 10.1007/s00442-011-1904-5

Zhang, Z. J., Elser, J. J., Cease, A. J., Zhang, X. M., Yu, Q., Han, X. G., et al. (2014). Grasshoppers regulate N:P stoichiometric homeostasis by changing phosphorus contents in their frass. *PLoS ONE* 9:e103697. doi: 10.1371/journal.pone.0103697

Conflict of Interest Statement: The authors declare that the research was conducted in the absence of any commercial or financial relationships that could be construed as a potential conflict of interest.

PERMISSIONS

All chapters in this book were first published in FEART, by Frontiers; hereby published with permission under the Creative Commons Attribution License or equivalent. Every chapter published in this book has been scrutinized by our experts. Their significance has been extensively debated. The topics covered herein carry significant findings which will fuel the growth of the discipline. They may even be implemented as practical applications or may be referred to as a beginning point for another development.

The contributors of this book come from diverse backgrounds, making this book a truly international effort. This book will bring forth new frontiers with its revolutionizing research information and detailed analysis of the nascent developments around the world.

We would like to thank all the contributing authors for lending their expertise to make the book truly unique. They have played a crucial role in the development of this book. Without their invaluable contributions this book wouldn't have been possible. They have made vital efforts to compile up to date information on the varied aspects of this subject to make this book a valuable addition to the collection of many professionals and students.

This book was conceptualized with the vision of imparting up-to-date information and advanced data in this field. To ensure the same, a matchless editorial board was set up. Every individual on the board went through rigorous rounds of assessment to prove their worth. After which they invested a large part of their time researching and compiling the most relevant data for our readers.

The editorial board has been involved in producing this book since its inception. They have spent rigorous hours researching and exploring the diverse topics which have resulted in the successful publishing of this book. They have passed on their knowledge of decades through this book. To expedite this challenging task, the publisher supported the team at every step. A small team of assistant editors was also appointed to further simplify the editing procedure and attain best results for the readers.

Apart from the editorial board, the designing team has also invested a significant amount of their time in understanding the subject and creating the most relevant covers. They scrutinized every image to scout for the most suitable representation of the subject and create an appropriate cover for the book.

The publishing team has been an ardent support to the editorial, designing and production team. Their endless efforts to recruit the best for this project, has resulted in the accomplishment of this book. They are a veteran in the field of academics and their pool of knowledge is as vast as their experience in printing. Their expertise and guidance has proved useful at every step. Their uncompromising quality standards have made this book an exceptional effort. Their encouragement from time to time has been an inspiration for everyone.

The publisher and the editorial board hope that this book will prove to be a valuable piece of knowledge for researchers, students, practitioners and scholars across the globe.

LIST OF CONTRIBUTORS

Georg F. Zellmer and **Anja Moebis**
Volcanic Risk Solutions, Institute of Agriculture and Environment, Massey University, Palmerston North, New Zealand

Naoya Sakamoto, Nozomi Matsuda and Hisayoshi Yurimoto
Isotope Imaging Laboratory, Department of Natural History Sciences, Hokkaido University, Sapporo, Japan

Yoshiyuki Iizuka
Institute of Earth Sciences, Academia Sinica, Taipei, Taiwan

Shyh-Lung Hwang
Department of Materials Science and Engineering, National Dong Hwa University, Hualien, Taiwan

Rivaldo R. Silva
Laboratório de Paleontologia da Universidade Luterana do Brasil, Torres, Brazil

Jorge Ferigolo
Museu de Ciências Naturais da Fundação Zoobotânica do Rio Grande do Sul, Porto Alegre, Brazil

Piotr Bajdek
Independent Researcher, Aleja Najs´wie tszej Maryi Panny 20/20A, Cze stochowa, Poland

Graciela Piñeiro
Facultad de Ciencias, Instituto de Ciencias Geológicas, Montevideo, Uruguay

Ellyn M. Enderlin and Gordon S. Hamilton
Climate Change Institute, University of Maine, Orono, ME, USA
School of Earth and Climate Sciences, University of Maine, Orono, ME, USA

Shad O' Neel
Alaska Science Center, US Geological Survey, Anchorage, AK, USA

Timothy C. Bartholomaus
Department of Geological Sciences, University of Idaho , Moscow, ID, USA

Mathieu Morlighem
Department of Earth System Science, University of California Irvine, Irvine, CA, USA

John W. Holt
Institute for Geophysics, University of Texas at Austin, Austin, TX, USA

Christian Vögeli and Mathias Bavay
Research Unit Snow and Permafrost, WSL Institute for Snow and Avalanche Research SLF, Davos, Switzerland

Michael Lehning and Nander Wever
Research Unit Snow and Permafrost, WSL Institute for Snow and Avalanche Research SLF, Davos, Switzerland
Civil and Environmental Engineering, CRYOS School of Architecture, École Polytechnique Fédéralede Lausanne, Lausanne, Switzerland

Andrii Shelestov, Mykola Lavreniuk and Nataliia Kussul
Department of Space Information Technologies and Systems, Space Research Institute (NASU-SSAU), Kyiv, Ukraine
Department of Information Security, National Technical University of Ukraine "Igor Sikorsky Kyiv Polytechnic Institute," Kyiv, Ukraine

Alexei Novikov
Department of Information Security, National Technical University of Ukraine "Igor Sikorsky Kyiv Polytechnic Institute," Kyiv, Ukraine

Sergii Skakun
Department of Geographical Sciences, University of Maryland, College Park, MD, USA
NASA Goddard Space Flight Centre, Greenbelt, MD, USA

Andrew L. Rose
School of Environment, Science and Engineering, Southern Cross University, Lismore, NSW, Australia

Lizz Ultee and Jeremy N. Bassis
Department of Climate and Space, University of Michigan, Ann Arbor, MI, USA

Alessandro Aiuppa
Dipartimento DiSTeM, Università di Palermo, Palermo, Italy
Istituto Nazionale di Geofisica e Vulcanologia, Palermo, Italy

Luca Fiorani
Fusion and Technology for Nuclear Safety and Security Department, ENEA, Frascati, Italy

Simone Santoro
Dipartimento DiSTeM, Università di Palermo, Palermo, Italy
ENEA Guest, Frascati, Italy

Stefano Parracino
ENEA Guest, Frascati, Italy
Department of Industrial Engineering, University of Rome "Tor Vergata", Rome, Italy

Roberto D'Aleo
Dipartimento DiSTeM, Università di Palermo, Palermo, Italy

Marco Liuzzo
Istituto Nazionale di Geofisica e Vulcanologia, Palermo, Italy

Giovanni Maio
ENEA Guest, Frascati, Italy
Vitrociset SpA, Roma, Italy

Marcello Nuvoli
Fusion and Technology for Nuclear Safety and Security Department, ENEA, Frascati, Italy

James A. Bradley
Department of Earth Sciences, University of Southern California, Los Angeles, CA, USA
School of Geographical Sciences, University of Bristol, Bristol, UK

Alexandre M. Anesio
School of Geographical Sciences, University of Bristol, Bristol, UK

Sandra Arndt
School of Geographical Sciences, University of Bristol, Bristol, UK
Department of Earth and Environmental Sciences, Université Libre de Bruxelles, Brussels, Belgium

Joan Martí, Stefania Bartolini, Laura Becerril and Adelina Geyer
Group of Volcanology, Institute of Earth Sciences Jaume Almera, Agencia Estatal Consejo Superior de Investigaciones Científicasn CSIC, Barcelona, Spain

Carmen López
Instituto Geográfico Nacional, Madrid, Spain

Kai Berglar, Dieter Franke, Rüdiger Lutz, Bernd Schreckenberger and Volkmar Damm
Federal Institute for Geosciences and Natural Resources (BGR), Hannover, Germany

Jaime Otero, Francisco J. Navarro, Javier J. Lapazaran and Roman Finkelnburg
Department of Applied Mathematics, Universidad Politécnica de Madrid, Madrid, Spain

Ethan Welty
Institute of Arctic and Alpine Research, University of Colorado Boulder, Boulder, CO, USA

Darek Puczko
Institute of Biochemistry and Biophysics, Polish Academy of Sciences, Warsaw, Poland

Eglantine Boulard, MaryM.Reagan and Wendy L. Mao
Geological Sciences, Stanford University, Stanford, CA, USA

Guillaume Morard
Centre National de la Recherche Scientifique, UMR Centre Nationalde la Recherche Scientifique 7590, Institut de Minéralogie, de Physique des Matériaux et de Cosmochimie, IRD Sorbonne Universités — Université Pierreet Marie Curie, Muséum National d'Histoire Naturelle, Paris, France

Mohamed Mezouar
European Synchrotron Radiation Facility (ESRF), Grenoble, France

Yijin Liu
Stanford Synchrotron Radiation Light source, SLAC National Accelerator Laboratory, Menlo Park, CA, USA

Ai L. Koh
Stanford Nano Shared Facilities, Stanford University, Stanford, CA, USA

JulienStodolna
EDF Lab Les Renardieres, Dpt MMC, MoretsurLoing, France

Nobuo Geshi and Teruki Oikawa
Institute of Earthquake and Volcano Geology, Geological Survey of Japan, National Institute of Advanced Industrial Science and Technology (AIST), Tsukuba, Japan

Aron Stubbins and Leticia M. Silva
Department of Marine Sciences, Skidaway Institute of Oceanography, University of Georgia, Savannah, GA, USA

Thorsten Dittmar
Research Group for Marine Geochemistry, Institute for Chemistry and Biology of the Marine Environment, University of Oldenburg, Oldenburg, Germany

John T. Van Stan
Department of Geology and Geography, Georgia Southern University, Statesboro, GA, USA

Bruno Massa
Dipartimento di Scienze e Tecnologie, Universitá degli Studi del Sannio, Benevento, Italy
Istituto Nazionale di Geofisica e Vulcanologia, Sezione di Napoli, Osservatorio Vesuviano, Napoli, Italy

Luca D'Auria
Istituto Nazionale di Geofisica e Vulcanologia, Sezione di Napoli, Osservatorio Vesuviano, Napoli, Italy Istituto per il Rilevamento Elettromagnetico dell'Ambiente, Consiglio Nazionale delle Ricerche, Napoli, Italy

Elena Cristiano
Istituto Nazionale di Geofisica e Vulcanologia, Sezione di Napoli, Osservatorio Vesuviano, Napoli, Italy

Ada De Matteo
Dipartimento di Scienze e Tecnologie, Universitá degli Studi del Sannio, Benevento, Italy

Judith Sitters
Ecology and Biodiversity, Department Biology, Vrije Universiteit Brussel, Brussels, Belgium, Department of Aquatic Ecology
Netherlands Institute of Ecology (NIOO-KNAW), Wageningen, Netherlands

Department of Terrestrial Ecology, Netherlands Institute of Ecology (NIOO-KNAW), Wageningen, Netherlands

Elisabeth S. Bakker
Department of Aquatic Ecology, Netherlands Institute of Ecology (NIOO-KNAW), Wageningen, Netherlands

Michiel P. Veldhuis
Faculty of Science and Engineering, Groningen Institute for Evolutionary Life Sciences, University of Groningen, Groningen, Netherlands

G. F. Veen
Department of Terrestrial Ecology, Netherlands Institute of Ecology (NIOO-KNAW), Wageningen, Netherlands

Harry Olde Venterink
Ecology and Biodiversity, Department Biology, Vrije Universiteit Brussel, Brussels, Belgium

Michael J. Vanni
Department of Biology, Miami University, Oxford, OH, USA

Index

A

Agricultural Monitoring, 53
Airborne Digital Sensor, 38-40
Allochthonous Nutrients, 101
Andesitic Magma, 169, 171-175
Andesitic Tephras, 1
Anthropocene, 179
Anthropogenic Change, 99-100, 110
Aragonite Structures, 165-166
Autotroph Productivity, 204

B

Baseline Simulation, 101, 105
Bathymetric Elevation, 135
Bed Topography, 36, 74-78, 80-81, 83, 85, 150
Big Data, 53-54, 62
Bioavailability, 63, 70-73, 110
Biogeochemical Processes, 63, 113
Biome, 100, 114, 189
Bioreactors, 100
Boundary Layers, 1-2, 4
Bromalites, 8-9, 17-18, 20, 25

C

Carbonate, 14, 64, 160-161, 163-167
Compressional Deformation, 135-137, 142, 144, 146
Condensed Aromatics, 178, 184-185, 187-188
Coprolites, 8-25
Cryospheric Ecosystems, 102
Cryptic Phase Zoning, 1, 5-6
Crystal Growth, 1, 3-4, 6-7
Crystal Nucleation, 1-4, 6-7

D

Dolomite, 13-14, 166-167

E

Earthquake Hypocenter, 192
Ecosystem Structure, 63
Electron Microscopy, 2-3, 5, 12-13, 15, 160-161, 164-165
Electronic Structure, 64
Eruption Fissures, 169-172, 174-175
Extracellular Environment, 63-64, 66-67, 69-72

F

Feeder Dike, 169, 173
Flowline, 31, 74, 84, 150-153, 155-156, 158
Focal Mechanism, 191-193, 200-202
Force Balance Analyses, 35
Fossil-lagerstätte, 8
Frontal Ablation, 149-150

G

Glacier Advance, 74, 81-83
Glacier Dynamics, 26-27, 34-37, 74, 83, 85, 158-159
Glacier Flow, 26-27, 36, 85, 150
Glacier Forefields, 99-101, 104, 106, 108-110, 113
Glacier Geometry, 27, 74, 150, 155
Glacier Substrate, 75
Gondwana, 8-9, 24-25
Growth Textures, 1

H

Heterogeneous Satellite Imagery, 53
Heterotrophic Biomass, 100, 104-105
Hydrofracturing, 125, 150
Hydrological Models, 38

I

Ice Age, 84, 100-101
Inter-annual Scaling, 38, 44, 48-50
Isotopography, 2

L

Labile Redox-active Compounds, 63
Land Use, 49-50, 53-54
Local Stress, 31-32, 118-119, 122, 130, 132, 169-170, 174-175, 200

M

Magnetic Anomaly, 136-138, 140, 144
Mangrullo Formation, 8-12, 14-22
Mean Absolute Error, 38, 44, 46
Mesosauridae, 8-9, 22, 24-25
Metamorphism, 39, 47
Monogenetic Fields, 118, 120, 129-131
Monogenetic Volcanism, 118-119, 129-130, 133
Mountain Precipitation, 38

P

Paroxysms, 87, 96-97

Permafrost, 38, 101, 189-190

Permian, 8-18, 20, 22-25

Phase Diagrams, 160

Phase Transition, 160-161, 164, 166-167

Photosynthetic Organisms, 64, 68

Phreatomagmatic Deposits, 118

Pizzo Peak, 92-93

Plastic Approximation, 74-75, 85

Plastic Network Approach, 74

Polymorphs, 160, 164, 166

Precipitation Scaling, 38-39, 42, 50

Preferential Deposition, 38

R

Redox Conditions, 63-64, 67-68, 71-72

Redox Thermodynamics, 63-64

Regression Tree, 57

Remote Sensing, 38-40, 50-51, 53-54, 58-59, 61-62, 86, 93, 95-97, 149, 159

S

Satellite Imagery, 36-37, 53-56, 60, 62, 85

Simulated Terminus, 83

Spatial Interpolation, 42

Stress Coupling, 28, 34

Stromboli, 86-87, 90-91, 93-97, 174, 176

Subducting Slab, 166

Submarine Melt, 149-150

Summit Vents, 86

Superoxide, 63-64, 72-73

Swallow Tail Textures, 1

T

Tephra Layers, 171

Terminus Position, 31, 75-78, 80, 83, 149-150, 153, 155, 157-158

Terrestrial Herbivores, 204-208

Thermogenic, 179, 187

Translocation, 208

Tundra Zone, 101

V

Vancori Scan, 90-93

Volcanic Plume, 86-94

Volcanic Susceptibility, 118-120, 129-131

Volcanic Tephras, 1, 6

W

Waste Stoichiometry, 204, 208